Studies in Systems, Decision and Control

Volume 491

The series "Studies in Systems, Decision and Control" (SSDC) covers both new developments and advances, as well as the state of the art, in the various areas of broadly perceived systems, decision making and control–quickly, up to date and with a high quality. The intent is to cover the theory, applications, and perspectives on the state of the art and future developments relevant to systems, decision making, control, complex processes and related areas, as embedded in the fields of engineering, computer science, physics, economics, social and life sciences, as well as the paradigms and methodologies behind them. The series contains monographs, textbooks, lecture notes and edited volumes in systems, decision making and control spanning the areas of Cyber-Physical Systems, Autonomous Systems, Sensor Networks, Control Systems, Energy Systems, Automotive Systems, Biological Systems, Vehicular Networking and Connected Vehicles, Aerospace Systems, Automation, Manufacturing, Smart Grids, Nonlinear Systems, Power Systems, Robotics, Social Systems, Economic Systems and other. Of particular value to both the contributors and the readership are the short publication timeframe and the world-wide distribution and exposure which enable both a wide and rapid dissemination of research output.

Indexed by SCOPUS, DBLP, WTI Frankfurt eG, zbMATH, SCImago.

All books published in the series are submitted for consideration in Web of Science.

Abdellatif Ben Makhlouf ·
Mohamed Ali Hammami · Omar Naifar
Editors

State Estimation and Stabilization of Nonlinear Systems

Theory and Applications

Springer

Editors
Abdellatif Ben Makhlouf
Department of Mathematics
University of Sfax
Sfax, Tunisia

Mohamed Ali Hammami
Department of Mathematics
University of Sfax
Sfax, Tunisia

Omar Naifar
Control and Energy Management
Laboratory
National School of Engineering of Sfax
University of Sfax
Sfax, Tunisia

ISSN 2198-4182 ISSN 2198-4190 (electronic)
Studies in Systems, Decision and Control
ISBN 978-3-031-37969-7 ISBN 978-3-031-37970-3 (eBook)
https://doi.org/10.1007/978-3-031-37970-3

This Springer imprint is published by the registered company Springer Nature Switzerland AG
The registered company address is: Gewerbestrasse 11, 6330 Cham, Switzerland

Contents

Practical $h-$Stability of Nonlinear Impulsive Systems: A Survey

B. Ghanmi

Abstract In this chapter, we introduce a new type of stability for nonlinear impulsive systems of differential equations namely practical $h-$stability. By using Lyapunov stability theory, some sufficient conditions which guarantee practical $h-$stability are established. Our original results generalize well-known fundamental stability results, practical stability, practical exponential stability and practical asymptotic stability for nonlinear time-varying impulsive systems. Then, two classes of nonlinear impulsive systems, namely perturbed and cascaded impulsive systems are concerned. Furthermore, the problem of practical $h-$stabilization for certain classes of nonlinear impulsive systems is devoted. Finally, two numerical examples are given to show the effectiveness of our theoretical results.

Keywords Impulsive systems · $h-$stability · Practical $h-$stability · Lyapunov theory · Practical $h-$stabilization

1 Introduction

The differential equations with impulse effects are a new branch of the theory of ordinary differential equations. They describe evolution processes which at certain moments change their state rapidly. In addition, it is well known, that impulsive differential equations appear as a natural description of observed evolution phenomena of several real world problems. For example, many biological phenomena involving thresholds, chemistry [8], engineering [6], bursting rhythm models in medicine and biology, optimal control models in economics, pharmacokinetics and frequency modulated systems, theoretical physics, radiophysics, mechanics, do exhibit impulsive effects.

B. Ghanmi (✉)
University of Gafsa, Gafsa, Tunisia
e-mail: bghanmi@gmail.com

Faculty of Sciences of Gafsa, Department of Mathematics, Sidi Ahmed Zarroug, 2112 Gafsa, Tunisia

A. Ben Makhlouf et al. (eds.), *State Estimation and Stabilization of Nonlinear Systems*, Studies in Systems, Decision and Control 491,
https://doi.org/10.1007/978-3-031-37970-3_1

The Lyapunov stability theory is well known and is widely used in concrete problems of the real world. There are various concepts of stability, such as absolute stability, asymptotical stability, conditional stability and practical stability, when the origin is not necessarily an equilibrium point. In recent years, the theory of practical stability has been developed very intensively and attracts much attention, LaSalle and Lefschetz first brought forth the concept of practical stability in 1961 [16]. Furthermore, Lakshmikantham et al. presented a systematic study of practical stability in 1990. The main results in this prospect are due to Martynyuk and his collaborators [18, 19, 21, 23, 24] and the references cited therein. Practical stability, being quite different from stability in the sense of Lyapunov, is a significant performance specification from an engineering point of view for the following reason: A system might be stable or asymptotically stable in theory, however it is actually unstable in practice because the stable domain or the domain of the desired attractor is not large enough and on the other hand, sometimes the desired state of a system may be mathematically unstable and yet the system may oscillate sufficiently near this state that its performance is acceptable, that is, it is stable in practice. For standard state-space systems, [16–18, 20] presented a systematic study of the theory of practical stability and collected most valuable results. In [7], an impulsive differential-difference equations is considered and by using the method of comparison and differential inequalities for piecewise continuous functions, sufficient conditions for practical stability of the solutions of such systems are obtained. Applications to population dynamics are also given. It is worth mentioning some interesting previous works on practical stability [29, 30] and some references cited therein.

In 1984 M. Pinto introduced the notion of $h-$stability with the intention of obtaining results about stability for a weakly stable system (at least, weaker than those given exponential asymptotic stability) under some perturbations. That is, Pinto extended the study of exponential asymptotic stability to a variety of reasonable systems called $h-$systems. The notion of $h-$stability is quite flexible because it includes the classical notions of uniform or exponential stability within one common framework. In [13], the author study a general variational stability introduced mainly for non-autonomous systems. Several works are published around the notion of $h-$stability [5, 12, 31–35].

Motivated by the existing literature on stability, practical stability and $h-$stability, the contribution of the paper is to introduce and study the notion of the practical $h-$stability of nonlinear impulsive systems of differential equations. That is, we will extend the study of practical exponential asymptotic stability to a variety of reasonable impulsive systems called practical $h-$systems (at least, for systems with stabilities weaker than those given by practical exponential asymptotic stability). For this purpose, we use Lyapunov theory to establish global uniform practical $h-$stability of nonlinear impulsive systems of differential equations. Then, it is used to obtain sufficient conditions that ensure the practical $h-$stability of nonlinear impulsive systems. The rest of this paper is organized as follows. In the next section, some definitions, notations and hypotheses are summarized and the system description is given. Our main results are stated in Sect. 3, sufficient conditions for practical $h-$stability are given using the Lyapunov theory. However, Sect. 4 establishes the main results concerning the practical $h-$stability of perturbed and cascade systems. Section 5 is devoted to control applications. Finally, some numerical examples are given to demonstrate the validity of the results.

2 Definitions, Notations and Hypotheses

In this section, we introduce notations, definitions, and preliminary facts which are used throughout this paper.

Let $\mathbf{R}_+ = [0, +\infty[$, $\mathbf{R}^n$, $n \in \mathbf{N}^*$ be the n-dimensional Euclidean space with elements $x = col(x_1, x_2, \ldots, x_n)$ and the Euclidian norm $\|x\| = \left(\sum_{i=1}^{n} x_i^2\right)^{\frac{1}{2}}$. For $\delta \geq 0$, $\mathcal{B}_\delta$ denotes the closed ball of $\mathbf{R}^n$ centered at zero, i.e., $\mathcal{B}_\delta := \{x \in \mathbf{R}^n : \|x\| \leq \delta\}$.

In this paper, we aim to consider the following impulsive system:

$$\begin{aligned} \dot{x}(t) &= f(t, x(t)), \quad t \neq \tau_k,\ k = 1, 2, \ldots \\ \Delta x(\tau_k) &= I_k(x(\tau_k^-)), \quad t = \tau_k,\ k = 1, 2, \ldots \end{aligned} \tag{1}$$

where $t \in \mathbf{R}_+$ and $x(t) = [x_1(t), x_2(t), \ldots x_n(t)]^T \in \mathbf{R}^n$ is the state vector, $I_k : \mathbf{R}^n \to \mathbf{R}^n$ are continuous functions and $f : \mathbf{R}_+ \times \mathbf{R}^n \to \mathbf{R}^n$ is a nonlinear function. The discrete set of impulsive moments $\{\tau_k\}_{k=1}^{+\infty}$ is a known sequence of real numbers which satisfies $0 < \tau_1 < \tau_2 < \cdots < \tau_k < \cdots$, and $\lim\limits_{k \longrightarrow +\infty} \tau_k = +\infty$, $\Delta x(\tau_k) = x(\tau_k^+) - x(\tau_k^-)$, denotes the jumps in the state variable at the time instants τ_k where $x(\tau_k^+) = \lim\limits_{t \to \tau_k^+} x(t)$ and $x(\tau_k^-) = \lim\limits_{t \to \tau_k^-} x(t)$. We shall assume that the origin is not required to be an equilibrium for the system under consideration. This indeed fails in many situations when studying practical stability see [3].

For given $t_0 \in \mathbf{R}_+$ and $x_0 \in \mathbf{R}^n$, with Eq. (1) one associates the initial condition

$$x(t_0^+) = x_0. \tag{2}$$

A function $x : [t_0, t_0 + a) \to \mathbf{R}^n$ with $a > 0$ is finite or infinite, is said to be a solution of the initial value problem (IVP) (1)–(2) if:

(i) $x(t)$ is left continuous on $[t_0, t_0 + a)$ and satisfies $x(t_0^+) = x_0$;
(ii) $x(t)$ is differentiable and $x'(t) = f(t, x(t))$ everywhere on $(t_0, t_0 + a)\backslash\{\tau_1, \tau_2, \ldots, \tau_k, \ldots\}$ and
(iii) for any $t = \tau_k \in [t_0, t_0 + a)$, $x(t^+) = \lim\limits_{s \to t^+} x(s)$ exists and satisfies $x(t^+) = x(t) + I_k(x(t))$.

In this work, without loss of generality, suppose that $0 \leq t_0 = \tau_0 < \tau_1 < \tau_2 < \cdots < \tau_k < \cdots$. Then, the condition (2) becomes $x(t_0) = x_0$ (see [28]).

Denote by $x(t) = x(t, t_0, x_0)$ the solution of system (1) satisfying the initial condition $x(t_0, t_0, x_0) = x_0$. We further assume that the functions f and I_k, $k \in \mathbf{N}^*$, are smooth enough on $\mathbf{R}_+ \times \mathbf{R}^n$ and $\mathbf{R}^n$, respectively, to guarantee existence, uniqueness and continuability of the solution $x(t)$ of the system (1) on the interval $[t_0, +\infty)$ for all suitable initial data $x_0 \in \mathbf{R}^n$ and $t_0 \in \mathbf{R}_+$.

We also introduce the following classes of functions

- If $1 \leq p < +\infty$ is a real number, $\mathcal{L}^p(\mathbf{R}_+) = \Big\{\varphi : \mathbf{R}_+ \to \mathbf{R};\ \Big(\int_0^{+\infty} |\varphi(t)|^p dt\Big)^{\frac{1}{p}} < +\infty\Big\}$, the space of functions $\varphi : \mathbf{R}_+ \to \mathbf{R}$ for which the pth power of the absolute value is Lebesgue integrable. The Lebesgue space $\mathcal{L}^p(\mathbf{R}_+)$ is equipped by the norm $\|.\|_p$ defined by, if $\varphi \in \mathcal{L}^p(\mathbf{R}_+)$, $\|\varphi\|_p \triangleq \Big(\int_0^{+\infty} |\varphi(t)|^p dt\Big)^{\frac{1}{p}}$.
- $\mathcal{PC}[\mathbf{R}_+, \mathbf{R}]$ the set of functions $\psi : \mathbf{R}_+ \to \mathbf{R}$ continuous everywhere except for the points τ_k, $k = 1, 2, \ldots$ at which $\psi(\tau_k^-)$ and $\psi(\tau_k^+)$, $k = 1, 2, \ldots$, exist and $\psi(\tau_k^-) = \psi(\tau_k)$
- $\mathcal{PC}^1[\mathbf{R}_+, \mathbf{R}]$ the set of functions $\psi : \mathbf{R}_+ \to \mathbf{R}$ continuously differentiable everywhere except for the points τ_k, $k = 1, 2, \ldots$ at which $\psi'(\tau_k^-)$ and $\psi'(\tau_k^+)$, $k = 1, 2, \ldots$, exist and $\psi'(\tau_k^-) = \psi'(\tau_k)$

Evidently, $\mathcal{PC}[\mathbf{R}_+, \mathbf{R}]$ is a Banach space with norm $\|\psi\|_{\mathcal{PC}} := \sup_{t\geq 0}\{|\psi(t)|\}$ and

$$\mathcal{PC}^1[\mathbf{R}_+, \mathbf{R}] = \{\psi : \mathbf{R}_+ \to \mathbf{R};\ \psi' \in \mathcal{PC}[\mathbf{R}_+, \mathbf{R}]\}.$$

Let $\tau_0 = t_0$ and introduce the following sets

$$G_k = \{(t, x) \in \mathbf{R}_+ \times \mathbf{R}^n;\ \tau_{k-1} < t < \tau_k\}, k = 1, 2, \ldots \text{ and } G = \bigcup_{k=1}^{+\infty} G_k.$$

Next, we shall define the class of Lyapunov functions that will be used in the qualitative investigations of nonlinear impulsive systems.

Definition 1 A function $V : \mathbf{R}_+ \times \mathbf{R}^n \to \mathbf{R}_+$ belongs to class $\mathcal{V}_0$ if

- V is continuous in G and for each $x \in \mathbf{R}^n$ and $k \in \mathbb{N}$, $\lim_{(t,y)\to(\tau_k^+,x)} V(t, y) = V(\tau_k^+, x)$ exists
- For all $t \geq t_0$, $V(t, 0) \geq 0$.

Definition 2 Given a class $\mathcal{V}_0$ function $V : \mathbf{R}_+ \times \mathbf{R}^n \to \mathbf{R}_+$.

- If $x(t)$ is a solution of (1), the upper right-hand derivative of V with respect to system (1) is defined by

$$D^+V(t, x) = \limsup_{\delta \longrightarrow 0^+} \frac{1}{\delta}\Big(V(t + \delta, x(t + \delta, t, x)) - V(t, x)\Big) \tag{3}$$

  for $(t, x) \in \mathbf{R}_+ \times \mathbf{R}^n$.

- The total derivative $D^+V_{(2.1)}$ with respect to system (1) of the function V is defined by

$$D^+V_{(2.1)}(t,x) = \limsup_{\delta \longrightarrow 0^+} \frac{1}{\delta}\Big(V(t+\delta, x+\delta f(t,x))) - V(t,x)\Big) \tag{4}$$

Remark 1 Let $V : \mathbf{R}_+ \times \mathbf{R}^n \to \mathbf{R}_+$ be a class $\mathcal{V}_0$ function. Then,

- if V is Lipschitz with respect to x, Yoshizawa [15] has proved that

$$D^+V_{(2.1)}(t,x) = D^+V(t,x). \tag{5}$$

- if $V \in \mathcal{C}^1[\mathbf{R}_+ \times \mathbf{R}^n, \mathbf{R}_+]$, then

$$D^+V_{(2.1)}(t,x) = D^+V(t,x) = \dot{V}_{(2.1)}(t,x) \triangleq \nabla_t V(t,x) + \nabla_x^T V(t,x).f(t,x). \tag{6}$$

Let $h : \mathbf{R}_+ \to \mathbf{R}_+^*$ be a positive and continuous function.

Next, we define the new notion of practical $h-$stability for the impulsive system (1).

Definition 3 The system (1) is called a practical $h-$system if there exist constants $c \geq 1$, $r \geq 0$ and for $t_0 \in \mathbf{R}_+$ there exists $\delta = \delta(t_0) > 0$ such that for all $t \geq t_0$ and all $\|x_0\| \leq \delta(t_0)$

$$\|x(t,t_0,x_0)\| \leq c\|x_0\|h(t)h^{-1}(t_0) + r,\ t \geq t_0 \tag{7}$$

where $h^{-1}(t) = \frac{1}{h(t)}$

If (1) is a practical $h-$system, then the function h as well as the parameters c and r depend only on f and are the same for all solutions of system (1).

Remark 2 The new concept of practical $h-$system is motivated by the notion of $h-$system given by M. Pinto in [14]. In particular, the notion of practical $h-$system coincide with that of $h-$system when $r = 0$.

Definition 4 Suppose that $h : \mathbf{R}_+ \to \mathbf{R}_+^*$ is positive, continuous and bounded function on $\mathbf{R}_+$. The system (1) is called:

(i) Practical $h-$Stable (PhS) if there exist constants $c \geq 1, r \geq 0$ and for all initial time $t_0 \in \mathbf{R}_+$ there exists $\delta = \delta(t_0) > 0$ such that for all initial state $x_0 \in \mathbf{R}^n$ with $\|x_0\| \leq \delta$, the solution $x(t,t_0,x_0)$ satisfies (7).

(ii) Uniformly Practically $h-$Stable (UPhS) if there exist constants $c \geq 1, r \geq 0$ and there exists $\delta > 0$ independent of t_0 such that for all $t_0 \in \mathbf{R}_+$, for all $x_0 \in \mathbf{R}^n$ with $\|x_0\| \leq \delta$, the solution $x(t,t_0,x_0)$ satisfies (7).

(iii) Globally Uniformly Practically $h-$Stable (GUPhS) if there exist constants $c \geq 1$, $r \geq 0$ such for all $t_0 \in \mathbf{R}_+$ and all $x_0 \in \mathbf{R}^n$, the solution $x(t,t_0,x_0)$ satisfies (7).

Remark 3 Definition 4 generalizes the notions of $h-$ stability. More precisely, when $r = 0$ we recover the usual definition of $h-$stability [4, 14, 25–27]. However, the reciprocal is false. Moreover, for $r > 0$ and for some special cases of h, the practical $h-$stability coincide with a known practical types of stability:

(a) If $h(t) = c$ for a positive constant c, then the system (1) is practically stable (see [9, 11]).
(b) If $h(t) = e^{-\lambda t}$ for positive constant λ, then (1) is practically exponentially stable (see [10]).
(c) If $h(t)$ is a strictly decreasing function such that $\lim_{t \longrightarrow +\infty} h(t) = 0$, then the ball $\mathcal{B}_r$ is uniformly asymptotically stable on $\mathcal{B}_\delta$. More precisely, the solutions of system (1) converge to the ball $\mathcal{B}_r$ of radius r (i.e., $\limsup_{t \to +\infty} \|x(t, t_0, x_0)\| \leq r$, for $x_0 \in \mathcal{B}_\delta$) (see [2]).

The following lemmas will also be required in our investigations.

Lemma 1 ([1]) *Suppose $v \in \mathcal{PC}^1[\mathbf{R}_+, \mathbf{R}]$ satisfies the following scalar impulsive differential inequality*

$$\begin{cases} D^+v(t) \leq a(t)v(t) + b(t); \; t \neq \tau_k, \quad k \in \mathbb{N}^* \\ v(\tau_k^+) \leq c_k v(\tau_k^-) + d_k; \; t = \tau_k, \; k \in \mathbb{N}^* \\ v(t_0) = v_0 \end{cases} \tag{8}$$

where $a(t), b(t) \in PC[\mathbf{R}_+, \mathbf{R}]$, $c_k \geq 0$ and d_k are constants. Then $v(.)$ satisfies

$$v(t) \leq v_0 \Big(\prod_{t_0 < \tau_k < t} c_k \Big) e^{\int_{t_0}^t a(s)ds} + \sum_{t_0 < \tau_k < t} \Big(\prod_{\tau_k < \tau_j < t} c_j \Big) e^{\int_{\tau_k}^t a(s)ds} d_k$$
$$+ \int_{t_0}^t \Big(\prod_{s < \tau_k < t} c_k \Big) e^{\int_s^t a(u)du} b(s)ds.$$

Lemma 2 *For all $p \geq 1$ and all $a, b \geq 0$*

$$(a + b)^{\frac{1}{p}} \leq a^{\frac{1}{p}} + b^{\frac{1}{p}}.$$

3 Main Results

3.1 Sufficient Conditions for Practical h−stability

Sufficient conditions on practical $h-$stability of system (1) will be discussed by Liapunov's second method.

Theorem 1 *Suppose that h is a positive bounded continuously differentiable function on $\mathbf{R}_+$. Furthermore, assume that there exist constants $c \geq 1$, $a \geq 0$ and there exists a function $V \in \mathcal{V}_0$, satisfying the following properties:*

(i) $\|x\| \le V(t,x) \le c_1\|x\| + a,\ t \in \mathbf{R}_+,\ x \in \mathbf{R}^n$,

(ii) $D^+V_{(2.1)}(t,x) \le \frac{h'(t)}{h(t)}\Big(V(t,x) - a\Big),\ (t,x) \in G$,

(iii) $V(\tau_k^+, x + I_k(x) \le V(\tau_k, x) + d_k,\ k \in \mathbf{N}^*,\quad x \in \mathbf{R}^n$ *such that* $\displaystyle\sum_{k=1}^{+\infty} \frac{d_k}{h(\tau_k)} < +\infty$.

Then, the system (1) is globally uniformly practically $h-$stable.

Remark 4 If $a = 0$ and $d_k = 0$ for all $k \in \mathbf{N}^*$, we recover the usual definition of $h-$stability (see [5]).

As a consequence of Theorem 1, we have the following result.

Corollary 1 *Suppose that h is a positive bounded continuously differentiable function on $\mathbf{R}_+$. Furthermore, assume that there exist constants $c \ge 1$, $a \ge 0$ and there exists a function $V \in \mathcal{V}_0$ satisfying the following properties:*

(i) $\|x\|^p \le V(t,x) \le c\|x\|^p + a,\ t \in \mathbf{R}_+,\ x \in \mathbf{R}^n$,

(ii) $D^+V_{(2.1)}(t,x) \le \frac{h'(t)}{h(t)}\Big(V(t,x) - a\Big),\ (t,x) \in G$,

(iii) $V(\tau_k^+, x + I_k(x) \le V(\tau_k, x(\tau_k)) + d_k,\ k \in \mathbf{N}^*,\ x \in \mathbf{R}^n$ *such that* $\displaystyle\sum_{k=1}^{+\infty} \frac{d_k}{h(\tau_k)} < +\infty$.

Then, the system (1) is globally uniformly practically $h^{\frac{1}{p}}-$stable.

In [22], Martynyuk studied the exponential x_1-stability by splitting the vector $x \in \mathbf{R}^n$ into two subvector $x_i \in \mathbf{R}^{n_i}$, $i = 1, 2$, $n_1 + n_2 = n$. A $h-$stability results are obtained when all conditions of Theorem 1 [22] are adopted (see [27]). Here, we suppose more general assumption on the Lyapunov function candidate $V(t,x)$ and we obtain the following result:

Corollary 2 *Assume that the function h is bounded and continuously differentiable on $\mathbf{R}_+$. In addition, suppose the existence of constants $N, \gamma > 0$, $a \ge 0$, a class $\mathcal{K}$ function α defined on $\mathbf{R}_+$ and a function $V \in \mathcal{V}_0$ satisfying the following properties*

(i) $N\|x\|^\gamma \le V(t,x) \le \alpha(\|x\|) + a,\ t \in \mathbf{R}_+,\ x \in \mathbf{R}^n$,

(ii) $D^+V_{(2.1)}(t,x) \le \frac{h'(t)}{h(t)}\Big(V(t,x) - a\Big),\ (t,x) \in G$,

(iii) $V(\tau_k^+, x + I_k(x) \le V(\tau_k, x) + d_k,\ k \in \mathbb{N}^*,\quad x \in \mathbf{R}^n$ *such that* $\displaystyle\sum_{k=1}^{+\infty} \frac{d_k}{h(\tau_k)} < +\infty$.

Then, the system (1) is uniformly practically $h^{\frac{1}{\gamma}}-$stable.

4 Practical $h-$stability of Perturbed and Cascaded System

4.1 Practical h−stability of Perturbed System

We shall be interested by the relations between the solutions of the unperturbed (nominal) system (1) and the solutions of the perturbed system

$$\begin{aligned} \dot{y}(t) &= f(t, y) + g(t, y), & t \neq \tau_k,\ k = 1, 2, \ldots \\ \Delta y(\tau_k) &= I_k(y(\tau_k^-)) + J_k(y(\tau_k^-)), & t = \tau_k,\ k = 1, 2, \ldots \end{aligned} \tag{1}$$

where $g : \mathbf{R}_+ \times \mathbf{R}^n \to \mathbf{R}^n$. We assume that functions f, g, I_k and J_k satisfy all required conditions for existence and uniqueness of the solutions of system (1) on the interval $[t_0, +\infty)$ for all suitable initial data $y_0 \in \mathbf{R}^n$ and $t_0 \in \mathbf{R}_+$. The unique solutions of (1) is denoted by $y(t, t_0, y_0)$, satisfying the initial conditions $y(t_0, t_0, y_0) = y_0$. Suppose the following assumptions hold

(A1) : There exists a Lyapunov function $V \in \mathcal{V}_0$ which satisfies all conditions (i)–(iii) of Theorem 1 such that for all $(t, x), (t, y) \in \mathbf{R}_+ \times \mathbf{R}^n$,

$$|V(t, x) - V(t, y)| \leq L \|x - y\|, L > 0.$$

(A2) : For all $t \in \mathbf{R}_+$ and $y \in \mathbf{R}^n$, the perturbation term satisfies

$$\|g(t, y)\| \leq \lambda(t)\|y\| + \beta(t) \tag{2}$$

where $\lambda, \beta : \mathbf{R}_+ \to \mathbf{R}_+$ such that $\lambda \in \mathcal{L}^1(\mathbf{R}_+)$, $\lambda h^{-1} \in \mathcal{L}^1(\mathbf{R}_+)$ and $\beta h^{-1} \in \mathcal{L}^1(\mathbf{R}_+)$.

(A3) : For all $k \in \mathbf{N}^*$ and $y \in \mathbf{R}^n$, $\|J_k(y)\| \leq \mu_k$ such that $\sum_{k=1}^{+\infty} \frac{\mu_k}{h(\tau_k)} < +\infty$.

Theorem 2 *Under assumptions (A1)–(A3), the perturbed system (1) is globally uniformly practically* $\tilde{h}-$*stable where* $\tilde{h}(t) = h(t)e^{L\int_0^t \lambda(s)ds}$.

4.2 Practical h−stability of Cascaded System

Consider now cascaded systems (Σ_1, Σ_2) of the form:

$$\Sigma_1 : \quad \begin{aligned} \dot{x}_1 &= f_1(t, x_1) + g(t, x_1, x_2), & t \neq \tau_k,\ k = 1, 2, \ldots \\ \Delta x_1(\tau_k) &= I_k^1(x_1(\tau_k^-)) + J_k(x_1(\tau_k^-), x_2(\tau_k^-)), & t = \tau_k,\ k = 1, 2, \ldots \end{aligned} \tag{3}$$

$$\Sigma_2: \quad \begin{aligned} \dot{x}_2 &= f_2(t, x_2), & t \neq \tau_k,\ k = 1, 2, \ldots \\ \Delta x_2(\tau_k) &= I_k^2(x_2(\tau_k^-)), & t = \tau_k,\ k = 1, 2, \ldots \end{aligned} \tag{4}$$

where $x_1 \in \mathbf{R}^{n_1}$, $x_2 \in \mathbf{R}^{n_2}$ and $x := [x_1, x_2]^T$, denotes the state of the system (3)–(4). We assume that functions f_1, f_2, g and I_k^1, I_k^2, J_k satisfy all required conditions for existence and uniqueness of the solutions of system (3)–(4) on the interval $[t_0, +\infty)$ for all suitable initial data $x_0 \in \mathbf{R}^n$ and $t_0 \in \mathbf{R}_+$.

The general topic of the Sect. 4.2 is to study the Lyapunov practical $h-$stability of the nonlinear time-varying cascaded impulsive systems (3)–(4). Roughly speaking this a system in open loop as illustrated in the figure below

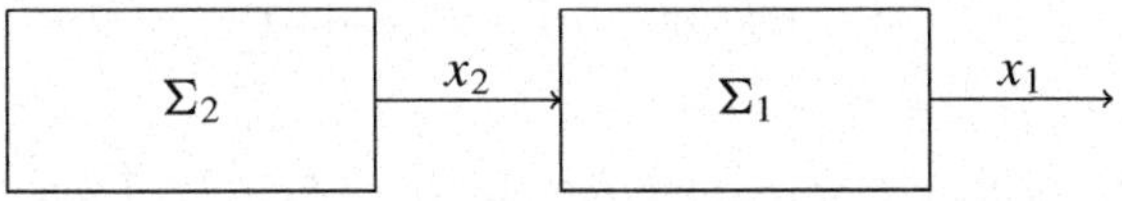

Cascade interonnetion of two dynamical impulsive systems Σ_1 and Σ_2.

Let us consider the following subsystems

$$\begin{aligned} \dot{x}_1 &= f_1(t, x_1), & t \neq \tau_k,\ k = 1, 2, \ldots \\ \Delta x_1(\tau_k) &= I_k^1(x_1(\tau_k^-)), & t = \tau_k,\ k = 1, 2, \ldots \end{aligned} \tag{5}$$

In this part, we give sufficient conditions to guarantee that if systems (4) and (5) are globally uniformly practically $h-$stable then the cascaded system (3)–(4) is also globally uniformly practically $h-$stable, if the interconnection term $g(t, x)$ satisfies some condition of boundedness. The solution of cascaded system (3)–(4) is denoted by $x(., t_0, x_0) = (x_1(., t_0, x_0),$
$x_2(., t_0, x_0))$, which satisfies $x(t_0, t_0, x_0) = x(t_0) = x_0$.

For $i = 1, 2$, we introduce the following sets

$$G_k^i = \{(t, x_i) \in \mathbf{R}_+ \times \mathbf{R}^{n_i};\ \tau_{k-1} < t < \tau_k\},\ k = 1, 2, \ldots \text{ and } G_i = \bigcup_{k=1}^{+\infty} G_k^i.$$

In the next, the global uniform practical $h-$stability of cascaded system (3)–(4) is studied. Before proposing our theorem, we introduce the following assumptions.

(A4) : There are two Lyapunov functions $V_1(t, x_1)$ and $V_2(t, x_2)$ such that

(i) $V_i \in \mathcal{V}_0$, $i = 1, 2$ such that $|V_1(t, x_1) - V_1(t, y_1)| \leq L_1 \|x_1 - y_1\|$, $x_1, y_1 \in \mathbf{R}^{n_1}$, $t \in \mathbf{R}_+$ and $L_1 > 0$.

(ii) $\|x_i\| \leq V_i(t, x_i) \leq c_i \|x_i\| + a_i$, $c_i > 0$, $a_i \geq 0$ and $(t, x_i) \in \mathbf{R}_+ \times \mathbf{R}^{n_i}$, $i = 1, 2$

(iii) $D^+ V_i(t, x_i) \leq \frac{h'(t)}{h(t)}\Big(V_i(t, x_i) - a_i\Big)$, $(t, x_i) \in G_i$, $i = 1, 2$

iv) $V_i(\tau_k^+, x_i + I_k^i(x_i) \leq V(\tau_k^-, x_i(\tau_k^-)) + d_k^i,\ t = \tau_k,\ k \in \mathbf{N}^*$ such that $\sum_{k=1}^{+\infty} \frac{d_k^i}{h(\tau_k)} < +\infty, i = 1, 2.$

(A5) : (Boundedness of the interconnection term) The function g satisfies

$$\|g(t, x_1, x_2)\| \leq \lambda(t)(\|x_1\| + \|x_2\| + 1),\ t \geq 0,\ x_1 \in \mathbf{R}^{n_1},\ x_2 \in \mathbf{R}^{n_2} \tag{6}$$

where $\lambda : \mathbf{R}_+ \to \mathbf{R}_+$ such that $\lambda \in \mathcal{L}^1(\mathbf{R}_+)$ and $\lambda h^{-1} \in \mathcal{L}^1(\mathbf{R}_+)$.

(A6) : For all $k \in \mathbf{N}^*$, $x_1 \in \mathbf{R}^{n_1}$ and $x_2 \in \mathbf{R}^{n_2}$

$$\|J_k(x_1, x_2)\| \leq \mu_k \text{ such that } \sum_{k=1}^{+\infty} \frac{\mu_k}{h(\tau_k)} < +\infty.$$

Theorem 3 *Assume that assumptions* **(A4)–(A6)** *hold. Then, the cascaded system (3)–(4) is globally uniformly practically $\tilde{h}-$stable where $\tilde{h}$ is a continuous, positive and bounded function defined on* $\mathbf{R}_+$.

5 Practical $h-$stabilization

In this part, we consider the following nonlinear impulsive system with delayed perturbations:

$$\begin{aligned} \dot{x} &= F(t, x) + G(t, x)\Big(\sum_{i=0}^{r} g_i(t, x(t - \delta_i(t))) + u\Big), && r \in \mathbf{N},\ t \neq \tau_k, k = 1, 2, \ldots \\ \Delta x(\tau_k) &= x(\tau_k^+) - x(\tau_k^-) = I_k(x(\tau_k^-), u(\tau_k^+)) + J_k(x(\tau_k^-)), && t = \tau_k,\ k = 1, 2, \ldots \end{aligned} \tag{1}$$

where $x \in \mathbf{R}^n$ is the state vector, $u \in \mathbf{R}^n$ is the control vector, $F : \mathbf{R}_+ \times \mathbf{R}^n \longrightarrow \mathbf{R}^n$, $G : \mathbf{R}_+ \times \mathbf{R}^n \longrightarrow \mathbf{R}^{n\times m}$ are known functions, and perturbations $g_i(t, x(t - \delta_i(t)) \in \mathbf{R}^n$ are time-varying nonlinear continuous unknown functions of the state. The time delays $\delta_i(t),\ i = 1, 2, \ldots, r$ are nonnegative, bounded and continuous functions, i.e., there exist positive constants $\overline{\delta}_i,\ i = 0, 1, \ldots, r$ such that $0 \leq \delta_i(t) \leq \overline{\delta}_i$. The initial condition function is given by $x(t) = \theta(t),\ t \in [-\overline{\delta}, 0]$ where $\overline{\delta} = \max_{0\leq i\leq r}\{\overline{\delta}_i\}$ and $\theta(t)$ is a piecewise continuous vectored-valued initial function on $[-\overline{\delta}, 0]$.

The corresponding system without perturbations, called the nominal system, is described by

$$\begin{aligned} \dot{x} &= F(t, x) + G(t, x)u, && t \neq \tau_k,\ k = 1, 2, \ldots \\ \Delta x(\tau_k) &= I_k(x(\tau_k^-), u(\tau_k^+)), && t = \tau_k,\ k = 1, 2, \ldots \end{aligned} \tag{2}$$

Moreover, the so-called unforced nominal system of system (1) is defined by

$$\begin{aligned} \dot{x} &= F(t,x), && t \neq \tau_k,\ k=1,2,\dots \\ \Delta x(\tau_k) &= I_k(x(\tau_k^-),0), && t=\tau_k,\ k=1,2,\dots \end{aligned} \tag{3}$$

The interest in this section is in finding feedback laws $u = u(t, x(t))$ which make the closed-loop impulsive system

$$\begin{aligned} \dot{x} =& F(t,x) + G(t,x)\Big(\sum_{i=0}^{r} g_i(t, x(t-\delta_i(t))) + u(t,x(t))\Big), && t \neq \tau_k,\ r \in \mathbf{N} \\ \Delta x(\tau_k) =& I_k(x(\tau_k^-), u(\tau_k^+)) + J_k(x(\tau_k^-)), && t=\tau_k,\ k=1,2,\dots \end{aligned} \tag{4}$$

globally uniformly practically $h-$stable.

Before giving our synthesis approach, we introduce for system (1) the following standard assumptions.

(A7) : There exists a continuously differentiable function $V : \mathbf{R}_+ \times \mathbf{R}^n \longrightarrow \mathbf{R}_+$ which satisfies

$$|V(t,x) - V(t,y)| \leq L\,\|x-y\| \ \text{for all}\ t \in \mathbf{R}_+, x, y \in \mathbf{R}^n \tag{5}$$

$$\|x\| \leq V(t,x) \leq c\|x\| + a, \ \text{for all}\ (t,x) \in \mathbf{R}_+ \times \mathbf{R}^n \tag{6}$$

$$\dot{V}_{(5.3)}(t,x) = \nabla_t V(t,x) + \nabla_x^T V(t,x).F(t,x) \leq \frac{h'(t)}{h(t)}\Big(V(t,x) - a\Big),\ t \neq \tau_k,\ x \in \mathbf{R}^n,\ a \geq 0 \tag{7}$$

$$V(\tau_k^+, x + I_k(x,0)) \leq c_k V(\tau_k, x) + d_k,\ t=\tau_k,\ k \in \mathbf{N}^*, 0 \leq c_k \leq 1 \tag{8}$$

such that $\displaystyle\sum_{k=1}^{+\infty} \frac{d_k}{h(\tau_k)} < +\infty$ and $h : \mathbf{R}_+ \rightarrow \mathbf{R}_+^*$ is a continuous and bounded function defined.

(A8) : The perturbations $g_i(t, x(t-\delta_i(t))$ are bounded in Euclidean norm as follows:

$$\text{for all}\ t \in \mathbf{R}_+,\ \|g_i(t, x(t-\delta_i(t))\| \leq \beta_i \|x(t-\delta_i(t))\|,\ \beta_i > 0,\ i = 0, 1, \dots, r. \tag{9}$$

(A9) : For all $k \in \mathbf{N}^*, x, y \in \mathbf{R}^n$ and all $u, v \in \mathbf{R}^n$, $\|I_k(x,u) - I_k(y,v)\| \leq \|x-y\|$ $+ \|u-v\|$ and $\|J_k(x)\| \leq \mu_k$ such that $\displaystyle\sum_{k=1}^{+\infty} \frac{\mu_k}{h(\tau_k)} < +\infty$.

Now, we have the following theorem which shows the practical $h-$stability of system (1) under the state feedback controller

Theorem 4 *Suppose assumptions **(A7)–(A9)** hold. Then, the system (1) is globally uniformly practically $h-$stable under the following controller given by (10)*

$$\begin{cases} u(t) &= \varphi(t,x(t)) = -\dfrac{\Big(\sum\limits_{i=0}^{r}\beta_i\chi_i(t)\Big)^2 G^T(t,x)\nabla_x V(t,x)}{\|\nabla_x^T V(t,x).G(t,x)\|\sum\limits_{i=0}^{r}\beta_i\chi_i(t)+\lambda(t)h(t)V(t,x(t))}, t\neq\tau_k,\ r\in\mathbf{N} \\ u(\tau_k^+) &= \dfrac{1-c_k}{L}x(\tau_k^-), t=\tau_k,\ k=1,2,\ldots \end{cases} \tag{10}$$

where $\lambda\in\mathcal{L}^1(\mathbf{R}_+)$ *and*

$$\chi_i(t) = \max_{s\in[t-\overline{\delta}_i,t]}\|x(s)\|,\ i=0,1,\ldots,r. \tag{11}$$

In practice, it is unavoidable to include uncertain parameters and perturbations in practical control systems due to modeling errors, measurement errors, approximations, slowly varying parameters, etc. Therefore, the stability problem and stabilization problem of uncertain dynamical systems with time-varying delays has attracted considerable attention of many researchers in recent years. From the above statements, we consider the problem of practical $h-$stabilization for nonlinear impulsive uncertain systems with time-varying delay of the form

$$\begin{aligned} &\dot{x} = F(t,x)+G(t,x)\sum_{i=0}^{r}g_i(t,x(t-\delta_i(t)))+G(t,x)(1+\delta(t,x(t)))u, && t\neq\tau_k,\ r\in\mathbf{N} \\ &\Delta x(\tau_k) = I_k(x(\tau_k^-),u(\tau_k^+))+J_k(x(\tau_k^-)), && t=\tau_k,\ k=1,2,\ldots \end{aligned} \tag{12}$$

where $\delta(t,x(t))$ represents the uncertainties.

Further assumption is given as follows:

(A10) : For all $x\in\mathbf{R}^n$, there exists a known function $\phi(t,x)$ such that

$$1+\delta(t,x)\geq\phi(t,x)>0. \tag{13}$$

Then, the following theorem can be obtained that provides the practical $h-$stability result of system (12).

Theorem 5 *Assume that* ***(A7)–(A10)*** *hold. Then, the system (12) is globally uniformly practically* $h-$*stable under the following controller*

$$\begin{cases} u(t) &= \zeta(t,x(t)) = -\dfrac{\Big(\sum\limits_{i=0}^{r}\beta_i\chi_i(t)\Big)^2 G^T(t,x)\nabla_x V(t,x)}{\phi(t,x(t))\|\nabla_x^T V(t,x).G(t,x)\|\sum\limits_{i=0}^{r}\beta_i\chi_i(t)+\lambda^*(t)h(t)V(t,x(t))}, t\neq\tau_k,\ r\in\mathbf{N} \\ u(\tau_k^+) &= \dfrac{1-c_k}{L}x(\tau_k^-), t=\tau_k,\ k=1,2,\ldots \end{cases} \tag{14}$$

where $\lambda^*(t)=\phi(t,x(t))\mu(t)$ *such that* $\mu\in\mathcal{L}^1(\mathbf{R}_+)$.

6 Examples

Example 1 To clarify better our contribution, let us consider the impulsive differential equation

$$\begin{aligned} \dot{x} &= f(t,x), && t \neq \tau_k,\ k = 1,2,\ldots \\ \Delta x(\tau_k) &= I_k(x(\tau_k^-)), && t = \tau_k,\ k = 1,2,\ldots \end{aligned} \tag{1}$$

where $x = [x_1, x_2, \ldots, x_n]^T \in \mathbf{R}^n$ and $f(t,x) = \Big[f_1(t,x), f_2(t,x), \ldots, f_n(t,x)\Big]^T$ with

(i) $f_m(t,x) = \lambda(t)\Big(x_m - e^{t^4+2t^2+1}x_m^5 \prod_{i\neq m} x_i^2 + \frac{\sin(\sqrt{\frac{3\alpha\beta}{n}}x_m)}{x_m}\Big),\ m \in \{1,2,\ldots,n\}$

$\alpha, \beta > 0$ and $\lambda : \mathbf{R}_+ \longrightarrow \mathbf{R}_+$ a continuous function which satisfies $\int_0^{+\infty} \lambda(s)ds < \frac{1}{e(1+\alpha)}$.

(ii) For all $k = 1,2,\ldots$, $I_k(x) = (\frac{1}{\sqrt{2}} - 1)x + \sqrt{\frac{\mu_k}{2}}\sin(x)$ such that $\mu_k \geq 0$ for all $k = 1,2,\ldots$, and

$$\sum_{k=1}^{+\infty} \mu_k e^{-3\int_0^{\tau_k}\lambda(s)ds} < +\infty.$$

Let us consider the Lyapunov function

$$V(t,x) = \|x\|^2 + \beta e^{-3\varepsilon_0 \int_0^t \lambda(s)ds}$$

where $\varepsilon_0 > 0$ is chosen such that the condition hold

$$\varepsilon_0 e^{-\varepsilon_0 \int_0^{+\infty}\lambda(s)ds} - \alpha \geq 1. \tag{2}$$

It is clear that $V(t,x)$ satisfy condition $i)$ of Theorem 1 with $c = 1$ and $a = \beta$. In addition $f(t,0) = \sqrt{\frac{3\alpha\beta}{n}}\lambda(t)\Big[1, 1, \ldots, 1\Big]^T \neq 0$.

The derivative of $V(t,x)$ along the solutions of system (1) is given by:

If $t \neq \tau_k,\ k = 1,2,\ldots$

$$\begin{aligned} \dot{V}_{(6.1)}(t,x) =& 2\sum_{k=1}^{n} x_k \dot{x}_k - 3\beta\varepsilon_0\lambda(t)e^{-3\varepsilon_0\int_0^t \lambda(s)ds}, \\ \leq& \lambda(t)\Big[3\|x\|^2 - 3\beta\Big(\varepsilon_0 e^{-\varepsilon_0\int_0^{+\infty}\lambda(s)ds} - \alpha\Big)\Big]. \end{aligned}$$

Then, under the condition given by (2), we obtain

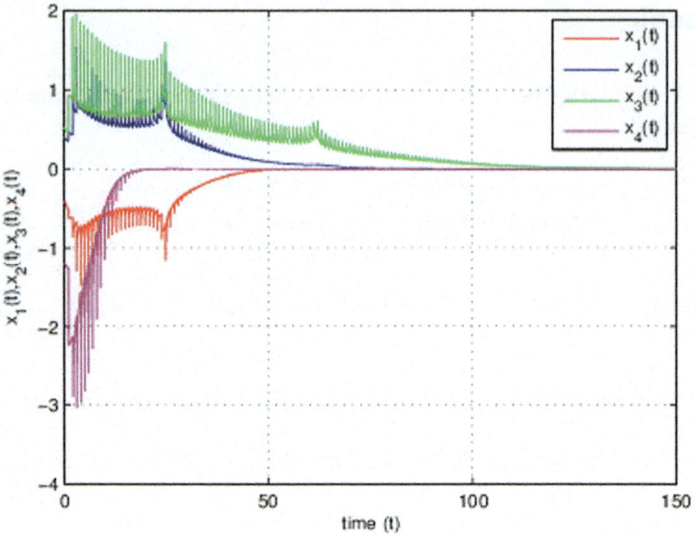

Fig. 1 Trajectories of system (1)

$$\dot{V}_{(6.1)}(t,x) \leq 3\lambda(t)\Big[\,\|x\|^2-\beta\Big] \leq 3\lambda(t)\Big(V(t,x)-\beta\Big) = \frac{h'(t)}{h(t)}\Big(V(t,x)-\beta\Big)$$

where $h(t)=e^{3\int_0^t \lambda(s)ds}$ is a positive, continuous and bounded function defined on $\mathbf{R}_+$.

If $t=\tau_k$, $k=1,2,\ldots$

$$\begin{aligned} V(\tau_k^+, x(\tau_k^+)) &= \|x(\tau_k^+)\|^2 + \beta e^{-3\varepsilon_0\int_0^{\tau_k}\lambda(s)ds}, \\ &= \|\frac{1}{\sqrt{2}}x(\tau_k^-) + \sqrt{\frac{\mu_k}{2}}\sin(x(\tau_k^-))\|^2 + \beta e^{-3\varepsilon_0\int_0^{\tau_k}\lambda(s)ds}, \\ &\leq \|x(\tau_k^-)\|^2 + \beta e^{-3\varepsilon_0\int_0^{\tau_k}\lambda(s)ds} + \mu_k, \\ &= V(\tau_k^-, x(\tau_k^-)) + \mu_k. \end{aligned}$$

Therefore, all hypothesis of Theorem 1 are satisfied. Then, the system (1) is globally uniformly practically $h-$stable. By using Matlab, a simulation is obtained (see Fig. 1).

Example 2 To illustrate Theorem 4, we consider the following nonlinear impulsive system with delayed perturbations:

$$\begin{aligned} \dot{x} &= F(t,x) + G(t,x)\Big(\sum_{i=1}^{3} g_i(t, x(t-\delta_i(t))) + u(t)\Big), \quad t \neq \tau_k,\ k=1,2,\ldots \\ \Delta x(\tau_k) &= I_k(x(\tau_k^-), u(\tau_k^+)) + J_k(x(\tau_k^-)), \quad t=\tau_k,\ k=1,2,\ldots \end{aligned} \tag{3}$$

where

- $F(t,x)=\lambda(t)\begin{bmatrix} -3sgn(x_1)+\frac{5}{2}x_1-x_1^3x_2^4+\frac{\sin(\sqrt{\frac{3}{2}}x_1)}{x_1(1+x_2^2)} \\ \\ -3sgn(x_2)+\frac{5}{2}x_2-x_2^7x_1^6+\frac{\sin(\sqrt{\frac{3}{2}}x_2)}{x_2(2+x_1^2)} \end{bmatrix}$ and

  $G(t,x)=\begin{bmatrix} 4x_2 & 0 \\ 0 & -2x_1^2 \end{bmatrix}$ with sgn is the signum function given by $sgn(t)=\begin{cases} 1, & t>0 \\ 0, & t=0 \\ -1, & t<0 \end{cases}$.

- $g_1(t,x(t-\delta_1(t)))=\begin{bmatrix} x_2(t-\delta_1(t)) \\ \\ 2\sin(x_1(t-\delta_1(t))) \end{bmatrix}$, $g_2(t,x(t-\delta_2(t)))=\begin{bmatrix} \sqrt{2|x_1(t-\delta_2(t))x_2(t-\delta_2(t))|} \\ \\ 0 \end{bmatrix}$ and $g_3(t,x(t-\delta_3(t)))=\begin{bmatrix} \sin(x_2(t-\delta_3(t))) \\ \\ \sqrt{6|x_1(t-\delta_3(t))x_2(t-\delta_3(t))|} \end{bmatrix}$.

- For all $k=1,2,\ldots$, $I_k(x,u)=(\frac{c_k}{\sqrt{2}}-1)x-\sqrt{\frac{2c_k^2-1}{2}}\cos(x)+\sin(u)$ such that

$$\frac{1}{\sqrt{2}}\le c_k\le 1 \text{ and } \sum_{k=1}^{+\infty} c_k e^{-3\int_0^{\tau_k}\lambda(s)ds}<+\infty.$$

- For all $k=1,2,\ldots$, $J_k(x)=\mu_k\cos(x)$ such that $\mu_k\ge 0$ and $\sum_{k=1}^{+\infty}\mu_k e^{-3\int_0^{\tau_k}\lambda(s)ds}<+\infty$.

In addition, suppose the condition hold:

$$\int_0^{+\infty}\lambda(s)ds=\frac{1}{9e}<\frac{2}{9e}.$$

The unforced nominal system of (3) is given by

$$\begin{aligned} &\dot{x}=F(t,x), && t\ne\tau_k,\ k=1,2,\ldots \\ &\Delta x(\tau_k)=I_k(x(\tau_k^-),0)=(\frac{c_k}{\sqrt{2}}-1)x(\tau_k^-)-\sqrt{\frac{2c_k^2-1}{2}}\cos(x(\tau_k^-)), && t=\tau_k,\ k=1,2,\ldots \end{aligned} \tag{4}$$

Choose a Lyapunov function $V(t,x)$ for the nominal system (4) in order to construct the robust control law in (10) which gives the global uniform practical $h-$stability of the system (3) as follows:

$$V(t,x) = \sqrt{W(t,x)}$$

such that

$$W(t,x) = \|x\|^2 + e^{-6\varepsilon_0 \int_0^t \lambda(s)ds} = x_1^2 + x_2^2 + e^{-6\varepsilon_0 \int_0^t \lambda(s)ds}$$

where $\varepsilon_0 > 0$ is chosen such that

$$\varepsilon_0 e^{-6\varepsilon_0 \int_0^{+\infty} \lambda(s)ds} - \frac{1}{2} \geq 1.$$

It is obvious that for all $t \in \mathbf{R}_+$, $x, y \in \mathbf{R}^n$

$$\|x\| \leq V(t,x) \leq \|x\| + 1 \text{ and } |V(t,x) - V(t,y)| \leq \|x-y\|.$$

For $t \neq \tau_k$, $k = 1, 2, \ldots$, we obtain

$$\begin{aligned}\dot{W}_{(6.4)}(t,x) =& \nabla_t W(t,x) + \nabla_x^T W(t,x).F(t,x),\\ \leq& \frac{h'(t)}{h(t)}\Big(2W(t,x) - 2 - 2\|x\|\Big) \leq \frac{h'(t)}{h(t)}\Big(2W(t,x) - 2\sqrt{W(t,x)}\Big),\end{aligned}$$

where $h(t) = e^{3\int_0^t \lambda(s)ds}$ is a positive, continuous and bounded function defined on $\mathbf{R}_+$.

If $t = \tau_k$, $k = 1, 2, \ldots$, then we have

$$\begin{aligned}W(\tau_k^+, x(\tau_k^+)) =& \|x(\tau_k^+)\|^2 + e^{-6\varepsilon_0 \int_0^{\tau_k} \lambda(s)ds}\\ \leq& c_k^2 W(\tau_k^-, x(\tau_k^-)) + (1 - c_k^2) + 2c_k^2 - 1\\ =& c_k^2 W(\tau_k^-, x(\tau_k^-)) + c_k^2.\end{aligned}$$

It follows that the Lyapunov function $V(t,x) = \sqrt{W(t,x)}$ satisfies:

$$\begin{cases} \dot{V}_{(6.4)}(t,x) = \frac{\dot{W}_{(6.4)}(t,x)}{2\sqrt{W(t,x)}} \leq \frac{h'(t)}{h(t)}\Big(V(t,x) - 1\Big), & t \neq \tau_k,\ k = 1, 2, \ldots \\ V(\tau_k^+, x(\tau_k^+)) = \sqrt{W(\tau_k^+, x(\tau_k^+))} \leq c_k V(\tau_k^-, x(\tau_k^-)) + c_k, & t = \tau_k,\ k = 1, 2, \ldots \end{cases}$$

Then, the feedback controller given by (10) guaranteeing the global uniform practical $h-$stability of the above system can be represented by

$$\begin{cases} u(t) \quad = -\frac{\Big(2\chi_1(t)+\chi_2(t)+2\chi_3(t)\Big)^2}{\sqrt{64x_1^2x_2^2+16x_1^4x_2^2}\Big(2\chi_1(t)+\chi_2(t)+2\chi_3(t)\Big)+\lambda(t)h(t)V(t,x(t))} \begin{bmatrix} 8x_1x_2 \\ -4x_1^2x_2 \end{bmatrix}, t \neq \tau_k,\ r \in \mathbf{N} \\ u(\tau_k^+) = (1 - c_k)x(\tau_k^-), t = \tau_k,\ k = 1, 2, \ldots \end{cases} \tag{5}$$

where $\chi_i(t)$, $i = 1, 2, 3$ is given by (11). Moreover, we obtain an estimate of the convergence for the uniform global practical $h-$stability of the closed-loop system (3)–(5) as

$$\|x(t, t_0, x_0)\| \leq c\|x_0\|H(t)H^{-1}(t_0) + \rho, \;\forall t \geq t_0$$

where $H(t) = h(t)e^{\int_0^t \lambda(s)h(s)ds}$ is a positive, continuous and bounded function defined on $\mathbf{R}_+$.

References

1. Lakshmikantham, V., Bainov, D., Simeonov, P.S.: Theory of Impulsive Differential Equations. Modern Applied Mathematics, vol. 6. World Scientific, New Jersy (1989)
2. Chaillet, A., Loría, A., Kelly, R.: Robustness of PID-controlled Manipulators vis-à-vis actuator dynamics and external disturbances. Eur. J. Control **6**, 563–576 (2007)
3. Byrnes, C.I., Celani, F., Isidori, A.: Omega-limit sets of a class of nonlinear systems that are semiglobally practically stabilized. Int. J. Robust Nonlinear Control **15**(7), 315–333 (2005)
4. Kulev, G.K., Bainov, D.D.: Lipschitz stability of impulsive systems of differential equations. Dyn. Stab. Syst.: Int. J. **8**(1), 1–17 (1993). https://doi.org/10.1080/02681119308806145
5. Choi, S.K., Koo, N., Ryu, C.: Stability of linear impulsive differential equations via t_∞-similarity. J. Chungcheong Math. Soc. **26**(4) (2013)
6. Dishliev, A., Bainov, D.D.: Dependence upon initial conditions and parameters of solutions of impulsive differential equations with variable structure. Int. J. Theor. Phys. **29**, 655–676 (1990)
7. Bainov, D.D., Dishliev, A.B., Stamova, I.M.: Practical stability of the solutions of impulsive systems of differential-difference equations via the method of comparison and some applications to population dynamics. ANZIAM J. **43**, 525–539 (2002)
8. Bainov, D.D., Simeonov, P.S.: Systems with Impulse Effect: Stability, Theory and Applications. Ellis Horwood Limited, Chichester (1989)
9. Bainov, D.D., Simeonov, P.S.: Impulsive Differential Equations: Asymptotic Propertites of the Solutions. World Scientific Publishing Co., Inc, River Edge, NJ (1995)
10. Dlala, M., Ghanmi, B., Hammami, M.A.: Exponential practical stability of nonlinear impulsive systems: converse theorem and applications. In: Dynamics of Continuous, Discrete and Impulsive Systems. Mathematical Analysis, vol. 21, pp. 37–64 (2014)
11. Dlala, M., Hammami, M.A.: Uniform exponential practical stability of impulsive perturbed systems. J. Dyn. Control Syst. **13**(3), 373–386 (2007)
12. Choi, S.K., Koo, N.: Variationally stable impulsive differential systems. Dyn. Syst. **30**(4), 435–449 (2015)
13. Pinto, M.: Stability of nonlinear differential systems. Appl. Anal.: Int. J. **43**(1–2), 1–20 (1992). https://doi.org/10.1080/00036819208840049
14. Pinto, M.: Perturbations of asymptotically stable differential systems. Analysis **4**(1–2), 161–175 (1984)
15. Yoshizawa, T.: Stability Theory by Lyapunov's Second Method. The Mathematical Society of Japan, Tokyo (1966)
16. Lasalle, J., Lefschetz, S.: Stability by Lyapunov Direct Method and Application. Academic Press, New York (1961)
17. La Salle, L.S.: Stability by Lyapunov's Direct Method. Academic Press, New York (1961)
18. Lakshmikantham, V., Leela, S., Martynyuk, A.A.: Practical Stability of Nonlinear Systems. World Scientific Press, Singapore (1990)
19. Lakshmikantham, V., Leela, S., Martynyuk, A.A.: Stability Analysis of Nonlinear Systems. Marcel Dekker, New York (1989)

20. Martynyuk, A.A., Sun, Z.: Practical Stability and Its Applications. Science Press, Beijing (2003). (in Chinese)
21. Martynyuk, A.A.: Stability and control: theory, methods and applications. In: Advances in Stability Theory at the End of the 20th Century, vol. 13. Taylor and Francis, London (2003)
22. Martynuk, A.A.: On exponential stability with respect to some of the variables. Russian Acad. Sci. Dokl. Math. **48**, 1720 (1994)
23. Garashchenko, G., Pichkur, V.: Properties of optimal sets of practical stability of differential inclusions: part I and II. J. Autom. Inf. Sci. **38**, 111 (2006)
24. Liu, B., Hill, D.J.: Uniform stability and ISS of discrete-time impulsive hybrid systems. Nonlinear Anal.: Hybrid Syst. **4**(2), 319–333 (2010)
25. Choi, S.K., Koo, N.J.: Asymptotic equivalence between two linear Volterra difference systems. Comput. Math. Appl. **47**, 461–471 (2004)
26. Choi, S.K., Goo, Y.H., Koo, N.J.: Lipschitz stability and exponential asymptotic stability for the nonlinear functional differential systems. Dyn. Syst. Appl. **6**, 397–410 (1997)
27. Choi, S.K., Koo, N.J., Ryu, H.S.: $h-$stability of differential systems via t_∞-similarity. Bull. Korean Math. Soc. **34**, 371–383 (1997)
28. Yang, T.: Impulsive Control Theory. Springer, Berlin, Germany (2001)
29. Caraballo, T., Hammami, M.A., Mchiri, L.: Practical asymptotic stability of nonlinear stochastic evolution equations. Stoch. Anal. Appl. **32**, 77–8 (2014)
30. Caraballo, T., Hammami, M.A., Mchiri, L.: Practical exponential stability in mean square of stochastic partial differential equations. Collect. Math. **66**(2), 261–271 (2015)
31. Pinto, M.: Asymptotic behavior of differential systems with impulse effect. Nonlinear Anal., Theory, Methods Appl. **30**(2), 1133–1140 (1997)
32. Medina, R., Pinto, M.: Variationally stable difference equations. Nonlinear Anal. **30**, 1141–1152 (1997)
33. Choi, S.K., Koo, N.J., Ryu, H.S.: $h-$stability of differential systems via t_∞-similarity. Bull. Korean Math. Soc. **34**, 371–383 (1997)
34. Choi, S.K., Koo, N.: A converse Theorem on $h-$stability via impulsive variational systems. J. Korean Math. Soc. **53**(5), 1115–1131 (2016)
35. Choi, S.K., Koo, N.: Variationally stable impulsive differential systems. Dyn. Syst. **30**(4), 435–449 (2015)

Practical Exponential Stabilization for Semi-Linear Systems in Hilbert Spaces

Hanen Damak

Abstract In this chapter, we investigate the practical exponential stabilization of semi-linear systems satisfying some relaxed conditions in Hilbert spaces. Sufficient conditions for stabilizability are established by solving a standard Lyapunov equation. We propose some classes of memoryless state linear and nonlinear feedback controllers. Numerical examples are given to validate the proposed theoretical results.

Keywords Semi-linear systems · Practical exponential stabilization · Partial differential equation · Feedback controller · Lyapunov method

2000 Mathematics Subject Classification: 35B35 · 35B40 · 35K58 · 93C10

1 Introduction

State feedback stabilization of infinite-dimensional nonlinear systems has been the focus of intensive work in the last decades, see [1–7, 10, 12, 18, 20] and the references cited therein. Moreover, there are many results about the stabilization of fractional-order systems, see [8, 9, 14–17]. In various situations, it is difficult to design a control law guaranteeing the stabilization for semi-linear systems in Hilbert spaces. Consequently, relaxing this stability issue is interesting. For instance, one can only ensure that the system trajectories converge within a ball centered at the origin. For this aim, a more general stability property called practical stability [2, 11] has been investigated. The practical stabilization is to find the state feedback candidate such that the solution of the closed-loop system is practically exponentially stable in the Lyapunov sense in which the origin was not supposed to be necessarily an equilibrium point of the system. In this case, the authors in [2] investigated the practical stabilization for a class of time-varying control systems in Hilbert spaces. Over recent years, the theory of robust stability analysis of partial differential equations (PDEs)

H. Damak (✉)
Faculty of Sciences of Sfax, University of Sfax, Road of Soukra, BP 1171, 3000 Sfax, Tunisia
e-mail: hanen.damak@yahoo.fr

A. Ben Makhlouf et al. (eds.), *State Estimation and Stabilization of Nonlinear Systems*, Studies in Systems, Decision and Control 491,
https://doi.org/10.1007/978-3-031-37970-3_2

has been extensively studied by many researchers in the qualitative theory of control dynamical systems, see [1, 2, 6]. This theory exploits a broad range of techniques and tools from such diverse fields as Lyapunov methods and semigroup, etc. Sufficient conditions are given that verify the practical stabilization of non-autonomous infinite-dimensional systems depending on a parameter, see [5].

The main contribution of this chapter is the study of the problem of feedback exponential stabilization for a class of semi-linear systems in Hilbert spaces based on Lyapunov functions. We refer the reader to [1, 5, 19] and the references therein for more information on this direction. We show, how under the assumption of stabilizability of the pair (A, B), we can construct a stabilizing two-controller feedback law: a linear controller and a nonlinear controller. A new approach for the stability analysis is proposed.

The rest of this chapter is organized as follows. In Sect. 2, the system description, notations and some preliminary results are presented. The required assumptions and the statement of the main results are provided in Sect. 3. Two simulation examples are carried out in Sect. 4 to show the correctness of the derived main results. A general conclusion is given in the final section of this chapter.

2 Preliminaries

We use the following notation throughout the paper. $\mathbb{R}_+$ denotes the set of all non-negative real numbers. Let X and U denote a state space and a space of input values respectively and let both of them be Hilbert, endowed with norms $\|\cdot\|_X$ and $\|\cdot\|_U$. For linear normed spaces X, Y let $L(X, Y)$ be the space of bounded linear operators from X to Y and $L(X) := L(X, X)$. A norm in these spaces we denote by $\|\cdot\|$. For an operator A, A^* is the adjoint, $Dom(A)$ is the domain and I is the identity operator. $C(X, Y)$ denotes the space of all continuous functions from X to Y. $L_p(0, \ell),\ p \geq 1$ is the space of p-th power integrable functions $f : (0, \ell) \to \mathbb{R}$ with the norm $\|f\|_{L_p(0,\ell)} = (\int_0^\ell |f(x)|^p dx)^{\frac{1}{p}}$. The space $H^k(0, \ell),\quad k \in \mathbb{N}$, is a Sobolev space of functions $u \in L_p(0, \ell)$, such that for each natural $i \leq k$, the weak derivative $u^{(i)}$ exists and belongs to $L_p(0, \ell)$. $H_0^k(0, \ell)$ denotes the closure of smooth functions with compact support in $(0, \ell)$ in the norm of $H^k(0, \ell),\quad k \in \mathbb{N}$.

Consider the following non-autonomous semi-linear evolution equation:

$$\begin{cases} \dot{x}(t) = Ax(t) + B[u(t) + \Gamma(t, x(t))], \\ x(0) = x_0, \end{cases} \tag{1}$$

where $t \geq 0$ is the time, $x(t) \in X$ is the system state, $u(t) \in U$ is the control input, x_0 is the initial condition, A is the infinitesimal generator of an analytic semigroup $T(t)$ and $B \in L(U, X)$ with $\|B\| \neq 0$. The operator $\Gamma : \mathbb{R}_+ \times X \longrightarrow X$ is continuous in t and is locally Lipschitz continuous in x, uniformly in t on bounded intervals.

We consider mild solutions of (1), i.e. solutions of the integral form

$$x(t) = T(t)x_0 + \int_0^t T(t-s)B[u(s) + \Gamma(s, x(s))]ds \tag{2}$$

belonging to the class $C([0, t_m], X)$ for some $t_m > 0$.

Under the condition that Γ is locally Lipschitz continuous in x, uniformly in t on bounded intervals, it is shown in [19, Theorem 1.4] that Eq. (3) has a unique mild solution on $[0, t_m]$. Moreover, if $t_m < \infty$, then $\lim_{t \to t_m} \|x(t)\|_X = \infty$.

The corresponding system without perturbations, called the nominal system, is described by

$$\dot{x}(t) = Ax(t), \quad x(0) = x_0, \quad t \geq 0. \tag{3}$$

Next, we recall the definition of the generator of an exponentially stable semi-group as well as that of the exponential stabilizability, see Curtain and Zwart [1] for more details.

Definition 1 The operator A generates an exponentially stable semigroup $T(t)$ if the initial value problem (3) has a unique solution $x(t) = T(t)x_0$, and $\|T(t)\| \leq Ce^{-\gamma t}$, for all $t \geq 0$ with some positive numbers C and γ.

Definition 2 The pair $\{A, B\}$ is said to be exponentially stabilizable if there exists a feedback operator $D \in L(X, U)$, such that the operator $A + BD$ generates an exponentially stable semigroup S_{BD}.

To study the stability properties of (1) with respect to external inputs, we use the notion of practical exponential stabilizability.

Definition 3 ([3, Definition 1]) System (1) is practically exponentially stabilizable if there exists a continuous feedback control $u : [0, \infty) \to U$, such that system (1) in closed-loop with $u(t)$ satisfies the following properties:

(i) For any initial condition $x_0 \in X$, there exists a unique mild solution $x(t)$ defined on $[0, \infty)$.

(ii) There exist positive scalars ω, c, η, such that the solution of the system (1) satisfies

$$\|x(t)\|_X \leq c\|x_0\|_X e^{-\omega t} + \eta, \quad \forall t \geq 0. \tag{4}$$

When (i) and (ii) are satisfied for (1), we say that (1) in closed-loop with $u(t)$ is globally practically uniformly exponentially stable, see [2, 3, 6] for more details.

Remark 1 The inequality (4) implies that the trajectory will be ultimately bounded. That is the solution is bounded and approaches a neighborhood of the origin for sufficiently large t.

Recall that a self-adjoint operator $P \in L(X)$ is positive if

$$\langle Px, x \rangle > 0 \text{ for all } x \in X \backslash \{0\}.$$

A positive operator $P \in L(X)$ is called coercive if there exists $c > 0$, such that

$$\langle Px, x\rangle \geq c\|x\|_X^2 \text{ for all } x \in Dom(P).$$

Proposition 1 ([1, Theorem 5.1.3]) *Suppose that A is the infinitesimal generator of the C_0−semigroup $T(t)$ on the Hilbert space X. Then, $T(t)$ is exponentially stable if and only if there exists a coercive positive self-adjoint operator $P \in L(X)$ and a constant $\xi > 0$, such that*

$$\langle Ax, Px\rangle + \langle Px, Ax\rangle \leq -\xi\langle x, x\rangle, \quad \forall x \in Dom(A). \tag{5}$$

3 Main Results

3.1 Exponential Stabilization by Linear State Controller

In this part, we establish the exponential stabilization in the practical sense for (1) via a linear state estimate controller under some restrictions on the nonlinearities. Let's consider the following assumptions.

($\mathcal{H}_1$) The pair $\{A, B\}$ is exponentially stabilizable, that is there exists a constant operator $D \in L(X, U)$, a constant $\mu > 0$, and a coercive positive self-adjoint operator P

$$\mu I \leq P \leq \|P\| I, \tag{6}$$

where $\mu > 0$, which satisfies

$$\langle A_D^* x, Px\rangle + \langle Px, A_D x\rangle \leq -\xi\langle x, x\rangle, \quad \forall x \in Dom(A_D), \tag{7}$$

with $A_D = A + BD$.

($\mathcal{H}_2$) The nonlinear operator $\Gamma(t, x)$ satisfies:

$$\|\Gamma(t, x)\|_X \leq \kappa\|x\|_X + \sigma(t), \quad \forall t \geq 0, \quad \forall x \in X, \tag{8}$$

where $\kappa > 0$ and $\sigma : \mathbb{R}_+ \longrightarrow \mathbb{R}_+$ is a continuous function satisfying

$$\left(\int_0^\infty \sigma(s)^2\, ds\right)^{\frac{1}{2}} \leq M_\sigma < +\infty.$$

Theorem 1 *Suppose that assumptions ($\mathcal{H}_1$) and ($\mathcal{H}_2$) hold, and the constant κ verifies*

$$\kappa < \frac{\xi}{2\|P\|\|B\|}$$

Then, the system (1) *in closed-loop with the linear feedback*

$$u(t) = Dx(t) \tag{9}$$

is globally uniformly practically exponentially stable.

Proof First, note that under the assumptions of the theorem, it is easy to see that Eq. (1) has a unique mild solution $x \in C([0, \infty), X)$ for any $x_0 \in X$ by applying the results of Pazy [19, Theorem 1.4] for any $x_0 \in X$ and this mild solution is even a classical solution which satisfies (2). We consider the following Lyapunov function:

$$V(t, x) = \langle Px, x \rangle.$$

Let us compute the derivative of V with respect to the system (1) in closed-loop with the controller (9). For $x(t) \in Dom(A_D) = Dom(A)$, we have

$$\begin{aligned}\dot{V}(t, x) &= \langle P\dot{x}, x \rangle + \langle Px, \dot{x} \rangle \\ &= \langle P[A_D x + B\Gamma(t, x)], x \rangle + \langle Px, [A_D x + B\Gamma(t, x)] \rangle.\end{aligned}$$

Since

$$\langle Px, A_D x \rangle = \langle A_D^* Px, x \rangle,$$

by applying the Lyapunov Eq. (7) and using Cauchy-Schwartz inequality, we obtain

$$\dot{V}(t, x) \leq -\xi \langle x, x \rangle + 2\|P\|\|B\|\|\Gamma(t, x)\|_X \|x\|_X.$$

It follows by (8) that

$$\dot{V}(t, x) \leq -\Lambda \|x\|_X^2 + 2\|P\|\|B\|\sigma(t)\|x\|_X, \tag{10}$$

where $\Lambda = \xi - 2\|P\|\|B\|\kappa$.
Then, we get from (10) that

$$\dot{V}(t, x) \leq -\frac{\Lambda}{\|P\|} V(t, x) + \frac{2\|P\|\|B\|\sigma(t)}{\sqrt{\mu}} \sqrt{V(t, x)}. \tag{11}$$

Since A generated an analytic semigroup, we can apply a density argument for the operator A to prove that (11) holds on the whole X. Let

$$\ell(t) = \sqrt{V(t, x)}.$$

The derivative of ℓ with respect to time is given by

$$\dot{\ell}(t) \leq -\frac{\Lambda}{2\|P\|} \ell(t) + \frac{\|P\|\|B\|\sigma(t)}{\sqrt{\mu}}. \tag{12}$$

Then, from (12) we have

$$\ell(t) \leq \ell(0)e^{-\frac{\Lambda}{2\|P\|}t} + \frac{\|P\|\|B\|}{\sqrt{\mu}}e^{-\frac{\Lambda t}{2\|P\|}}\int_0^t \sigma(s)e^{\frac{\Lambda s}{2\|P\|}}ds.$$

Thus, by using the Hölder inequality one obtains

$$\begin{aligned}
\ell(t) &\leq \ell(0)e^{-\frac{\Lambda}{2\|P\|}t} + \frac{\|P\|\|B\|}{\sqrt{\mu}}e^{-\frac{\Lambda t}{2\|P\|}}((\int_0^\infty \sigma(s)^2 ds)^{\frac{1}{2}}.(\int_0^t e^{\frac{\Lambda s}{\|P\|}})^{\frac{1}{2}}) \\
&\leq \ell(0)e^{-\frac{\Lambda}{2\|P\|}t} + \frac{M_\sigma\|P\|\|B\|}{\sqrt{\mu}}e^{-\frac{\Lambda t}{2\|P\|}}\left(\frac{\|P\|}{\Lambda}(e^{\frac{\Lambda t}{\|P\|}t} - 1)\right)^{\frac{1}{2}} \\
&\leq \ell(0)e^{-\frac{\Lambda}{2\|P\|}t} + \frac{M_\sigma\|P\|\|B\|}{\sqrt{\mu}}e^{-\frac{\Lambda t}{2\|P\|}}\left(\frac{\|P\|}{\Lambda}e^{\frac{\Lambda t}{\|P\|}t}\right)^{\frac{1}{2}}.
\end{aligned}$$

Therefore, by (6), it follows that

$$\|x(t)\|_X \leq \sqrt{\frac{\|P\|}{\mu}}\|x_0\|_X e^{-\frac{\Lambda}{2\|P\|}t} + \frac{M_\sigma\|P\|^{\frac{3}{2}}\|B\|}{\mu\sqrt{\Lambda}}.$$

Hence, the system (1) in closed-loop with the linear feedback (9) is globally uniformly practically exponentially stable.

To investigate the practical stabilization of the semi-linear system (1) we shall suppose the following assumption on the nonlinearities more than considered in Theorem 1.

($\mathcal{H}_3$) Assume that,

$$\|\Gamma(t,x)\|_X \leq \epsilon\|x\|_X + \nu(t)\|x\|_X^\alpha, \quad 0 < \alpha < 1, \quad \forall x \in X, \quad \forall t \geq 0, \tag{13}$$

where $\epsilon > 0$ and $\nu : \mathbb{R}_+ \longrightarrow \mathbb{R}_+$ is a continuous function satisfying

$$(\int_0^\infty \nu(s)^2\, ds)^{\frac{1}{2}} \leq M_\nu < +\infty. \tag{14}$$

We have the following theorem.

Theorem 2 *Suppose that assumptions ($\mathcal{H}_1$) and ($\mathcal{H}_3$) hold, and the constant ϵ verifies*

$$\epsilon < \frac{\xi}{2\|P\|\|B\|}$$

Then, the system (1) in closed-loop with the linear feedback (9) is globally uniformly practically exponentially stable.

Proof First, note that under the assumptions of the theorem, it is easy to see that Eq. (1) has a unique global mild solution $x(t)$ by applying the results of Pazy [19, Theorem 1.4] for any $x_0 \in X$ and this mild solution is even a classical solution. Define the Lyapunov function:

$$V(t,x) = \langle Px, x\rangle.$$

Let us compute the derivative of V with respect to the system (1) in closed-loop with the controller (9).

$$\begin{aligned}\dot{V}(t,x) &= \langle P\dot{x}, x\rangle + \langle Px, \dot{x}\rangle \\ &= \langle P[A_D x + B\Gamma(t,x)], x\rangle + \langle Px, [A_D x + B\Gamma(t,x)]\rangle.\end{aligned}$$

By using Cauchy Schwartz inequality and (13), we have

$$\dot{V}(t,x) \le -(\xi - 2\epsilon\|P\|\|B\|)\|x\|_X^2 + 2\|P\|\|B\|\nu(t)\|x\|_X^{\alpha+1}.$$

Since

$$\mu\|x\|_X^2 \le V(t,x) \le \|P\|\|x\|_X^2,$$

we get

$$\dot{V}(t,x) \le -\varsigma V(t,x) + \widetilde{\nu}(t)V(t,x)^{\frac{\alpha+1}{2}}, \tag{15}$$

where $\varsigma = \frac{\xi - 2\epsilon\|P\|\|B\|}{\|P\|}$ and $\widetilde{\nu}(t) = \frac{2\|P\|}{\mu^{\frac{\alpha+1}{2}}}$. By using a density argument for the operator A, to prove that (15) hold on the whole X.
Let

$$\theta(t) = V(t,x)^{\frac{1-\alpha}{2}}.$$

The derivative of θ with respect to time satisfies:

$$\dot{\theta}(t) \le -\varsigma\frac{1-\alpha}{2}\theta(t) + \frac{1-\alpha}{2}\widetilde{\nu}(t).$$

Then it follows:

$$\theta(t) \le \theta(0)e^{-\varsigma\frac{1-\alpha}{2}t} + \frac{1-\alpha}{2}e^{-\varsigma\frac{1-\alpha}{2}t}\int_0^t \widetilde{\nu}(s)e^{\varsigma\frac{1-\alpha}{2}s}ds.$$

Using Hölder's Inequality, we obtain

$$\begin{aligned}\theta(t) &\le \theta(0)e^{-\varsigma\frac{1-\alpha}{2}t} + \frac{1-\alpha}{2}e^{-\varsigma\frac{1-\alpha}{2}t}\Big(\int_0^\infty \widetilde{\nu}(s)^2 ds\Big)^{\frac{1}{2}}\Big(\Big(\int_0^t e^{2\varsigma\frac{1-\alpha}{2}s}ds\Big)^{\frac{1}{2}}\Big) \\ &\le \theta(0)e^{-\varsigma\frac{1-\alpha}{2}t} + \frac{1-\alpha}{2}e^{-\varsigma\frac{1-\alpha}{2}t}\frac{2M_\nu}{(1-\alpha)\varsigma}(e^{\varsigma(1-\alpha)t} - 1)^{\frac{1}{2}}.\end{aligned}$$

Thus, by (6), it follows that

$$\|x(t)\|^{1-\alpha} \leq \left(\sqrt{\frac{\|P\|}{\mu}}\right)^{1-\alpha} \|x_0\|_X^{1-\alpha} e^{-\varsigma\frac{1-\alpha}{2}t} + \frac{M_\nu}{\varsigma\mu^{\frac{1-\alpha}{2}}}$$

Using $a^p + b^p \leq \frac{1}{p}(a+b)^p$, for any $a, b \in \mathbb{R}_+$ and any $p \in]0, 1[$, one has

$$\|x(t)\|_X \leq \left(\frac{1}{1-\alpha}\right)^{\frac{1}{1-\alpha}} \sqrt{\frac{\|P\|}{\mu}} \|x_0\|_X e^{-\frac{\varsigma}{2}t} + \frac{M_\nu^{\frac{1}{1-\alpha}}}{\varsigma\sqrt{\mu}}.$$

We deduce that, the system (1) in closed-loop with the linear feedback (9) is globally uniformly practically exponentially stable.

As a special case of Theorem 2, the following result can be easily obtained.

Corollary 1 *Suppose that Assumptions* ($\mathcal{H}_1$) *and* ($\mathcal{H}_3$) *hold and the operator* $\Gamma(t, x)$ *satisfies for all* $t \geq t_0 \geq 0$, *and all* $x \in X$

$$\|\Gamma(t, x)\|_X \leq \nu(t)\|x\|_X^\alpha, \quad 0 < \alpha < 1, \quad \forall x \in X, \quad \forall t \geq 0, \tag{16}$$

such that $\nu : \mathbb{R}_+ \longrightarrow \mathbb{R}_+$ *is a continuous function satisfying (14). Then, the system* (1) *in closed-loop with the linear feedback (9) is globally uniformly practically exponentially stable.*

3.2 Exponential Stabilization by Nonlinear State Controller

In this subsection, we give a sufficient condition, in terms of a perturbation term, for the non-autonomous evolution Eq. (1) to be uniformly practically exponentially stabilizable. We shall construct a nonlinear feedback law that makes the system (1) globally uniformly practically exponentially stable.
For this purpose, the following assumptions are considered.

($\mathcal{H}_4$) The operator A is exponentially stable, there exists a coercive positive self-adjoint operator P, and a positive constant κ, such that

$$\mu I \leq P \leq \|P\| I,$$

where $\mu > 0$, which satisfies

$$A^*P + PA \leq -\kappa I. \tag{17}$$

($\mathcal{H}_5$) There exists a continuous function $\omega : X \to \mathbb{R}_+$, such that

$$\|\Gamma(t,x)\|_X \leq \omega(x),$$

for all $t \in \mathbb{R}_+$ and all $x \in X$.

We are now in a position to present the following theorem.

Theorem 3 *Suppose that Assumptions* ($\mathcal{H}_3$) *and* ($\mathcal{H}_4$) *hold. Then, the control system* (1) *is uniformly practically exponentially stabilizable by the nonlinear feedback*

$$u(t) = -\frac{B^* P x \omega(x)^2}{\|B^* P x\| \omega(x) + \rho}, \tag{18}$$

where ρ *is a positive constant.*

Proof First, note that under the assumptions of the theorem, the mild solution $x(t)$ of (1) exists and is unique, according to a classical existence and uniqueness theorem ([19, Theorem 1.4]). Let $t_m \in (0, \infty]$ be such that $[0, t_m)$ is the maximal time interval over which a system (1) admits a solution. We consider the following Lyapunov function:

$$V(t,x) = \langle Px, x \rangle.$$

Let us compute the derivative of V with respect to the system (1) in closed-loop with the controller (18). For $x(t) \in Dom(A)$, we obtain

$$\begin{aligned} \dot{V}(t,x) &= \langle P\dot{x}, x\rangle + \langle Px, \dot{x}\rangle \\ &= \langle P(Ax(t) + Bu(t) + B\Gamma(t, x(t)), x\rangle + \langle Px, Ax(t) + Bu(t) + B\Gamma(t, x(t))\rangle. \end{aligned}$$

Since

$$\langle Px, Ax\rangle = \langle A^* Px, x\rangle,$$

by applying the Lyapunov Eq. (17) in the form

$$\langle A^* Px, x\rangle + \langle PAx, x\rangle + \leq -\kappa \|x\|_X^2,$$

and using Cauchy-Schwartz inequality, we obtain

$$\dot{V}(t,x) \leq -\kappa\|x\|_X^2 + 2\|B^* Px\| \|\Gamma(t,x)\|_X - \frac{2\|B^* Px\|^2 \omega(x)^2}{\|B^* Px\|\omega(x) + \rho}$$

Together with Assumption ($\mathcal{H}_5$), we get

$$\begin{aligned} \dot{V}(t,x) &\leq -\frac{\kappa}{\|P\|} V(t,x) + \frac{2\rho \|B^* Px\|\omega(x)}{\|B^* Px\|\omega(x) + \rho} \\ &\leq -\frac{\kappa}{\|P\|} V(t,x) + 2\rho. \end{aligned}$$

Then, the system (1) has a unique mild solution, defined on $[0, \infty)$, and this system in closed-loop with the nonlinear feedback (18) is globally uniformly practically exponentially stable.

4 Examples

In this section, we illustrate our results with several examples of partial differential equations (PDEs) by semi-linear evolution equations.

Example 1 We Consider the controlled heat equation

$$\begin{cases} \dfrac{\partial x(l,t)}{\partial t} = \dfrac{\partial^2 x(l,t)}{\partial^2 l} + \mathbf{1}_{[\frac{1}{2},1]}(l)u(l,t) + \dfrac{1}{\sqrt{1+t^2}}\sqrt{\|x(l,t)\|}, \\ \dfrac{\partial x}{\partial l}(0,t) = 0 = \dfrac{\partial x}{\partial l}(1,t), \quad x(l,0) = x_0(l), \quad t \geq 0, \end{cases} \tag{19}$$

We take $X = L_2(0,1)$ as state space and $U = \mathbb{C}$ as input space. Let A be the operator defined by $A = \dfrac{\partial^2}{\partial^2 l}$, where $D(A) = \{h \in L_2(0,1), h, \frac{\partial h}{\partial l}$ are absolutely continuous, $\frac{\partial^2 h}{\partial^2 l} \in L_2(0,1)$ and $\dfrac{dh}{dl}(0) = 0 = \dfrac{dh}{dl}(1)\}$. The input operator $B = \mathbf{1}_{[\frac{1}{2},1]}(l)$ and the nonlinear operator is defined by $\Gamma(t,x) = \dfrac{1}{\sqrt{1+t^2}}\sqrt{\|x(l,t)\|}$, where $\mathbf{1}_{[\frac{1}{2},1]}(l) = \begin{cases} 1, \text{ if } \frac{1}{2} \leq l \leq 1 \\ 0, \text{ elsewhere.} \end{cases}$

On the other hand, A has the eigenvalues $0, -n^2\pi^2,\ n \geq 1$ and the corresponding eigenvectors $\{1, \sqrt{2}\cos(n\pi l),\ n \geq 1\}$. It follows from [1] that, A is the infinitesimal generator of the C_0-semigroup. We choose a stabilizing feedback

$$u(t) = Dx, \tag{20}$$

with $Dx = -3\langle x, 1\rangle$, where $\langle , \rangle$ is the inner product on $L_2(0,1)$. It is easy to verify that $A + BD$ has the eigenvalues $-3, -(n\pi)^2, n \geq 1$. Thus, the pair $\{A, B\}$ is exponentially stabilizable. Moreover, the condition (16) is satisfied with $\nu(t) = \frac{1}{\sqrt{1+t^2}}$ and $\alpha = \frac{1}{2}\cdot$ From Corollary 1 one can conclude that the system (19) in closed-loop with the linear feedback (20) is globally uniformly practically exponentially stable.

Figure 1 show the evolution of the state $x(l,t)$ of the closed-loop system (19) and (20) with the initial condition $x_0(l) = l(1-l)$.

Example 2 Let $c > 0$ and $L > 0$. Consider the following reaction-diffusion equation

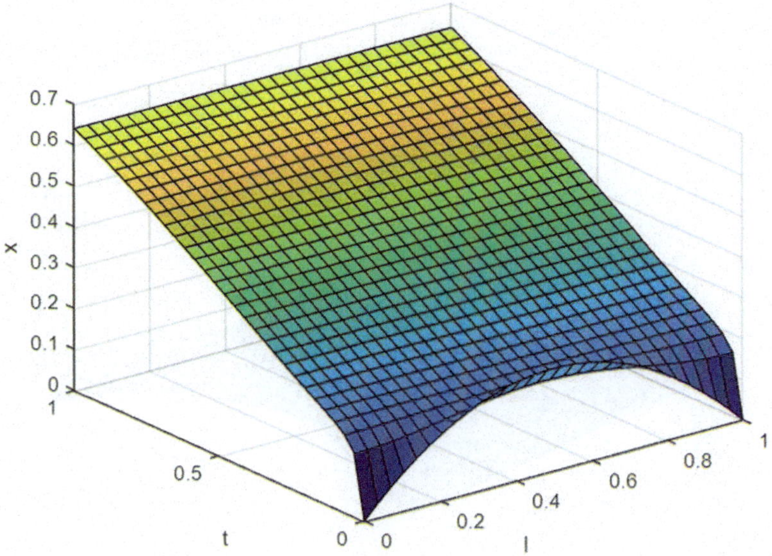

Fig. 1 Evolution of the state $x(l, t)$ of the closed-loop system (19) and (20)

$$\begin{cases} \dfrac{\partial x(\zeta, t)}{\partial t} = c\dfrac{\partial^2 x(\zeta, t)}{\partial^2 \zeta} + u(t) + e^{-t}\cos(x(\zeta, t))x(\zeta, t), \\ \\ x(0, t) = 0 = x(L, t), \quad x(\zeta, 0) = x_0(\zeta), \end{cases} \tag{21}$$

on the region $(\zeta, t) \in (0, L) \times (0, \infty)$ of the $\mathbb{R}-$ valued functions $x(\zeta, t)$ and $u(\zeta, t)$.

We denote $X = L_2(0, L)$, $U = C([0, L])$ and $Ah = c\dfrac{\partial^2 h}{\partial^2 \zeta}$, with domain $D(A) = H_0^1(0, L) \cap H^2(0, L)$ is a generator of an analytic semigroup. We choose the norm on $H_0^1(0, L)$ as

$$\|s\|_{H_0^1(0,L)} = \left(\int_0^L \left(\frac{\partial s}{\partial x}\right)^2 dx\right)^{\frac{1}{2}}.$$

Moreover, $Spec(A) = \{-c(\frac{\pi n}{L})^2 \backslash n = 1, 2, ...\}$. Then, the operator A is exponentially stable.

Let B be the input operator operator defined by $B = I$ and the nonlinear operator is defined by

$$\Gamma(t, x) = e^{-t}\cos(x(\zeta, t))x(\zeta, t)\cdot$$

Take $P = \frac{1}{2c}(\frac{L}{\pi})^2 I$. We have

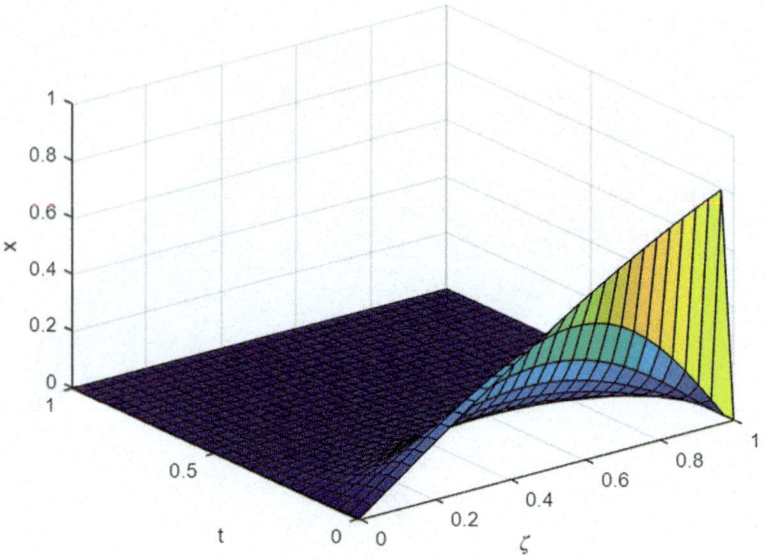

Fig. 2 Evolution of the state $x(\zeta, t)$ of the closed-loop system (21) and (22)

$$\langle Ax, Px\rangle + \langle Px, Ax\rangle = \frac{1}{c}\left(\frac{L}{\pi}\right)^2 \langle Ax, x\rangle = \left(\frac{L}{\pi}\right)^2 \int_0^L \frac{\partial^2 x}{\partial^2 \zeta} x d\zeta.$$

Applying the Friedrich's inequality (see [13]), we obtain

$$\begin{aligned}\langle Ax, Px\rangle + \langle Px, Ax\rangle &\leq -\left(\frac{L}{\pi}\right)^2 \int_0^L \left(\frac{\partial x}{\partial \zeta}\right)^2 d\zeta \\ &\leq -\|x\|^2_{L_2(0,L)}.\end{aligned}$$

Hence, the Assumption $(\mathcal{H}_3)$ is fulfilled with $\mu = \|P\| = \frac{1}{2c}(\frac{L}{\pi})^2$ and $\kappa = 1$.

One can see that Assumption $(\mathcal{H}_5)$ is satisfied with $\omega(x) = x$. From (18) the proposed stabilizing nonlinear feedback law is given by

$$u(t) = -\frac{L^2 x(\zeta, t)^3}{L^2 x(\zeta, t)\|x(\zeta, t)\| + 2c\pi^2}. \tag{22}$$

Then, all hypotheses of Theorem 3 are satisfied. Hence, we obtain the practical uniform exponential stability of the closed-loop system (21) with the controller (22).

For simulation, let's take $c = L = 1$. Figure 2 show the evolution of the state $x(\zeta, t)$ of the closed-loop system (21) and (22) with the initial condition $x_0(\zeta) = \sin(\zeta)$.

5 Conclusion

In this chapter, some novel results related to the practical exponential stabilization problem for semi-linear systems have been given in Hilbert spaces by using Lyapunov techniques. The nonlinearities are supposed uniformly bounded by some known functions where the nominal systems are linear. Finally, some simulation results are given to validate the theoretical results.

References

1. Curtain, R.F., Zwart, H.J.: An Introduction to Infinite Dimensional Linear Systems Theory. Springer, New York (1995)
2. Damak, H., Hammami, M.A.: Stabilization and practical asymptotic stability of abstract differential equations. Numer. Funct. Anal. Optim. **37**, 1235–1247 (2016)
3. Damak, H.: On the practical compensator design of time-varying perturbed systems in Hilbert spaces. Numer. Funct. Anal. Optim. **34**, 650–666 (2022)
4. Damak, H.: Compensator Design Via the Separation Principle for a Class of Nonlinear Uncertain Evolution Equations on a Hilbert Space. Stud. Syst. Decis. Control. **410**, 87–100 (2022)
5. Damak, H., Hammami, M.A.: Stabilization of non-autonomous infinite-dimensional systems depending on a parameter. Int. J. Comput. Math. **100**(3), 666–680 (2023)
6. Damak, H., Hammami, M.A., Maaloul, K.: Stability analysis for non-autonomous semilinear evolution equations in Hilbert spaces: a practical approach. Oper. Matrices. **16**, 1045–1062 (2023)
7. Damak, H.: Compensator Design Via the Separation Principle for a Class of Semilinear Evolution Equations. Ukr. Math. J. (2023). https://doi.org/10.1007/s11253-023-02131-8
8. Jmal, A., Naifar, O., Ben Makhlouf, A., Derbel, N., Hammami, M.A.: On observers design for nonlinear Caputo fractional-order systems. Asian J. Control. **20**, 1–8 (2018)
9. Jmal, A., Elloumi, M., Naifar, O., Ben Makhlouf, A., Hammami, M.A.: State estimation for nonlinear conformable fractional-order systems: a healthy operating case and a faulty operating case. Asian J. Control. *22*(5), 1870–1879 (2020)
10. Karafyllis, I., Jiang, Z.P.: Stability and Stabilization of Nonlinear Systems. Springer, London (2011)
11. Lakshmikantham, V., Leela, S., Martynyuk, A.A.: Practical Stability of Nonlinear Systems. World Scientific, Singapore (1998)
12. Luo, Z.H., Guo, B.Z., Morgul, O.: Stability and Stabilization of Infinite Dimensional Systems With Applications. Springer, London (1999)
13. Mitrinovic, D.S., Pecaric, J.E., Fink, A.M.: Inequalities Involving Functions and Their Integrals and Derivatives. Kluwer Academic Publishers Group, Dordrecht (1991)
14. Naifar, O., Ben, Makhlouf A., Hammami, M.A., Ouali, A.: State feedback control law for a class of nonlinear time-varying system under unknown time-varying delay. Nonlinear Dyn. **82**, 349–355 (2015)
15. Naifar, O., Ben, Makhlouf A., Hammami, M.A., Ouali, A.: On observer design for a class of nonlinear systems including unknown time-delay. Mediterr. J. Math. **13**, 2841–2851 (2016)
16. Naifar, O., Rebiai, G., Ben, Makhlouf A., Hammami, M.A., Guezane-Lakoud, A.: Stability analysis of conformable fractional-order nonlinear systems depending on a parameter. J. Appl. Anal. **26**, 287–296 (2020)
17. Naifar, O., Ben Makhlouf, A.: Fractional Order Systems-Control Theory and Applications. Studies in Systems, Decision and Control. Springer (2022)
18. Orlov, Y., Lou, Y., Christofides, P.D.: Robust stabilization of infinite-dimensional systems using sliding-mode output feedback control. Int. J. Control. **77**, 1115–1136 (2004)

19. Pazy, A.: Semigroups of Linear Operators and Applications to Partial Differential Equations. Springer, New York (1983)
20. Phat, V.N., Kiet, T.T.: On the Lyapunov equation in Banach spaces and applications to control problems. Int. J. Math. Math. Sci. **29**, 155–166 (2002)

An Observer Controller for Delay Impulsive Switched Systems

Imen Ellouze

Abstract In this paper the asymptotic stability and the observer-based controller problems for a new class of delayed-impulsive switched systems perturbed by a nonlinearity satisfying the sector condition are studied. By building appropriate Lyapunov-Krasovskii functions that take into account all delay informations, namely the lower and upper bounds and the upper bound of its first derivative, a new stability criteria is derived and formulated in the form of linear matrix inequalities. The first criterion is used to construct delay-dependent dynamic observer-based output feedback control. In order to reduce conservatism, some relaxation matrices which have made it possible to extend the field of feasibility of the LMIs and to tolerate fast delay are introduced. Furthermore, a numerical example is given to prove effectiveness of the proposed criteria.

Keywords Impulsive switched systems · Time-varying delay · Linear matrix inequality · Stability analysis · Observer-based control

Mathematics Subject Classification: 93C10 · 93C30 · 93D05 · 93D20

1 Introduction

A switched system is a hybrid dynamical system consisting of a family of continuous-time or discrete-time subsystems and a switching law that orchestrates the switching between them. Such systems have received considerable attention for several decades due to their numerous applications in many fields, such as mechanical systems, automotive industry, aircraft and air traffic control. Up until the present moment, many available results on the stability analysis and controller synthesis of switched systems have been established, for example [15, 19]. However, these systems can not cover some useful switched systems existing in the real world such as optimal control

I. Ellouze (✉)
Department of Mathematics, Faculty of Sciences of Sfax, Sfax, Tunisia
e-mail: imen_ellouz@yahoo.fr

A. Ben Makhlouf et al. (eds.), *State Estimation and Stabilization of Nonlinear Systems*, Studies in Systems, Decision and Control 491,
https://doi.org/10.1007/978-3-031-37970-3_3

models in economics, bursting rhythm models in pathology, and flying object motions that display a certain kind of dynamics with the impulsive jump phenomena at the switching points. Such kind of switched systems which are called impulsive switched systems, cannot be well described by using pure continuous or pure discrete models. Recently, there is an increasing interest among the control community in terms of stability analysis, stabilization and observer design so as to achieve a required stability performance of impulsive switched systems [1, 8–10, 13, 14, 16–18, 22].

Furthermore, time delays are ubiquitous in many industrial and natural processes because of measurement and computational delays and transmission and transport lags. They may cause instability or undesirable system performance in feedback systems such as chaos. Therefore, it is of significant importance to consider time delay effects for the impulsive switched system. A large number of available results have been derived for time delay impulsive systems [2–7, 11, 12] as well as delayed switched systems [18]. However, fewer results have been reported for stability analysis and control of impulsive switched systems with time-varying delay in a range. Most of the papers have studied the delay-independent stability criteria. To the best of the authors knowledge, some results on stability of impulsive switched system with time delays have been reported via different approaches. In particular, Xu et al. [20] investigated the asymptotic stability and robust stabilization of a class of impulsive switched systems with constant delay by using a Lyapunov-Krasovskii technique. But those results cannot be applied to impulsive switched systems with time-varying delays. In [9], Liu and Feng investigated the uniform stability and the uniform asymptotical stability of impulsive delayed switched systems by employing the method of Razumikhin-Lyapunov functional. And Yang and Zhu [22] studied the exponential stability of impulsive switched systems with time-varying delays by establishing some impulsive delay differential inequalities. However, to construct a suitable Lyapunov function to satisfy those conditions in [9] and to apply the designed switching law in [22] to real switched systems are very difficult and ineffective in some cases. Moreover, compared with the widely studied impulsive delayed differential systems the corresponding theory for impulsive delayed switched systems has not been fully systematically developed, and there exists enough room for further investigation and improvement.

Motivated by the above discussions, in this paper, we deal with the stability problem of a class of nonlinear impulsive switched systems with time-varying delay in a range which can be represented as a feedback connection of a linear dynamical system and a nonlinearity satisfying a sector constraint. New Delay-dependent stability criterion is derived by employing some integral inequalities. We use this new criterion to construct a delay-dependent dynamic observer-based output feedback control such that the observer error system is presented as the feedback interconnection of a linear system and a state-dependent sector-bounded nonlinearity. Those criteria are obtained and formulated in the form of linear matrix inequalities (LMIs) which are robustness tools provided by Matlab.

Motivated by the conservatism of the existing results, we managed and overcome the complexity of the system and its components, namely the variable delay, the non-linearity which depends on the output and is often a source of instability and degradation in system performance, the pulses and the mode change after each pulse. This is made possible by choosing appropriate Lyapunov-Krasovskii functions, LMI approach introducing some relaxation matrices which allowed us to avoid any restriction on the first time-delay's derivative function. Thus obtained criteria are useful for delays with fast or slow variations. It should be noted that the mathematical model discussed in this paper combines all the complexities mentioned above and is not yet treated in the literature.

This paper is organized as follows. In Sect. 2, system under consideration is described. In Sect. 3, a necessary lemma is given and a criterion of asymptotic stability is derived. In Sect. 4, a sufficient condition for feedback controller design is provided. In Sect. 5, observer-based control approach is developed. An illustrative example and concluding remarks are given in Sects. 6 and 7 respectively.

2 System Description

Let $\mathbb{R}$ denote the set of real numbers, $\mathbb{N}$ the set of positive integers, $\mathbb{R}^n$ the n-dimensional Euclidean space, $\mathbb{R}^{n\times m}$ the set of all $n \times m$ real matrices, I is the identity matrix of the set $\mathbb{R}^{n\times n}$, $\mathcal{C}_{n,h_M} := \mathcal{C}([-h_M, 0], \mathbb{R}^n)$ the space of continuous functions mapping the interval $[-h_M, 0]$ to $\mathbb{R}^n$, where h_M is a positive constant. Let $diag\{a_1, \dots, a_n\}$ denotes diagonal with the diagonal elements a_i, $1 \leq i \leq n$ and $A > 0$ a positive definite matrix A whose symmetric terms are denoted by $*$.

Consider the following impulsive switched system with two-additive time-varying-delays

$$\begin{cases} \dot{x}(t) &= A_{0,i_k}x(t) + A_{1,i_k}x(t - h_1(t) - h_2(t)) + B_{i_k}\omega^{i_k}(t) + G_{i_k}u_k(t), & t \neq t_k, \\ \Delta x(t) &= D_{i_k}x(t), & t = t_k, \\ y(t) &= C_{0,i_k}x(t) + C_{1,i_k}x(t - h_1(t) - h_2(t)), & t \neq t_k, \\ \omega^{i_k}(t) &= -\varphi^{i_k}(t, y(t)), & \\ x(t) &= \phi(t), & t \in [-h_M, 0]. \end{cases} \quad (1)$$

where $x(t) \in \mathbb{R}^n$ is the system state, $y(t) \in \mathbb{R}^p$ the measured output, $u_k(t) \in \mathbb{R}^p$ the control input, the matrices $A_{j,i_k} \in \mathbb{R}^{n\times n}$, $C_{j,i_k} \in \mathbb{R}^{p\times n}$, $1 \leq j \leq 2$, $B_{i_k} \in \mathbb{R}^{n\times p}$ and $D_{i_k} \in \mathbb{R}^{n\times n}$ are constant real matrices. $x(t) = \phi \in \mathcal{C}_{n,h_M}$ is the initial condition of system (1). $\Delta x(t) = x(t^+) - x(t^-)$, $x(t^+) = \lim_{h\to 0^+} x(t + h)$, $x(t^-) = \lim_{h\to 0^+} x(t - h)$, and $x(t_k^-) = \lim_{h\to 0^+} x(t_k - h) = x(t_k)$, which implies that the solution of (1) is left continuous at t_k. $i_k \in \{1, ...m\}$, $m \in \mathbb{N}$. Moreover, $h_1(t)$, $h_2(t)$ represent the two delay components in the state and $(t_k)_{k\in\mathbb{N}}$ a sequence of impulsive switching times, for which the following assumptions are satisfied

Assumption 1 • The sequence $(t_k)_{k\in\mathbb{N}}$ has the property that $t_0 = 0$, $t_k \in \mathbb{R}^+$ and $t_k < t_{k+1}$, $\forall k \in \mathbb{N}$.

- At the time point t_k, the system switches to the i_k subsystem from the i_{k-1} subsystem.

Assumption 2 The time delays $h_1(t)$ and $h_2(t)$ are bounded time-varying differentiable functions that satisfy

$$0 \le h_{1,m} \le h_1(t) < h_{1,M} \;\; and \;\; \dot{h_1}(t) < \mu_1 < \infty,$$
$$0 \le h_{2,m} \le h_2(t) < h_{2,M} \;\; and \;\; \dot{h_2}(t) < \mu_2 < \infty.$$

With $h_M = h_{1,M} + h_{2,M}$, $h_m = \min\{h_{1,m}, h_{2,m}\}$ and $\mu = \mu_1 + \mu_2$, one can write

$$0 \le h_m \le h(t) = h_1(t) + h_2(t) < h_M \;\; and \;\; \dot{h}(t) < \mu, \tag{2}$$

where h_m, h_M and μ are known constant reals. Note that h_m may not be equal to 0.

Assumption 3 The nonlinear function $\varphi^{i_k} : \mathbb{R}_+ \times \mathbb{R}^p \to \mathbb{R}^p$ is continuous and belongs to the sector $[0, K_{i_k}]$, i.e φ^{i_k} satisfies

$$(\varphi^{i_k})^\top(t, y)\left[\varphi^{i_k}(t, y) - K_{i_k} y\right] \le 0, \;\; \forall (t, y) \in \mathbb{R}_+ \times \mathbb{R}^p, \tag{3}$$

where $K_{i_k} \in \mathbb{R}^{p \times p}$ is a positive definite matrix.

3 Asymptotic Stability Result

In this section, stability criterion for impulsive switched systems with time-varying delay is derived. Before this, the following lemma is needed.

Lemma 1 ([21]) *Let $x(t) \in \mathbb{R}^n$ be a vector-valued function with first-order continuous-derivative entries. Then, the following integral inequality holds for any matrices $M_1, M_2 \in \mathbb{R}^{n \times n}$ and $X = X^\top > 0$, and a scalar function $h(t) \ge 0$:*

$$-\int_{t-h(t)}^{t} \dot{x}^\top(s) X \dot{x}(s) ds \le \xi^\top(t) \Upsilon \xi(t) + h(t)\xi^\top(t)\Gamma^\top X^{-1}\Gamma\xi(t), \tag{1}$$

where

$$\Upsilon := \begin{bmatrix} M_1^\top + M_1 & -M_1^\top + M_2 \\ * & -M_2^\top - M_2 \end{bmatrix}, \; \Gamma^\top := \begin{bmatrix} M_1^\top \\ M_2^\top \end{bmatrix}, \; \xi(t) := \begin{bmatrix} x(t) \\ x(t - h(t)) \end{bmatrix}.$$

Remark 1 The inequality given in the Lemma 1 is called an integral inequality. It plays a key role in the derivation of a robust stability criterion in this paper. Note that thanks to these extras matrices M_1 and M_2 that allow a wide freedom of choice one can expand the field of feasibility of the LMIs provided in this paper. Indeed, in Theorem 1, we have six matrices.

Under the sector condition (3), a sufficient condition for absolute stability of system (1) is given and the following theorem is obtained.

Theorem 1 *Under Assumptions 1, 2 and 3, for given scalars* $0 \leq h_m < h_M$*, if there exist a scalar* $\epsilon_k > 0$*, positive definite matrices* $P_{i_k} > 0$, $Q_{j,i_k} > 0$, $R_{j,i_k} > 0$, $j \in \{1, 2, 3\}$*, and real matrices* $M_{j,i_k}, N_{j,i_k}, S_{j,i_k} \in \mathbb{R}^{n \times n}$, $j \in \{1, 2\}$ *such that for* $i_k = 1, 2, ..., m$*, the following LMIs hold:*

$$\Xi_1^{i_k} < 0, \tag{2}$$

and

$$\Theta_1^{i_k} = \begin{bmatrix} P_{i_{k-1}} & (I + D_k)^T P_{i_k} \\ * & P_{i_k} \end{bmatrix} > 0. \tag{3}$$

where

$$\Xi_1^{i_k} = \begin{bmatrix} \Xi_{11} & \Xi_{12} & 0 & \Xi_{14} & \Xi_{15} & h_M A_{0,i_k}^T R_{1,i_k} & h_M M_{1,i_k}^T & \Xi_{18} & 0 & h_M A_{0,i_k}^T R_{3,i_k} & h_M S_{1,i_k}^T \\ * & \Xi_{22} & 0 & 0 & \Xi_{25} & h_M A_{1,i_k}^T R_{1,i_k} & 0 & \Xi_{28} & 0 & h_M A_{1,i_k}^T R_{3,i_k} & h_M S_{2,i_k}^T \\ * & * & \Xi_{33} & \Xi_{34} & 0 & 0 & 0 & 0 & \Xi_{39} & 0 & 0 \\ * & * & * & \Xi_{44} & 0 & 0 & h_M M_{2,i_k}^T & 0 & \Xi_{49} & 0 & 0 \\ * & * & * & * & \Xi_{55} & h_M B_{i_k}^T R_{1,i_k} & 0 & \Xi_{58} & 0 & h_M B_{i_k}^T R_{3,i_k} & 0 \\ * & * & * & * & * & -h_M R_{1,i_k} & 0 & 0 & 0 & 0 & 0 \\ * & * & * & * & * & * & -h_M R_{1,i_k} & 0 & 0 & 0 & 0 \\ * & * & * & * & * & * & * & \Xi_{88} & 0 & 0 & 0 \\ * & * & * & * & * & * & * & * & \Xi_{99} & 0 & 0 \\ * & * & * & * & * & * & * & * & * & -h_M R_{3,i_k} & 0 \\ * & * & * & * & * & * & * & * & * & * & -h_M R_{3,i_k} \end{bmatrix}$$

with

$$\begin{aligned}
\Xi_{11} &= A_{0,i_k}^\top P_{i_k} + P_{i_k} A_{0,i_k} + Q_{1,i_k} + Q_{2,i_k} + Q_{3,i_k} + M_{1,i_k}^\top + M_{1,i_k} + S_{1,i_k}^\top + S_{1,i_k}, \\
\Xi_{12} &= P_{i_k} A_{1,i_k} - S_{1,i_k}^\top + S_{2,i_k}, \\
\Xi_{14} &= -M_{1,i_k}^\top + M_{2,i_k}, \\
\Xi_{15} &= P_{i_k} B_{i_k} - \epsilon\, C_{0,i_k}^\top K_{i_k}, \\
\Xi_{22} &= -(1 - \mu) Q_{3,i_k} - S_{2,i_k}^\top - S_{2,i_k}, \\
\Xi_{25} &= -\epsilon_k\, C_{1,i_k}^\top K_{i_k}, \\
\Xi_{33} &= -Q_{1,i_k} + N_{1,i_k}^\top + N_{1,i_k}, \\
\Xi_{34} &= -N_{1,i_k}^\top + N_{2,i_k}, \\
\Xi_{44} &= -Q_{2,i_k} - M_{2,i_k}^\top - M_{2,i_k} - N_{2,i_k}^\top - N_{2,i_k}, \\
\Xi_{55} &= -2\epsilon_k\, I, \\
\Xi_{18} &= (h_M - h_m) A_{0,i_k}^T R_{2,i_k}, \\
\Xi_{28} &= (h_M - h_m) A_{1,i_k}^T R_{2,i_k},
\end{aligned}$$

$$\Xi_{58} = (h_M - h_m)B_{i_k}^T R_{2,i_k},$$
$$\Xi_{88} = -(h_M - h_m)R_{2,i_k},$$
$$\Xi_{39} = (h_M - h_m)N_{1,i_k}^T,$$
$$\Xi_{49} = (h_M - h_m)N_{2,i_k}^T,$$
$$\Xi_{99} = -(h_M - h_m)R_{2,i_k}.$$

Then the trivial solution of the impulsive switched system (1) *with the control input* $u_k(t) = 0$ *is globally uniformly asymptotically stable.*

Proof When $t \in [t_k, t_{k+1})$, consider the Lyapunov-Krasovskii function candidate

$$\begin{aligned} V_{i_k}(x(t)) = {} & x^\top(t)P_{i_k}x(t) + \int_{t-h_m}^{t} x^\top(s)Q_{1,i_k}x(s)ds + \int_{t-h_M}^{t} x^\top(s)Q_{2,i_k}x(s)ds \\ & + \int_{t-h(t)}^{t} x^\top(s)Q_{3,i_k}x(s)ds + \int_{-h_M}^{0}\int_{t+\theta}^{t} \dot{x}^\top(s)R_{1,i_k}\dot{x}(s)dsd\theta \\ & + \int_{-h_M}^{-h_m}\int_{t+\theta}^{t} \dot{x}^\top(s)R_{2,i_k}\dot{x}(s)dsd\theta \\ & + \int_{-h(t)}^{0}\int_{t+\theta}^{t} \dot{x}^\top(s)R_{3,i_k}\dot{x}(s)dsd\theta \end{aligned} \tag{4}$$

The derivative of the function V defined by (4) along the solution of the impulsive switched system (1) is given by

$$\begin{aligned} D^+V_{i_k}(x(t)) = {} & 2\dot{x}^\top(t)P_{i_k}x(t) + x^\top(t)Q_{1,i_k}x(t) - x^\top(t-h_m)Q_{1,i_k}x(t-h_m) \\ & + x^\top(t)Q_{2,i_k}x(t) - x^\top(t-h_M)Q_{2,i_k}x(t-h_M) \\ & + x^\top(t)Q_{3,i_k}x(t) - (1-\dot{h}(t))x^\top(t-h(t))Q_{3,i_k}x(t-h(t)) \\ & + h_M\dot{x}^\top(t)R_{1,i_k}\dot{x}(t) - \int_{t-h_M}^{t} \dot{x}^\top(s)R_{1,i_k}\dot{x}(s)ds \\ & + (h_M - h_m)\dot{x}^\top(t)R_{2,i_k}\dot{x}(t) - \int_{t-h_M}^{t-h_m} \dot{x}^\top(s)R_{2,i_k}\dot{x}(s)ds \\ & + h(t)\dot{x}^\top(t)R_{3,i_k}\dot{x}(t) - \int_{t-h(t)}^{t} \dot{x}^\top(s)R_{3,i_k}\dot{x}(s)ds. \end{aligned}$$

Using Assumption 2 and Lemma 1 yields

$$\begin{aligned} D^+V_{i_k}(x(t)) \le {} & 2\dot{x}^\top(t)P_{i_k}x(t) + x^\top(t)[Q_{1,i_k} + Q_{2,i_k} + Q_{3,i_k}]x(t) - x^\top(t-h_m)Q_{1,i_k}x(t-h_m) \\ & - x^\top(t-h_M)Q_{2,i_k}x(t-h_M) - (1-\mu)x^\top(t-h(t))Q_{3,i_k}x(t-h(t)) \\ & + \dot{x}^\top(t)[h_M R_{1,i_k} + (h_M - h_m)R_{2,i_k} + h_M R_{3,i_k}]\dot{x}(t) \\ & + \xi_1^\top(t)\Upsilon_{1,i_k}\xi_1(t) + h_M\xi_1^\top(t)\Gamma_{1,i_k}^\top R_{1,i_k}^{-1}\Gamma_{1,i_k}\xi_1(t) \end{aligned}$$

$$+\xi_2^\top(t)\Upsilon_{2,i_k}\xi_2(t) + (h_M - h_m)\xi_2^\top(t)\Gamma_{2,i_k}^\top R_{2,i_k}^{-1}\Gamma_{2,i_k}\xi_2(t)$$
$$+\xi_3^\top(t)\Upsilon_{3,i_k}\xi_3(t) + h_M\xi_3^\top(t)\Gamma_{3,i_k}^\top R_{3,i_k}^{-1}\Gamma_{3,i_k}\xi_3(t) \tag{5}$$

with

$$\xi_1(t) = \begin{bmatrix} x(t) \\ x(t-h_M) \end{bmatrix}; \Gamma_{1,i_k}^\top = \begin{bmatrix} M_{1,i_k}^\top \\ M_{2,i_k}^\top \end{bmatrix}; \Upsilon_{1,i_k} = \begin{bmatrix} M_{1,i_k}^\top + M_{1,i_k} & -M_{1,i_k}^\top + M_{2,i_k} \\ * & -M_{2,i_k}^\top - M_{2,i_k} \end{bmatrix},$$
$$\xi_2(t) = \begin{bmatrix} x(t-h_m) \\ x(t-h_M) \end{bmatrix}; \Gamma_{2,i_k}^\top = \begin{bmatrix} N_{1,i_k}^\top \\ N_{2,i_k}^\top \end{bmatrix}; \Upsilon_{2,i_k} = \begin{bmatrix} N_{1,i_k}^\top + N_{1,i_k} & -N_{1,i_k}^\top + N_{2,i_k} \\ * & -N_{2,i_k}^\top - N_{2,i_k} \end{bmatrix},$$
$$\xi_3(t) = \begin{bmatrix} x(t) \\ x(t-h(t)) \end{bmatrix}; \Gamma_{3,i_k}^\top = \begin{bmatrix} S_{1,i_k}^\top \\ S_{2,i_k}^\top \end{bmatrix}; \Upsilon_{3,i_k} = \begin{bmatrix} S_{1,i_k}^\top + S_{1,i_k} & -S_{1,i_k}^\top + S_{2,i_k} \\ * & -S_{2,i_k}^\top - S_{2,i_k} \end{bmatrix},$$

Then the inequality (5) can be rewritten as:

$$D^+V_{i_k}(x(t)) \leq \eta^\top(t)\ \Pi^{i_k}\ \eta(t),\ \forall\eta(t) \neq 0, \tag{6}$$

where

$$\Pi^{i_k} := \begin{bmatrix} \Pi_{11} & \Pi_{12} & 0 & \Pi_{14} & \Pi_{15} \\ * & \Pi_{22} & 0 & 0 & \Pi_{25} \\ * & * & \Pi_{33} & \Pi_{34} & 0 \\ * & * & * & \Pi_{44} & 0 \\ * & * & * & * & \Pi_{55} \end{bmatrix}, \eta(t) := \begin{bmatrix} x(t) \\ x(t-h(t)) \\ x(t-h_m) \\ x(t-h_M) \\ \omega^{i_k}(t) \end{bmatrix}$$

with

$$\begin{aligned}
\Pi_{11} &= A_{0,i_k}^\top P_{i_k} + P_{i_k}A_{0,i_k} + Q_{1,i_k} + Q_{2,i_k} + Q_{3,i_k} + h_M A_{0,i_k}^\top R_{1,i_k}A_{0,i_k} \\
&\quad +(h_M - h_m)A_{0,i_k}^\top R_{2,i_k}A_{0,i_k} + h_M A_{0,i_k}^\top R_{3,i_k}A_{0,i_k} + M_{1,i_k}^\top + M_{1,i_k} \\
&\quad +h_M M_{1,i_k}^\top R_{1,i_k}^{-1}M_{1,i_k} + h_M S_{1,i_k}^\top R_{3,i_k}^{-1}S_{1,i_k} + S_{1,i_k}^\top + S_{1,i_k}, \\
\Pi_{12} &= P_{i_k}A_{1,i_k} + h_M A_{0,i_k}^\top R_{1,i_k}A_{1,i_k} + (h_M - h_m)A_{0,i_k}^\top R_{2,i_k}A_{1,i_k} + h_M A_{0,i_k}^\top R_{3,i_k}A_{1,i_k} \\
&\quad -S_{1,i_k}^\top + S_{2,i_k} + h_M S_{1,i_k}R_{3,i_k}^{-1}S_{2,i_k}, \\
\Pi_{14} &= -M_{1,i_k}^\top + M_{2,i_k} + h_M M_{1,i_k}^\top R_{1,i_k}^{-1}M_{2,i_k}, \\
\Pi_{15} &= P_{i_k}B_{i_k} + h_M A_{0,i_k}^\top R_{1,i_k}B_{i_k} + (h_M - h_m)A_{0,i_k}^\top R_{2,i_k}B_{i_k} + h_M A_{0,i_k}^\top R_{3,i_k}B_{i_k}, \\
\Pi_{22} &= -(1-\mu)Q_{3,i_k} - S_{2,i_k}^\top - S_{2,i_k} + h_M A_{1,i_k}^\top R_{1,i_k}A_{1,i_k} + (h_M - h_m)A_{1,i_k}^\top R_{2,i_k}A_{1,i_k} \\
&\quad +h_M A_{1,i_k}^\top R_{3,i_k}A_{1,i_k} + h_M S_{2,i_k}^\top R_{3,i_k}^{-1}S_{2,i_k}, \\
\Pi_{25} &= h_M A_{1,i_k}^\top R_{1,i_k}B_{i_k} + (h_M - h_m)A_{1,i_k}^\top R_{2,i_k}B_{i_k} + h_M A_{1,i_k}^\top R_{3,i_k}B_{i_k}, \\
\Pi_{33} &= -Q_{1,i_k} + N_{1,i_k}^\top + N_{1,i_k} + (h_M - h_m)N_{1,i_k}^\top R_{2,i_k}^{-1}N_{1,i_k}, \\
\Pi_{34} &= -N_{1,i_k}^\top + N_{2,i_k} + (h_M - h_m)N_{1,i_k}^\top R_{2,i_k}^{-1}N_{1,i_k}, \\
\Pi_{44} &= -Q_{2,i_k} - M_{2,i_k}^\top - M_{2,i_k} + h_M M_{2,i_k}^\top R_{1,i_k}^{-1}M_{2,i_k} + (h_M - h_m)N_{2,i_k}^\top R_{2,i_k}^{-1}N_{2,i_k} \\
&\quad -N_{2,i_k}^\top - N_{2,i_k}, \\
\Pi_{55} &= h_M B_{i_k}^\top R_{1,i_k}B_{i_k} + (h_M - h_m)B_{i_k}^\top R_{2,i_k}B_{i_k} + h_M B_{i_k}^\top R_{3,i_k}B_{i_k}.
\end{aligned}$$

If the LMI (2) is satisfied, then using the Shur complement formula, gives

$$\Xi_1^{i_k} < 0 \Longleftrightarrow \Sigma^{i_k} < 0,$$

where

$$\Sigma^{i_k} = \begin{bmatrix} \Sigma_{11} & \Sigma_{12} & \Sigma_{13} & \Sigma_{14} & \Sigma_{15} \\ * & \Sigma_{22} & \Sigma_{23} & \Sigma_{24} & \Sigma_{25} \\ * & * & \Sigma_{33} & \Sigma_{34} & \Sigma_{35} \\ * & * & * & \Sigma_{44} & \Sigma_{45} \\ * & * & * & * & \Sigma_{55} \end{bmatrix},$$

with

$$\begin{aligned}
\Sigma_{ij} &= \Pi_{ij}, (i, j \in \{1, 2, 3, 4\}),\\
\Sigma_{15} &= \Pi_{15} - \epsilon_k\, C_{0,i_k}^{\top} K_{i_k},\\
\Sigma_{25} &= \Pi_{25} - \epsilon_k\, C_{1,i_k}^{\top} K_{i_k},\\
\Sigma_{35} &= \Pi_{35},\\
\Sigma_{45} &= \Pi_{45},\\
\Sigma_{55} &= \Pi_{55} - 2\epsilon_k\, I.
\end{aligned}$$

Then, it is easy to verify that

$$\begin{aligned}
\eta^{\top}(t)\Sigma^{i_k}\eta(t) = \eta^{\top}(t)\Pi^{i_k}\eta(t) - 2\epsilon_k\omega^{i_k\top}(t)\omega^{i_k}(t) - 2\epsilon_k\omega^{i_k\top}(t)[K_{i_k}C_{0,i_k}x(t)\\
+K_{i_k}C_{1,i_k}x(t - h(t))].
\end{aligned}$$

Using Assumption 1 and the $\mathcal{S}$-procedure, leads to following inequality

$$\eta^{\top}(t)\ \Pi^{i_k}\ \eta(t) < 0. \tag{7}$$

Thus, $D^{+}V_{i_k}(x(t)) \leq 0$. It means that the impulsive switched system is asymptotically stable, except possibly at the impulsive and switching points. Now, let us look at these time points. Note that at time point $t_k, k = 1, 2, ...,$ the system switches from the i_{k-1} subsystem to the i_k subsystem. To ensure the asymptotic stability, the following condition is required to be satisfied:

$$\begin{aligned}
V_{i_k}(x(t_k^{+})) - V_{i_k}(x(t_k)) &= x(t_k^{+})^T P_{i_k} x(t_k^{+}) - x(t_k)^T P_{i_{k-1}} x(t_k)\\
&\leq x(t_k)^T[(I + D_k)^T P_{i_k}(I + D_k) - P_{i_{k-1}}]x(t_k) < 0.
\end{aligned}$$

This means that

$$(I + D_k)^T P_{i_k}(I + D_k) - P_{i_{k-1}} < 0,$$

or, equivalently,

$$P_{i_{k-1}} - (I + D_k)^T P_{i_k} (I + D_k) > 0. \tag{8}$$

From Schur complement, one can see that the inequality (8) is equivalent to condition (3). Thus, by (2) and (3), the impulsive switched system is globally uniformly asymptotically stable. □

4 Design of Feedback Controller

The main idea of this section is to focus on the design of feedback controllers of the form $u_k(t) = L_{i_k} x(t)$, for impulsive switched systems with time-varying delays. Thus, the objective of this section is to provide a computational procedure to construct an appropriate matrix L_{i_k} such that the corresponding closed-loop system is asymptotically stable.

Theorem 2 *Under Assumptions 1, 2 and 3, for given scalars* $0 \leq h_m < h_M$, λ_{j,i_k}, α_{j,i_k}, $\beta_{j,i_k} \in \mathbb{R}$, $j \in \{1, 2\}$, *if there exist a scalar* $\epsilon_k > 0$, *positive definite matrices* $P_{i_k} > 0$, $\overline{P}_{i_k} > 0$, $\overline{Q}_{j,i_k} > 0$, $\overline{R}_{j,i_k} > 0$, $j \in \{1, 2, 3\}$, *and a matrix* $Y_{i_k} \in \mathbb{R}^{p \times n}$ *such that for* $i_k = 1, 2, ..., m$, *the following conditions are satisfied:*

$$\Xi_2^{i_k} < 0, \tag{1}$$

and

$$\Theta_1 i_k = \begin{bmatrix} P_{i_{k-1}} & (I + D_k)^T P_{i_k} \\ * & P_{i_k} \end{bmatrix} > 0. \tag{2}$$

Where

$$\Xi_2^{i_k} = \begin{bmatrix} \Xi_{11} & \Xi_{12} & 0 & \Xi_{14} & \Xi_{15} & \Xi_{16} & \Xi_{17} & \Xi_{18} & 0 & \Xi_{110} & \Xi_{111} \\ * & \Xi_{22} & 0 & 0 & \Xi_{25} & \Xi_{26} & 0 & \Xi_{28} & 0 & \Xi_{210} & \Xi_{211} \\ * & * & \Xi_{33} & \Xi_{34} & 0 & 0 & 0 & 0 & \Xi_{39} & 0 & 0 \\ * & * & * & \Xi_{44} & 0 & 0 & \Xi_{47} & 0 & \Xi_{49} & 0 & 0 \\ * & * & * & * & -2\epsilon_k.I & h_M B_{i_k}^T & 0 & \Xi_{58} & 0 & h_M B_{i_k}^T & 0 \\ * & * & * & * & * & -h_M \overline{R}_{1,i_k} & 0 & 0 & 0 & 0 & 0 \\ * & * & * & * & * & * & -h_M \overline{R}_{1,i_k} & 0 & 0 & 0 & 0 \\ * & * & * & * & * & * & * & \Xi_{88} & 0 & 0 & 0 \\ * & * & * & * & * & * & * & * & \Xi_{99} & 0 & 0 \\ * & * & * & * & * & * & * & * & * & -h_M \overline{R}_{3,i_k} & 0 \\ * & * & * & * & * & * & * & * & * & * & -h_M \overline{R}_{3,i_k} \end{bmatrix}$$

with

$$\begin{aligned}
\Xi_{11} &= \overline{P}_{i_k}(A_{0,i_k} + (\lambda_{1,i_k} + \alpha_{1,i_k})I)^\top + (A_{0,i_k} + (\lambda_{1,i_k} + \alpha_{1,i_k})I)\overline{P}_{i_k} + G_{i_k}Y_{i_k} + Y_{i_k}^\top G_{i_k}^\top \\
&\quad + \overline{Q}_{1,i_k} + \overline{Q}_{2,i_k} + \overline{Q}_{3,i_k} \\
\Xi_{12} &= A_{1,i_k}\overline{P}_{i_k} + (\alpha_{2,i_k} - \alpha_{1,i_k})\overline{P}_{i_k}, \\
\Xi_{14} &= (\lambda_{2,i_k} - \lambda_{1,i_k})\overline{P}_{i_k}, \\
\Xi_{15} &= B_{i_k} - \epsilon_k \overline{P}_{i_k} C_{0,i_k}^\top K_{i_k}, \\
\Xi_{16} &= h_M \overline{P}_{i_k} A_{0,i_k}^\top + h_M Y_{i_k}^\top G_{i_k}^\top, \\
\Xi_{17} &= \lambda_{1,i_k} h_M \overline{R}_{1,i_k}, \\
\Xi_{18} &= (h_M - h_m)\overline{P}_{i_k} A_{0,i_k}^\top + (h_M - h_m) Y_{i_k}^\top G_{i_k}^\top, \\
\Xi_{110} &= h_M \overline{P}_{i_k} A_{0,i_k}^\top + h_M Y_{i_k}^\top G_{i_k}^\top, \\
\Xi_{111} &= \alpha_{1,i_k} h_M \overline{R}_{3,i_k}, \\
\Xi_{22} &= -(1-\mu)\overline{Q}_{3,i_k} - 2\alpha_{2,i_k}\overline{P}_{i_k}, \\
\Xi_{25} &= -\epsilon_k \overline{P}_{i_k} C_{1,i_k}^\top K_{i_k}, \\
\Xi_{26} &= h_M \overline{P}_{i_k} A_{1,i_k}^\top, \\
\Xi_{28} &= (h_M - h_m)\overline{P}_{i_k} A_{1,i_k}^\top, \\
\Xi_{210} &= h_M \overline{P}_{i_k} A_{1,i_k}^\top, \\
\Xi_{211} &= \alpha_{2,i_k} h_M \overline{R}_{3,i_k}, \\
\Xi_{33} &= -\overline{Q}_{1,i_k} + 2\beta_{1,i_k}\overline{P}_{i_k}, \\
\Xi_{34} &= (\beta_{2,i_k} - \beta_{1,i_k})\overline{P}_{i_k}, \\
\Xi_{39} &= \beta_{1,i_k}(h_M - h_m)\overline{R}_{2,i_k}, \\
\Xi_{44} &= -\overline{Q}_{2,i_k} - 2(\lambda_{2,i_k} + \beta_{1,i_k})\overline{P}_{i_k}, \\
\Xi_{47} &= \lambda_{2,i_k} h_M \overline{R}_{1,i_k}, \\
\Xi_{49} &= \beta_{2,i_k}(h_M - h_m)\overline{R}_{2,i_k}, \\
\Xi_{58} &= (h_M - h_m) B_{i_k}^T, \\
\Xi_{88} &= -(h_M - h_m)\overline{R}_{2,i_k}, \\
\Xi_{99} &= -(h_M - h_m)\overline{R}_{2,i_k}.
\end{aligned}$$

Then the trivial solution of the impulsive switched system (1) *is globally uniformly asymptotically stable by the linear state feedback* $u_k(t) = L_{i_k}x(t)$ *where*

$$L_{i_k} = Y_{i_k}\overline{P_{i_k}}^{-1}.$$

Proof Substitute $u_k(t) = L_{i_k}x(t)$ in Eq. (1). Then, the corresponding impulsive switched system (1) becomes

$$\begin{cases} \dot{x}(t) &= \overline{A}_{0,i_k}x(t) + A_{1,i_k}x(t-h(t)) + B_{i_k}\omega^{i_k}(t), & t \neq t_k, \\ \Delta x(t) &= D_{i_k}x(t), & t = t_k, \\ y(t) &= C_{0,i_k}x(t) + C_{1,i_k}x(t-h(t)), & t \neq t_k, \\ \omega^{i_k}(t) &= -\varphi^{i_k}(t, y(t)), & \\ x(t) &= \phi(t), & t \in [-h_M, 0]. \end{cases} \tag{3}$$

where $\overline{A}_{0,i_k} = A_{0,i_k} + G_{i_k}L_{i_k}$.
Using Theorem 1, the origin of system (3) is globally uniformly asymptotically stable, if there exist positive definite matrices $P_{i_k} > 0$, $Q_{1,i_k} > 0$, $Q_{2,i_k} > 0$, $Q_{3,i_k} > 0$, $R_{1,i_k} > 0$, $R_{2,i_k} > 0$, $R_{3,i_k} > 0$, and M_{1,i_k}, M_{2,i_k}, N_{1,i_k}, N_{2,i_k}, S_{1,i_k}, $S_{2,i_k} \in \mathbb{R}^{n\times n}$ such that the LMIs (3) and (2) with replacing A_{0,i_k} by $A_{0,i_k} + G_{i_k}L_{i_k}$ hold. Furthermore, it can be seen that inequality (2) is equivalent to the feasibility of the following LMI

$$T^{i_k\top}\ \Xi_L^{i_k}\ T^{i_k} < 0,$$

where $\Xi_L^{i_k}$ is the matrix given in the LMI (2) with replacing A_{0,i_k} by $\overline{A}_{0,i_k}$, and

$$T^{i_k} = diag\{P_{i_k}^{-1}, P_{i_k}^{-1}, P_{i_k}^{-1}, P_{i_k}^{-1}, I, R_{1,i_k}^{-1}, R_{1,i_k}^{-1}, R_{2,i_k}^{-1}, R_{2,i_k}^{-1}, R_{3,i_k}^{-1}, R_{3,i_k}^{-1}\}.$$

Denoting

$$\overline{P}_{i_k} = P_{i_k}^{-1}, \overline{Q}_{1,i_k} = P_{i_k}^{-1}Q_{1,i_k}P_{i_k}^{-1}, \overline{Q}_{2,i_k} = P_{i_k}^{-1}Q_{2,i_k}P_{i_k}^{-1}, \overline{Q}_{3,i_k} = P_{i_k}^{-1}Q_{3,i_k}P_{i_k}^{-1},$$

$$\overline{R}_{1,i_k} = R_{1,i_k}^{-1}, \overline{R}_{2,i_k} = R_{2,i_k}^{-1}, \overline{R}_{3,i_k} = R_{3,i_k}^{-1}, L_{i_k}P_{i_k}^{-1} = Y_{i_k},$$

and picking $M_{j,i_k} = \lambda_{j,i_k}P_{i_k}$, $N_{j,i_k} = \beta_{j,i_k}P_{i_k}$, $S_{j,i_k} = \alpha_{j,i_k}P_{i_k}$, $j \in \{1, 2\}$, leads to the desired LMI (9). The rest of the proof is similar to that of Theorem 1 and will be omitted. □

5 Observer-Based Control Design

In this section, observer-based control approach is developed to perform the stabilization of the system proposed in the previous section where the state information is not available.

5.1 System Under Consideration

Consider the following nonlinear impulsive switched system

$$\begin{cases} \dot{x}(t) &= A_{0,i_k}x(t) + A_{1,i_k}x(t-h(t)) + B_{i_k}\omega^{i_k}(t), & t \neq t_k, \\ \Delta x(t) &= D_{i_k}x(t), & t = t_k, \\ y(t) &= C_{0,i_k}x(t) + C_{1,i_k}x(t-h(t)), & \\ \omega^{i_k}(t) &= -\gamma^{i_k}(H_{0,i_k}x(t) + H_{1,i_k}x(t-h(t))), & \\ x(t) &= \phi(t), & t \in [-h_M, 0], \end{cases} \tag{1}$$

where $h(t)$ is the delay defined as in Sect. 1 and satisfies assumption (2.3) and $H_{0,i_k}, H_{1,i_k} \in \mathbb{R}^{p\times n}$. The multivariate nonlinearity $\gamma^{i_k}\colon \mathbb{R}^p \to \mathbb{R}^p$ is assumed to verify the following assumption:

Assumption 4 The nonlinearity is decentralized in the sense that $\gamma_j^{i_k}(v)$, $(v \in \mathbb{R}^p, 1 \leq j \leq p)$, depends only on v_j; that is

$$\gamma^{i_k}(v) = \begin{bmatrix} \gamma_1^{i_k}(v_1) \\ \vdots \\ \vdots \\ \gamma_p^{i_k}(v_p) \end{bmatrix} \tag{2}$$

and the conditions that γ^{i_k} verifies $\gamma^{i_k}(0) = 0$ and

$$a_j \leq \frac{\gamma_j^{i_k}(t) - \gamma_j^{i_k}(t')}{t - t'} \leq b_j \;\; \forall t,\, t' \in \mathbb{R},\; t \neq t',\;\; \forall 1 \leq j \leq p. \tag{3}$$

are also supposed. If $\gamma_j^{i_k}$, $1 \leq j \leq p$, is continuous differentiable, then its slope is restricted by

$$\frac{d\gamma_j^{i_k}}{dv_j}(v_j) \in [a_j, b_j], \forall\, 1 \leq j \leq p, v_j \in \mathbb{R}.$$

In the following, it is assumed further that the scalar function $\gamma_j^{i_k}$ satisfies (3) with $a_j = 0,\ \forall 1 \leq j \leq p$.

Remark 2 If $a_j \neq 0$, a new function $\widetilde{\gamma_j^{i_k}}(t) := \gamma_j^{i_k}(t) - a_j t$ which satisfies (3) with $\widetilde{a_j} = 0, \widetilde{b_j} = b_j - a_j$, can be defined and so system (4) can be rewritten as

$$\begin{cases} \dot{x}(t) &= \widetilde{A}_{0,i_k}x(t) + \widetilde{A}_{1,i_k}x(t-h(t)) + B_{i_k}\widetilde{\omega^{i_k}}(t), & t \neq t_k, \\ \Delta x(t) &= D_{i_k}x(t), & t = t_k, \\ y(t) &= C_{0,i_k}x(t) + C_{1,i_k}x(t-h(t)), & \\ \widetilde{\omega}^{i_k}(t) &= -\gamma^{i_k}(H_{0,i_k}x(t) + H_{1,i_k}x(t-h(t))), & \\ x(t) &= \phi(t), & t \in [-h_M, 0], \end{cases}$$

with

$$\widetilde{A}_{0,i_k} = A_{0,i_k} - B_{i_k}diag\{a_1, ..., a_p\}H_{0,i_k}$$
$$\widetilde{A}_{1,i_k} = A_{1,i_k} - B_{i_k}diag\{a_1, ..., a_p\}H_{1,i_k}.$$

5.2 *Observer Design*

In this subsection, observer is designed for a class of impulsive switched systems with time-varying delays in a range. The approach is to represent the observer error system as the feedback interconnection of a linear system and a state-dependent multivariate sector-bounded nonlinearity.

With the Assumption 3, the observer has the following form:

$$\begin{cases} \dot{\widehat{x}}(t) &= A_{0,i_k}\widehat{x}(t) + A_{1,i_k}\widehat{x}(t-h(t)) + B_{i_k}\widehat{\omega^{i_k}}(t) + L_{i_k}(\widehat{y}(t) - y(t)), & t \neq t_k, \\ \Delta\widehat{x}(t) &= D_{i_k}\widehat{x}(t), & t = t_k, \\ \widehat{y}(t) &= C_{0,i_k}\widehat{x}(t) + C_{1,i_k}\widehat{x}(t-h(t)), & \\ \widehat{\omega^{i_k}}(t) &= -\gamma^{i_k}(E_{i_k}(\widehat{y}(t) - y(t)) + H_{0,i_k}\widehat{x}(t) + H_{1,i_k}\widehat{x}(t-h(t))), & \\ \widehat{x}(t) &= \phi(t), & t \in [-h_M, 0], \end{cases} \tag{4}$$

The objective of this part is to design the matrices $E_{i_k} \in \mathbb{R}^{p\times p}$ and $L_{i_k} \in \mathbb{R}^{n\times p}$ to make the observer error $e(t) = x(t) - \widehat{x}(t)$ approach to zero. At this point, it is assumed that the solution of (4) does not escape to infinity in finite time. From (4) and (4), for all $t \neq t_k$ the dynamics of the observer error $e(t)$ is governed by

$$\dot{e}(t) = (A_{0,i_k} + L_{i_k}C_{0,i_k})e(t) + (A_{1,i_k} + L_{i_k}C_{1,i_k})e(t-h(t)) - B_{i_k}[\gamma^{i_k}(v(t)) - \gamma^{i_k}(w(t))], \tag{5}$$

where

$$v(t) = H_{0,i_k}x(t) + H_{1,i_k}x(t-h(t)),$$
$$w(t) = E_{i_k}(\widehat{y}(t) - y(t)) + H_{0,i_k}\widehat{x}(t) + H_{1,i_k}\widehat{x}(t-h(t)).$$

To represent the observer error system (5) as the feedback interconnection of a linear system and multivariate sector nonlinearity, the function $\gamma^{i_k}(v(t)) - \gamma^{i_k}(w(t))$ can be considered as a function of t and

$$z(t) := v(t) - w(t) = (H_{0,i_k} + E_{i_k}C_{0,i_k})e(t) + (H_{1,i_k} + E_{i_k}C_{1,i_k})e(t-h(t)).$$

That is, a state-dependent multivariate nonlinearity in t and $z(t)$:

$$\psi^{i_k}(t, z(t)) := \gamma^{i_k}(v(t)) - \gamma^{i_k}(w(t)). \tag{6}$$

Then, the observer error (5) can be rewritten as

$$\begin{cases} \dot{e}(t) &= (A_{0,i_k} + L_{i_k}C_{0,i_k})e(t) + (A_{1,i_k} + L_{i_k}C_{1,i_k})e(t-h(t)) + B_{i_k}\Omega^{i_k}(t), \ t \neq t_k, \\ \Delta e(t) &= D_{i_k}e(t), \qquad t = t_k, \\ z(t) &= (H_{0,i_k} + E_{i_k}C_{0,i_k})e(t) + (H_{1,i_k} + E_{i_k}C_{1,i_k})e(t-h(t)), \\ \Omega^{i_k}(t) &= -\psi^{i_k}(t, z(t)) \end{cases} \tag{7}$$

Beginning to show that under the Assumption 3, the nonlinearity $\psi^{i_k}(t, z(t))$ satisfies a multivariate sector property. Indeed, it can be seen that

$$\begin{aligned} \psi_j^{i_k}(t, z(t)) &= \gamma_j^{i_k}(v(t)) - \gamma_j^{i_k}(w(t)) \\ &= \gamma_j^{i_k}(v_j(t)) - \gamma_j^{i_k}(w_j(t)) \\ &= \psi_j^{i_k}(t, z_j(t)) \\ &= \int_0^1 [\frac{\partial \gamma_j^{i_k}}{\partial s}]_{s=v_j+\lambda(w_j(t)-v_j(t))}(v_j(t) - w_j(t))d\lambda, \ \forall 1 \leq j \leq p. \end{aligned}$$

Thus, from Assumption 3, each component $(\psi_j^{i_k})_{1 \leq j \leq p}$ satisfies the following sector condition

$$\psi_j^{i_k}(t, z_j(t))[\psi_j^{i_k}(t, z_j(t)) - b_j z_j(t)] \leq 0. \tag{8}$$

Taking

$$K_{i_k} = diag\{b_1, ..., b_p\} > 0,$$

it can be seen that

$$(\psi^{i_k})^T(t, z(t))[\psi^{i_k}(t, z(t)) - K_{i_k}z(t)] \leq 0. \tag{9}$$

So, the following result is derived.

Theorem 3 *Under Assumptions 1, 2 and 4, for given scalars* $0 \leq h_m < h_M$, λ_{j,i_k}, α_{j,i_k}, $\beta_{j,i_k} \in \mathbb{R}$, $j \in \{1, 2\}$, *if there exist a scalar* $\epsilon_k > 0$, *positive definite matrices* $P_{i_k} > 0$, $Q_{j,i_k} > 0$, $\overline{R}_{j,i_k} > 0$, $j \in \{1, 2, 3\}$, *and real matrices* $E_{i_k} \in \mathbb{R}^{p \times p}$, *and* $Y_{i_k} \in \mathbb{R}^{n \times p}$, *such that for* $i_k = 1, 2, ..., m$, *the following LMIs are satisfied:*

$$\Delta^{i_k} < 0, \tag{10}$$

and

$$\Theta_1^{i_k} = \begin{bmatrix} P_{i_{k-1}} & (I + D_k)^T P_{i_k} \\ * & P_{i_k} \end{bmatrix} > 0. \tag{11}$$

where

$$
\Delta^{i_k} = \left[\begin{array}{ccccccccccc}
\Delta_{11} & \Delta_{12} & 0 & \Delta_{14} & \Delta_{15} & h_M(A_{0,i_k}^\top P_{i_k} + C_{0,i_k}^T Y_{i_k}^T) & h_M \lambda_{1,i_k} \overline{R}_{1,i_k} & (h_M - h_m)(A_{0,i_k}^\top P_{i_k} + C_{0,i_k}^T Y_{i_k}^T) & 0 & h_M(A_{0,i_k}^\top P_{i_k} + C_{0,i_k}^T Y_{i_k}^T) & h_M \alpha_{1,i_k} \overline{R}_{3,i_k} \\
* & \Delta_{22} & 0 & 0 & \Delta_{25} & h_M(A_{1,i_k}^\top P_{i_k} + C_{1,i_k}^T Y_{i_k}^T) & 0 & (h_M - h_m)(A_{1,i_k}^\top P_{i_k} + C_{1,i_k}^T Y_{i_k}^T) & 0 & h_M(A_{1,i_k}^\top P_{i_k} + C_{1,i_k}^T Y_{i_k}^T) & h_M \alpha_{2,i_k} \overline{R}_{3,i_k} \\
* & * & \Delta_{33} & \Delta_{34} & 0 & 0 & 0 & 0 & (h_M - h_m)\beta_{1,i_k} \overline{R}_{2,i_k} & 0 & 0 \\
* & * & * & \Delta_{44} & 0 & 0 & h_M \lambda_{2,i_k} \overline{R}_{1,i_k} & 0 & (h_M - h_m)\beta_{2,i_k} \overline{R}_{2,i_k} & 0 & 0 \\
* & * & * & * & -2\varepsilon_k I & h_M B_{i_k}^T P_{i_k} & 0 & (h_M - h_m) B_{i_k}^\top P_{i_k} & 0 & h_M B_{i_k}^\top P_{i_k} & 0 \\
* & * & * & * & * & -h_M \overline{R}_{1,i_k} & 0 & 0 & 0 & 0 & 0 \\
* & * & * & * & * & * & -h_M \overline{R}_{1,i_k} & 0 & 0 & 0 & 0 \\
* & * & * & * & * & * & * & -(h_M - h_m)\overline{R}_{2,i_k} & 0 & 0 & 0 \\
* & * & * & * & * & * & * & * & -(h_M - h_m)\overline{R}_{2,i_k} & 0 & 0 \\
* & * & * & * & * & * & * & * & * & -h_M \overline{R}_{3,i_k} & 0 \\
* & * & * & * & * & * & * & * & * & * & -h_M \overline{R}_{3,i_k}
\end{array}\right]
$$

with

$$
\begin{aligned}
\Delta_{11} &= [A_{0,i_k} + (\lambda_{1,i_k} + \alpha_{1,i_k})I]^T P_{i_k} + P_{i_k}[A_{0,i_k} + (\lambda_{1,i_k} + \alpha_{1,i_k})I] + Q_{1,i_k} + Q_{2,i_k} \\
&\quad + Q_{3,i_k} + C_{0,i_k}^T Y_{i_k}^T + Y_{i_k} C_{0,i_k}, \\
\Delta_{12} &= P_{i_k}[A_{1,i_k} + (\alpha_{2,i_k} - \alpha_{1,i_k})I] + Y_{i_k} C_{1,i_k}, \\
\Delta_{14} &= -(\lambda_{1,i_k} - \lambda_{2,i_k}) P_{i_k}, \\
\Delta_{15} &= P_{i_k} B_{i_k} - \epsilon_k (H_{0,i_k} + E_{i_k} C_{0,i_k})^\top K_{i_k}, \\
\Delta_{22} &= -(1 - \mu) Q_{3,i_k} - 2\alpha_{2,i_k} P_{i_k}, \\
\Delta_{25} &= -\epsilon_k (H_{1,i_k} + E_{i_k} C_{1,i_k})^\top K_{i_k}, \\
\Delta_{33} &= -Q_{1,i_k} + 2\beta_{1,i_k} P_{i_k}, \\
\Delta_{34} &= -(\beta_{1,i_k} - \beta_{2,i_k}) P_{i_k}, \\
\Delta_{44} &= -Q_{2,i_k} - 2(\lambda_{2,i_k} + \beta_{2,i_k}) P_{i_k}.
\end{aligned}
$$

Then the system (4) *is an observer for the impulsive switched system* (4). *The gain matrix* E_{i_k} *is given by the resolution of the LMI* (10) *and* $L_{i_k} = P_{i_k}^{-1} Y_{i_k}$.

Proof Sufficient conditions for the convergence of the observer error are given by Theorem 1. Indeed, let $\Lambda_{E,L}^{i_k}$ be the matrix $\Xi_1^{i_k}$ given in the LMI (2) with replacing A_{0,i_k} by $(A_{0,i_k} + L_{i_k} C_{0,i_k})$, A_{1,i_k} by $(A_{1,i_k} + L_{i_k} C_{1,i_k})$, C_{0,i_k} by $(H_{0,i_k} + E_{i_k} C_{0,i_k})$, and C_{1,i_k} by $(H_{1,i_k} + E_{i_k} C_{1,i_k})$.

Then, it is easy to verify that we can have the following equivalence:

$$\Lambda_{E,L}^{i_k} < 0 \Longleftrightarrow T^{i_k T} \Lambda_{E,L}^{i_k} T^{i_k} < 0,$$

where

$$T^{i_k} = diag\{I, I, I, I, I, R_{1,i_k}^{-1} P_{i_k}, R_{1,i_k}^{-1} P_{i_k}, R_{2,i_k}^{-1} P_{i_k}, R_{2,i_k}^{-1} P_{i_k}, R_{3,i_k}^{-1} P_{i_k}, R_{3,i_k}^{-1} P_{i_k}\}.$$

Then, if the following matrices are chosen such that $\overline{R}_{j,i_k} = P_{i_k} R_{j,i_k}^{-1} P_{i_k}$, $1 \leq j \leq 3$, $Y_{i_k} = P_{i_k} L_{i_k}$, $M_{j,i_k} = \lambda_{j,i_k} P_{i_k}$, $N_{j,i_k} = \beta_{j,i_k} P_{i_k}$ and $S_{j,i_k} = \alpha_{j,i_k} P_{i_k}$ then the desired LMI (10) will be attained. □

5.3 *Observer-Based Control*

In this subsection, observers-based feedback controllers of the form $u_k(t) = F_{i_k} \widehat{x}(t)$ which robustly stabilize the following system are designed.

$$\begin{cases} \dot{x}(t) &= A_{0,i_k} x(t) + A_{1,i_k} x(t-h(t)) + B_{i_k} \omega^{i_k}(t) + G_{i_k} u_k(t), \ t \neq t_k, \\ \Delta x(t) &= D_{i_k} x(t), \qquad t = t_k, \\ y(t) &= C_{0,i_k} x(t) + C_{1,i_k} x(t-h(t)), \\ \omega^{i_k}(t) &= -\gamma^{i_k}(H_{0,i_k} x(t) + H_{1,i_k} x(t-h(t))), \end{cases} \tag{12}$$

where the nonlinearity γ^{i_k} satisfies Assumption 3. with $a_j = 0$. The dynamic of the observers is defined by

$$\begin{cases} \dot{\widehat{x}}(t) &= (A_{0,i_k} + G_{i_k} F_{i_k}) \widehat{x}(t) + A_{1,i_k} \widehat{x}(t-h(t)) + L_{i_k}(\widehat{y}(t) - y(t)) + B_{i_k} \widehat{\omega}_{i_k}(t), \ t \neq t_k, \\ \Delta \widehat{x}(t) &= D_{i_k} \widehat{x}(t), \qquad t = t_k, \\ \widehat{y}(t) &= C_{0,i_k} \widehat{x}(t) + C_{1,i_k} \widehat{x}(t-h(t)), \\ \widehat{\omega}_{i_k}(t) &= -\gamma^{i_k}(H_{0,i_k} \widehat{x}(t) + H_{1,i_k} \widehat{x}(t-h(t)) + E_{i_k}(\widehat{y}(t) - y(t))), \end{cases} \tag{13}$$

Consequently, the observer error is

$$\begin{cases} \dot{e}(t) &= (A_{0,i_k} + L_{i_k} C_{0,i_k}) e(t) + (A_{1,i_k} + L_{i_k} C_{1,i_k}) e(t-h(t)) + B_{i_k} \Omega^{i_k}(t), \ t \neq t_k, \\ \Delta e(t) &= D_{i_k} e(t), \qquad t = t_k, \\ z(t) &= (H_{0,i_k} + E_{i_k} C_{0,i_k}) e(t) + (H_{1,i_k} + E_{i_k} C_{1,i_k}) e(t-h(t)), \\ \Omega^{i_k}(t) &= -\psi^{i_k}(t, z(t)) \qquad , \end{cases} \tag{14}$$

with

$$\begin{aligned} \psi^{i_k}(t, z(t)) &= \gamma^{i_k}(v(t)) - \gamma^{i_k}(w(t)), \\ v(t) &= H_{0,i_k} x(t) + H_{1,i_k} x(t-h(t)), \\ w(t) &= E_{i_k}(\widehat{y}(t) - y(t)) + H_{0,i_k} \widehat{x}(t) + H_{1,i_k} \widehat{x}(t-h(t)). \end{aligned}$$

Here, the objective of this subsection is to develop a procedure to design observer-based controllers for the impulsive switched system (12), such that the resulting closed loop subsystem given by

$$\begin{cases} \dot{x}(t) &= (A_{0,i_k} + G_{i_k} F_{i_k})x(t) + A_{1,i_k} x(t - h(t)) - B_{i_k} \gamma^{i_k}(v(t)) - G_{i_k} F_{i_k} e(t), \ t \neq t_k, \\ \Delta x(t) &= \qquad D_{i_k} x(t), \qquad t = t_k, \end{cases} \tag{15}$$

is globally asymptotically stable.

Theorem 4 *Under Assumptions 1, 2 and 4, for given scalars* $0 \leq h_m < h_M$, λ_{j,x,i_k}, λ_{j,e,i_k}, α_{j,x,i_k}, α_{j,e,i_k}, β_{j,x,i_k}, $\beta_{j,e,i_k} \in \mathbb{R}$, $j \in \{1, 2\}$ *if there exist two scalars* $\epsilon_{e,k} > 0$ *and* $\epsilon_{x,k} > 0$, *positive definite matrices* $P_{x,i_k} > 0$, $Q_{j,x,i_k} > 0$, $R_{j,x,i_k} > 0$, $P_{e,i_k} > 0$, $Q_{j,e,i_k} > 0$, $R_{j,e,i_k} > 0$, $j \in \{1, 2, 3\}$, *and real matrices* $E_{i_k} \in \mathbb{R}^{p \times p}$, $Y_{i_k} \in \mathbb{R}^{p \times n}$ *and* $\overline{Y}_{i_k} \in \mathbb{R}^{n \times p}$ *such that for* $i_k = 1, 2, ..., m$, *the following LMIs hold:*

$$\bigwedge_{e}^{i_k} < 0, \tag{16}$$

$$\bigvee_{x}^{i_k} < 0, \tag{17}$$

$$\Theta_{1,x}^{i_k} = \begin{bmatrix} P_{x,i_{k-1}} & (I + D_k)^T P_{x,i_k} \\ * & P_{x,i_k} \end{bmatrix} > 0, \tag{18}$$

and

$$\Theta_{2,e}^{i_k} = \begin{bmatrix} P_{e,i_{k-1}} & (I + D_k)^T P_{e,i_k} \\ * & P_{e,i_k} \end{bmatrix} > 0. \tag{19}$$

where

$$\bigwedge_{e}^{i_k} = \begin{bmatrix} \Delta_{11} & \Delta_{12} & 0 & \Delta_{14} & \Delta_{15} & h_M(A_{0,i_k}^{\top} P_{e,i_k} + C_{0,i_k}^T \overline{Y}_{i_k}^T) & h_M \lambda_{1,e,i_k} R_{1,e,i_k} \\ * & \Delta_{22} & 0 & 0 & \Delta_{25} & h_M(A_{1,i_k}^{\top} P_{e,i_k} + C_{1,i_k}^T \overline{Y}_{i_k}^T) & 0 \\ * & * & \Delta_{33} & \Delta_{34} & 0 & 0 & 0 \\ * & * & * & \Delta_{44} & 0 & 0 & h_M \lambda_{2,e,i_k} R_{1,e,i_k} \\ * & * & * & * & -2\varepsilon_k I & h_M B_{i_k}^T P_{e,i_k} & 0 \\ * & * & * & * & * & -h_M R_{1,e,i_k} & 0 \\ * & * & * & * & * & * & -h_M R_{1,e,i_k} \\ * & * & * & * & * & * & * \\ * & * & * & * & * & * & * \\ * & * & * & * & * & * & * \\ * & * & * & * & * & * & * \end{bmatrix}$$

$$
\begin{matrix}
(h_M - h_m)(A_{0,i_k}^\top P_{e,i_k} + C_{0,i_k}^T \overline{Y}_{i_k}^T) & 0 & h_M(A_{0,i_k}^\top P_{e,i_k} + C_{0,i_k}^T \overline{Y}_{i_k}^T) & h_M \alpha_{1,e,i_k} R_{3,e,i_k} \\
(h_M - h_m)(A_{1,i_k}^\top P_{e,i_k} + C_{1,i_k}^T \overline{Y}_{i_k}^T) & 0 & h_M(A_{1,i_k}^\top P_{e,i_k} + C_{1,i_k}^T \overline{Y}_{i_k}^T) & h_M \alpha_{2,e,i_k} R_{3,e,i_k} \\
0 & (h_M - h_m)\beta_{1,e,i_k} R_{2,e,i_k} & 0 & 0 \\
0 & (h_M - h_m)\beta_{2,e,i_k} R_{2,e,i_k} & 0 & 0 \\
(h_M - h_m) B_{i_k}^\top P_{e,i_k} & 0 & h_M B_{i_k}^\top P_{e,i_k} & 0 \\
0 & 0 & 0 & 0 \\
0 & 0 & 0 & 0 \\
-(h_M - h_m) R_{2,e,i_k} & 0 & 0 & 0 \\
* & -(h_M - h_m) R_{2,e,i_k} & 0 & 0 \\
* & * & -h_M R_{3,e,i_k} & 0 \\
* & * & * & -h_M R_{3,e,i_k}
\end{matrix}
\Bigg],
$$

$$
\bigvee_x^{i_k} = \begin{bmatrix}
\bigvee_{11} & \bigvee_{12} & 0 & \bigvee_{14} & \bigvee_{15} & \bigvee_{16} & \bigvee_{17} & \bigvee_{18} & 0 & \bigvee_{110} & \bigvee_{111} \\
* & \bigvee_{22} & 0 & 0 & \bigvee_{25} & \bigvee_{26} & 0 & \bigvee_{28} & 0 & \bigvee_{210} & \bigvee_{211} \\
* & * & \bigvee_{33} & \bigvee_{34} & 0 & 0 & 0 & 0 & \bigvee_{39} & 0 & 0 \\
* & * & * & \bigvee_{44} & 0 & 0 & \bigvee_{47} & 0 & \bigvee_{49} & 0 & 0 \\
* & * & * & * & -2\epsilon_{x,k}.I & h_M B_{i_k}^T & 0 & \bigvee_{58} & 0 & h_M B_{i_k}^T & 0 \\
* & * & * & * & * & -h_M R_{1,x,i_k} & 0 & 0 & 0 & 0 & 0 \\
* & * & * & * & * & * & -h_M R_{1,x,i_k} & 0 & 0 & 0 & 0 \\
* & * & * & * & * & * & * & \bigvee_{88} & 0 & 0 & 0 \\
* & * & * & * & * & * & * & * & \bigvee_{99} & 0 & 0 \\
* & * & * & * & * & * & * & * & * & -h_M R_{3,x,i_k} & 0 \\
* & * & * & * & * & * & * & * & * & * & -h_M R_{3,x,i_k}
\end{bmatrix}
$$

with

$$
\begin{aligned}
\Lambda_{11} &= [A_{0,i_k} + (\lambda_{1,e,i_k} + \alpha_{1,e,i_k})I]^T P_{e,i_k} + P_{e,i_k}[A_{0,i_k} + (\lambda_{1,e,i_k} + \_1, e, i_k)I] + Q_{1,e,i_k} \\
&\quad + Q_{2,e,i_k} + Q_{3,e,i_k} + C_{0,i_k}^T \overline{Y}_{i_k}^T + \overline{Y}_{i_k} C_{0,i_k} \\
\Lambda_{12} &= P_{e,i_k}[A_{1,i_k} + (\alpha_{2,e,i_k} - \alpha_{1,e,i_k})I] + \overline{Y}_{i_k} C_{1,i_k}, \\
\Lambda_{14} &= -(\lambda_{1,e,i_k} - \lambda_{2,e,i_k}) P_{e,i_k}, \\
\Lambda_{15} &= P_{e,i_k} B_{i_k} - \epsilon_k (H_{0,i_k} + E_{i_k} C_{0,i_k})^\top K_{i_k}, \\
\Lambda_{22} &= -(1-\mu) Q_{3,e,i_k} - 2\alpha_{2,e,i_k} P_{e,i_k}, \\
\Lambda_{25} &= -\epsilon_k (H_{1,i_k} + E_{i_k} C_{1,i_k})^\top K_{i_k}, \\
\Lambda_{33} &= -Q_{1,e,i_k} + 2\beta_{1,e,i_k} P_{e,i_k}, \\
\Lambda_{34} &= -(\beta_{1,e,i_k} - \beta_{2,e,i_k}) P_{e,i_k}, \\
\Lambda_{44} &= -Q_{2,e,i_k} - 2(\lambda_{2,e,i_k} + \beta_{2,e,i_k}) P_{e,i_k}, \\
\bigvee_{11} &= P_{x,i_k}(A_{0,i_k} + (\lambda_{1,x,i_k} + \alpha_{1,x,i_k})I)^\top + (A_{0,i_k} + (\lambda_{1,x,i_k} + \alpha_{1,x,i_k})I) P_{x,i_k} \\
&\quad + G_{i_k} Y_{i_k} + Y_{i_k}^\top G_{i_k}^\top + Q_{1,x,i_k} + Q_{2,x,i_k} + Q_{3,x,i_k}, \\
\bigvee_{12} &= A_{1,i_k} P_{x,i_k} + (\alpha_{2,x,i_k} - \alpha_{1,x,i_k}) P_{x,i_k}, \\
\bigvee_{14} &= (\lambda_{2,x,i_k} - \lambda_{1,x,i_k}) P_{x,i_k},
\end{aligned}
$$

$$\bigvee_{15} = B_{i_k} - \epsilon_{x,k} P_{x,i_k} C_{0,i_k}^{\top} K_{i_k},$$

$$\bigvee_{16} = h_M P_{x,i_k} A_{0,i_k}^{\top} + h_M Y_{i_k}^{\top} G_{i_k}^{\top},$$

$$\bigvee_{17} = \lambda_{1,x,i_k} h_M R_{1,x,i_k},$$

$$\bigvee_{18} = (h_M - h_m) P_{x,i_k} A_{0,i_k}^{\top} + (h_M - h_m) Y_{i_k}^{\top} G_{i_k}^{\top},$$

$$\bigvee_{110} = h_M P_{x,i_k} A_{0,i_k}^{\top} + h_M Y_{i_k}^{\top} G_{i_k}^{\top},$$

$$\bigvee_{111} = \alpha_{1,x,i_k} h_M R_{3,x,i_k},$$

$$\bigvee_{22} = -(1 - \mu) Q_{3,x,i_k} - 2\alpha_{2,x,i_k} P_{x,i_k},$$

$$\bigvee_{25} = -\epsilon_k P_{x,i_k} C_{1,i_k}^{\top} K_{i_k},$$

$$\bigvee_{26} = h_M P_{x,i_k} A_{1,i_k}^{\top},$$

$$\bigvee_{28} = (h_M - h_m) P_{x,i_k} A_{1,i_k}^{\top},$$

$$\bigvee_{210} = h_M P_{x,i_k} A_{1,i_k}^{\top},$$

$$\bigvee_{211} = \alpha_{2,x,i_k} h_M R_{3,x,i_k},$$

$$\bigvee_{33} = -Q_{1,x,i_k} + 2\beta_{1,x,i_k} P_{x,i_k},$$

$$\bigvee_{34} = (\beta_{1,x,i_k} - \beta_{1,x,i_k}) P_{x,i_k},$$

$$\bigvee_{39} = \beta_{1,x,i_k} (h_M - h_m) R_{2,x,i_k},$$

$$\bigvee_{44} = -Q_{2,x,i_k} - 2(\lambda_{2,x,i_k} + \beta_{1,x,i_k}) P_{x,i_k},$$

$$\bigvee_{47} = \lambda_{2,x,i_k} h_M R_{1,x,i_k},$$

$$\bigvee_{49} = \beta_{2,x,i_k} (h_M - h_m) R_{2,x,i_k},$$

$$\bigvee_{58} = (h_M - h_m) B_{i_k}^{T},$$

$$\bigvee_{88} = -(h_M - h_m)R_{2,x,i_k},$$

$$\bigvee_{99} = -(h_M - h_m)R_{2,x,i_k}.$$

Then the impulsive switched system (12) *is globally uniformly asymptotically stable by the dynamic feedback* $u_k(t) = F_{i_k}\widehat{x}(t)$ *and* (13), *where the gain matrices are given by*

$$L_{i_k} = P_{e,i_k}^{-1}\overline{Y}_{i_k} \quad and \quad F_{i_k} = Y_{i_k}P_{x,i_k}^{-1}.$$

Proof When $t \in [t_k, t_{k+1})$, define the Lyapunov-Krasovskii functional V_{i_k} as follows

$$V_{i_k}(t) := aV_{x,i_k}(t, x_t) + V_{e,i_k}(t, e_t), \tag{20}$$

where

$$\begin{aligned} V_{x,i_k}(t, x_t) &= x^\top(t)P_{x,i_k}x(t) + \int_{t-h_m}^{t} x^\top(s)Q_{1,x,i_k}x(s)ds + \int_{t-h_M}^{t} x^\top(s)Q_{2,x,i_k}x(s)ds \\ &+ \int_{t-h(t)}^{t} x^\top(s)Q_{3,x,i_k}x(s)ds + \int_{-h_M}^{0}\int_{t+\theta}^{t} \dot{x}^\top(s)R_{1,x,i_k}\dot{x}(s)dsd\theta \\ &+ \int_{-h_M}^{-h_m}\int_{t+\theta}^{t} \dot{x}^\top(s)R_{2,x,i_k}\dot{x}(s)dsd\theta \\ &+ \int_{-h(t)}^{0}\int_{t+\theta}^{t} \dot{x}^\top(s)R_{3,x,i_k}\dot{x}(s)dsd\theta, \end{aligned}$$

and

$$\begin{aligned} V_{e,i_k}(t, e_t) &= x^\top(t)P_{e,i_k}x(t) + \int_{t-h_m}^{t} x^\top(s)Q_{1,e,i_k}x(s)ds + \int_{t-h_M}^{t} x^\top(s)Q_{2,e,i_k}x(s)ds \\ &+ \int_{t-h(t)}^{t} x^\top(s)Q_{3,e,i_k}x(s)ds + \int_{-h_M}^{0}\int_{t+\theta}^{t} \dot{x}^\top(s)R_{1,e,i_k}\dot{x}(s)dsd\theta \\ &+ \int_{-h_M}^{-h_m}\int_{t+\theta}^{t} \dot{x}^\top(s)R_{2,e,i_k}\dot{x}(s)dsd\theta \\ &+ \int_{-h(t)}^{0}\int_{t+\theta}^{t} \dot{x}^\top(s)R_{3,e,i_k}\dot{x}(s)dsd\theta, \end{aligned}$$

$x_t, e_t \in \mathcal{C}_{n,\overline{h}}$, $P_{x,i_k} > 0$, $P_{e,i_k} > 0$, $Q_{j,x,i_k} > 0$, $Q_{j,e,i_k} > 0$, $R_{j,x,i_k} > 0$, $R_{j,e,i_k} > 0$; $1 \leq j \leq 3$ are matrices to be determined, and a is a positive real to be suitably chosen.

Firstly, the derivative of V_{x,i_k} along the trajectories of system (15) is given by

$$
\begin{aligned}
D^+V_{x,i_k}(t,x_t) &= 2\dot{x}^\top(t)P_{x,i_k}x(t) + x^\top(t)Q_{1,x,i_k}x(t) - x^\top(t-h_m)Q_{1,x,i_k}x(t-h_m) \\
&+x^\top(t)Q_{2,x,i_k}x(t) - x^\top(t-h_M)Q_{2,x,i_k}x(t-h_M) \\
&+x^\top(t)Q_{3,x,i_k}x(t) - (1-\dot{h}(t))x^\top(t-h(t))Q_{3,x,i_k}x(t-h(t)) \\
&+h_M\dot{x}^\top(t)R_{1,x,i_k}\dot{x}(t) - \int_{t-h_M}^{t}\dot{x}^\top(s)R_{1,x,i_k}\dot{x}(s)ds \\
&+(h_M-h_m)\dot{x}^\top(t)R_{2,x,i_k}\dot{x}(t) - \int_{t-h_M}^{t-h_m}\dot{x}^\top(s)R_{2,x,i_k}\dot{x}(s)ds \\
&+h(t)\dot{x}^\top(t)R_{3,x,i_k}\dot{x}(t) - \int_{t-h(t)}^{t}\dot{x}^\top(s)R_{3,x,i_k}\dot{x}(s)ds
\end{aligned}
$$

Applying Assumption 2 and Lemma 1 to the terms of the right-hand side of the previous equality for any matrices $M_{j,x,i_k}, N_{j,x,i_k}, S_{j,x,i_k} \in \mathbb{R}^{n\times n}$, $j \in \{1,2\}$, leads to the following inequality

$$
\begin{aligned}
D^+V_{x,i_k}(t,x_t) &\le 2\dot{x}^\top(t)P_{x,i_k}x(t) + x^\top(t)[Q_{1,x,i_k}+Q_{2,x,i_k}+Q_{3,x,i_k}]x(t) - x^\top(t-h_m) \\
&\times Q_{1,x,i_k}x(t-h_m) - x^\top(t-h_M)Q_{2,x,i_k}x(t-h_M) - (1-\mu)x^\top(t-h(t)) \\
&\times Q_{3,x,i_k}x(t-h(t)) + \dot{x}^\top(t)[h_MR_{1,x,i_k} + (h_M-h_m)R_{2,x,i_k} + h_MR_{3,x,i_k}]\dot{x}(t) \\
&+\xi_{1,x}^\top(t)\Upsilon_{1,x,i_k}\xi_{1,x}(t) + h_M\xi_{1,x}^\top(t)\Gamma_{1,x,i_k}^\top R_{1,x,i_k}^{-1}\Gamma_{1,x,i_k}\xi_{1,x}(t) \\
&+\xi_{2,x}^\top(t)\Upsilon_{2,x,i_k}\xi_{2,x}(t) + (h_M-h_m)\xi_{2,x}^\top(t)\Gamma_{2,x,i_k}^\top R_{2,x,i_k}^{-1}\Gamma_{2,x,i_k}\xi_{2,x}(t) \\
&+\xi_{3,x}^\top(t)\Upsilon_{3,x,i_k}\xi_{3,x}(t) + h_M\xi_{3,x}^\top(t)\Gamma_{3,x,i_k}^\top R_{3,x,i_k}^{-1}\Gamma_{3,x,i_k}\xi_{3,x}(t)
\end{aligned}
$$

with

$$
\xi_{1,x}(t) = \begin{bmatrix} x(t) \\ x(t-h_M) \end{bmatrix}; \Gamma_{1,x,i_k}^\top = \begin{bmatrix} M_{1,x,i_k}^\top \\ M_{2,x,i_k}^\top \end{bmatrix}; \Upsilon_{1,x,i_k} = \begin{bmatrix} M_{1,x,i_k}^\top + M_{1,x,i_k} & -M_{1,x,i_k}^\top + M_{2,x,i_k} \\ * & -M_{2,x,i_k}^\top - M_{2,x,i_k} \end{bmatrix},
$$

$$
\xi_{2,x}(t) = \begin{bmatrix} x(t-h_m) \\ x(t-h_M) \end{bmatrix}; \Gamma_{2,x,i_k}^\top = \begin{bmatrix} N_{1,x,i_k}^\top \\ N_{2,x,i_k}^\top \end{bmatrix}; \Upsilon_{2,x,i_k} = \begin{bmatrix} N_{1,x,i_k}^\top + N_{1,x,i_k} & -N_{1,x,i_k}^\top + N_{2,x,i_k} \\ * & -N_{2,x,i_k}^\top - N_{2,x,i_k} \end{bmatrix},
$$

$$
\xi_{3,x}(t) = \begin{bmatrix} x(t) \\ x(t-h(t)) \end{bmatrix}; \Gamma_{3,x,i_k}^\top = \begin{bmatrix} S_{1,x,i_k}^\top \\ S_{2,x,i_k}^\top \end{bmatrix}; \Upsilon_{3,x,i_k} = \begin{bmatrix} S_{1,x,i_k}^\top + S_{1,x,i_k} & -S_{1,x,i_k}^\top + S_{2,x,i_k} \\ * & -S_{2,x,i_k}^\top - S_{2,x,i_k} \end{bmatrix}.
$$

In order to simplify the expressions, this notation can be adopted

$$
\overline{Z}_{i_k} := h_MR_{1,x,i_k} + (h_M-h_m)R_{2,x,i_k} + h_MR_{3,x,i_k}.
$$

For any $b > 0$, the following inequalities are obtained

$$
\begin{aligned}
-2e^T(t)(G_{i_k}F_{i_k})^T\overline{Z}_{i_k}\Gamma_{i_k}\eta(t) &\le be^T(t)(G_{i_k}F_{i_k})^T\overline{Z}_{i_k}(G_{i_k}F_{i_k})e(t) + \frac{1}{b}\eta^T(t)\Gamma_{i_k}^T\overline{Z}_{i_k}\Gamma_{i_k}\eta(t) \\
-2e^T(t)(G_{i_k}F_{i_k})^TP_{x,i_k}x(t) &\le be^T(t)(G_{i_k}F_{i_k})^TP_{x,i_k}(G_{i_k}F_{i_k})e(t) + \frac{1}{b}x^T(t)P_{x,i_k}x(t).
\end{aligned}
$$

Then, it follows that

$$
\begin{aligned}
D^+V_{x,i_k}(t,x_t) &\le \delta^T(t)\Theta_{i_k}\delta(t) + \frac{1}{b}\zeta^T(t)\Gamma_{i_k}^T\overline{Z}_{i_k}\Gamma_{i_k}\zeta(t) + \frac{1}{b}x^T(t)P_{x,i_k}x(t) \\
&+e^T(t)(G_{i_k}F_{i_k})^T(bP_{x,i_k} + (1+b)\overline{Z}_{i_k})(G_{i_k}F_{i_k})e(t), \qquad (21)
\end{aligned}
$$

where

$$\Theta_{i_k} := \begin{bmatrix} \Theta_{11} & \Theta_{12} & 0 & \Theta_{14} & \Theta_{15} \\ * & \Theta_{22} & 0 & 0 & \Theta_{25} \\ * & * & \Theta_{33} & \Theta_{34} & 0 \\ * & * & * & \Theta_{44} & 0 \\ * & * & * & * & \Theta_{55} \end{bmatrix}, \delta(t) := \begin{bmatrix} x(t) \\ x(t-h(t)) \\ x(t-h_m) \\ x(t-h_M) \\ -\gamma^{i_k}(v(t)) \end{bmatrix}$$

$$\Gamma_{i_k}^T := \begin{bmatrix} (A_{0,i_k} + G_{i_k}F_{i_k})^T \\ A_{1,i_k}^T \\ B_{i_k}^T \end{bmatrix}, \zeta(t) := \begin{bmatrix} x(t) \\ x(t-h(t)) \\ -\gamma^{i_k}(v(t)) \end{bmatrix}$$

with

$$\begin{aligned}
\Theta_{11} =\ & (A_{0,i_k} + G_{i_k}F_{i_k})^\top P_{x,i_k} + P_{x,i_k}(A_{0,i_k} + G_{i_k}F_{i_k}) + Q_{1,x,i_k} + Q_{2,x,i_k} + Q_{3,x,i_k} \\
& + h_M(A_{0,i_k} + G_{i_k}F_{i_k})^\top R_{1,x,i_k}(A_{0,i_k} + G_{i_k}F_{i_k}) + (h_M - h_m)(A_{0,i_k} + G_{i_k}F_{i_k})^\top \\
& R_{2,x,i_k}(A_{0,i_k} + G_{i_k}F_{i_k}) + h_M(A_{0,i_k} + G_{i_k}F_{i_k})^\top R_{3,x,i_k}(A_{0,i_k} + G_{i_k}F_{i_k}) + M_{1,x,i_k}^\top \\
& + M_{1,x,i_k} + h_M M_{1,x,i_k}^\top R_{1,x,i_k}^{-1} M_{1,x,i_k} + h_M S_{1,x,i_k}^\top R_{3,x,i_k}^{-1} S_{1,x,i_k} + S_{1,x,i_k}^\top + S_{1,x,i_k}, \\
\Theta_{12} =\ & P_{x,i_k}A_{1,i_k} + h_M(A_{0,i_k} + G_{i_k}F_{i_k})^\top R_{1,x,i_k}A_{1,i_k} + (h_M - h_m)(A_{0,i_k} + G_{i_k}F_{i_k})^\top \\
& R_{2,x,i_k}A_{1,i_k} + h_M(A_{0,i_k} + G_{i_k}F_{i_k})^\top R_{3,x,i_k}A_{1,i_k} - S_{1,x,i_k}^\top + S_{2,x,i_k} + h_M S_{1,x,i_k} \\
& R_{3,x,i_k}^{-1} S_{2,x,i_k}, \\
\Theta_{14} =\ & -M_{1,x,i_k}^\top + M_{2,x,i_k} + h_M M_{1,x,i_k}^\top R_{1,x,i_k}^{-1} M_{2,x,i_k}, \\
\Theta_{15} =\ & P_{x,i_k}B_{i_k} + h_M(A_{0,i_k} + G_{i_k}F_{i_k})^\top R_{1,x,i_k}B_{i_k} + (h_M - h_m)(A_{0,i_k} + G_{i_k}F_{i_k})^\top R_{2,x,i_k}B_{i_k} \\
& + h_M(A_{0,i_k} + G_{i_k}F_{i_k})^\top R_{3,x,i_k}B_{i_k}, \\
\Theta_{22} =\ & -(1-\mu)Q_{3,x,i_k} - S_{2,x,i_k}^\top - S_{2,x,i_k} + h_M A_{1,i_k}^\top R_{1,x,i_k}A_{1,i_k} + (h_M - h_m)A_{1,i_k}^\top R_{2,x,i_k}A_{1,i_k} \\
& + h_M A_{1,i_k}^\top R_{3,x,i_k}A_{1,i_k} + h_M S_{2,x,i_k}^\top R_{3,x,i_k}^{-1} S_{2,x,i_k}, \\
\Theta_{25} =\ & h_M A_{1,i_k}^\top R_{1,x,i_k}B_{i_k} + (h_M - h_m)A_{1,i_k}^\top R_{2,x,i_k}B_{i_k} + h_M A_{1,i_k}^\top R_{3,x,i_k}B_{i_k}, \\
\Theta_{33} =\ & -Q_{1,x,i_k} + N_{1,x,i_k}^\top + N_{1,x,i_k} + (h_M - h_m)N_{1,x,i_k}^\top R_{2,x,i_k}^{-1} N_{1,x,i_k}, \\
\Theta_{34} =\ & -N_{1,x,i_k}^\top + N_{2,x,i_k} + (h_M - h_m)N_{1,x,i_k}^\top R_{2,x,i_k}^{-1} N_{1,x,i_k}, \\
\Theta_{44} =\ & -Q_{2,x,i_k} - M_{2,x,i_k}^\top - M_{2,x,i_k} + h_M M_{2,x,i_k}^\top R_{1,x,i_k}^{-1} M_{2,x,i_k} + (h_M - h_m)N_{2,x,i_k}^\top \\
& R_{2,x,i_k}^{-1} N_{2,x,i_k} - N_{2,x,i_k}^\top - N_{2,x,i_k}, \\
\Theta_{55} =\ & h_M B_{i_k}^\top R_{1,x,i_k}B_{i_k} + (h_M - h_m)B_{i_k}^\top R_{2,x,i_k}B_{i_k} + h_M B_{i_k}^\top R_{3,x,i_k}B_{i_k}.
\end{aligned}$$

Subsequently, using the proof of Theorem 1 for any free matrices M_{j,e,i_k}, N_{j,e,i_k}, S_{j,e,i_k}; $j \in \{1, 2\}$ the derivative of the functional $V_{e,i_k}(t, e_t)$ along the trajectories of (7) yields the following inequality

$$\dot{V}_{e,i_k}(t, e_t) \leq \mu^T(t)\chi_{i_k}\mu(t), \tag{22}$$

where

$$\chi_{i_k} := \begin{bmatrix} \chi_{11} & \chi_{12} & 0 & \chi_{14} & \chi_{15} \\ * & \chi_{22} & 0 & 0 & \chi_{25} \\ * & * & \chi_{33} & \chi_{34} & 0 \\ * & * & * & \chi_{44} & 0 \\ * & * & * & * & \chi_{55} \end{bmatrix}, \mu(t) := \begin{bmatrix} e(t) \\ e(t-h(t)) \\ e(t-h_m) \\ e(t-h_M) \\ \omega^{i_k}(t) \end{bmatrix}$$

with

$$\begin{aligned}
\chi_{11} &= (A_{0,i_k} + L_{i_k}C_{0,i_k})^\top P_{e,i_k} + P_{e,i_k}(A_{0,i_k} + L_{i_k}C_{0,i_k}) + Q_{1,e,i_k} + Q_{2,e,i_k} + Q_{3,e,i_k} \\
&\quad + h_M(A_{0,i_k} + L_{i_k}C_{0,i_k})^\top R_{1,e,i_k}(A_{0,i_k} + L_{i_k}C_{0,i_k}) + (h_M - h_m)(A_{0,i_k} + L_{i_k}C_{0,i_k})^\top R_{2,e,i_k} \\
&\quad (A_{0,i_k} + L_{i_k}C_{0,i_k}) + h_M(A_{0,i_k} + L_{i_k}C_{0,i_k})^\top R_{3,e,i_k}(A_{0,i_k} + L_{i_k}C_{0,i_k}) + M_{1,e,i_k}^\top + M_{1,e,i_k} \\
&\quad + h_M M_{1,e,i_k}^\top R_{1,e,i_k}^{-1} M_{1,e,i_k} + h_M S_{1,e,i_k}^\top R_{3,e,i_k}^{-1} S_{1,e,i_k} + S_{1,e,i_k}^\top + S_{1,e,i_k}, \\
\chi_{12} &= P_{e,i_k}(A_{1,i_k} + L_{i_k}C_{1,i_k}) + h_M(A_{0,i_k} + L_{i_k}C_{0,i_k})^\top R_{1,e,i_k}(A_{1,i_k} + L_{i_k}C_{1,i_k}) + (h_M - h_m) \\
&\quad (A_{0,i_k} + L_{i_k}C_{0,i_k})^\top R_{2,e,i_k}(A_{1,i_k} + L_{i_k}C_{1,i_k}) + h_M(A_{0,i_k} + L_{i_k}C_{0,i_k})^\top R_{3,e,i_k} \\
&\quad (A_{1,i_k} + L_{i_k}C_{1,i_k}) - S_{1,e,i_k}^\top + S_{2,e,i_k} + h_M S_{1,e,i_k} R_{3,e,i_k}^{-1} S_{2,e,i_k}, \\
\chi_{14} &= -M_{1,e,i_k}^\top + M_{2,e,i_k} + h_M M_{1,e,i_k}^\top R_{1,e,i_k}^{-1} M_{2,e,i_k}, \\
\chi_{15} &= P_{e,i_k}B_{i_k} + h_M(A_{0,i_k} + L_{i_k}C_{0,i_k})^\top R_{1,e,i_k}B_{i_k} + (h_M - h_m)(A_{0,i_k} + L_{i_k}C_{0,i_k})^\top \\
&\quad R_{2,e,i_k}B_{i_k} + h_M(A_{0,i_k} + L_{i_k}C_{0,i_k})^\top R_{3,e,i_k}B_{i_k}, \\
\chi_{22} &= -(1-\mu)Q_{3,e,i_k} - S_{2,e,i_k}^\top - S_{2,e,i_k} + h_M(A_{1,i_k} + L_{i_k}C_{1,i_k})^\top R_{1,e,i_k}(A_{1,i_k} + L_{i_k}C_{1,i_k}) \\
&\quad + (h_M - h_m)(A_{1,i_k} + L_{i_k}C_{1,i_k})^\top R_{2,e,i_k}(A_{1,i_k} + L_{i_k}C_{1,i_k}) \\
&\quad + h_M(A_{1,i_k} + L_{i_k}C_{1,i_k})^\top R_{3,e,i_k}(A_{1,i_k} + L_{i_k}C_{1,i_k}) + h_M S_{2,e,i_k}^\top R_{3,e,i_k}^{-1} S_{2,e,i_k}, \\
\chi_{25} &= h_M(A_{1,i_k} + L_{i_k}C_{1,i_k})^\top R_{1,e,i_k}B_{i_k} + (h_M - h_m)(A_{1,i_k} + L_{i_k}C_{1,i_k})^\top R_{2,e,i_k}B_{i_k} \\
&\quad + h_M(A_{1,i_k} + L_{i_k}C_{1,i_k})^\top R_{3,e,i_k}B_{i_k}, \\
\chi_{33} &= -Q_{1,e,i_k} + N_{1,e,i_k}^\top + N_{1,e,i_k} + (h_M - h_m)N_{1,e,i_k}^\top R_{2,e,i_k}^{-1} N_{1,e,i_k}, \\
\chi_{34} &= -N_{1,e,i_k}^\top + N_{2,e,i_k} + (h_M - h_m)N_{1,e,i_k}^\top R_{2,e,i_k}^{-1} N_{1,e,i_k}, \\
\chi_{44} &= -Q_{2,e,i_k} - M_{2,e,i_k}^\top - M_{2,e,i_k} + h_M M_{2,e,i_k}^\top R_{1,e,i_k}^{-1} M_{2,e,i_k} + (h_M - h_m)N_{2,e,i_k}^\top R_{2,e,i_k}^{-1} \\
&\quad N_{2,e,i_k} - N_{2,e,i_k}^\top - N_{2,e,i_k}, \\
\chi_{55} &= h_M B_{i_k}^\top R_{1,e,i_k}B_{i_k} + (h_M - h_m)B_{i_k}^\top R_{2,e,i_k}B_{i_k} + h_M B_{i_k}^\top R_{3,e,i_k}B_{i_k}.
\end{aligned}$$

Using (21) and (22), leads to this inequality

$$\begin{aligned}
\dot{V}_{i_k}(t) &\leq \mu^T(t)\chi_{i_k}\mu(t) + a\delta^T(t)\Theta_{i_k}\delta(t) + \frac{a}{b}\zeta^T(t)\Gamma_{i_k}^T\overline{Z}_{i_k}\Gamma_{i_k}\zeta(t) + \frac{a}{b}x^T(t)P_{x,i_k}x(t) \\
&\quad + e^T(t)(G_{i_k}F_{i_k})^T(abP_{x,i_k} + a(1+b)\overline{Z}_{i_k})(G_{i_k}F_{i_k})e(t).
\end{aligned}$$

It is clear that for the negativity of the function V_{i_k}, it suffices to have the following two conditions verified:

$$\mu^T(t)\chi_{i_k}\mu(t) < 0 \text{ and } \delta^T(t)\Theta_{i_k}\delta(t) < 0.$$

Indeed, in this case by applying the S-procedure the real $b > 0$ (which is large enough) can be chosen such that

$$\delta^T(t)\Theta_{i_k}\delta(t) + \frac{1}{b}\zeta^T(t)\Gamma_{i_k}^T\overline{Z}_{i_k}\Gamma_{i_k}\zeta(t) + \frac{1}{b}x^T(t)P_{x,i_k}x(t) < 0,$$

and then, the real $a > 0$ may be made sufficiently small such that

$$\mu^T(t)\chi_{i_k}\mu(t) + e^T(t)(G_{i_k}F_{i_k})^T(abP_{x,i_k} + a(1+b)\overline{Z}_{i_k})(G_{i_k}F_{i_k})e(t) < 0.$$

Now, using Theorem 3, $\mu^T(t)\chi_{i_k}\mu(t) < 0$ is equivalent to the LMI (16) and the gain matrix L_{i_k} is given by $L_{i_k} = P_{e,i_k}^{-1}\overline{Y}_{i_k}$.
To establish a sufficient condition in term of LMI, the sector condition

$$(\gamma^{i_k})^T(v(t))[\gamma^{i_k}(v(t)) - K_{i_k}(v(t))] \leq 0, \tag{23}$$

and the S-procedure imply

$$\delta^T(t)\Theta_{i_k}\delta(t) - 2\epsilon_{x,k}(\gamma^{i_k})^T(v(t))\gamma^{i_k}(v(t)) - 2\epsilon_{x,k}(\gamma^{i_k})^T(v(t))K_{i_k}[H_{0,i_k}x(t) + H_{1,i_k}x(t-h(t))] < 0, \tag{24}$$

for all $\delta(t) \neq 0$. This gives the following LMI

$$\bigsqcup_x^{i_k} = \left[\begin{matrix} \sqcup_{11} & \sqcup_{12} & 0 & \sqcup_{14} & \sqcup_{15} & h_M(A_{0,i_k} + G_{i_k}F_{i_k})^T R_{1,x,i_k} & h_M M_{1,x,i_k}^T \\ * & \sqcup_{22} & 0 & 0 & \sqcup_{25} & h_M(A_{0,i_k} + G_{i_k}F_{i_k})^T R_{1,x,i_k} & 0 \\ * & * & \sqcup_{33} & \sqcup_{34} & 0 & 0 & 0 \\ * & * & * & \sqcup_{44} & 0 & 0 & h_M M_{2,x,i_k}^T \\ * & * & * & * & \sqcup_{55} & h_M B_{i_k}^T R_{1,x,i_k} & 0 \\ * & * & * & * & * & -h_M R_{1,x,i_k} & 0 \\ * & * & * & * & * & * & -h_M R_{1,x,i_k} \\ * & * & * & * & * & * & * \\ * & * & * & * & * & * & * \\ * & * & * & * & * & * & * \\ * & * & * & * & * & * & * \end{matrix}\right.$$

$$\left.\begin{matrix} \sqcup_{18} & 0 & h_M(A_{0,i_k} + G_{i_k}F_{i_k})^T R_{3,x,i_k} & h_M S_{1,x,i_k}^T \\ \sqcup_{28} & 0 & h_M A_{1,i_k}^T R_{3,x,i_k} & h_M S_{2,x,i_k}^T \\ 0 & \sqcup_{39} & 0 & 0 \\ 0 & \sqcup_{49} & 0 & 0 \\ \sqcup_{58} & 0 & h_M B_{i_k}^T R_{3,x,i_k} & 0 \\ 0 & 0 & 0 & 0 \\ 0 & 0 & 0 & 0 \\ \sqcup_{88} & 0 & 0 & 0 \\ * & \sqcup_{99} & 0 & 0 \\ * & * & -h_M R_{3,x,i_k} & 0 \\ * & * & * & -h_M R_{3,x,i_k} \end{matrix}\right]$$

with

$$\bigsqcup_{11} = (A_{0,i_k} + G_{i_k} F_{i_k})^\top P_{x,i_k} + P_{x,i_k}(A_{0,i_k} + G_{i_k} F_{i_k}) + Q_{1,x,i_k} + Q_{2,x,i_k} + Q_{3,x,i_k}$$
$$+ M^\top_{1,x,i_k} + M_{1,x,i_k} + S^\top_{1,x,i_k} + S_{1,x,i_k},$$
$$\bigsqcup_{12} = P_{x,i_k} A_{1,i_k} - S^\top_{1,x,i_k} + S_{2,x,i_k},$$
$$\bigsqcup_{14} = -M^\top_{1,x,i_k} + M_{2,x,i_k},$$
$$\bigsqcup_{15} = P_{x,i_k} B_{i_k} - \epsilon_{x,k} H^\top_{0,i_k} K_{i_k},$$
$$\bigsqcup_{22} = -(1-\mu) Q_{3,i_k} - S^\top_{2,x,i_k} - S_{2,x,i_k},$$
$$\bigsqcup_{25} = -\epsilon_{x,k} H^\top_{1,i_k} K_{i_k},$$
$$\bigsqcup_{33} = -Q_{1,x,i_k} + N^\top_{1,x,i_k} + N_{1,x,i_k},$$
$$\bigsqcup_{34} = -N^\top_{1,x,i_k} + N_{2,x,i_k},$$
$$\bigsqcup_{44} = -Q_{2,x,i_k} - M^\top_{2,x,i_k} - M_{2,x,i_k} - N^\top_{2,x,i_k} - N_{2,x,i_k},$$
$$\bigsqcup_{55} = -2\epsilon_{x,k} I,$$
$$\bigsqcup_{18} = (h_M - h_m)(A_{0,i_k} + L_{i_k} C_{0,i_k})^T R_{2,x,i_k},$$
$$\bigsqcup_{28} = (h_M - h_m) A^T_{1,i_k} R_{2,x,i_k},$$
$$\bigsqcup_{58} = (h_M - h_m) B^T_{i_k} R_{2,x,i_k},$$
$$\bigsqcup_{88} = -(h_M - h_m) R_{2,x,i_k},$$
$$\bigsqcup_{39} = (h_M - h_m) N^T_{1,x,i_k},$$
$$\bigsqcup_{49} = (h_M - h_m) N^T_{2,x,i_k},$$
$$\bigsqcup_{99} = -(h_M - h_m) R_{2,x,i_k}.$$

Finally, as in the proof of Theorem 2, for given scalars $\lambda_{j,x,i_k}, \alpha_{j,x,i_k}, \beta_{j,x,i_k} \in \mathbb{R}$, $j \in \{1, 2\}$, this LMI is equivalent to the LMI (17) and the gain matrix F_{i_k} is given by

$$F_{i_k} = Y_{i_k} P_{x,i_k}^{-1}.$$ □

Remark 3 Two steps to solve the LMIs in Theorem 4. The first step is to determine the gain matrix F_{i_k} by solving the LMIs (17) and (18) . The second step is to solve the LMIs (16) and (19) in order to determine the gain matrices E_{i_k} and L_{i_k}.

6 Illustrative Example

As an illustration, we consider a two-dimension system in the form of (1). Without loss of generality, we assume that there are two subsystems, that is $i_k \in \{1, 2\}$ between which the dynamical system alternates. We consider robust performance of the system using Theorem 2. The parameters of the system are specified as follows:

Subsystem 1 The nonlinearity has the expression $\varphi^{i_k}(t, y(t)) = \sin(y(t)) + y(t)$, $i_k = 1, 2$ which verifies the sector condition $[0, 2]$ and the time-delay function is $h(t) = h_1(t) + h_2(t) = 0.3 + 0.3\sin(2t)$

$$A_{0,1} = \begin{pmatrix} -3 & 0 \\ 0 & -3 \end{pmatrix}, \quad A_{1,1} = \begin{pmatrix} -1 & 0 \\ 0 & 1 \end{pmatrix}, \quad B_1 = \begin{pmatrix} -1 & 0 \\ 0 & -1 \end{pmatrix},$$

$$C_{0,1} = \begin{pmatrix} 1 & 0 \\ 0 & 1 \end{pmatrix}, \quad C_{1,1} = \begin{pmatrix} 0 & -1 \\ -1 & -2 \end{pmatrix}, \quad K_1 = \begin{pmatrix} 0.005 & 0 \\ 0 & 0.005 \end{pmatrix},$$

$$G_1 = \begin{pmatrix} -1 \\ -1 \end{pmatrix}, \quad D_1 = \begin{pmatrix} 2 & 0 \\ 0 & 2 \end{pmatrix}$$

Subsystem 2

$$A_{0,2} = \begin{pmatrix} -2.02 & -1 \\ 0.5 & 0.22 \end{pmatrix}, \quad A_{1,2} = \begin{pmatrix} 0.49 & 1 \\ -0.1, & -0.68 \end{pmatrix}, \quad B_2 = \begin{pmatrix} -0.5 & 1 \\ 0.2 & 0 \end{pmatrix},$$

$$C_{0,2} = \begin{pmatrix} 0.4 & 0 \\ 0 & 0.5 \end{pmatrix}, \quad C_{1,2} = \begin{pmatrix} 0.2 & 0 \\ 0 & 0.3 \end{pmatrix}, \quad K_2 = \begin{pmatrix} 0.005 & 0 \\ 0 & 0.005 \end{pmatrix},$$

$$G_2 = \begin{pmatrix} -0.8 \\ -0.8 \end{pmatrix}, \quad D_2 = \begin{pmatrix} 2 & 0 \\ 0 & 2 \end{pmatrix}$$

Choosing $h_m = 10^{-4}$, $h_M = 0.088$, $\mu = 0.01$, $\lambda_{1,1} = -1$, $\lambda_{2,1} = -1.2$, $\lambda_{1,2} = -1$, $\lambda_{2,2} = -1.2$, $\alpha_{1,1} = -0.2$, $\alpha_{2,1} = 0$, $\alpha_{1,2} = -0.2$, $\alpha_{2,2} = 0$, $\beta_{1,1} = 0$, $\beta_{2,1} = 0$, $\beta_{1,2} = 0$, $\beta_{2,2} = 0$ and solving the inequalities (9) and (2) in Theorem 2 by using the feasp command within the MATLAB environment to calculate the positive define symmetric matrices, we get $\epsilon_1 = \epsilon_2 = 0.6809$,

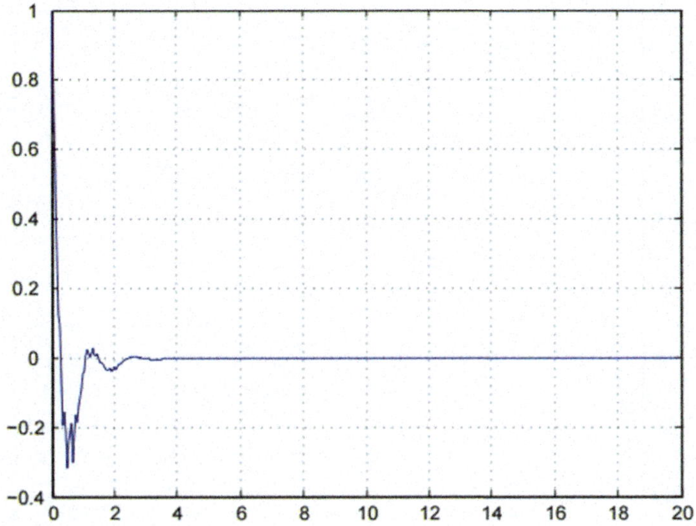

Fig. 1 x_1 trajectory

$P_1 = P_2 = e^{-10} * \begin{pmatrix} 0.1354 & 0 \\ 0 & 0.1354 \end{pmatrix}$,
$\overline{P}_1 = \begin{pmatrix} 16.4098 & -4.0844 \\ -4.0844 & 16.4098 \end{pmatrix}$, $\overline{P}_2 = \begin{pmatrix} 2.5180 & -1.0237 \\ -1.0237 & 0.8354 \end{pmatrix}$
$\overline{Q}_{1,1} = \begin{pmatrix} 33.8349 & 1.9364 \\ 1.9364 & 31.9154 \end{pmatrix}$, $\overline{Q}_{2,1} = \begin{pmatrix} 63.7211 & -7.3456 \\ -1.0237 & 61.2726 \end{pmatrix}$, $\overline{Q}_{3,1} = \begin{pmatrix} 35.6985 & 0.9898 \\ 0.9898 & 35.8650 \end{pmatrix}$
$\overline{Q}_{1,2} = \begin{pmatrix} 3.0175 & 0.7218 \\ 0.7218 & 16.6849 \end{pmatrix}$, $\overline{Q}_{2,2} = \begin{pmatrix} 8.3727 & -1.6938 \\ -1.6938 & 3.4005 \end{pmatrix}$, $\overline{Q}_{3,2} = \begin{pmatrix} 3.0727 & 0.7120 \\ 0.7120 & 1.7078 \end{pmatrix}$
$\overline{R}_{1,1} = \begin{pmatrix} 51.4636 & 0.7354 \\ 0.7354 & 51.2020 \end{pmatrix}$, $\overline{R}_{2,1} = \begin{pmatrix} 61.0741 & 0.0457 \\ 0.0457 & 61.0457 \end{pmatrix}$, $\overline{R}_{3,1} = \begin{pmatrix} 60.9241 & 0.0600 \\ 0.0600 & 60.8880 \end{pmatrix}$,
$\overline{R}_{1,2} = \begin{pmatrix} 5.2115 & 0.4931 \\ 0.4931 & 4.4892 \end{pmatrix}$, $\overline{R}_{2,2} = \begin{pmatrix} 6.9575 & 0.0510 \\ 0.0510 & 6.8833 \end{pmatrix}$, $\overline{R}_{3,2} = \begin{pmatrix} 6.9246 & 0.0707 \\ 0.0707 & 6.8152 \end{pmatrix}$ and matrices Y_1
$= 10^3 \times \begin{pmatrix} 1.7922 & -1.7729 \end{pmatrix}$, $Y_2 = \begin{pmatrix} -221.4804 & 225.2574 \end{pmatrix}$.
The state trajectories of the closed-loop system are shown in Figs. 1 and 2.

7 Conclusion

In this paper, the observer-based control problem for a novel class of impulsive delayed switched systems was studied. New criteria are established and expressed as a set of linear matrix inequalities (LMIs). By constructing an appropriate Lyapunov-Krasovskii function and introducing some extra matrices, we could manage the parameters that can disturb and affect the stability of the dynamic system in consideration and we managed to stabilize it by an observer-based control.

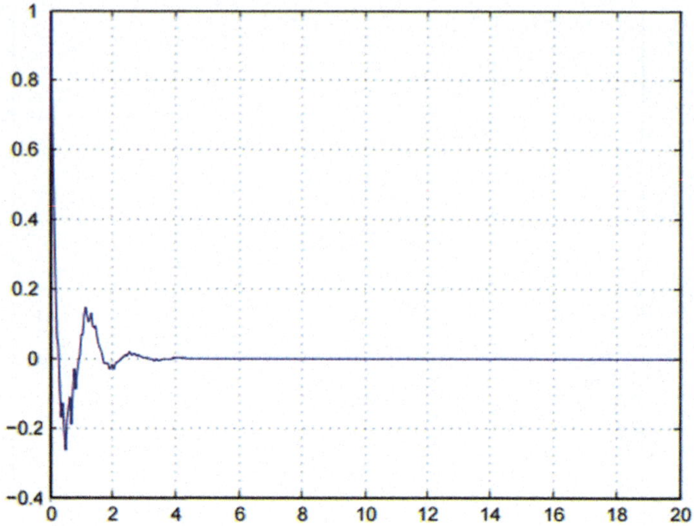

Fig. 2 x_2 trajectory

References

1. Binbin, D., Xiaojie, Z.: Delay-dependent stability analysis and synthesis for uncertain impulsive switched system with mixed delays. Discret. Dyn. Nat. Soc. Hindawi publishing corporation, New York (2011). https://doi.org/10.1155/2011/381571
2. Caraballo, T., Hammami, M.A., Mchiri, L.: Practical exponential stability of impulsive stochastic functional differential equations. Syst. Control. Lett. **109**, 43–48 (2017)
3. Ellouze, I., Benhamed, B., Hammami, M.A.: Practical uniform stability of nonlinear differential delay equations. Mediterr. J. Math. **8**(4), 607–620 (2011)
4. Ellouze, I., Hammami, M.A.: On the practical stability of impulsive control systems with multiple time delay. Dyn. Contin. Discret. Impuls. Syst. Ser. Math. Anal. **20**, 341–356 (2013)
5. Ellouze, I.: Practical observer for impulsive systems. J. Korean Math. Soc. **51**(1), 99–111 (2014)
6. Ellouze, I.: On the practical stabilization of nonlinear systems via hybrid feedback control. Math. Notes. **107**(4), 589–594 (2020)
7. Ellouze, I.: Robust stabilization of delay systems: hybrid control. Bull. Malays. Math. Sci. Soc. **44**, 467–478 (2021)
8. Ellouze, I., Vivalda, J.C., Hammami, M.A.: A separation principle for linear impulsive systems. Eur. J. Control. **20**, 105–110 (2014)
9. Feng, G., Cao, J.: Stability analysis of impulsive switched singular systems. IET Control. Theory Appl. **9**(6), 4–13 (2015)
10. Gao, L., Wang, D., Wang, G.: Further results on exponential stability for impulsive switched nonlinear time-delay systems with delayed impulse effects. Appl. Math. Comput. **268**, 186–200 (2015)
11. Gu, K., Kharitonov, V.L., Chen, J.: Stability of Time Delay Systems. Birkauser, Boston (2003)
12. Hale, J.K., Lunel, S.M.V.: Introduction to Functional Differential Equations. Springer-Verlag, Applied Mathematical Sciences (1993)
13. Li, X., Xiang, Z.: Observer design of discrete-time impulsive switched nonlinear systems with time-varying delays. Appl. Math. Comput. **229**, 327–339 (2014)

14. Li, Z., Soh, Y., Wen, C.: Switched and Impulsive Systems Analysis, Design, and Applications Neutral Systems with Time-Varying Delays and Nonlinear Perturbations. Springer-Verlag, Berlin, Heidelberg (2005)
15. Liberzon, D., Morse, A.S.: Basic problems in stability and design of switched systems. IEEE Control Syst. Mag. **19**(5), 59–70 (1999)
16. Mahmoudi, A., Momeni, A., Aghdam, A.G., Gohari, P.: On observer design for a class of impulsive switched systems. Proc. Am. Control. Conf. 4633–4639 (2008)
17. Naifer, O., Ben Makhlouf, A.: Advances in Observer Design and Observation for Nonlinear Systems. Springer (2022). https://doi.org/10.1007/978-3-030-92731-8
18. Wang, L., Li, X.: Stability analysis of impulsive delayed switched systems and applications. Math. Methods Appl. Sci. **35**(10), 1161–1174 (2012)
19. Xiang, Z., Sun, Y.N., Chen, Q.: Stabilization for a class of switched neutral systems under asynchronous switching. Trans. Inst. Meas. Control. **34**(7), 793–801 (2011)
20. Xu, H., Liu, X., Teo, K.: A LMI approach to stability analysis and synthesis of impulsive switched systems with time delays. Nonlinear Anal. Hybrid Syst. **2**, 38–50 (2008)
21. Zhang, X.M., Wu, M., She, J.H., He, Y.: Delay-dependent stabilization of linear systems with time-varying state and input delays. Autom. **41**, 1405–1412 (2005)
22. Yang, C., Zhu, W.: Stability analysis of impulsive switched systems with time delays. Math. Comput. Model. **4**(3), 608–617 (2009)

Stabilization of TS Fuzzy Systems via a Practical Observer

N. Hadj Taieb, M. A. Hammami, and F. Delmotte

Abstract This paper focuses of the analysis and study of the advanced approach for the asymptotic behaviors of perturbed fuzzy systems via an observer design. We introduce the notion of global practical exponential stability for Takagi-Sugeno fuzzy systems and by using common quadratic Lyapunov function and parallel distributed compensation controller techniques we study the asymptotic stability of the solutions of such systems by optimizing the number of rules. The proposed approach for stability analysis is based on the determination of the bounds of perturbations that characterize the asymptotic convergence of the solutions to a closed ball centered at the origin.

Keywords Takagi-Sugeno fuzzy systems · perturbations Practical exponential stabilization · Practical observer

1 Introduction

Approximating a function using fuzzy systems is the problem of identifying a fuzzy model by training it to fit a certain set of data points. Multiple approaches have been used to solve this problem. Takagi-Sugeno ([25]) rules and fuzzy inference methods proposed by many authors such as ([13, 14, 19–23, 25–35]) are utilized in function approximation problems using fuzzy logic. A small number of simple rules can be used to approximate such functions using the Takagi-Sugeno system. The issue that

N. Hadj Taieb (✉)
University of Sfax, IPEIS, Sfax, Tunisia
e-mail: nizar.hadjtaieb@yahoo.fr

M. A. Hammami
University of Sfax, Faculty of Sciences of Sfax, Sfax, Tunisia
e-mail: MohamedAli.Hammami@fss.rnu.tn

F. Delmotte
University of Artois, Bethune, France
e-mail: franccois.delmotte@univ-artois.fr

A. Ben Makhlouf et al. (eds.), *State Estimation and Stabilization of Nonlinear Systems*, Studies in Systems, Decision and Control 491,
https://doi.org/10.1007/978-3-031-37970-3_4

we encounter while using the TS rules is that the consequent part does not cover most of the function space and the rules are hard to interpret, while the Takagi-Sugeno Fuzzy system has a better perspective in presenting the rules and a better interpretability. If we consider a given function space, and if we conduct a good analysis on the function, we are able to generate rules that are specific to that space which will lead to a smooth approximation of the curve. The Lyapunov stability theory is the main approach for these kinds of problems, among them, the simplest approaches consists in looking for a common quadratic Lyapunov function (CQLF) by using the concept of the parallel distributed compensation (PDC) technique ([20, 24, 26]) to design a stabilizing controller. However, another important issue in stability analysis of nonlinear systems is to study the behavior of the solutions in practical sense, it means that the trajectories converge to a small neighborhood of the origin. Stability is one of the most important concepts concerning the properties of control systems ([10–12, 22]). In [32], fuzzy reliability analysis method for multidisciplinary systems was developed to decouple the fuzzy reliability analysis from the optimization. In practical engineering systems, external disturbances tend to introduce oscillation, even causes instability. In these cases, the systems are usually not stable in the sense of Lyapunov stability, but sometimes their performance may be acceptable in practice just because they oscillate sufficiently near a mathematically unstable course. Furthermore, in some cases, though a system is stable, it can not be acceptable in practice engineering just because the attraction domain is not large enough ([5–7]). To deal with these situation, the concept of practical stability ([1–4], [8, 9] [15–18]), which is derived from the so called, finite time stability, is more useful. The goal of this paper is to present the practical stability concept for a class of Takagi-Sugeno fuzzy systems via an observer design.

2 Fuzzy Control Systems

Consider a class of the continuous-time T-S fuzzy control system which can be described by the following fuzzy rules,

Rule i : If $z_1(t)$ is M_{i1} and $z_2(t)$ is M_{i2} ... and $z_p(t)$ is M_{ip}, then

$$\dot{x}(t) = A_i x(t) + B_i u(t), \ i = 1, 2, ..., r,$$

where $x(t) \in \mathbb{R}^n$ is the state vector, $u(t) \in \mathbb{R}^m$ is the control input vector, $A_i \in \mathbb{R}^{n\times n}$ and $B_i \in \mathbb{R}^{n\times m}$ are the system matrix input matrix, $i = 1, ..., r$ is the number of fuzzy rules, M_{ij} are the inputs fuzzy sets, $z(t) = [z_1(t), ..., z_p(t)]^T$ are measurable variables, i.e., the premise variables. Using weighted average defuzzifiers, the aggregated fuzzy model is given by

$$\dot{x}(t) = \frac{\sum_{i=1}^{r} w_i(z)\big(A_i x(t) + B_i u(t)\big)}{\sum_{i=1}^{r} w_i(z)},$$

where

$$w_i(z) = \prod_{j=1}^{r} M_{ij}(z_j).$$

Let $\mu_i(z)$ be the membership functions that belong to class C^1, i,e., they are continuous differentiable and defined as

$$\mu_i(z) = \frac{w_i(z)}{\sum_{i=1}^{r} w_i(z)}.$$

Then the fuzzy system has the state-space form

$$\dot{x}(t) = \sum_{i=1}^{r} \mu_i(z)\big(A_i x(t) + B_i u(t)\big). \tag{1}$$

μ_i are such that $\mu_i(z) \geq 0$ for $i = 1, 2, ..., r$ and $\sum_{i=1}^{r} \mu_i(z) = 1$.

Many published results, concerning the control of the fuzzy system, are based on the PDC principle. The design of the fuzzy controller shares the same antecedent as the fuzzy system and employs a linear state feedback control in the consequent part. For each local dynamics the controller is defined as

Rule i : If $z_1(t)$ is M_{i1} and $z_2(t)$ is M_{i2} ... and $z_p(t)$ is M_{ip}, then

$$u(t) = -K_i x(t), \ i = 1, 2, ..., r, \tag{2}$$

where K_i is the local state feedback gain. Consequently, the defuzzified result is

$$u(t) = -\sum_{i=1}^{r} \mu_i(z) K_i x(t). \tag{3}$$

The system (1) in closed-loop with the fuzzy controller (3) yields the following fuzzy system,

$$\dot{x}(t) = \sum_{i=1}^{r}\sum_{j=1}^{r} \mu_i(z)\mu_j(z)\big(A_i - B_i K_j\big)x(t). \tag{4}$$

A sufficient condition for the stability is deduced using Lyapunov's direct method. Suppose that a common positive definite matrix P exists, so that the following conditions are satisfied:

$$(A_i - B_i K_i)^T P + P(A_i - B_i K_i) < 0, \; i = 1, 2, ..., r,$$

and

$$\frac{1}{2}(A_i - B_i K_j + A_j - B_j K_i)^T P + \frac{1}{2} P(A_i - B_i K_j + A_j - B_j K_i) < 0, \; 1 \leq i < j \leq r.$$

When these conditions are satisfied, the fuzzy system (4) is then asymptotically stable. The design work can be transformed into a convex problem [24], which is efficiently solved by linear matrix inequalities optimization. If the solution is feasible, meaning that the stabilization constraints are met, then local state feedback gains are obtained.

3 Control of Perturbed Fuzzy Systems

Motivated by the results of the above section concerning the control of fuzzy-model, we will extend the T-S fuzzy system with the presence of external disturbances. Consider the following T-S fuzzy uncertain model,

Rule i : If $z_1(t)$ is M_{i1} and $z_2(t)$ is M_{i2} ... and $z_p(t)$ is M_{ip}, then

$$\dot{x}(t) = A_i x(t) + B_i u(t) + \Omega_i(t, x(t)), \; i = 1, 2, ..., r. \tag{1}$$

$\Omega_i : \mathbb{R}^+ \times \mathbb{R}^n \longrightarrow \mathbb{R}^n$ are continuous functions which represent the perturbations terms for $i = 1, 2, ..., r$.

We shall suppose the following assumption required for the stabilization problem.

($\mathcal{H}_1$) The pairs (A_i, B_i), $i = 1, ..., r$, are controllable, that is each nominal local model is controllable.

The fuzzy system is then inferred to be

$$\dot{x}(t) = \sum_{i=1}^{r} \mu_i(z)\Big(A_i x(t) + B_i u(t) + \Omega_i(t, x(t))\Big). \tag{2}$$

As is mentioned in the above section, μ_i are the membership functions that belong to class $\mathcal{C}^1$, i.e., they are continuously differentiable, which ensures the existence and uniqueness of solutions, and are subject to the following conditions: $0 \leq \mu_i(z) \leq 1$ and $\sum_{i=1}^{r} \mu_i(z) = 1$.

The functions Ω_i, $i = 1, 2, ..., r$, represent the uncertain external disturbances of each fuzzy subsystem and are time-varying satisfying the following condition.

($\mathcal{H}_2$) Assume that,

$$\|\Omega_i(t, x(t))\| \leq \pi_i(t), \quad i = 1, 2, ..., r, \tag{3}$$

for all $t \geq 0$ and $x \in \mathbb{R}^n$, where π_i are known nonnegative continuous functions for $i = 1, ..., r$, with

$$\left(\int_0^{+\infty} \pi(t)^2 \, dt\right)^{\frac{1}{2}} \leq \tilde{\pi} < +\infty,$$

where

$$\pi(t) := \left(\sum_{i=1}^{r} \pi_i(t)^2\right)^{\frac{1}{2}},$$

$\tilde{\pi}$ is a nonnegative constant.

The fuzzy control rule is defined as above and we will consider the fuzzy uncertain system (2). Therefore, the closed-loop system with respect the fuzzy control (2) and (3) is given by

$$\dot{x}(t) = \sum_{i=1}^{r}\sum_{j=1}^{r} \mu_i(z)\mu_j(z)\big(A_i - B_i K_j\big)x(t) + \sum_{i=1}^{r} \mu_i(z)\Omega_i(t, x(t)). \tag{4}$$

Thus,

$$\dot{x}(t) = \sum_{i=1}^{r} \mu_i^2 G_{ii} x(t) + 2\sum_{i<j} \mu_i \mu_j G_{ij} x(t) + \sum_{i=1}^{r} \mu_i \Omega_i(t, x(t)),$$

where

$$G_{ii} = A_i - B_i K_i$$

and

$$G_{ij} = \frac{1}{2}(A_i - B_i K_j + A_j - B_j K_i).$$

The controller synthesis initially considers the stability of the local fuzzy dynamics. That is, the stable feedback gains are determined for every subsystem. Suppose that there exist positive symmetric and definite matrices P, Q_i, and Q_{ij} $(i < j)$, and some matrices $K_i, i = 1, ..., r$, such that the following inequalities [21] hold,

$$G_{ii}^T P + PG_{ii} < -Q_i,\ i = 1, 2, ..., r, \tag{5}$$

and

$$G_{ij}^T P + PG_{ij} < -Q_{ij},\ 1 \leq i < j \leq r. \tag{6}$$

Based on this assumption, each nominal local model is controllable and a suitable feedback gain can be obtained.

3.1 Practical Stabilization

As a first step, we need to recall what is meant by uniformly ultimately bounded and uniform global practical exponential stability of dynamic systems ([1–3]). Consider a system described by

$$\dot{x} = F(t, x) \tag{7}$$

with $t \in \mathbb{R}_+$ is the time and $x \in \mathbb{R}^n$ is the state.

Definition 1 The system (7) is said uniformly ultimately bounded if there exists $R > 0$, such that for all $R_1 > 0$, there exists a $T = T(R_1) > 0$ such that

$$\|x(t_0)\| \leq R_1 \Rightarrow \|x(t)\| \leq R \quad \text{for all} \quad t \geq t_0 + T \quad \text{and} \quad t_0 \geq 0.$$

Definition 2 The system (7) is said uniformly globally practically exponentially stable, if there exists a ball

$$\mathcal{B}_r = \{x \in \mathbb{R}^n \ / \ \|x\| \leq r\},$$

such that $\mathcal{B}_r$ is uniformly globally practically exponentially stable, it means that, there exists $r > 0$ such that, for all $\varepsilon > r$, there exists $\epsilon = \epsilon(\varepsilon) > 0$ such that, for all $t_0 \geq 0$, $\|x(t_0)\| \leq \epsilon$, we have

$$\|x(t)\| \leq \eta \|x(t_0)\| e^{-\upsilon(t-t_0)} + r, \quad \text{for all} \ t \geq t_0,$$

with $\eta > 0$, $\upsilon > 0$.

In the case where $t = t_0$, the solution of (7) with respect the initial condition $(0, x(0))$, satisfies:

$$\|x(t)\| \leq \gamma\|x(0)\|e^{-\upsilon t} + r, \quad \text{for all} \ \ t \geq 0, \tag{8}$$

with $\eta > 0$, $\upsilon > 0$.

Our objective is to find some conditions on the functions $\pi_i(t)$ such that the fuzzy system (4) is globally uniformly practically exponentially stable. If $\pi_i(t) = 0$ for all $i = 1, ..., r$, the fuzzy uncertain system (4) has an equilibrium point at the origin. In this case, we can analyze the stability of the closed loop system behavior for the origin as an equilibrium point. If $\pi_i(t) \neq 0$, for a certain $i = 1, ..., r$, then the origin can will not be an equilibrium point of the fuzzy uncertain system (4). In this case, we can study the uniform ultimate boundedness of the solutions of the fuzzy uncertain system, also, the convergence of the solutions toward a neighborhood of the origin.

Now, one can state the following theorem.

Theorem 1 *Suppose that the assumptions* $(\mathcal{H}_1)$, $(\mathcal{H}_2)$ *hold and there exist a common positive definite matrix* P *and some feedback gain matrices* K_i, $i = 1, ..., r$, *such that the stability conditions (5) and (6) are satisfied, then the fuzzy closed-loop system (4) with the control laws (2) and (3) is guaranteed to be globally uniformly practically exponentially stable.*

Proof Consider the Lyapunov function candidate $V(t, x) = x^T Px$. It's derivative with respect to time is given by,

$$\dot{V}(t,x) = \sum_{i=1}^{r} \mu_i^2 x^T(G_{ii}^T P + PG_{ii})x + 2\sum_{i<j}^{r} \mu_i\mu_j x^T(G_{ij}^T P + PG_{ij})x + 2x^T P \sum_{i=1}^{r} \mu_i \Omega_i(t, x(t)).$$

The first two terms on the right-hand side constitute the derivative of the Lyapunov function $V(x)$ with respect the nominal system, while the third term is the effect of the perturbation. On the one hand, we have

$$x^T(G_{ii}^T P + PG_{ii})x \leq -\lambda_{min}(Q_i)\|x\|^2, \ i = 1, 2, ..., r,$$

and

$$x^T(G_{ij}^T P + PG_{ij})x \leq -\lambda_{min}(Q_{ij})\|x\|^2, \ 1 \leq i < j \leq r.$$

It follows that,

$$\dot{V}(t,x) \leq -\sum_{i=1}^{r} \mu_i^2 \lambda_{min}(Q_i)\|x\|^2 - 2\sum_{i<j}^{r} \mu_i\mu_j\lambda_{min}(Q_{ij})\|x\|^2 + 2x^T P \sum_{i=1}^{r} \mu_i \Omega_i(t, x(t)).$$

Thus,

$$\dot{V}(t,x) \leq -(\sum_{i=1}^{r} \mu_i^2 \lambda_{min}(Q_i) + 2\sum_{i<j}^{r} \mu_i\mu_j\lambda_{min}(Q_{ij}))\|x\|^2 + 2x^T P \sum_{i=1}^{r} \mu_i \Omega_i(t, x(t)).$$

Then, one gets

$$\dot{V}(t,x) \leq -\lambda_0 \|x\|^2 \sum_{i=1}^{r} \sum_{i=1}^{r} \mu_i \mu_j + 2x^T P \sum_{i=1}^{r} \mu_i \Omega_i(t, x(t)),$$

where $\lambda_0 = \inf\{(\lambda_{\min}(Q_i); i = 1, ..., r); (\lambda_{\min}(Q_{ij}); 1 \leq i < j \leq r)\}$, $\lambda_{\min(\max)}$ denotes the smallest (largest) eigenvalue of the matrix.

Since,

$$\sum_{i=1}^{r} \sum_{j=1}^{r} \mu_i \mu_j = 1,$$

then, we have

$$\dot{V}(t,x) \leq -\lambda_0 \|x\|^2 + 2x^T P \sum_{i=1}^{r} \mu_i \Omega(t, x(t)).$$

Taking into account the assumption $(\mathcal{H}_2)$, we have

$$\| \sum_{i=1}^{r} \mu_i \Omega_i(t, x(t)) \| \leq \sum_{i=1}^{r} \mu_i \pi_i(t).$$

Taking into account the above expressions, it follows that

$$\dot{V}(t,x) \leq -\lambda_0 \|x\|^2 + 2\|x\| \|P\| \sum_{i=1}^{r} \mu_i \pi_i(t).$$

Thus, by using the Cauchy-Schwartz inequality, one has

$$\dot{V}(t,x) \leq -\lambda_0 \|x\|^2 + 2\|x\| \|P\| ((\sum_{i=1}^{r} \mu_i^2)^{\frac{1}{2}} (\sum_{i=1}^{r} \pi_i(t)^2)^{\frac{1}{2}}).$$

It follows that,

$$\dot{V}(t,x) \leq -\lambda_0 \|x\|^2 + 2\|P\| \pi(t) \|x\|.$$

Since,

$$\lambda_{min}(P)\|x\|^2 \leq V(t,x) = x^T P x \leq \lambda_{max}(P)\|x\|^2,$$

then, by taking $\|P\| = \lambda_{max}(P)$, yields

$$\dot{V}(t,x) \leq -\frac{\lambda_0}{\lambda_{max}(P)} V(t,x) + 2\frac{\lambda_{max}(P)}{\lambda_{min}^{\frac{1}{2}}(P)} \pi(t) V(t,x)^{\frac{1}{2}}.$$

Let,

$$\eta = \frac{\lambda_0}{\lambda_{max}(P)} > 0,$$

and

$$\rho(t) = 2\frac{\lambda_{max}(P)}{\lambda_{min}^{\frac{1}{2}}(P)}\pi(t).$$

With the previous notations, it follows that

$$\dot{V}(t,x) \leq -\eta V(t,x) + \rho(t)V(t,x)^{\frac{1}{2}}.$$

In the last expression, we make the following change of variable, $w(t) = V(t,x)^{\frac{1}{2}}$. The derivative with respect to time is given by

$$\dot{w}(t) = \frac{\dot{V}(t,x)}{2V(t,x)^{\frac{1}{2}}}.$$

This implies that,

$$\dot{w}(t) \leq -\frac{1}{2}\eta w(t) + \frac{1}{2}\rho(t).$$

Thus,

$$w(t) \leq w(0)e^{-\frac{1}{2}\eta t} + \frac{1}{2}e^{-\frac{1}{2}\eta t}.\int_0^t \rho(s)e^{\frac{1}{2}\eta s}ds.$$

It follows that,

$$\lambda_{min}^{\frac{1}{2}}(P)\|x(t)\| \leq \lambda_{max}^{\frac{1}{2}}(P)\|x(0)\|e^{-\frac{1}{2}\eta t} + \frac{1}{2}e^{-\frac{1}{2}\eta t}.((\int_0^t \rho(s)^2 ds)^{\frac{1}{2}}.(\int_0^t (e^{\frac{1}{2}\eta s})^2 ds)^{\frac{1}{2}}).$$

So,

$$\lambda_{min}^{\frac{1}{2}}(P)\|x(t)\| \leq \lambda_{max}^{\frac{1}{2}}(P)\|x(0)\|e^{-\frac{1}{2}\eta t} + \frac{1}{2}e^{-\frac{1}{2}\eta t}.((\int_0^{+\infty} \rho(s)^2 ds)^{\frac{1}{2}}.(\int_0^t e^{\eta s} ds)^{\frac{1}{2}}).$$

One gets,

$$\lambda_{min}^{\frac{1}{2}}(P)\|x(t)\| \leq \lambda_{max}^{\frac{1}{2}}(P)\|x(0)\|e^{-\frac{1}{2}\eta t} + 2\tilde{\pi}(\frac{\lambda_{max}(P)}{\lambda_{min}^{\frac{1}{2}}(P)})^2 e^{-\frac{1}{2}\eta t}.(\frac{1}{\eta}(e^{\frac{1}{\eta t}} - 1))^{\frac{1}{2}}.$$

Hence,

$$\|x(t)\| \leq \frac{\lambda_{max}^{\frac{1}{2}}(P)}{\lambda_{min}^{\frac{1}{2}}(P)}\|x(0)\|e^{-\frac{1}{2}\eta t} + 2\tilde{\pi}\frac{\lambda_{max}^2(P)}{\lambda_{min}^{\frac{3}{2}}(P)}e^{-\frac{1}{2}\eta t}.(\frac{1}{\eta}e^{\frac{1}{\eta t}})^{\frac{1}{2}}.$$

Then,

$$\|x(t)\| \leq \frac{\lambda_{max}^{\frac{1}{2}}(P)}{\lambda_{min}^{\frac{1}{2}}(P)}\|x(0)\|e^{-\frac{1}{2}\eta t} + 2\frac{\lambda_{max}^{2}(P)}{\lambda_{min}^{\frac{3}{2}}(P)}\frac{\tilde{\pi}}{\eta^{\frac{1}{2}}}.$$

Therefore, $\mathcal{B}_r$, with $r = 2\frac{\lambda_{max}^{2}(P)}{\lambda_{min}^{\frac{3}{2}}(P)}\frac{\tilde{\pi}}{\eta^{\frac{1}{2}}}$, is globally uniformly practically exponentially stable. □

Remark 1 Note that the last inequality implies that system (4) is uniformly ultimately bounded in the sense of Definition 3.1.

3.2 Practical Stabilization via an Observer Design

Takagi-Sugeno (T-S) fuzzy models are usually used to describe nonlinear systems by a set of IF-THEN rules that gives local linear representations of subsystems. The overall model of the system is then formed as a fuzzy blending of these subsystems. It is important to study their stability or the synthesis of stabilizing controllers. The stability of TS models has been derived by means of several methods: Lyapunov approach, switching systems theory, linear system with modeling uncertainties, etc.

Let now consider the following T.S fuzzy dynamic model:

$$\dot{x} = \sum_{i=1}^{r} \mu_i(z)\big(A_i x + B_i u\big) \tag{9}$$

$$y = \sum_{i=1}^{r} \mu_i(z) C_i x \tag{10}$$

where $x \in \mathbb{R}^n$ is the state, $u \in \mathbb{R}^m$ is the control input, and $y \in \mathbb{R}^q$ is the output. The matrices A_i, B_i and C_i are of appropriate dimension, $r \geq 2$ is the number of rules, z is the premise vector which may include unmeasurable variables. It is assumed that $\mu_i(z) \geq 0$, for all $i = 1, ..., r$ and $\sum_{i=1}^{r} \mu_i(z) = 1$, for all $t \geq 0$.

In many practical control problems, the physical state variables of systems are partially or fully unavailable for measurement, since the state variables are not accessible by sensing devices and transducers are not available or very expensive. In such cases, observer based control schemes should be designed to estimate the state for (9). Taking $\hat{y}$ defined by

$$\hat{y} = \sum_{i=1}^{r} \mu_i(z) C_i \hat{x}.$$

In this case, an observer can be designed which has the form:

$$\dot{\hat{x}} = \sum_{i=1}^{r} \mu_i(z) \big(A_i \hat{x} + B_i u\big) - \sum_{i=1}^{r} \mu_i(z) L_i (\hat{y} - y). \tag{11}$$

Next, in order to prove a result of stabilization by means of an observer one can summarize this fact by the following steps.

We want to estimate the state x from the available signals u and y. We assume that we know the matrices $(A;\ B;\ C)$. Suppose that the pair $(A;\ B)$ is controllable and the pair $C;\ A)$ is observable. On the one hand, the controller feedback $u = Kx$ stabilizes the liner system in presence of perturbations and therefore the hole system under some restrictions on the nonlinearity part, here the matrix K is chosen such that $\mathcal{R}eel\lambda(A + BK) < 0$ this is holds because the pair $(A;\ B)$ is controllable. On the other hand, since the pair $(C;\ A)$ is observable, the gain matrix L can be chosen such that $\mathcal{R}eel\lambda(A - LC) < 0$. Hence, we can consider the following observer (estimator):

$$\text{Observer} \begin{cases} \dot{\hat{x}} = A\hat{x} + Bu + \Omega(t, x(t)) - L\left(C\hat{x} - y\right) \\ y = Cx \end{cases}$$

The error $e = \hat{x} - x$ satisfies

$$\dot{e} = (A - LC)\, e + (\Omega(t, \hat{x}) - \Omega(t, x))$$

and the estimated controller is given by: $u = K\hat{x}$.

The closed-loop system under state estimate feedback:

$$\dot{x} = Ax + BK\hat{x} + \Omega(t, x).$$

Let us consider the linear input-output system:

$$\begin{cases} \dot{x} = Ax + Bu + \Omega(t, x) \\ y = Cx \end{cases}$$

On has the following general system "observer based-controller"

$$\begin{cases} \dot{\hat{x}} = A\hat{x} + BK\hat{x} + \Omega(t, \hat{x}) - LCe \\ \\ \dot{e} = (A - LC)e + \Omega(t, \hat{x}) - \Omega(t, x) \end{cases}$$

In order to study the stabilization problem via an observer, we consider the system

$$\dot{\hat{x}} = f(\hat{x}, e) + R(\hat{x}, e)$$

$$\dot{e} = g(\hat{x}, e)$$

where,

$$f(\hat{x}, e) = (A + BK)\hat{x} + \Omega(t, \hat{x})$$

$$R(\hat{x}, e) = -LCe$$

and

$$g(\hat{x}, e) = (A - LC)e + \Omega(t, \hat{x}) - \Omega(t, x).$$

We have, both subsystems: $\dot{\hat{x}} = f(\hat{x}, 0)$ and $\dot{e} = g(\hat{x}, e)$ are globally practically exponentially stable, then a Lyapunov function candidate associated to the cascaded system can be taken as

$$V(\hat{x}, e) = V_1(\hat{x}) + \sigma V_2(e), \ \sigma > 0.$$

V_1 and V_2 are respectively the associated Lyapunov functions for $\dot{\hat{x}} = f(\hat{x}, 0)$ and $\dot{e} = g(\hat{x}, e)$. This yields, with $\sigma > 0$ small enough, that the following estimation holds:

$$\|(\hat{x}, e)\| \leq \lambda e^{-\gamma t}\|(\hat{x}(0), e(0))\| + \rho, \ \rho > 0, \ \lambda > 0 \ \gamma > 0, \ \forall t \geq 0.$$

Design of a practical observer Let consider the following output for the system (2): $y = \sum_{i=1}^{r} \mu_i(z)C_i x$, $y \in \mathbb{R}^q$ and C_i has an appropriate dimension. Taking $\hat{y}$ defined by $\hat{y} = \sum_{i=1}^{r} \mu_i(z)C_i\hat{x}$. In this case, an observer can be designed, where the premise variables are measurable, which has the form:

$$\dot{\hat{x}} = \sum_{i=1}^{r} \mu_i(z)\big(A_i\hat{x} + B_i u + \Omega_i(t, \hat{x})\big) - L_i(\hat{y} - y).$$

We wish to find the gain matrices L_i such that the error $e(t) = \hat{x}(t) - x(t)$ converge exponentially to a small ball centered at the origin as t goes to infinity. We assume that the following rules are given concerning the observer of each subsystem.

Rule l : If $z_1(t)$ is M_{i1} and $z_2(t)$ is M_{i2} ... and $z_p(t)$ is M_{ip}, then

$$\dot{\hat{x}} = \big(A_i\hat{x} + B_i u + \Omega_i(t, \hat{x})\big) - L_i(\hat{y} - y).$$

It suffices to take $L = \sum_{i=1}^{r} \eta_i(z) L_i$. Thus,

$$\dot{\hat{x}} = \sum_{i=1}^{r} \mu_i(z)\big(A_i\hat{x} + B_i u + \Omega_i(t,\hat{x})\big) - \sum_{i=1}^{r} \mu_i(z) L_i(\hat{y} - y).$$

Taking into account the above construction, the system error is given by:

$$\dot{e} = \sum_{i=1}^{r}\sum_{j=1}^{r} \mu_i(z)\eta_j(z)(A_i - L_j C_i)e + \sum_{i=1}^{r} \mu_i\big(\Omega_i(t,\hat{x}) - \Omega_i(t,\hat{x}-e)\big).$$

The above equation can also be written as:

$$\dot{e} = \sum_{i=1}^{r} \mu_i(z)^2 (A_i - L_j C_i)e + 2\sum_{i<j}^{r} \mu_i(z)\mu_j(z)\Big(\frac{1}{2}(A_i - L_j C_i)e + \frac{1}{2}(A_j - L_i C_j)e\Big)$$

$$+ \sum_{i=1}^{r} \mu_i\big(\Omega_i(t,\hat{x}) - \Omega_i(t,\hat{x}-e)\big).$$

The error system is exponentially stable if there exists a common symmetric positive definite matrix $\bar{P}$, such that

$$(A_i - L_j C_i)^T \bar{P} + \bar{P}(A_j - L_i C_j) = -\bar{Q}_{ij},\ i, j = 1, \ldots, r,$$

where $\bar{Q}_{ij}$ are positive definite matrices. In other hand, one can reach the same result if we suppose that:

$$(A_i - L_j C_i)^T \bar{P} + \bar{P}(A_j - L_i C_j) = -\bar{Q}_{ij},\ i, j = 1, \ldots, r.$$

Therefore, the Lyapunov function candidate can be taken as $\bar{V}(e) = e^T \bar{P} e$ to show the global exponential stability of the error equation.

In this situation, one gets

$$\dot{\bar{V}}(e) \leq -\bar{\lambda}_0\|e\|^2 \sum_{i=1}^{r}\sum_{j=1}^{r} \mu_i\mu_j + 2e^T\bar{P}\sum_{i=1}^{r} \mu_i\big(\Omega_i(t,\hat{x}) - \Omega_i(t,\hat{x}-e)\big),$$

for a certain non-negative constant $\bar{\lambda}_0$. It follows that,

$$\dot{\bar{V}}(e) \leq -\bar{\lambda}_0\|e\|^2 + 2e^T\bar{P}\sum_{i=1}^{r} \mu_i\big(\Omega_i(t,\hat{x}) - \Omega_i(t,\hat{x}-e)\big).$$

Since,

$$\lambda_{min}(\bar{P})\|e\|^2 \le V(t,e) = e^T(\bar{P})e \le \lambda_{max}((\bar{P}))\|e\|^2,$$

then, by taking $\|(\bar{P})\| = \lambda_{max}((\bar{P}))$, yields, as in the proof of Theorem 3.1, the following estimation:

$$\dot{V}(t,e) \le -\bar{\eta}V(t,e) + \bar{\rho}(t)V(t,e)^{\frac{1}{2}},$$

for $\bar{\eta} > 0$ and a bounded function $\bar{\rho}(t)$. Thus, one can obtain the following estimation, (see [10]), on the trajectories of the error equation:

$$\|e(t)\| \le \bar{\lambda}\|e(0)\|e^{-\bar{\gamma}t} + \bar{r},\ \bar{\lambda} > 0,\ \bar{\gamma} > 0,\ \bar{r} > 0.$$

Hence, the estimated fuzzy controller: $u(\hat{x}) = \sum_{j=1}^{r} \eta_j(z)K_j\hat{x}(t)$ ensures the stability of the closed-loop cascade system:

$$\begin{cases} \dot{\hat{x}} = \sum_{i=1}^{r} \mu_i\big(A_i\hat{x} + B_i u(\hat{x}) + \Omega_i(t,\hat{x})\big) - \sum_{i=1}^{r}\sum_{j=1}^{r} \mu_i\mu_j L_j C_i e \\ \dot{e} = \sum_{i=1}^{r}\sum_{j=1}^{r} \mu_i\mu_j(A_i - L_jC_i)e + \sum_{i=1}^{r} \mu_i\big(\Omega_i(t,\hat{x}) - \Omega_i(t,\hat{x}-e)\big) \end{cases}$$

in the sense that it is globally uniformly exponentially stable toward a small ball B_ρ centered at the origin of $\mathbb{R}^{2n}$.

In the modeling of uncertain nonlinear systems, two types of uncertainties may occur. The first one, named structural uncertainty, is referred to as parametric uncertainties that are due to formalized unknown nonlinearities. The second one, known as unstructured uncertainty, is often due to non-formalized modeling errors and external disturbances. Let us quote that, taking into account these uncertainties that can be understood as more practical applicability. Indeed, with the growing complexity of nonlinear systems, it is often necessary to make approximations in the dynamical modeling process. Therefore, the main objective is now to provide stability conditions, in terms of LMI, that ensure the practical stabilization for uncertain T-S models.

The fuzzy system is then inferred to be

$$\dot{x}(t) = \sum_{i=1}^{r} \mu_i(z)\Big(A_i x(t) + B_i u(t) + B_i\Omega_i(t, x(t))\Big). \tag{12}$$

The functions Ω_i, $i = 1, 2, ..., r,$, with respect the dimension of the matrices B_i, represent the uncertain external disturbances of each fuzzy subsystem and are time-varying satisfying the following condition.

$(\mathcal{H}_2')$ Assume that,

$$\|\Omega(t, x(t))\| \leq \zeta_i(t, x), \quad i = 1, 2, ..., r, \tag{13}$$

for all $t \geq 0$ and $x \in \mathbb{R}^n$, where ζ_i are known nonnegative continuous functions for $i = 1, ..., r$, with $\zeta_i(t, x) \leq \zeta(t, x)$, for all (t, x) and $i = 1, ..., r$.
We will use the following composite fuzzy controller:

$$u(t) = \sum_{j=1}^{r} \mu_j K_j x + \tilde{u}, \tag{14}$$

$\tilde{u}$ is related to the uncertainties, which is chosen in the following form:

$$\tilde{u} = \begin{cases} -\dfrac{B^T P x \zeta(t, x)}{\|B^T P x\| + \varepsilon \|x\|^2} & \text{if } x \neq 0 \\ 0 & \text{if } x = 0 \end{cases}$$

for a certain $\varepsilon > 0$.

From, (4) and (5), regarding each matrix $(G_{ii}^T P + P G_{ii})$, one has:

$$\lambda_{min}(G_{ii}^T P + P G_{ii})\|x\|^2 \leq^T x(G_{ii}^T P + P G_{ii})x \leq \lambda_{max}(G_{ii}^T P + P G_{ii})\|x\|^2,$$

$\lambda_{min}(.)(resp.\lambda_{max}(.))$ denotes the smallest (resp. the largest) eigenvalue of the matrix.

Define

$$\alpha = \max_{i,j} \lambda_{max}(G_{ii}^T P + P G_{ii})$$

for $1 \leq i < j \leq r$. A relaxed condition concerning the coupling effect is expressed as:

$$\sum_{i<j}^{r} \mu_i \mu_j x^T (G_{ij}^T P + P G_{ij}) x \leq k \|x\|^2,$$

where $k = \frac{r(r-1)}{2}\alpha$. Indeed, we have

$$\sum_{i<j}^{r} \mu_i \mu_j x^T (G_{ij}^T P + P G_{ij}) x \leq \sum_{i<j}^{r} \mu_i \mu_j \lambda_{max}(G_{ij}^T P + P G_{ij}) \|x\|^2.$$

It follows that,

$$\sum_{i<j}^{r} \mu_i \mu_j x^T (G_{ij}^T P + P G_{ij}) x \leq \sum_{i<j}^{r} \mu_i \mu_j \max_{i,j} \lambda_{max}(G_{ij}^T P + P G_{ij}) \|x\|^2.$$

Hence,

$$\sum_{i<j}^{r} \mu_i \mu_j x^T (G_{ij}^T P + P G_{ij}) x \leq \alpha \sum_{i<j}^{r} \mu_i \mu_j \|x\|^2 = \alpha \frac{r(r-1)}{2} \|x\|^2.$$

The representation of the global nonlinearities denoted by the following bound positive continuous function, for all $t \geq 0$,

$$\max_{x \in \mathbb{R}^n} \zeta(t,x) \leq \frac{1}{2\varepsilon} (\inf_{i=1,\ldots,r} \lambda_{\min}(Q_i) \sum_{i=1}^{r} \mu_i^2 - 2k) \tag{15}$$

with $\varepsilon > 0$ and

$$k < \frac{1}{2} \inf_{i=1,\ldots,r} \lambda_{\min}(Q_i) \sum_{i=1}^{r} \mu_i^2.$$

Now, one can state the following theorem.

Theorem 2 *Suppose that the assumptions $(\mathcal{H}_1)$, $(\mathcal{H}_2')$ hold and there exist a common positive definite matrix P and some feedback gain matrices K_i, $i = 1, \ldots, r$, such that the stability conditions (5) and (6) are satisfied, then the fuzzy closed-loop system (12) with the control laws (14) is guaranteed to be globally uniformly practically exponentially stable provided that $\zeta(t,x)$ satisfies (15).*

Note that, given $\zeta(t,x)$, ε can be chosen small enough with in such away (4.6) holds.

Proof Taking into account the fact that $B = \sum_{l=1}^{r} \mu_l B_l$, the derivative of $V(x) = x^T P x$ is given by:

$$\dot{V}(x) = \sum_{i=1}^{r} \mu_i^2 x^T (G_{ii}^T P + P G_{ii}) x + 2 \sum_{i<j}^{r} \mu_i \mu_j x^T (G_{ij}^T P + P G_{ij}) x + 2x^T \tilde{P} B \tilde{u} + 2x^T P \sum_{i=1}^{r} \mu_i B_i \Omega_i(t,x).$$

Thus, for $x \neq 0$, one gets

$$\dot{V}(x) \leq -\sum_{i=1}^{r} \mu_i^2 \lambda_{\min}(Q_i) \|x\|^2 + 2k\|x\|^2 + 2\|x^T P B\| \zeta(t,x) - 2 \frac{x^T P B B^T P x \zeta(t,x)}{\|B^T P x\| + \varepsilon \|x\|^2}.$$

It follows that

$$\dot{V}(x) \leq - \inf_{i=1,\ldots,r} \lambda_{\min}(Q_i) \sum_{i=1}^{r} \mu_i^2 \|x\|^2 + 2k\|x\|^2 + 2 \frac{\|B^T P x\| \zeta(t,x) \varepsilon \|x\|^2}{\|B^T P x\| + \varepsilon \|x\|^2}.$$

Hence,

$$\dot{V}(x) \leq - \inf_{i=1,\ldots,r} \lambda_{\min}(Q_i) \sum_{i=1}^{r} \mu_i^2 \|x\|^2 + 2k\|x\|^2 + 2\zeta(t,x)\varepsilon\|x\|^2.$$

and so,

$$\dot{V}(x) \leq \Big[- \inf_{i=1,\ldots,r} \lambda_{\min}(Q_i) \sum_{i=1}^{r} \mu_i^2 + 2k + 2\zeta(t,x)\varepsilon \Big] \|x\|^2.$$

Since $\zeta(t, x)$ satisfies (15), then one can obtain an estimation on $\dot{V}(x)$ as $\dot{V}(x) \leq -l\|x\|^2$, $l > 0$. Hence, the fuzzy system (12) in closed-loop with (14) is globally uniformly exponentially stable. ◇

Note that, given $\zeta(t, x)$ we can choose $\varepsilon > 0$ small enough such that (15) holds provided that

$$k < \frac{1}{2} \inf_{i=1,\ldots,r} \lambda_{\min}(Q_i) \sum_{i=1}^{r} \mu_i^2.$$

According to the above analysis, the design procedure for T-S fuzzy systems is summarized as follows.
Step 1: Verify that assumption (A_l, B_l) are controllable for $l = 1, \ldots, r$.
Step 2: Verify that assumption $(\mathcal{H}_1)$, $(\mathcal{H}_2)$ are satisfied (resp. $(\mathcal{H}_1)$, $(\mathcal{H}_2')$).
Step 3: Solve the Lyapunov equations (5) and (6) to obtain P, K_i, Q_i, $i = 1, \ldots, r$. Note that, for simplicity one can choose $Q_i = I$.
Step 4: By using the control toolbox, execute the nonlinear program based on equations $G_{ii}^T P + PG_{ii})$, $1 \leq i < j \leq r$, to determine k. The nonlinear programming is expressed in the following form: Determine $\lambda_{max}(G_{ii}^T P + PG_{ii})$ and then

$$\alpha = \max_{i,j} \lambda_{max}(G_{ii}^T P + PG_{ii})$$

for $1 \leq i < j \leq r$.
Step 5: Construct the fuzzy controller (2) and (3) (resp. (2.14)) for (4) (respectively for (12)).
Step 6: Verify the condition (15) imposed on $\zeta(t, x)$ for a suitable choice of ε.

References

1. Ben Abdallah, A., Ellouze, I., Hammami, M. A.: Practical stability of nonlinear time-varying cascade systems. J. Dyn. Control. Syst. **15**, 45–62 (2009)
2. Benabdallah, A., Ellouze, I., Hammami, M.A.: Practical exponential stability of perturbed triangular systems and a separation principle. Asian J. Control. **13**(3), 445–448 (2011)
3. Ben Hamed, B., Ellouze, I., Hammami, M.A.: Practical uniform stability of nonlinear differential delay equations. Mediterr. J. Math. **6**, 139–150 (2010)
4. Ben Hamed, B., Hammami, M.A.: Practical stabilization of a class of uncertain time-varying nonlinear delay systems. J. Control. Theory Appl. **7**, 175–180 (2009)
5. Ben Makhlouf, A., Hammami, M.A.: A nonlinear inequality and application to global asymptotic stability of perturbed systems. Math. Methods Appl. Sci. **38**(12), 2496–2505 (2015)
6. Ben Makhlouf, A., Hammami, M.A., Sioud, K.: Stability of fractional-order nonlinear systems depending on a parameter. Bull. Korean Math. Soc. **54**(4), 1309–1321 (2017)
7. Nasser, B.B., Boukerrioua, K., Defoort, M., Djemai, M., Hammami, M.A., Laleg-Kirati, T.M.: Sufficient conditions for uniform exponential stability and h-stability of some classes of dynamic equations on arbitrary time scales. Nonlinear Anal. Hybrid Syst. **32**, 54–64 (2019)
8. Caraballo, T., Hammami, M.A., Mchiri, L.: Practical stability of stochastic delay evolution equations. Acta Applicandae Mathematicae **142**, 91–105 (2016)
9. Caraballo, T., Hammami, M.A., Mchiri, L.: Practical exponential stability of impulsive stochastic functional differential equations. Syst. Control. Lett. **109**, 43–48 (2017)
10. Delmotte, F., Hammami, M.A., Jellouli, A.: Exponential stabilization of fuzzy systems with perturbations by using state estimation. Int. J. Gen. Syst. **50**(4), 388–408 (2021)
11. Dlala, M., Hammami, M.A.: Uniform exponential practical stability of impulsive perturbed systems. J. Dyn. Control. Syst. **13**(3), 373–386 (2007)
12. Ellouze, I., Hammami, M.A.: A separation principle of time-varying dynamical systems: a practical stability approach. Math. Model. Anal. **12**(3), 297–308 (2007)
13. Hadj Taieb, N., Hammami, M.A., Delmotte, F.: A separation principle for Takagi-Sugeno control fuzzy systems. Arch. Control. Sci. **29**(2), 227–245 (2019)
14. Hadj Taieb, N., Hammami, M.A., Delmotte, F.: Stabilization of a certain class of fuzzy control systems with uncertainties. Arch. Control. Sci. **27**(3), 453–481 (2017)
15. Hammami, M., Hammami, M.A., De La Sen, M.: Exponential stability of time-varying systems subject to discrete time-varying delays and nonlinear delayed perturbations. Math. Probl. Eng. **2015**(1), 12 (2015). Article ID 641268. https://doi.org/10.1155/2015/641268.
16. Hadj Taieb, N., Hammami, M.A.: Some new results on the global uniform asymptotic stability of time-varying dynamical systems. IMA J. Math. Control. Inf. **32**, 1–22 (2017)
17. Hadj Taieb, N.: Stability an analysis for time-varying nonlinear systems. Int. J. Control. (2020). https://doi.org/10.1080/00207179.2020.1861332
18. Hamzaoui, A., Taieb, N.H., Hammami, M.A.: Practical partial stability of time-varying systems. Discret. Contin. Dyn. Syst. B. (2021). https://doi.org/10.3934/dcdsb.2021197
19. Huang, D., Nguang, S.K.: Robust H_∞ static output feedback control of fuzzy systems: an ILMI approach. IEEE Trans. Syst. Man Cybern. Part B Cybern. **36**(1), 216–222 (2006)
20. Lee, H.J., Park, J.B., Chen, G.: Robust fuzzy control of nonlinear systems with parameter uncertainties. IEEE Trans. Fuzzy Syst. **9**, 369–379 (2001)
21. Liu, Z.: New approach to H_∞ controller designs based on fuzzy observers for T-S fuzzy systems via LMI. Autom. **39**, 1571–1582 (2003)
22. Chunyan, L., Lu, W., Jingzhe, L.: Importance analysis of different components in a multicomponent system under fuzzy inputs. Struct. Multidiscip. Optim. **65**, Article number: 93 (2022)
23. Nian, X., Feng, J.: Guaranteed-cost control of a linear uncertain system with multiple time-varying delays: an LMI approach. IEE Proc. Part D Control. Theory Appl. **150**(1), 17–22 (2003)
24. Park, J., Kim, J., Park, D.: LMI-based design of stabilizing fuzzy controllers for nonlinear systems described by Takagi-Sugeno fuzzy model. Fuzzy Sets Syst. **122**, 73–82 (2003)

25. Takagi, T., Sugeno, M.: Fuzzy identification of systems and its applications to modeling and control. IEEE Trans. Syst. Man Cybern. Part B Cybern. **15**, 116–132 (1985)
26. Tseng, C.S., Chen, B.S., Uang, H.J.: Fuzzy tracking control design for nonlinear dynamics systems via T-S fuzzy model. IEEE Trans. Fuzzy Syst. **9**(3), 381–392 (2001)
27. Tanaka, K., Ikeda, T., Wang, H.O.: Robust stabilization of a class of uncertain nonlinear systems via fuzzy control: quadratic stabilizability, H_∞ control theory and linear matrix inequalities. IEEE Trans. Fuzzy Syst. **4**, 1–13 (1996)
28. Tanaka, K., Sugeno, M.: Stability analysis and design of fuzzy control systems. Fuzzy Sets Syst. **45**, 135–156 (1992)
29. Tong, R.M.: A control engineering review of fuzzy systems. Autom. **13**, 559–568 (1977)
30. Tong, S., Wang, T., Li, H.X.: Fuzzy robust tracking control for uncertain nonlinear systems. Int. J. Approx. Reason. **30**, 73–90 (2002)
31. Wang, Y., Liu, J., Sun, C.M., Sun, Q.: On the stability and con-vergence rate analysis for the nonlinear uncertain systems based upon active disturbance rejection control. Int. J. Robust Nonlinear Control. **30**(2020), 5728–5750 (2020)
32. Wang, L., Xiong, C., Wang, X., Liu, G., Shi, Q.: Sequential optimization and fuzzy reliability analysis for multidisciplinary systems. Struct. Multidiscip. Optim. **60**, 1079–1095 (2019)
33. Zadeh, L.A.: Outline of a new approach to the analysis of complex systems and decision process. IEEE Trans. Syst. Man Cybern. Part B Cybern. **3**, 28–44 (1973)
34. Zhang, J.M., Li, R.H., Zhang, P.A.: Stability analysis and systematic design of fuzzy control systems. Fuzzy Sets Syst. **120**, 65–72 (2001)
35. Zhang, H., Liu, M., Shi, Y., Sheng, J.: Extended LMI representatives for stability and stabilization of discrete-time Takagi-Sugeno fuzzy systems. Optim. Control. Appl. Methods. **35**(6), 647–655 (2014)

Observer-Based Robust Tracking Controller Design of Nonlinear Dynamic Systems Represented by Bilinear T-S Fuzzy Systems

Chekib Ghorbel and Naceur Benhadj Braiek

Abstract This chapter presents an observer-based tracking control strategy of nonlinear dynamic systems described as bilinear Takagi-Sugeno fuzzy system in the presence of parameter uncertainties and external disturbances. Firstly, a bilinear fuzzy model is constructed, based on which the dynamics of fuzzy observer, observer error, and tracking error are designed. Secondly, a Lyapunov's direct method and a parallel distributed compensation concept are recalled. By using Schur complement, separation lemma, and linear matrix inequality tools, sufficient stability conditions are established. Finally, a numerical simulation is carried out to demonstrate the effectiveness of the proposed control scheme on tracking of a nonlinear mechanical system.

Keywords Bilinear fuzzy model · Observer design · Tracking · Robust control · Linear matrix inequality

1 Introduction

During these last decades, fuzzy techniques have been successfully used in nonlinear system modelling, identification, and control [3, 7, 16, 28]. However, the fuzzy logic control study has received considerable attention in numerous fields, such as robotics, power systems, chemical engineering, image processing, vehicular technology, industrial automation, electronics, civil engineering, and automobiles. Particularly, the well-known Takagi-Sugeno (T-S) fuzzy model is an effective approach to develop controllers for nonlinear dynamic systems. It represents them as fuzzy models by a set of If-then rules. Notably, based on the sector nonlinearity concept

C. Ghorbel (✉) · N. Benhadj Braiek
Laboratory of Advanced Systems, Polytechnic High School of Tunisia, University of Carthage, BP 743, 2078 La Marsa, Tunisia
e-mail: chekib.ghorbel@enicarthage.rnu.tn

N. Benhadj Braiek
e-mail: naceur.benhadj@ept.rnu.tn

A. Ben Makhlouf et al. (eds.), *State Estimation and Stabilization of Nonlinear Systems*, Studies in Systems, Decision and Control 491,
https://doi.org/10.1007/978-3-031-37970-3_5

[13, 30], the overall T-S model is achieved by interpolating those models through nonlinear fuzzy membership functions. In this case, a Parallel Distributed Compensation (PDC) concept has been employed to overcome different stability, stabilization, and tracking problems of fuzzy systems [12, 32, 36]. Indeed, each fuzzy control rule is designed from the corresponding fuzzy rule. The resulting controller is a blending of local controllers. Subsequently, the obtained stability conditions are in general established in terms of Linear Matrix Inequalities (LMI), which can be solved efficiently by using convex programming tools.

Bilinear systems have been widely studied, both from the theoretical and the practical point of view (see for example [15, 17, 23, 31] and relative bibliography. Indeed, many nonlinear dynamic systems cannot be adequately approximated by linear models. A good example of a bilinear process is the population of biological species [22, 25], which its mathematical model is described by $\dot{x} = x\,u$ where x is the population and u denotes the birth rate minus death rate. As well, it is possible to generalize to bilinear systems many analysis methods originally developed in the linear theory [8, 10, 20]. For example, the authors of [26, 28, 33, 34] presented robust H_∞ fuzzy controllers of fuzzy uncertain bilinear systems. Saoudi et al. [27] proposed a fault tolerant control methodology of bilinear systems. The authors of [19] studied the problem of robust synchronization of chaotic systems modeled as bilinear T-S fuzzy systems. More recently, the authors of [6] explored sufficient stability conditions for the synthesis of robust unknown input observers. The authors of [11] presented fuzzy control strategies for classes of linear and bilinear fuzzy systems.

In addition, in practical control applications, some state variables may not be available. Therefore, the state estimation is necessary to achieve guaranteed control objectives [1, 5, 29, 35]. Over the past decades, many research works have been carried out to investigate the problem of the observer synthesis and its application [1, 14, 18]. For example, the T-S fuzzy observer problem for dynamic systems has been addressed in [4, 14]. Several works as in [2, 37] dealing with observer design in explicit form proportional integral observer.

The main contributions can be summarized as follows:

- Compared with the previous research works, an observer-based robust tracking controller design of bilinear T-S fuzzy systems with decay rate and attenuation of the disturbances effect is designed;
- Sufficient stability conditions described on the LMI form are proposed. Following the Introduction, this chapter is structured as follows: In Sect. 2, the problem formulation is exposed. Section 3 presents an observer-based robust tracking control scheme. In Sect. 4, a simulation study is provided to illustrate the merit of the purpose. The chapter finalizes with Conclusion.

  Throughout this work, M^T denotes the transpose of M, $M + (*)$ stands for $M + M^T$, $\begin{pmatrix} A & (*) \\ B & C \end{pmatrix}$ represents $\begin{pmatrix} A & B^T \\ B & C \end{pmatrix}$, and $\chi\,(z) = \chi\,(z\,(t))$.

2 Problem Formulation

Let us consider a nonlinear dynamic system:

$$\begin{cases} \dot{x} = f(x) + g(x, u) \\ y = h(x) \end{cases} \tag{1}$$

where $f(x)$, $g(x, u)$, and $h(x)$ are nonlinear functions of the vector state $x \in \mathbb{R}^{n\times n}$, $u \in \mathbb{R}^m$ is the input vector, and $y \in \mathbb{R}^p$ is the output vector. By using the fuzzy inference method [30], the nonlinear dynamic system (1) can be represented by the following bilinear plant rules:

$$\begin{aligned} &\text{If } (z_1 \text{ is } M_{1i}) \text{ and } (z_2 \text{ is } M_{2i}) \text{ and } \ldots \text{ and } (z_k \text{ is } M_{ki}) \\ &\text{then } \dot{x} = \overline{A}_i\, x + \overline{B}_i\, u + \overline{N}_i\, x\, u + D_i\, w \text{ and } y = C_i\, x \end{aligned} \tag{2}$$

for $i = 1, 2, \ldots, r$ and $k \leq n$ where $w \in \mathbb{R}^n$ is the disturbances vector, M_{ji} is the fuzzy set, $z_1, z_2, \ldots, z_k$ are the known premise variables, $M_{ji}(z_j)$ is the degree of membership of z_j, $i = 1, 2, \ldots, r$ and $j = 1, 2, \ldots, k$, and r is the number of If-then rules. $A_i \in \mathbb{R}^{n\times n}$, $B_i \in \mathbb{R}^{n\times m}$, $C_i \in \mathbb{R}^{p\times n}$, $D_i \in \mathbb{R}^n$, and $N_i \in \mathbb{R}^{n\times n}$ are known. Let us assume that the matrices $\overline{A}_i$, $\overline{B}_i$, and $\overline{N}_i$ are given in the form below:

$$\begin{aligned} \overline{A}_i &= A_i + \Delta A_i(t) = A_i + H_i\, \psi(t)\, E_{Ai} \\ \overline{B}_i &= B_i + \Delta B_i(t) = B_i + H_i\, \psi(t)\, E_{Bi} \\ \overline{N}_i &= N_i + \Delta N_i(t) = N_i + H_i\, \psi(t)\, E_{Ni} \end{aligned} \tag{3}$$

where H_i, E_{Ai}, E_{Bi}, and E_{Ni} are known real matrices of appropriate dimensions and $\psi(t)$ is a time-varying matrix satisfying $\psi^T(t)\, \psi(t) \leq I$.
Then, the resulting dynamic equations are expressed as:

$$\begin{cases} \dot{x} = \sum\limits_{i=1}^{r} h_i \left(\overline{A}_i x + B_i u + \overline{N}_i x u + D_i\, w\right) \\ y = \sum\limits_{i=1}^{r} h_i\, C_i\, x \end{cases} \tag{4}$$

where $h_i = \frac{w_i}{\sum\limits_{i=1}^{r} w_i}$, $w_i = \prod\limits_{j=1}^{k} M_{ji}(z_j)$, and $z = \left(z_1\ z_2\ \ldots\ z_k\right)^T$. Recall that $\sum\limits_{i=1}^{r} h_i = 1$ and $h_i \geq 0$.
In practical situations, not all the system's states are available to be measured, thus it is necessary to design a fuzzy observer, which is in several cases given by the following state equation:

$$\dot{\hat{x}} = \sum_{i=1}^{r} h_i\, \left(A_i\, \hat{x} + B_i\, u + N_i\, \hat{x}\, u + L_i\, (y - \hat{y})\right) \tag{5}$$

where $L_i \in \mathbb{R}^{n\times p}$ are the observer gains to be determined.
The objective of this chapter is to design an observer-based robust tracking controller that ensures a good tracking between the estimated state variables of the controlled fuzzy system and the following reference model's ones:

$$\dot{x}_e = A_e\, x_e + B_e\, e \tag{6}$$

where $A_e \in \mathbb{R}^{n\times n}$ is an asymptotically stable matrix, $B_e \in \mathbb{R}^n$ is an input matrix, $x_e \in \mathbb{R}^n$ is a reference state, and $e \in \mathbb{R}$ is a reference input.

3 Observer-Based Control Scheme Design

The bilinear fuzzy model is firstly introduced in this section. Based on which, the dynamics of fuzzy observer, observer error, and tracking error are secondly determined. Thirdly, sufficient stability conditions are established to ensure the asymptotic stability of the controlled system with guaranteed control objectives.
Let us consider that the pairs $(A_i,\ B_i)$ are locally controllable and $(A_i,\ C_i)$ are locally observable, for $i = 1, 2, \ldots, r$. Then, the ith control law is defined as:

$$\begin{aligned} &\text{If } (z_1 \text{ is } M_{1i}) \text{ and } (z_2 \text{ is } M_{2i}) \text{ and } \ldots \text{ and } (z_k \text{ is } M_{ki}) \\ &\text{then } u = \rho\, \frac{K_i\, \varepsilon}{\sqrt{1+\varepsilon^T K_i^T K_i\, \varepsilon}} \end{aligned} \tag{7}$$

where $\varepsilon = x_e - \hat{x}$ is the tracking error, $K_i \in \mathbb{R}^{m\times n}$ are controller gains. Thus, the resulting fuzzy controller is described as:

$$\begin{aligned} u &= \sum_{i=1}^{r} h_i\, \rho\, \frac{K_i\, \varepsilon}{\sqrt{1+\varepsilon^T K_i^T K_i\, \varepsilon}} \\ &= \sum_{i=1}^{r} h_i\, \rho\, \sin(\theta_i) \\ &= \sum_{i=1}^{r} h_i\, \rho\, K_i\, \varepsilon\, \cos(\theta_i) \end{aligned} \tag{8}$$

where $\sin(\theta_i) = \frac{K_i\, \varepsilon}{\sqrt{1+\varepsilon^T K_i^T K_i\, \varepsilon}}$ and $\cos(\theta_i) = \frac{1}{\sqrt{1+\varepsilon^T K_i^T K_i\, \varepsilon}}$ for $\theta_i \in \left[-\frac{\pi}{2},\ \frac{\pi}{2}\right]$.
By assuming that $\varepsilon_o = x - \hat{x}$ and $\varepsilon_e = x - x_e$, the dynamics of the observer error and the tracking error are defined as follows:

$$\dot{\varepsilon}_o = \sum_{i=1}^{r}\sum_{j=1}^{r} h_i\, h_j\, \left(\upsilon_{1,ij}\, \varepsilon_o + \upsilon_{2,ij}\, \varepsilon_e + \upsilon_{3,ij}\, x_e + D_i\, w\right) \tag{9}$$

$$\dot{\varepsilon}_e = \sum_{i=1}^{r}\sum_{j=1}^{r} h_i\, h_j\left(\xi_{1,ij}\,\varepsilon_o + \xi_{2,ij}\,\varepsilon_e + \xi_{3,ij}\,x_e + D_i\,w - B_e\,e\right) \tag{10}$$

where:

- $\upsilon_{1,ij} = A_i - L_i\,C_i + \rho N_i\,\sin(\theta_j) + \rho\,\Delta B_i\,K_j\cos(\theta_j)$;
- $\upsilon_{2,ij} = \Delta A_i + \rho\,\Delta N_i\,\sin(\theta_j) - \rho\,\Delta B_i\,K_j\,\cos(\theta_j)$;
- $\upsilon_{3,ij} = \Delta A_i + \rho\,\Delta N_i\,\sin(\theta_j)$;
- $\xi_{1,ij} = \rho\,\overline{B}_i\,K_j\,\cos(\theta_j)$;
- $\xi_{2,ij} = \overline{A}_i + \rho\,\overline{N}_i\,\sin(\theta_j) - \rho\,\overline{B}_i\,K_j\,\cos(\theta_j)$;
- $\xi_{3,ij} = \overline{A}_i - A_e + \rho\,\overline{N}_i\,\sin(\theta_j)$.

In what follows, the augmented system is defined as:

$$\dot{x}_a = \sum_{i=1}^{r}\sum_{j=1}^{r} h_i\, h_j\left(\Upsilon_{1,ij}\,x_a + \Upsilon_{2,i}\,\varphi\right) \tag{11}$$

where $x_a = \begin{pmatrix}\varepsilon_o\\ \varepsilon_e\\ x_e\end{pmatrix}$, $\varphi = \begin{pmatrix} w\\ e\end{pmatrix}$, $\overline{\Upsilon}_{ij} = \begin{pmatrix}\upsilon_{1,ij} & \upsilon_{2,ij} & \upsilon_{3,ij}\\ \xi_{1,ij} & \xi_{2,ij} & \xi_{3,ij}\\ 0 & 0 & A_e\end{pmatrix}$, $\Upsilon_{2,i} = \begin{pmatrix} D_i & 0\\ D_i & -B_e\\ 0 & B_e\end{pmatrix}$.
Mainly, the control objective is to determine the gains K_i and L_i, for $i = 1, 2, \ldots, r$, that guarantee the asymptotic stability of the augmented system (11) and ensure the H_∞ tracking control performance:

$$\left\| H_{x_a\varphi}\right\|_\infty = \sup_{\|\varphi\|_2\neq 0}\frac{\|x_a\|_2}{\|\varphi\|_2} < \gamma \tag{12}$$

where $\gamma > 0$ is a prescribed disturbances attenuation level. Consequently, the following theorem presents the result.

Theorem The equilibrium of the augmented system (11) is globally asymptotically stable with the decay rate $\alpha > 0$ and satisfying the performance control objective (12) if there exist positive scalars γ, τ_i, for $i = 1, 2, \ldots, 10$, $\delta_{2,3,4} = \tau_2 + \tau_3 + \tau_4$, $\delta_{1,\rho} = \tau_5\rho^2 + \tau_8\left(1 + 2\rho^2\right) + \tau_9\left(1 + \rho^2\right)$, $\delta_{2,\rho} = \tau_6\left(1 + 2\rho^2\right) + \tau_7\,\rho^2 + \tau_{10}\left(1 + \rho^2\right)$, $\delta_{5,7} = \tau_5^{-1} + \tau_7^{-1}$, $\delta_{6,8} = \tau_6^{-1} + \tau_8^{-1}$, and $\delta_{9,10} = \tau_9^{-1} + \tau_{10}^{-1}$ verifying the following LMI formulation:

$$\begin{aligned}&\underset{P_1, X_2, P_3, R_i, V_i}{\text{maximize}}\ \alpha\\ &\text{subject to}:\\ &\begin{pmatrix} LM_{ij}^{11} & (*) & 0_{5\times 7}\\ LM_{ij}^{21} & LM^{22} & (*)\\ 0_{7\times 5} & LM_{ij}^{32} & LM^{33}\end{pmatrix} < 0,\\ &\text{for } i,\ j = 1, 2, \ldots, r\end{aligned} \tag{13}$$

where:

- $LM_{ij}^{11} = \begin{pmatrix} \zeta_i^{11} & 0 & 0 & P_1 D_i & 0 \\ 0 & \zeta_i^{22} & (A_i - A_e)\, X_2 & D_i X_2 & -B_e X_2 \\ 0 & (*) & \zeta_i^{33} & 0 & P_3 B_e \\ (*) & (*) & 0 & -\gamma^2 I & 0 \\ 0 & (*) & (*) & 0 & -\gamma^2 I \end{pmatrix}$;
- $\zeta_i^{11} = P_1 A_i - V_i\, C_i \,+ (*) + 2\,\alpha\, P_1$;
- $\zeta_i^{22} = A_i X_2 + (*) + 2\,\alpha X_2 + \,\delta_{2,3,4}\, \rho^2 I + \delta_{2,\rho} H_i H_i^T$;
- $\zeta_i^{33} = P_3 A_e + (*) + 2\,\alpha\, P_3 + \,\tau_3^{-1} N_i^T N_i + \delta_{9,10}\, E_{Ai}^T E_{Ai} + \,\delta_{9,10}\, E_{Ni}^T E_{Ni}$;
- $LM_{21}^{ij} = \begin{pmatrix} \rho P_1 & 0\,0\,0\,0 \\ B_i K_j & 0\,0\,0\,0 \\ H_i^T P_1 & 0\,0\,0\,0 \\ E_{B_i} K_j & 0\,0\,0\,0 \end{pmatrix}$;
- $LM^{22} = diag\,([-\tau_1^{-1} I \; - \tau_4 I \; -\delta_{1,\rho}^{-1} I \; - \delta_{5,7}^{-1} I\,])$;
- $LM_{32}^{ij} = \begin{pmatrix} X_2 & 0\,0\,0 \\ \rho K_j^T B_i & 0\,0\,0 \\ N_i X_2 & 0\,0\,0 \\ R_j & 0\,0\,0 \\ X_2\, E_{A_i} & 0\,0\,0 \\ E_{N_i} X_2 & 0\,0\,0 \\ E_{B_i}\, R_j & 0\,0\,0 \end{pmatrix}$;
- $LM^{33} = diag\,([-\,Q^{-1} \; - \tau_2^{-1} I \; - \tau_2\, I \; - \tau_4\, I \; - \delta_{6,8}^{-1}\, I \; - \delta_{6,8}^{-1}\, I \; - \delta_{6,8}^{-1}\, I\,])$;

Then, the feedback and observer gains are calculated from expressions $K_i = R_i\, P_2$ and $L_i = V_i\, P_1^{-1}$, respectively.

Proof Let us consider the quadratic Luapunov function:

$$V\,(x_a) = x_a^T\; P\; x_a \tag{14}$$

such that:

$$\dot{V}\,(x_a) \leq -\,2\;\alpha\; V\,(x_a) \tag{15}$$

where $P = P^T > 0$ and $\alpha > 0$.
In the following, the controlled system (11) is globally asymptotically stable with a decay rate α if:

$$\dot{V}\,(x_a) \,+\, 2\,\alpha\, V\,(x_a) \,+\, x_a^T\; Q^+\, x_a \;-\; \gamma^2\, \varphi^T\, \varphi \;<\; 0 \tag{16}$$

where $Q^+ = diag\,(0,\;\; Q,\;\; 0)$.

From relations (11), (14), and (15), the constraint (16) is equivalent to:

$$\sum_{i=1}^{r}\sum_{j=1}^{r} h_i\, h_j\, x_a^T \begin{pmatrix} P\,\overline{\Upsilon}_{1,ij} + (*) + 2\,\alpha\, P + Q^+ & (*) \\ \Upsilon_{2,i}^T P & -\gamma^2 I \end{pmatrix} x_a < 0 \tag{17}$$

for $i,\ j = 1, 2, \ldots, r$.
As all nonlinearities $h_i\ \in\ [0,\ \ 1]$, the matrix inequality (17) allows to write:

$$\begin{pmatrix} P\,\overline{\Upsilon}_{1,ij} + (*) + 2\,\alpha\, P + Q^+ & (*) \\ \Upsilon_{2,i}^T P & -\gamma^2 I \end{pmatrix} <\ 0, \text{for } i,\ j = 1, 2, \ldots, r \tag{18}$$

By choosing $P = diag\,(P_1,\ P_2,\ P_3)$, (18) becomes:

$$\begin{pmatrix} \mu_{ij}^{11} + \overline{\mu}_{ij}^{11} & \mu_{ij}^{12} + \overline{\mu}_{ij}^{12} & \overline{\mu}_{ij}^{13} & P_1 D_i & 0 \\ (*) & \mu_{ij}^{22} + \overline{\mu}_{ij}^{22} & \mu_{ij}^{23} + \overline{\mu}_{ij}^{23} & P_2 D_i & -P_2 B_e \\ 0 & (*) & P_3 A_e + (*) + 2\,\alpha\, P_3 & 0 & P_3 B_e \\ (*) & (*) & 0 & -\gamma^2 I & 0 \\ 0 & (*) & (*) & 0 & -\gamma^2 I \end{pmatrix} < 0 \tag{19}$$

where:

- $\mu_{ij}^{11} = P_1\left(A_i\ -\ L_i\, C_i + \rho N_i\ \sin\left(\theta_j\right)\right) + (*) + 2\,\alpha P_1;$
- $\mu_{ij}^{12} = \rho\, K_j^T\, B_i^T\, P_2 \cos\left(\theta_j\right);$
- $\mu_{ij}^{22} = P_2\left(A_i + \rho\, N_i\ \sin\left(\theta_j\right) - \rho\, B_i\, K_j\ \cos\left(\theta_j\right)\right) + (*) + 2\,\alpha P_2 + Q;$
- $\mu_{ij}^{23} = P_2\left(\overline{A}_i - A_e + \rho\, N_i\ \sin\left(\theta_j\right)\right);$
- $\overline{\mu}_{ij}^{11} = \rho\, P_1\, \Delta B_i K_j \cos\left(\theta_j\right) +\ (*);$
- $\overline{\mu}_{ij}^{12}{=}P_1\left(\Delta A_i +\ \rho\, \Delta N_i\ \sin\left(\theta_j\right)\ - \rho\, \Delta B_i\, K_j\ \cos\left(\theta_j\right)\right) + \rho\, K_j^T\, \Delta B_i^T\, P_2 \cos\left(\theta_j\right);$
- $\overline{\mu}_{ij}^{22} = P_2\left(\Delta A_i +\ \ \rho\, \Delta N_i\ \sin\left(\theta_j\right) - \rho\, \Delta B_i\, K_j\ \cos\left(\theta_j\right)\right) +\ \ (*);$
- $\overline{\mu}_{ij}^{13} = P_1\,(\Delta A_i +\ \rho\, \Delta N_i\ \sin\left(\theta_i\right));$
- $\overline{\mu}_{ij}^{23} = \Delta A_i + \rho\, P_2\, \Delta N_i\ \sin\left(\theta_j\right).$

It is pointing out that (19) contains certain parts Θ_{ij} and uncertain parts $\Delta\Theta_{ij}$. However, it leads to transform it on the following form:

$$\Theta_{ij} + \Delta\Theta_{ij}\ < 0, \text{for } i,\ j = 1, 2, \ldots, r \tag{20}$$

where:

- $\Theta_{ij} = \begin{pmatrix} \mu_{ij}^{11} & \mu_{ij}^{12} & 0 & P_1 D_i & 0 \\ (*) & \mu_{ij}^{22} & \mu_{ij}^{23} & P_2 D_i & -P_2 B_e \\ 0 & (*) & P_3 A_e + (*) + 2\,\alpha\, P_3 & 0 & P_3 B_e \\ (*) & (*) & 0 & -\gamma^2 I & 0 \\ 0 & (*) & (*) & 0 & -\gamma^2 I \end{pmatrix};$

- $\Delta\Theta_{ij} = \begin{pmatrix} \overline{\mu}_{ij}^{11} & \overline{\mu}_{ij}^{12} & \overline{\mu}_{ij}^{13} & 0 & 0 \\ (*) & \overline{\mu}_{ij}^{22} & \overline{\mu}_{ij}^{23} & 0 & 0 \\ (*) & (*) & 0 & 0 & 0 \\ 0 & 0 & 0 & 0 & 0 \\ 0 & 0 & 0 & 0 & 0 \end{pmatrix}.$

The next step is to transform Θ_{ij} and $\Delta\Theta_{ij}$ into diagonal matrices using lemma 1. Then, by replacing ΔA_i, ΔB_i, and ΔN_i by their respective quantities $H_i\ \psi\ (t)\ E_{Ai}$, $H_i\ \psi\ (t)\ E_{Bi}$, and $H_i\ \psi\ (t)\ E_{Ni}$ and considering positive scalars τ_i, for $i = 1, 2, \ldots, 10$, $\delta_{2,3,4} = \tau_2 + \tau_3 + \tau_4$, $\delta_{1,\rho} = \tau_5\rho^2 + \tau_8\left(1 + 2\rho^2\right) + \tau_9\left(1 + \rho^2\right)$, $\delta_{2,\rho} = \tau_6\left(1 + 2\rho^2\right) + \tau_7\,\rho^2 + \tau_{10}\left(1 + \rho^2\right)$, $\delta_{5,7} = \tau_5^{-1} + \tau_7^{-1}$, $\delta_{6,8} = \tau_6^{-1} + \tau_8^{-1}$, and $\delta_{9,10} = \tau_9^{-1} + \tau_{10}^{-1}$, this leads to obtain:

$$\Theta_{ij} \leq \begin{pmatrix} a_{ij}^{11} & 0 & 0 & P_1 D_i & 0 \\ 0 & a_{ij}^{22} & P_2\left(A_i - A_e\right) & P_2 D_i & -P_2 B_e \\ 0 & (*) & a_i^{33} & 0 & P_3 B_e \\ (*) & (*) & 0 & -\gamma^2 I & 0 \\ 0 & (*) & (*) & 0 & -\gamma^2 I \end{pmatrix} \tag{21}$$

$$\Delta\Theta_{ij} \leq \begin{pmatrix} b_{ij}^{11} & 0 & 0 & 0 & 0 \\ 0 & b_{ij}^{22} & 0 & 0 & 0 \\ 0 & 0 & b_i^{33} & 0 & 0 \\ 0 & 0 & 0 & 0 & 0 \\ 0 & 0 & 0 & 0 & 0 \end{pmatrix} \tag{22}$$

where:

- $a_{ij}^{11} = P_1\left(A_i - L_i\,C_i\right) + (*) + 2\,\alpha\,P_1 + \tau_1\,\rho^2 P_1^2 + \tau_1^{-1} N_i^T N_i + \tau_4^{-1} K_j^T B_i^T B_i K_j$;
- $a_{ij}^{22} = P_2 A_i + (*) + 2\,\alpha\,P_2 + Q + \delta_{2,3,4}\,\rho^2 P_2^2 + \tau_2\,\rho^2 P_2\,B_i\,K_j K_j^T B_i P_2 + \tau_2^{-1}\left(N_i^T N_i + K_j^T K_j\right)$;
- $a_i^{33} = P_3 A_e + (*) + 2\,\alpha\,P_3 + \tau_3^{-1} N_i^T N_i$;
- $b_{ij}^{11} = \delta_{1,\rho} P_1 H_i H_i^T P + \delta_{5,7} K_j^T E_{Bi}^T E_{Bi} K_j$;
- $b_{ij}^{22} = \delta_{2,\rho} P_2 H_i H_i^T P_2 + \delta_{6,8} E_{Ai}^T E_{Ai} + \delta_{6,8} E_{Ni}^T E_{Ni} + \delta_{6,8} K_j^T E_{Bi}^T E_{Bi} K_j$;
- $b_{ij}^{33} = \delta_{9,10} E_{Ai}^T E_{Ai} + \delta_{9,10} E_{Ni}^T E_{Ni}$.

From (21) and (22), inequality (20) is equivalent to:

$$\begin{pmatrix} c_{ij}^{11} & 0 & 0 & P_1 D_i & 0 \\ 0 & c_{ij}^{22} & P_2\left(A_i - A_e\right) & P_2 D_i & -P_2 B_e \\ 0 & (*) & c_i^{33} & 0 & P_3 B_e \\ (*) & (*) & 0 & -\gamma^2 I & 0 \\ 0 & (*) & (*) & 0 & -\gamma^2 I \end{pmatrix} < 0 \tag{23}$$

where:

- $c_{ij}^{11} = P_1 (A_i - L_i C_i) + (*) + 2\alpha P_1 + \tau_1 \rho^2 P_1^2 + \tau_1^{-1} N_i^T N_i + \tau_4^{-1} K_j^T B_i^T B_i K_j + \delta_{1,\rho} P_1 H_i H_i^T P_1 + \delta_{5,7} K_j^T E_{Bi}^T E_{Bi} K_j;$
- $c_{ij}^{22} = P_2 A_i + (*) + 2\alpha P_2 + Q + \delta_{2,3,4}\, \rho^2 P_2^2 + \tau_2\, \rho^2 \left(P_2 B_i K_j K_j^T B_i P_2\right) + \tau_2^{-1}\left(N_i^T N_i + K_j^T K_j\right) + \delta_{2,\rho} P_2 H_i H_i^T P_2 + \delta_{6,8} E_{Ai}^T E_{Ai} + \delta_{6,8} E_{Ni}^T E_{Ni} + \delta_{6,8} K_j^T E_{Bi}^T E_{Bi} K_j;$
- $c_i^{33} = P_3 A_e + (*) + 2\alpha P_3 + \tau_3^{-1} N_i^T N_i + \delta_{9,10} E_{Ai}^T E_{Ai} + \delta_{9,10} E_{Ni}^T E_{Ni}.$

Then, by multiplying (23) on the right and on the left by $diag\,(I,\ X_2,\ I,\ I,\ I)$ and assuming that $R_i = K_i X_2$ and $V_i = P_1 L_i$, it becomes:

$$\begin{pmatrix} d_{ij}^{11} & 0 & 0 & P_1 D_i & 0 \\ 0 & d_{ij}^{22} & (A_i - A_e) X_2 & D_i X_2 & -B_e X_2 \\ 0 & (*) & d_i^{33} & 0 & P_3 B_e \\ (*) & (*) & 0 & -\gamma^2 I & 0 \\ 0 & (*) & (*) & 0 & -\gamma^2 I \end{pmatrix} < 0 \tag{24}$$

where:

- $d_{ij}^{11} = P_1 A_i - V_i C_i + (*) + 2\alpha P_1 + \tau_1^{-1} N_i^T N_i + \tau_1 \rho^2 P_1^2 + \tau_4^{-1} K_j^T B_i^T B_i K_j + \delta_{1,\rho} P_1 H_i H_i^T P_1 + \delta_{5,7} K_j^T E_{Bi}^T E_{Bi} K_j;$
- $d_{ij}^{22} = A_i X_2 + (*) + 2\alpha X_2 + \delta_{2,3,4}\, \rho^2 I + \delta_{2,\rho} H_i H_i^T + X_2 Q X_2 + \tau_2\, \rho^2 B_i K_j K_j^T B_i + \tau_2^{-1}\left(X_2 N_i^T N_i X_2 + R_j^T R_j\right) + \delta_{6,8} X_2 E_{Ai}^T X_2 E_{Ai} + \delta_{6,8} X_2 E_{Ni}^T E_{Ni} X_2 + \delta_{6,8} R_j^T E_{Bi}^T E_{Bi} R_j;$
- $d_i^{33} = P_3 A_e + (*) + 2\alpha P_3 + \tau_3^{-1} N_i^T N_i + \delta_{9,10}\, E_{Ai}^T E_{Ai} + \delta_{9,10}\, E_{Ni}^T E_{Ni}.$

Therefore, the largest lower bound on the decay rate α is found be solving the optimization problem involving the LMI formulation, which is presented in (13).

4 Numerical Example: A Nonlinear Mechanical System

A nonlinear mass-spring-damper mechanical system is considered in this study to evaluate the applicability of the proposed control scheme. Its mathematical model is modeled as [9]:

$$\ddot{y} + g\,(y,\ \dot{y}) = h\,(y,\ \dot{y})\, u \tag{25}$$

where $g\,(y,\ \dot{y}) = 3.5y + 2.8 \sin(\dot{y}) - 0.65\dot{y}^2 \sin(\dot{y})$ and $h\,(y,\ \dot{y}) = 1 + 3\dot{y} - 0.85\dot{y}^2 \sin(\dot{y})$.

For $x_1 = y$, $x_2 = \dot{y}$, $x_2 \in [-x_{20},\ x_{20}]$, $v_2 = \max\left({x_2}^2\right)$, and $|\psi\,(t)| < 1$, the nonlinearities $x_2 \mapsto \sin(x_2)$ and $x_2 \mapsto {x_2}^2$ are decomposed as follows:

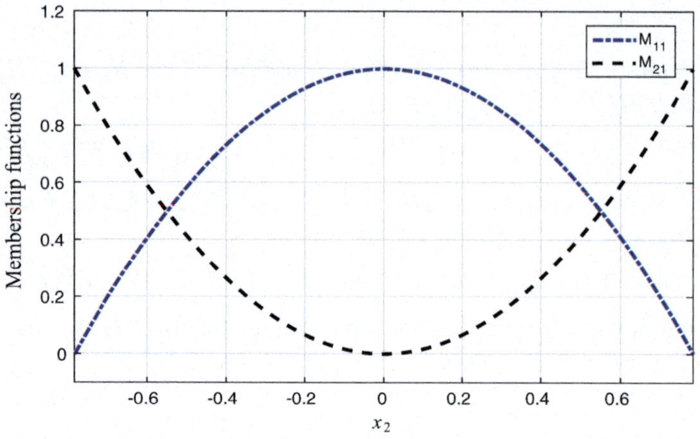

Fig. 1 Evolution of membership functions

$$\sin(x_2) = M_{11}(x_2)\,x_2 + M_{12}(x_2)\,\frac{\sin(x_{20})}{x_{20}}\,x_2 \tag{26}$$

$$x_2{}^2 = 0 + v_2\,\psi(t) \tag{27}$$

where $M_{11}(x_2) = \frac{x_{20}\sin(x_2) - x_2\sin(x_{20})}{x_2(x_{20}-\sin(x_{20}))}$ and $M_{12}(x_2) = \frac{x_{20}x_2 - x_{20}\sin(x_2)}{x_2(x_{20}-\sin(x_{20}))}$.
Figure 1 depicts the evolution of the membership functions $M_{11}(x_2)$ and $M_{12}(x_2)$ for $x_2 \in \left[-\frac{\pi}{4},\ \frac{\pi}{4}\right]$.

The considered system is described by the following two rules:

$$\text{If } (x_1 \text{ is } M_{11}(x_2)) \text{ then } \dot{x} = \overline{A}_1 x + B_1 u + \overline{N}_1 x u + D_1 w \tag{28}$$

$$\text{If } (x_1 \text{ is } M_{12}(x_2)) \text{ then } \dot{x} = \overline{A}_2 x + B_2 u + \overline{N}_2 x u + D_2 w \tag{29}$$

where:
$A_1 = \begin{pmatrix} 0 & 1 \\ -3.5 & -2.8 \end{pmatrix}$, $A_2 = \begin{pmatrix} 0 & 1 \\ -3.5 & -2.52 \end{pmatrix}$,
$B_1 = B_2 = \begin{pmatrix} 0 \\ 1 \end{pmatrix}$, $N_1 = N_2 = \begin{pmatrix} 0 & 0 \\ 0 & 3 \end{pmatrix}$, $D_1 = D_2 = \begin{pmatrix} 1 \\ 1 \end{pmatrix}$,
$\Delta A_1(t) = \begin{pmatrix} 0 \\ 0.65 \end{pmatrix} \psi(t) \begin{pmatrix} 0 & v_2 \end{pmatrix}$, $\Delta A_2(t) = \begin{pmatrix} 0 \\ 0.01 \end{pmatrix} \psi(t) \begin{pmatrix} 0 & v_2 \end{pmatrix}$,
$\Delta N_1(t) = \begin{pmatrix} 0 \\ 0.65 \end{pmatrix} \psi(t) \begin{pmatrix} 0 & -0.85\,v_2 \end{pmatrix}$, $\Delta N_2(t) = \begin{pmatrix} 0 \\ 0.01 \end{pmatrix} \psi(t) \begin{pmatrix} 0 & -0.015\,v_2 \end{pmatrix}$,
$\psi(t) = \sin(0.1\,t)$, $v_2 = 0.62$, and $w(t) = \begin{pmatrix} 0 \\ 0.08\,\sin(t) \end{pmatrix}$.
Using positive scalars $\gamma = 0.65$, $\rho = 0.85$, $\tau_1 = 0.72$, $\tau_2 = 0.28$, $\tau_3 = 0.81$, $\tau_4 = 0.66$, $\tau_5 = 0.80$, $\tau_6 = 0.47$, $\tau_7 = 0.88$, $\tau_8 = 0.50$, $\tau_9 = 0.62$, $\tau_{10} = 0.70$, $A_e =$

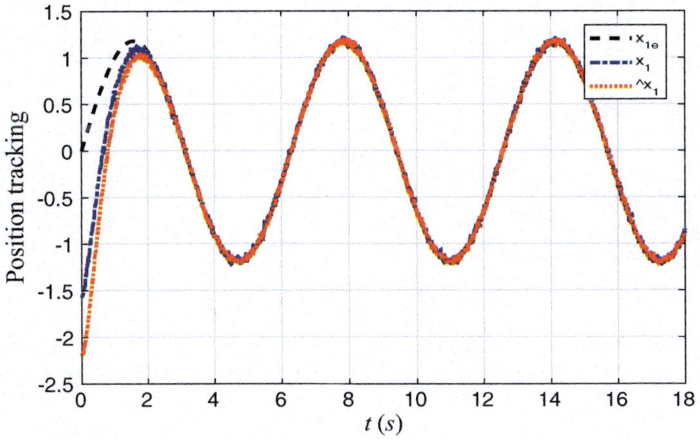

Fig. 2 Position tracking trajectory

$\begin{pmatrix} 0 & 1 \\ -8 & -6 \end{pmatrix}$, $B_e = \begin{pmatrix} 0 \\ 1 \end{pmatrix}$, and $Q = \begin{pmatrix} 0.085 & 0 \\ 0 & 0.006 \end{pmatrix}$, the solutions are: $\alpha = 0.806$, $P_1 = \begin{pmatrix} 53.125 & -8.162 \\ -8.162 & 23.045 \end{pmatrix}$, $P_2 = \begin{pmatrix} 27.361 & 8.157 \\ 8.157 & 13.607 \end{pmatrix}$, $P_3 = \begin{pmatrix} 108.734 & -16.560 \\ -16.560 & 92.066 \end{pmatrix}$, $K_1 = \begin{pmatrix} 72.106 & 6.480 \end{pmatrix}$, $K_2 = \begin{pmatrix} -42.063 & 28.145 \end{pmatrix}$, $L_1 = \begin{pmatrix} 28.316 \\ 112.230 \end{pmatrix}$, and $L_2 = \begin{pmatrix} 25.306 \\ 71.176 \end{pmatrix}$.

Figures 2 and 3 depict respectively the tracking trajectories of the position and the velocity under initial conditions $x\,(0) = \begin{pmatrix} -1.6 & \frac{\pi}{8} \end{pmatrix}^T$, $\hat{x}\,(0) = \begin{pmatrix} -2.2 & -\frac{\pi}{6} \end{pmatrix}^T$, and $x_e\,(0) = \begin{pmatrix} 0 & 0 \end{pmatrix}^T$. Figure 4 shows the control signal.

As a result, the designed controller ensures a good robust tracking control despite the presence of uncertainties and disturbances.

5 Conclusion

This chapter has presented an observer-based controller for nonlinear systems represented by bilinear Takagi-Sugeno fuzzy models with uncertainties and disturbances. Based on the parallel distributed compensation concept and the quadratic Lyapunov function with decay rate, sufficient stability conditions in terms of linear matrix inequalities that ensure good tracking control objectives have been provided. Finally, the simulation result shows the effectiveness of the proposed control scheme.

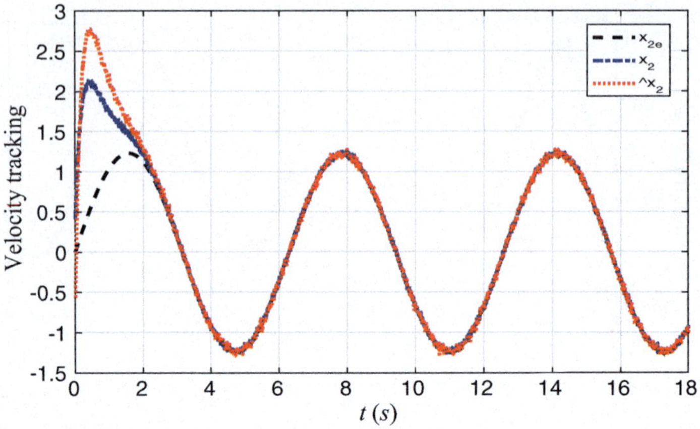

Fig. 3 Velocity tracking trajectory

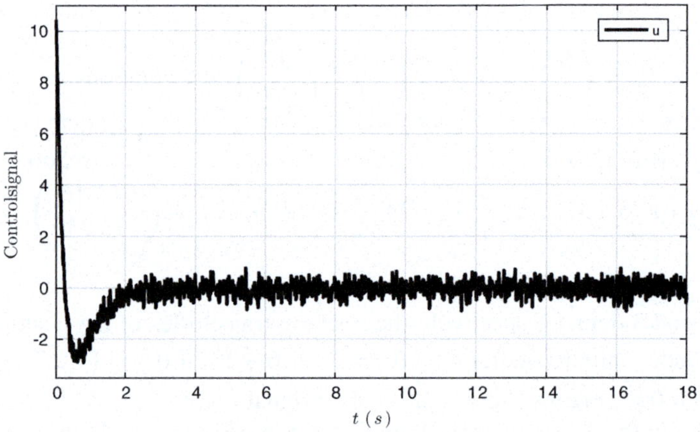

Fig. 4 Control signal

Appendix 1: Schur Complement Lemma

Given matrices A and B with appropriate dimensions and a positive scalar τ, the following inequality holds [21, 36, 38]:

$$A^T B + B^T A \leq \tau A^T A + \tau^{-1} B^T B \tag{30}$$

Appendix 2: Separation Lemma

Given matrices $M = M^T$, $Q = Q^T$, and L with appropriate dimensions, these statements are equivalent [3, 24, 36]:

$$\begin{array}{l} (i) \begin{pmatrix} M & (*) \\ L^T & Q \end{pmatrix} < 0 \\ (ii)\ Q < 0,\ M - L Q^{-1} L^T < 0 \\ (iii)\ M < 0,\ Q - L^T M^{-1} L < 0 \end{array} \tag{31}$$

References

1. Bernard, P., Andrieu, V., Astolfi, D.: Observer design for continuous-time dynamical systems. Annu. Rev. Control. **53**, 224–248 (2022)
2. Chang, J.-L., Wu, C.-P., Shih, C.-L.: Output feedback high-gain proportional integral control for minimum phase uncertain systems. J. Frankl. Inst. **352**(6), 2314–2328 (2015)
3. Chang, X.-H., Yang, G.-H.: Nonfragile H_∞ filter design for T-S fuzzy systems in standard form. IEEE Trans. Ind. Electron. **61**(7), 3448–3458 (2014)
4. Echreshavi, Z., Shasadeghi, M., Asemani, M.H.: H_∞ dynamic observer-based fuzzy integral sliding mode control with input magnitude and rate constraints. J. Frankl. Inst. **358**(1), 575–605 (2021)
5. Ellis, G.: Chapter 1—control systems and the role of observers. In: Ellis, G. (ed.) Observers in Control Systems, pp. 1–4. Academic Press, San Diego (2002)
6. Etienne, L., Langueh, K.A., Karkaba, H., Iovine, A.: Robust observer synthesis for bilinear parameter varying system. In: 2022 IEEE 61st Conference on Decision and Control, pp. 1900–1905 (2022)
7. Filev, D.P., Yager, R.R.: On the analysis of fuzzy logic controllers. Fuzzy Sets Syst. **68**(1), 39–66 (1994)
8. Gao, H., Lam, J., Wang, Z.: Discrete bilinear stochastic systems with time-varying delay: stability analysis and control synthesis. Chaos Solitons Fractals. **34**(2), 394–404 (2007)
9. Ghorbel, C., Benhadj Braiek, N.: PPDC design-based H_∞ non-fragile controller for a class of bilinear systems. In: 18th International Multi-Conference on Systems, Signals and Devices, pp. 1302–1307 (2021)
10. Ghorbel, C., Braiek, N.B.: PPDC design-based H_∞ non-fragile controller for a class of bilinear systems. In: 2021 18th International Multi-Conference on Systems, Signals, and Devices, pp. 1302–1307 (2021)
11. Ghorbel, C., Braiek, N.B.: Nonfragile H_∞ tracking control strategies for classes of linear and bilinear uncertain Takagi-Sugeno fuzzy systems. Trans. Inst. Meas. Control. **44**(11), 2166–2176 (2022)
12. Ghorbel, C., Tiga, A., Benhadj Braiek, N.: Proportional PDC design-based robust stabilization and tracking control strategies for uncertain and disturbed T-S model. Complex. **2020**, 1–9 (2020)
13. Huang, S.-J., He, X.-Q., Zhang, N.-N.: New results on H_∞ filter design for nonlinear systems with time delay via T-S fuzzy models. IEEE Trans. Fuzzy Syst. **19**(1), 193–199 (2011)
14. Huo, X., Ma, L., Zhao, X., Niu, B., Zong, G.: Observer-based adaptive fuzzy tracking control of MIMO switched nonlinear systems preceded by unknown backlash-like hysteresis. Inf. Sci. **490**, 369–386 (2019)
15. Karachalios, D.S., Gosea, I.V., Antoulas, A.C.: A framework for fitting quadratic-bilinear systems with applications to models of electrical circuits. IFAC-PapersOnLine. **55**(20), 7–12 (2022)

16. Kickert, W., Mamdani, E.: Analysis of a fuzzy logic controller. Fuzzy Sets Syst. **1**(1), 29–44 (1978)
17. Korovin, S.K., Fomichev, V.V.: Chapter 6. Observers for bilinear systems, pp. 165–190. De Gruyter, Berlin, New York (2009)
18. Ku, C.-C., Chang, W.-J., Tsai, M.-H., Lee, Y.-C.: Observer-based proportional derivative fuzzy control for singular Takagi-Sugeno fuzzy systems. Inf. Sci. **570**, 815–830 (2021)
19. Lee, S., Park, M., Baek, J.: Robust adaptive synchronization of a class of chaotic systems via fuzzy bilinear observer using projection operator. Inf. Sci. **402**, 182–198 (2017)
20. Li, T.-H.S., Tsai, S.-H.: T-S fuzzy bilinear model and fuzzy controller design for a class of nonlinear systems. IEEE Trans. Fuzzy Syst. **15**(3), 494–506 (2007)
21. Liu, J., Zhang, J., Liu, Y.: The Schur complement of strictly doubly diagonally dominant matrices and its application. Linear Algebr. Appl. **437**(1), 168–183 (2012)
22. Molina, M., Mota, M., Ramos, A.: Estimation of parameters in biological species with several mating and reproduction alternatives. Math. Biosci. **329**, 108471 (2020)
23. Penny, W., Ghahramani, Z., Friston, K.: Bilinear dynamical systems. Philos. Trans. R. Soc. B Biol. Sci. **360**(1457), 983–993 (2005)
24. Portman, V.: Robot stiffness evaluability problem: solution by Schur complements and collinear stiffness values. Mech. Mach. Theory. **161**, 1–16 (2021)
25. Rafikov, M., Balthazar, J., von Bremen, H.: Mathematical modeling and control of population systems: applications in biological pest control. Appl. Math. Comput. **200**(2), 557–573 (2008)
26. Rayouf, Z., Ghorbel, C., Benhadj Braiek, N.: Nonfragile H_∞ stabilizing nonlinear systems described by multivariable Hammerstein models. Complex. **2021**, 1–12 (2021)
27. Saoudi, D., Chadli, M., Benhadj Braiek, N.: Fault tolerant control for uncertain fuzzy bilinear models. IFAC-PapersOnLine. **48**(21), 1256–1261 (2015)
28. Sun, X., Zhang, H., Han, J., Wang, Y.: Non-fragile control for interval type-2 TSK fuzzy logic control systems with time-delay. J. Frankl. Inst. **354**(18), 7997–8014 (2017)
29. Sundarapandian, V.: Local observer design for nonlinear systems. Math. Comput. Model. **35**(1), 25–36 (2002)
30. Takagi, T., Sugeno, M.: Fuzzy identification of systems and its applications to modeling and control. IEEE Trans. Syst. Man Cybern. **15**(1), 116–132 (1985)
31. Tang, G.-Y., Zhao, Y.-D., Ma, H.: Optimal output tracking control for bilinear systems. Trans. Inst. Meas. Control. **28**(4), 387–397 (2006)
32. Tlili, A.S.: Proportional integral observer-based decentralized stabilization design scheme for nonlinear complex interconnected systems. Trans. Inst. Meas. Control. **41**(7), 1811–1823 (2019)
33. Tsai, S.-H., Chen, W.-H., Hsiao, M.-Y.: State-feedback stabilisation for fuzzy bilinear uncertain system with disturbance via fuzzy control approach. Int. J. Syst. Sci. **43**(8), 1386–1395 (2012)
34. Tsai, S.-H., Li, T.-H.S.: Robust H_∞ fuzzy control of a class of fuzzy bilinear systems with time-delay. Chaos Solitons Fractals. **39**(5), 2028–2040 (2008)
35. Tsinias, J.: Observer design for nonlinear systems. Syst. Control. Lett. **13**(2), 135–142 (1989)
36. Wei, Y., Peng, X., Qiu, J.: Robust and non-fragile static output feedback control for continuous-time semi-Markovian jump systems. Trans. Inst. Meas. Control. **38**(9), 1136–1150 (2016)
37. Wen, P., Dong, H., Huo, F., Li, J., Lu, X.: Observer-based PID control for actuator-saturated systems under binary encoding scheme. Neurocomputing. **499**, 54–62 (2022)
38. Zhang, S., Wang, Z., Ding, D., Dong, H., Alsaadi, F.E., Hayat, T.: Nonfragile H_∞ fuzzy filtering with randomly occurring gain variations and channel fadings. IEEE Trans. Fuzzy Syst. **24**(3), 505–518 (2016)

H_∞ Filter Design for Discrete-Time Switched Interconnected Systems with Time Delays

G. Arthi and M. Antonyronika

Abstract In this paper, H_∞ filter design for a class of uncertain discrete-time switched interconnected system is investigated. N-linear discrete-time subsystems with coupled states which has time-varying delays and external disturbances are considered. A suitable filter is designed for considered system with the help of Lyapunov-Krasovski theory and linear matrix inequality (LMI) approach. Sufficient conditions are developed to ensure the exponential mean-square stability of the error system. The filter parameters can be easily determined by using MATLAB LMI toolbox. Finally, a numerical example is provided to ensure the effectiveness of the obtained results.

Keywords Interconnected systems · Mean square stability · Discrete-time · Time-varying delays · Filter

1 Introduction

Switched systems are hybrid dynamical system which are composed of various subsystems with its own parameter, which are subjected to the switching laws between N subsystems. There are many approaches are proposed to analysis the switching systems, out of that the dwell-time and average dwell-time approaches are mostly used and found to be more useful. Because, switching systems based on average dwell-time are the most attractive and the most flexible. It has a wide range of applications in both the physical and man-made systems like power electronic, switched-capacitors network, chaos-generator, automated highway systems and air-traffic management system and full envelopes which are used in formulating the switched systems. There has been considerable interest on analysis of switched systems in the existing literature, see [8, 11, 16] and references therein.

G. Arthi (✉) · M. Antonyronika
Department of Mathematics, PSGR Krishnammal College for Women, Coimbatore 641004, India
e-mail: arthimath@gmail.com

A. Ben Makhlouf et al. (eds.), *State Estimation and Stabilization of Nonlinear Systems*, Studies in Systems, Decision and Control 491,
https://doi.org/10.1007/978-3-031-37970-3_6

Further, interconnected system is one which connects various subsystems with the coupling states, that is, it has the direct connection of two or more information systems used for sharing data and other information resources. It has a wide range of real-life application, such as electrical power system, economical system, process controller system, and many computer networks. Also, it has the main advantage that the switched and coupling terms are handled simultaneously [13, 22]. Moreover, when the real world system consist of coupled units or subsystems that are directly interacted with each other in a predictable and simple fashion to ensure a common pool of objectives, but the resulting overall systems displays complex behaviours [1, 6, 12].

Time delays which happens usually due to the system process and information's flows to a particular part of dynamical-systems and time delays which are unexpectable dynamical behaviour that is, oscillations and poor performances. Generally, delay independent criteria which leads to less conservative results than that of delay dependent ones, which are explained in [2, 9, 15]. In [5] authors discussed some applications of time varying delays which are taken into consideration in a realistic system designing i,e., chemical processing and thermal processing. The literatures [3, 14, 19, 21] explains the presence of time-varying delays will result in instability of the designed system.

Disturbance signal indicates unwanted input, it affects the designed systems output, results in increasing the system errors. There are many performance analysis used in the literature where the noises and disturbances can be handled, among them H_∞ performance have the advantages to overcome classical control techniques, which is used to synthesis the controller for achieving the required robust performance. The main idea of H_∞ filter designing is to estimate the state and to filter out the noises in the systems. H_∞ performance helps to deal with disturbance in the system. To deal with this disturbance, many performance has been established. Some researchers [4, 7, 10, 20] have studied that the H_∞ performance deals with the state estimation in-order to reduce the disturbance in the system. In addition, H_∞ theory for both linear and nonlinear systems have attracted considerable attention and important progress. But most of the above mentioned papers has not discussed about discrete time interconnected system where the switching signal and the coupling term are handled simultaneously with the time varying delays.

From the best of our knowledge, when the time delays appear in interconnected system, the problem of filter design will be still remains to be unsolved. With this motivation, we have focused on interconnected system and first we study the stability of interconnected system in the absence of disturbance. Further, we contribute to the development of H_∞ filter design for discrete-time linear interconnected systems. The main contributions of our work are:

1. A suitable filter is designed in-order to filter out the noise and estimate the error system and system state. By confining the LKF along with average dwell time, we illustrate the delay dependent sufficient condition to ensure the existence of a full order filter.

2. $\mathcal{H}_\infty$ performance index is studied for discrete-time switched interconnected system to reduce the disturbance. The set of LMI based sufficient conditions are derived in-order to ensure the stability of the interconnected system.
3. The coupling term is handled simultaneously along with the switched system to establish the effectiveness of the designed interconnected system.

These contributions of this work are expected to be widely useful for furthermore development of discrete-time interconnected systems. The arrangement of the sections presented are: problem formations and preliminaries, main result which are followed by numerical example to illustrate the effectiveness of the approached method.

2 Problem Formation and Preliminaries

Notations & preliminaries

The following notations are used in this paper. The superscript T and $'-1'$ used to denote matrix transposition and its inverse. $\mathbb{R}^{n\times n}$ is the $n \times n$-dimensional-Euclidean space, I and 0 is used to indicate identity matrices and zero matrices, diagonal matrices are denoted as $diag\{.\}$, $(*)$ is used to indicate the symmetric terms.

Lemma 1 ([23]) *Given matrices A, $Q = Q^T$ and $P > 0$, then $A^T PA - Q < 0$ holds if and only if there exists a matrix S such that*

$$\begin{bmatrix} -Q & A^T S^T \\ * & P - S - S^T \end{bmatrix} < 0.$$

Lemma 2 ([17]) *Assume that Ω, M and N_a are real matrices with appropriate dimensions and F(k) is a matrix function satisfying $F^T(k)F(k) \leq I$. Then,*

$$\Omega + MF(k)N_a + [MF(k)N_a]^T < 0.$$

holds if and only if there exist a scalar $\epsilon > 0$ satisfying

$$\Omega + \epsilon^{-1}MM^T + \epsilon N_a^T N_a < 0.$$

Definition 1 The system (3) is said to be mean-square with gauranteed H_∞ performance level $\gamma_i > 0$ if it is mean square stable under the zero initial condition and it is satisfies the following inequality,

$$\sum_{i\in\Omega}\{\sum_{s=k_0}^{\infty} \tilde{z}_i^T(s)\tilde{z}_i(s)\} \leq \sum_{i\in\Omega}\{\gamma_i^2 \sum_{s=k_0}^{\infty} d_i^T(s)d_i(s)\}.$$

Definition 2 For any $k \geq k_0$, a given switching signal i and $k_0 \leq \tau \leq k$, let N_i denote the switching numbers of i during the interval $[k_0, \text{k}]$. If there exist $T_a > 0$ and $N_0 \geq 0$ such that $N_i(k_0, k) \leq N_0 + (k - k_0)/\, T_a$, then T_a and N_0 are called the average dwell-time and the chatter bound respectively. As commonly used in the literature, we choose $N_0 = 0$.

Remark 1 The following equations are obvious

$$\sum_{i=1}^{N}\sum_{j=1}^{N}\tilde{x_j}^T(k)Q_{ij}\tilde{x_j}(k) = \sum_{i=1}^{N}\tilde{x_i}^T(k)\Big(\sum_{j=1}^{N}Q_{ji}\Big)\tilde{x_j}(k) \text{ (or)}$$

$$\sum_{j\in N_i{}^{in}}\tilde{x_j}^T(k)P_{ij}\tilde{x_j}(k) = \tilde{x_i}^T(k)\Big(\sum_{j\in N_i{}^{out}}P_{ji}\Big)\tilde{x_j}(k).$$

2.1 System Description

Consider the specially interconnected systems consisting of N-linear discrete-time subsystems with coupled states and time varying delays, whose ith system is described as

$$\begin{aligned}
x_i(k+1) &= A_i x_i(k) + A_{di}x_i(k-\tau_i(k)) + \sum_{j\in N_i{}^{in}} A_{dij}x_j(k-\tau_{ij}(k)) + B_i d_i(k),\\
y_i(k) &= C_i x_i(k) + D_i d_i(k), \qquad (1)\\
x_i(j) &= \phi_i(j), \forall j \in [-\tau, 0], \tau = max[-\tau_i(M) - \tau_{ij}(M)].
\end{aligned}$$

Where $x_i(k) \in \mathbb{R}^{n_i}$, $y_i(k) \in \mathbb{R}^{m_i}$ and $d_i(k) \in \mathbb{R}^{d_i}$ are state, measurement output and disturbance input of the ith subsystem respectively. The symbol $N_i{}^{in} = \{j \in \mathbb{N} - \{i\} | A_{dij} \neq 0\}$ and $N_i{}^{out} = \{j \in \mathbb{N} - \{i\} | A_{dji} \neq 0\}$ denotes the in-neighbour and out-neighbour set of ith subsystem. In particular $A_{dij} > 0$ describes, there is a connection between ith and jth subsystem. If $A_{dij} = 0$, then there is no connection between them. τ_i, τ_{ij} are unknown time-delay factors satisfying $0 \leq \tau_{im} \leq \tau_i(k) \leq \tau_{iM}$ and $0 \leq \tau_{ijm} \leq \tau_{ij}(k) \leq \tau_{ijM}$. The bounds τ_{im}, τ_{iM}, τ_{ijm} and τ_{ijM} are constant values. The disturbance vector $d_i(k) = \left[d_1(k)\, d_2(k) \ldots d_N(k)\right]^T$, can satisfies the condition $\sum_{k=1}^{\infty}[d^T(k)d(k) + d^T(k)d(k)] \leq v$, $v > 0$ to describe the external inputs are bounded.

2.2 Filter Description

We assume the measurement output $y_i(k)$ and output signal to be estimated $z_i(k)$ are described as follows:

$$y_i(k) = C_i x_i(k) + D_i d_i(k),$$
$$z_i(k) = M_i x_i(k),$$

where C_i, D_i, M_i are known constant matrices with appropriate dimensions. The filtering problem consists of obtaining an estimate $\hat{z}_i(k)$ of the signal $z_i(k)$ with an estimation error $\hat{z}_i(k) - z_i(k)$, consider the following full order filter for the estimation of $z_i(k)$:

$$\begin{aligned} \hat{x}_i(k+1) &= A_{fi}\hat{x}_i(k) + B_{fi}y_i(k), \\ \hat{z}_i(k) &= M_{fi}\hat{x}_i(k), \\ \hat{x}_i(k_0) &= 0, \end{aligned} \tag{2}$$

where $\hat{x}_i(k) \in \mathbb{R}^{n_i}$ is the state vector, $\hat{z}_i(k) \in \mathbb{R}^{p_i}$ is the output signal of the filter and A_{fi}, B_{fi},M_{fi} are the filter parameters to be designed.

Define the new state vector $\tilde{x}_i(k) = [{x_i}^T(k)\ {\hat{x}_i}^T(k)]^T$ and the filtering error vector $\tilde{z}_i(k) = z_i(k) - \hat{z}_i(k)$. Then, it follows from (1) and (2) that, the augmented system can be obtained in the following form:

$$\tilde{x}_i(k+1) = \bar{A}_i\tilde{x}_i(k) + \bar{A}_{di}\varepsilon\tilde{x}_i(k-\tau_i(k)) + \sum_{j\in {N_i}^{in}} \bar{A}_{dij}\varepsilon\tilde{x}_j(k-\tau_{ij}(k)) + \bar{B}_i d_i(k),$$

$$\begin{aligned} \tilde{z}_i(k) &= \mathbb{M}_i\tilde{x}_i(k), \\ \tilde{x}_i(k_0) &= \tilde{x}_0, \end{aligned} \tag{3}$$

where

$$\bar{A}_i = \begin{bmatrix} A_i & 0 \\ B_{fi}C_i & A_{fi} \end{bmatrix}, \bar{A}_{di} = \begin{bmatrix} A_{di} \\ 0 \end{bmatrix}, \bar{A}_{dij} = \begin{bmatrix} A_{dij} \\ 0 \end{bmatrix}, \bar{B}_i = \begin{bmatrix} B_i \\ B_{fi}D_i \end{bmatrix}, \varepsilon = \begin{bmatrix} I & 0 \end{bmatrix}, \mathbb{M}_i = \begin{bmatrix} M_i & -M_{fi} \end{bmatrix}.$$

Next we need to check our system (1) is mean square exponentially stable without disturbance.

3 Main Results

In this section, we first check the designed system is mean square exponentially stable with $d_i(k) = 0$. Then we study the sufficient condition to analyse exponential H_∞ filtering for the designed system (3).

Theorem 1 *For a given scalar $0 < \alpha < 1$, the system (3) with $d_i(k) = 0$ is said to be mean-square exponentially stable, if there exist $n \times n$ positive definite matrices P_i, Q_{1i}, Q_{2i}, Q_{3i}, R_{1ij}, R_{2ij}, R_{3ij} and any matrices S_{1i}, S_{2i} of appropriate dimensions, such that for all i, $j \in N$ and $i \neq j$, subject to LMIs*

$$\begin{bmatrix} \psi_{1i} & \psi_{2i}^T \\ * & P_i - S_i - S_i^T \end{bmatrix} < 0, \tag{4}$$

where

$$\psi_{1i} = diag\{\psi_{(1,1)}, -(1-\alpha)^{\tau_{im}} Q_{1i}, -(1-\alpha)^{\tau_{im}} Q_{2i}, -(1-\alpha)^{\tau_{iM}} Q_{3i}, M_1, M_2, M_3\},$$

$$M_1 = \sum_{j \in N_i^{in}} [-(1-\alpha)^{\tau_{im}} R_{1ij}], M_2 = \sum_{j \in N_i^{in}} [-(1-\alpha)^{\tau_{im}} R_{2ij}], M_3 = \sum_{j \in N_i^{in}} [-(1-\alpha)^{\tau_{iM}} R_{3ij}],$$

$$\psi_{(1,1)} = -(1-\alpha)P_i + Q_{1i} + Q_{3i} + (\tau_{iM} - \tau_{im} + 1)Q_{2i} + \sum_{j \in N_i^{out}} \left[(R_{1ji} + R_{3ji}) + (\tau_{iM} - \tau_{im} + 1)R_{2ji}\right],$$

$$\psi_{2i} = \left[\, S_{1i}\bar{A}_i \;\; 0 \;\; S_{1i}\bar{A}_{di} \;\; 0 \;\; 0 \;\; S_{1i}\sum_{j \in N_i^{in}} \bar{A}_{dij} \;\; 0 \,\right].$$

The estimate of state decay is $\sum_{i\in\Omega} \|\tilde{x}_i(k)\|^2 \leq \frac{\beta_2}{\beta_1}(1-\alpha)^{k-k_0} \sum_{i\in\Omega} \|\phi_i(k)\|_l^2$ *with* $\beta_1 = \min_{\forall i\in\Omega} \lambda_{\min}(P_i)$ *and* $\beta_2 = \max_{\forall i\in\Omega} \lambda_{\max}(P_i) + \max_{\forall i\in\Omega} \lambda_{\max}(Q_{1i}) + (1 + \tau_{iM} - \tau_{im}) \max_{\forall i\in\Omega} \lambda_{\max}(Q_{2i}) + \max_{\forall i\in\Omega} \lambda_{\max}(Q_{3i}) + \max_{\forall i\in\Omega} \sum_{j\in N_i^{in}} \times \lambda_{\max}(R_{1ij}) + (1 + \tau_{iM} - \tau_{im}) \max_{\forall i\in\Omega} \sum_{j\in N_i^{in}} \lambda_{\max}(R_{2ij}) + \max_{\forall i\in\Omega} \sum_{j\in N_i^{in}} \lambda_{\max}(R_{3ij})$.

Proof In order to prove the required result for the given system (3), we define the following LKF to present LMI based sufficient-condition

$$V(k) = \sum_{i\in\Omega} V_i(k) = \sum_{i\in\Omega} \left[V_{1i} + V_{2i} + V_{3i} + V_{4i} + V_{5i}\right], \tag{5}$$

where

$$V_{1i}(k) = \tilde{x}_i^T(k) P_i \tilde{x}_i(k),$$

$$V_{2i}(k) = \sum_{s=k-\tau_{im}}^{k-1} (1-\alpha)^{k-s-1} \tilde{x}_i^T(s)\varepsilon^T Q_{1i}\varepsilon\tilde{x}_i(s) + \sum_{s=k-\tau_i(k)}^{k-1} (1-\alpha)^{k-s-1} \tilde{x}_i^T(s)\varepsilon^T Q_{2i}\varepsilon\tilde{x}_i(s)$$
$$+ \sum_{s=k-\tau_{iM}}^{k-1} (1-\alpha)^{k-s-1} \tilde{x}_i^T(s)\varepsilon^T Q_{3i}\varepsilon\tilde{x}_i(s),$$

$$V_{3i}(k) = \sum_{r=-\tau_{iM}+1}^{-\tau_{im}} \sum_{s=k+r}^{k-1} (1-\alpha)^{k-s-1} \tilde{x}_i^T(s)\varepsilon^T Q_{2i}\varepsilon\tilde{x}_i(s),$$

$$V_{4i}(k) = \sum_{j\in N_i^{in}} \Bigg[\sum_{s=k-\tau_{ijm}}^{k-1} (1-\alpha)^{k-s-1} \tilde{x}_i^T(s)\varepsilon^T R_{1ij}\varepsilon\tilde{x}_i(s) + \sum_{s=k-\tau_{ij}(k)}^{k-1} (1-\alpha)^{k-s-1} \tilde{x}_i^T(s)\varepsilon^T R_{2ij}\varepsilon\tilde{x}_i(s)$$
$$\sum_{s=k-\tau_{ijM}}^{k-1} (1-\alpha)^{k-s-1} \tilde{x}_i^T(s)\varepsilon^T R_{3ij}\varepsilon\tilde{x}_i(s) \Bigg],$$

$$V_{5i}(k) = \sum_{j\in N_i^{in}} \Bigg[\sum_{r=-\tau_{ijM}+1}^{-\tau_{ijm}} \sum_{s=k+r}^{k-1} (1-\alpha)^{k-s-1} \tilde{x}_i^T(s)\varepsilon^T R_{2ij}\varepsilon\tilde{x}_i(s) \Bigg].$$

Let us denote the forward difference $\Delta V(k) = V(k+1) - V(k)$, we have the following

$$\begin{aligned}
\Delta V_1(k+1) + \alpha V_1(k) &= V_1(k+1) - (1-\alpha)V_1(k) \\
&= \tilde{x_i}^T(k+1)P_i\tilde{x_i}(k+1) - (1-\alpha)\tilde{x_i}^T(k)P_i\tilde{x_i}(k) \\
&= \Big[\bar{A}_i\tilde{x}_i(k) + \bar{A}_{di}\varepsilon\tilde{x}_i(k-\tau_i(k)) + \sum_{j\in N_i^{in}} \bar{A}_{dij}\varepsilon\tilde{x}_j(k-\tau_{ij}(k)) \\
&\quad +\bar{B}_i d_i(k)\Big]^T P_{1i}\Big[\bar{A}_i\tilde{x}_i(k) + \bar{A}_{di}\varepsilon\tilde{x}_i(k-\tau_i(k)) \\
&\quad + \sum_{j\in N_i^{in}} \bar{A}_{dij}\varepsilon\tilde{x}_j(k-\tau_{ij}(k)) + \bar{B}_i d_i(k)\Big] \\
&\quad -(1-\alpha)\tilde{x}^T(k)P_i\tilde{x}(k).
\end{aligned} \tag{6}$$

$$\begin{aligned}
\Delta V_2(k+1) + \alpha V_2(k) &= V_2(k+1) - (1-\alpha)V_2(k) \\
&= \tilde{x_i}^T(k)\varepsilon^T(Q_{1i} + Q_{2i} + Q_{3i})\varepsilon\tilde{x_i}(k) \\
&\quad -(1-\alpha)^{\tau_{im}}\tilde{x_i}^T(k-\tau_{im})\varepsilon^T Q_{1i}\varepsilon\tilde{x_i}(k-\tau_{im}) \\
&\quad -(1-\alpha)^{\tau_{im}}\tilde{x_i}^T(k-\tau_i(k))\varepsilon^T Q_{2i}\varepsilon\tilde{x_i}(k-\tau_i(k)) \\
&\quad -(1-\alpha)^{\tau_{iM}}\tilde{x_i}^T(k-\tau_{iM})\varepsilon^T Q_{3i}\varepsilon\tilde{x_i}(k-\tau_{iM}) \\
&\quad + \sum_{s=k+1-\tau_{iM}}^{k-\tau_{im}} (1-\alpha)^{k-s}\tilde{x}_i^T(s)\varepsilon^T Q_{2i}\varepsilon\tilde{x}_i(s).
\end{aligned} \tag{7}$$

$$\begin{aligned}
\Delta V_3(k+1) + \alpha V_3(k) &= V_3(k+1) - (1-\alpha)V_3(k) \\
&= (\tau_{iM} - \tau_{im})\tilde{x_i}^T(k)\varepsilon^T Q_{2i}\varepsilon\tilde{x_i}(k) - \sum_{r=k+1-\tau_{iM}}^{k-\tau_{im}} (1-\alpha)^{k-r}\tilde{x}_i^T(r)\varepsilon^T Q_{2i}\varepsilon\tilde{x}_i(r).
\end{aligned} \tag{8}$$

$$\begin{aligned}
\Delta V_4(k+1) + \alpha V_4(k) &= V_4(k+1) - (1-\alpha)V_4(k) \\
&= \sum_{j\in N_i^{in}} \Big[\tilde{x_j}^T(k)\varepsilon^T(R_{1ij} + R_{2ij} + R_{3ij})\varepsilon\tilde{x_j}(k) \\
&\quad -(1-\alpha)^{\tau_{ijm}}\tilde{x_j}^T(k-\tau_{ijm})\varepsilon^T R_{1ij}\varepsilon\tilde{x_j}(k-\tau_{ijm}) \\
&\quad -(1-\alpha)^{\tau_{ijm}}\tilde{x_j}^T(k-\tau_{ij}(k))\varepsilon^T R_{2ij}\varepsilon\tilde{x_j}(k-\tau_{ij}(k)) \\
&\quad -(1-\alpha)^{\tau_{ijM}}\tilde{x_j}^T(k-\tau_{ijM})\varepsilon^T R_{3ij}\varepsilon\tilde{x_j}(k-\tau_{ijM}) \\
&\quad + \sum_{s=k+1-\tau_{ijM}}^{k-\tau_{ijm}} (1-\alpha)^{k-s}\tilde{x}_j^T(s)\varepsilon^T R_{2ij}\varepsilon\tilde{x}_j(s)\Big].
\end{aligned} \tag{9}$$

$$\begin{aligned}
\Delta V_5(k+1) + \alpha V_5(k) &= V_5(k+1) - (1-\alpha)V_5(k) \\
&= \sum_{j\in N_i^{in}} \Big[(\tau_{ijM} - \tau_{ijm})\tilde{x_j}^T(k)\varepsilon^T R_{2ij}\varepsilon\tilde{x_j}(k) \\
&\quad - \sum_{r=k+1-\tau_{ijM}}^{k-\tau_{ijm}} (1-\alpha)^{k-r}\tilde{x}_j^T(r)\varepsilon^T R_{2ij}\varepsilon\tilde{x}_j(r)\Big].
\end{aligned} \tag{10}$$

By using Remark 1, (9) and (10) can be written as follows

$$\begin{aligned}\Delta V_4(k+1)+\alpha V_4(k) &= V_4(k+1)-(1-\alpha)V_4(k)\\ &= \tilde{x_i}^T(k)\varepsilon^T\Big[\sum_{j\in N_i{}^{out}} R_{1ji}+R_{2ji}+R_{3ji}\Big]\varepsilon\tilde{x_i}(k)\\ &\quad +\tilde{x_j}^T(k-\tau_{ijm})\varepsilon^T M_1\varepsilon\tilde{x_j}(k-\tau_{ijm})+\tilde{x_j}^T(k-\tau_{ij}(k))\varepsilon^T\\ &\quad \times M_2\varepsilon\tilde{x_j}(k-\tau_{ij}(k))+\tilde{x_j}^T(k-\tau_{ijM})\varepsilon^T M_3\varepsilon\tilde{x_j}(k-\tau_{ijM})\\ &\quad +\sum_{s=k+1-\tau_{ijM}}^{k-\tau_{ijm}}(1-\alpha)^{k-s}\tilde{x}_j^T(s)\varepsilon^T R_{2ij}\varepsilon\tilde{x}_j(s). \end{aligned} \tag{11}$$

$$\begin{aligned}\Delta V_5(k+1)+\alpha V_5(k) &= V_5(k+1)-(1-\alpha)V_5(k)\\ &= \tilde{x}^T(k)\varepsilon^T\Big[\sum_{j\in N_i{}^{out}}(\tau_{ijM}-\tau_{ijm})R_{2ij}\Big]\varepsilon\tilde{x}(k)\\ &\quad -\sum_{r=k+1-\tau_{ijM}}^{k-\tau_{ijm}}(1-\alpha)^{k-r}\tilde{x}_j^T(r)\varepsilon^T R_{2ij}\varepsilon\tilde{x}_j(r). \end{aligned} \tag{12}$$

To prove the stability results when $d_i(k)=0$, combining (6), (7), (8), (11) and (12), we have

$$\Delta V_i(k)+\alpha V_i(k)\le \zeta_k^T\psi_i\zeta_k, \tag{13}$$

where $\psi_i=\psi_{1i}+\psi_{2i}^T P_i\psi_{2i}$,

$$\begin{aligned}\zeta_k=\big[&\tilde{x_i}^T(k)\ \ \varepsilon^T\tilde{x_i}^T(k-\tau_m)\ \ \varepsilon^T\tilde{x_i}^T(k-\tau(k))\ \ \varepsilon^T\tilde{x_i}^T(k-\tau_M)\\ &\varepsilon^T\tilde{x_j}^T(k-\tau_{ijm})\ \ \varepsilon^T\tilde{x_j}^T(k-\tau_{ij}(k))\ \ \varepsilon^T\tilde{x_j}^T(k-\tau_{ijM})\big].\end{aligned}$$

Now by applying Lemma 1, in view of LMI (4) there exist a matrix S_i, one can get $\psi_i<0$. So that it is easy to verify that $\sum_{i\in\Omega}\big[\Delta V_i(k)+\alpha V_i(k)\big]\le 0$. Then we can get the following

$$\sum_{i\in\Omega}\big[\Delta V_i(k+1)-V_i(k)\big]\le\sum_{i\in\Omega}-\alpha V_i(k),$$

which implies that

$$\begin{aligned}\sum_{i\in\Omega}V_{ik}(k) &\le (1-\alpha)^{k-k_t}\sum_{i\in\Omega}V_{ik_t}(k_t)\\ &\le (1-\alpha)^{k-k_t}\mu\sum_{i\in\Omega}V_{ik_{t-1}}(k_t)\\ &\le \mu(1-\alpha)^{k-k_t}(1-\alpha)^{k_t-k_{t-1}}\sum_{i\in\Omega}V_{ik_{t-1}}(k_{t-1}) \end{aligned} \tag{14}$$

$$= \mu(1-\alpha)^{k-k_{t-1}} \sum_{i\in\Omega} V_{ik_{t-1}}(k_{t-1})$$

$$\vdots$$

$$\leq \mu^{N_i(k_0,k)}(1-\alpha)^{k-k_0} \sum_{i\in\Omega} V_{ik_0}(k_0).$$

Then by Definition 2 $N_i(k_0, k) \leq (k-k_0)/T_a$, (14) becomes

$$\sum_{i\in\Omega} V_{ik}(k) \leq \left((1-\alpha)\mu^{\frac{1}{Ta}}\right)^{k-k_0} \sum_{i\in\Omega} V_{ik_0}(k_0) \tag{15}$$

It can be verified from (5) that $V_{ik}(k) \geq \beta_1\|\tilde{x_i}\|^2$ and $V_{ik_0}(k_0) \leq \beta_2\|\tilde{\phi_i}\|_l^2$. Then we have

$$\sum_{i\in\Omega} \beta_1\|\tilde{x_i}\|^2 \leq \left((1-\alpha)\mu^{\frac{1}{Ta}}\right)^{k-k_0} \sum_{i\in\Omega} \beta_2\|\tilde{\phi_i}\|_l^2$$

$$\sum_{i\in\Omega} \|\tilde{x_i}\|^2 \leq \frac{\beta_2}{\beta_1}\vartheta^{k-k_0} \sum_{i\in\Omega} \|\tilde{\phi_i}\|_l^2. \tag{16}$$

Here $\vartheta = \left((1-\alpha)\mu^{\frac{1}{Ta}}\right)$, we can easily obtain $\vartheta < 1$ by using T_a. This proves the exponential stability of (3) with $d_i(k) = 0$. ■

Theorem 2 *For a given scalar* $0 < \alpha < 1$, *the system* (3) *is said to be mean-square exponentially stable with* H_∞ *performance level* $\gamma_i > 0$, *if there exist* $n \times n$ *positive definite matrices* P_i, Q_{1i}, Q_{2i}, Q_{3i}, R_{1ij}, R_{2ij}, R_{3ij} *and any matrices* S_{1i}, S_{2i} *of appropriate dimensions, such that for all* $i, j \in N$ *and* $i \neq j$, *the convex optimization problem:*

$$min_{P_i;Q_{1i};Q_{2i};Q_{3i};R_{1ij};R_{2ij};R_{3ij}}\ \rho_i \quad with \quad \rho_i = \gamma_i^2, \tag{17}$$

subject to LMIs

$$\begin{bmatrix} \bar{\psi_i} & 0 & \bar{\psi_{1i}}^T & \bar{\psi_{2i}}^T & \bar{\psi_{3i}}^T \\ * & -\gamma_i^2 I & B_i^T S_{1i}^T & D_i^T Bf_i^T S_{2i}^T & 0 \\ * & * & P_{1i} - S_{1i} - S_{1i}^T & P_{2i} & 0 \\ * & * & * & P_{3i} - S_{2i} - S_{2i}^T & 0 \\ * & * & * & * & -I \end{bmatrix} < 0, \tag{18}$$

where

$$\bar{\psi_i} = diag\{\bar{\psi}_{(1,1)}, -(1-\alpha)P_{2i}, -(1-\alpha)^{\tau_{im}} Q_{1i}, -(1-\alpha)^{\tau_{im}} Q_{2i}, -(1-\alpha)^{\tau_{iM}} Q_{3i}, M_1, M_2, M_3\},$$
$$\bar{\psi_{1i}} = \begin{bmatrix} S_{1i}A_i & 0 & 0 & S_{1i}A_{di} & 0 & 0 & S_{1i}\sum_{j\in N_i^{in}} A_{dij} & 0 \end{bmatrix},$$
$$\bar{\psi_{2i}} = \begin{bmatrix} S_{2i}Bf_iC_i & S_{1i}Af_i & 0 & 0 & 0 & 0 & 0 & 0 \end{bmatrix},$$
$$\bar{\psi_{3i}} = \begin{bmatrix} M_i & -M_{fi} & 0 & 0 & 0 & 0 & 0 & 0 \end{bmatrix}.$$

Then the optimal H_∞ performance level $\gamma_i > 0$. Further the filter parameters are $A_{fi} = S_{1i}^{-T} F_{1i}$, $B_{fi} = S_{2i}^{-T} F_{2i}$ and $M_{fi} = M_{Fi}$.

Proof First the matrix P_i is defined as

$$P_i = \begin{bmatrix} P_{1i} & P_{2i} \\ 0 & P_{3i} \end{bmatrix}.$$

Now we will discuss H_∞ performance for the system (3). For that we consider the following performance index $J = \sum_{i\in\Omega}\{\tilde{z_i}^T(k)\tilde{z}_i(k) - \gamma_i^2 w_i^T(k)w_i(k)\}$ for all non-zero $w_i(k)$, using LKF (5) together with (4) we can calculate

$$\Delta V_i(k) + \alpha V_i(k) + \sum_{i\in\Omega}\{\tilde{z_i}^T(k)\tilde{z}_i(k) - \gamma_i^2 w_i^T(k)w_i(k)\} \le \sum_{i\in\Omega} \zeta_{1k}^T \tilde{\psi}_i \zeta_{1k}, \quad (19)$$

where $\zeta_{1k}^T = \begin{bmatrix} \zeta_k^T & w_i(k) \end{bmatrix}$,

$$\tilde{\psi_i} = \begin{bmatrix} \bar{\psi_i} + \tilde{\psi}_{3i}^T \tilde{\psi}_{3i} & 0 & \bar{\psi_{1i}}^T & \bar{\psi_{2i}}^T \\ * & -\gamma_i^2 I & B_i^T S_{1i}^T & D_i^T Bf_i^T S_{2i}^T \\ * & * & P_{1i} - S_{1i} - S_{1i}^T & P_{2i} \\ * & * & * & P_{3i} - S_{2i} - S_{2i}^T \end{bmatrix}, \quad (20)$$

$\tilde{\psi}_{3i}^T = \begin{bmatrix} \mathbb{M}_i & 0_{n,7n} \end{bmatrix}$. In view of Schur-complement, taking matrices $S_i = diag\{S_{1i}, S_{2i}\}$ and $F_{1i} = S_{1i}A_{fi}$, $F_{2i} = S_{2i}B_{fi}$, it is easy to get (20) is equivalent to (18). Hence if LMI (18) holds for $0 < \alpha < 1$, then it follows (19) we have,

$$\Delta V_i(k) + \alpha V_i(k) + \sum_{i\in\Omega}\{\tilde{z_i}^T(k)\tilde{z}_i(k) - \gamma_i^2 w_i^T(k)w_i(k)\} \le 0. \quad (21)$$

Now to prove the H_∞ performance for given system, using (20) we get the following

$$V_i(k_1) \le (1-\alpha)V_i(k_0) - \left[\sum_{i\in\Omega}\{\tilde{z_i}^T(k_0)\tilde{z}_i(k_0) - \gamma_i^2 w_i^T(k_0)w_i(k_0)\}\right].$$

Iterating the above mentioned inequality, we have the following

$$V_i(k) \leq (1-\alpha)^{k-k_0} V_i(k_0) - \left[\sum_{i \in \Omega} \left\{ \sum_{s=k_0}^{k-1} (1-\alpha)^{k-s-1} \left[\tilde{z}_i{}^T(s)\tilde{z}_i(s) - \gamma_i^2 w_i^T(s) w_i(s) \right] \right\} \right].$$

$$\leq (1-\alpha)^{k-k_0} V_i(k_0) - \sum_{i\in\Omega} \sum_{s=k_0}^{k-1} (1-\alpha)^{k-s-1} \tilde{z}_i{}^T(s)\tilde{z}_i(s) + \sum_{i\in\Omega} \sum_{s=k_0}^{k-1} (1-\alpha)^{k-s-1} \gamma 2_i w_i^T(s) w_i(s)$$

Therefore we have

$$V_{ik}k \leq (1-\alpha)^{k-k_t} V_{ik}(k_t) - \sum_{i\in\Omega} \sum_{s=k_t}^{k-1} (1-\alpha)^{k-s-1} \tilde{z}_i{}^T(s)\tilde{z}_i(s) + \sum_{i\in\Omega} \sum_{s=k_t}^{k-1} (1-\alpha)^{k-s-1} \gamma 2_i w_i^T(s) w_i(s)$$

$$\leq (1-\alpha)^{k-k_t} \mu V_{ik_{t-1}}(k_t) - \sum_{i\in\Omega} \sum_{s=k_t}^{k-1} (1-\alpha)^{k-s-1} \tilde{z}_i{}^T(s)\tilde{z}_i(s) + \sum_{i\in\Omega} \sum_{s=k_t}^{k-1} (1-\alpha)^{k-s-1} \gamma 2_i w_i^T(s) w_i(s)$$

$$= (1-\alpha)^{k-k_0} \mu^{N(k_0,k)} V_{ik_0}(k_0) - \sum_{i\in\Omega} \sum_{s=k_0}^{k-1} \mu^{N(s,k)} (1-\alpha)^{k-s-1} [\tilde{z}_i{}^T(s)\tilde{z}_i(s) - \gamma_i^2 w_i^T(s) w_i(s)]$$

Under the zero initial condition, we have -$\sum_{s=k_0}^{k-1} \mu^{N(s,k)} (1-\alpha)^{k-s-1} \leq 0$, Then $\mu^{-N_i(0,k)}$ is multiplied on both sides the above inequality will become

$$\mu^{-N_i(0,k)} \sum_{i\in\Omega} \sum_{s=k_0}^{k-1} \mu^{N_i(s,k)} (1-\alpha)^{k-s-1} \tilde{z}_i{}^T(s)\tilde{z}_i(s) \leq \mu^{-N_i(0,k)} \sum_{i\in\Omega} \sum_{s=k_0}^{k-1} \mu^{N_i(s,k)} (1-\alpha)^{k-s-1} \gamma_i^2 w_i^T(s) w_i(s)$$

$$\sum_{i\in\Omega} \sum_{s=k_0}^{k-1} \mu^{-N_i(0,s)} (1-\alpha)^{k-s-1} \tilde{z}_i{}^T(s)\tilde{z}_i(s) \leq \sum_{i\in\Omega} \sum_{s=k_0}^{k-1} \mu^{-N_i(0,s)} (1-\alpha)^{k-s-1} \gamma_i^2 w_i^T(s) w_i(s).$$

By Definition 2, $N_i(0,s) \leq \frac{s}{T_a} \leq \frac{-s\ln(1-\alpha)}{\ln\mu}$ using this it is easy to obtain that

$$\sum_{i\in\Omega} \sum_{s=k_0}^{k-1} \mu^{\frac{-s\ln(1-\alpha)}{\ln\mu}} (1-\alpha)^{k-s-1} \tilde{z}_i{}^T(s)\tilde{z}_i(s) \leq \sum_{i\in\Omega} \sum_{s=k_0}^{k-1} (1-\alpha)^{k-s-1} \gamma_i^2 w_i^T(s) w_i(s)$$

$$\sum_{i\in\Omega} \sum_{s=k_0}^{k-1} (1-\alpha)^s (1-\alpha)^{k-s-1} \tilde{z}_i{}^T(s)\tilde{z}_i(s) \leq \gamma_i^2 \sum_{i\in\Omega} \sum_{s=k_0}^{k-1} (1-\alpha)^{k-s-1} w_i^T(s) w_i(s)$$

$$\sum_{i\in\Omega} \left\{ \sum_{s=k_0}^{\infty} (1-\alpha)^s \tilde{z}_i{}^T(s)\tilde{z}_i(s) \right\} \leq \sum_{i\in\Omega} \gamma_i^2 \left\{ \sum_{s=k_0}^{\infty} w_i^T(s) w_i(s) \right\}.$$

From the Definition 1 it is concluded that the system (3) is mean-square stable with guaranteed H_∞ performance index $\gamma_i > 0$. This proves the theorem. ∎

4 Numerical Example

In this section, we establish a numerical example to illustrate the usefulness and effectiveness of the filter design developed in this paper.

Example 1 We consider linear coupled system with time-varying delays consisting of three subsystems.

Case 1 First we consider the system whose parameters of each mode are:
Subsystem 1

$$A_1 = \begin{bmatrix} -0.01 & -0.02 & 0.02 \\ 0.02 & 0.03 & -0.02 \\ -0.02 & -0.01 & 0.02 \end{bmatrix}, A_{d1} = \begin{bmatrix} 0.02 & 0.1 & 0 \\ 0.4 & -0.05 & -0.1 \\ 0.1 & 0 & -0.09 \end{bmatrix}, B_1 = \begin{bmatrix} 1 \\ -0.1 \\ 0.5 \end{bmatrix},$$

$$A_{d12} = \begin{bmatrix} -0.1 & 0.1 & -0.2 \\ 0.2 & 0.2 & -0.1 \\ 0.1 & 0.1 & 0.1 \end{bmatrix}, C_1 = \begin{bmatrix} 0.6 \\ 0.3 \\ 0.1 \end{bmatrix}^T, A_{d13} = \begin{bmatrix} 0.2 & 0.1 & -0.2 \\ 0.2 & 0.1 & -0.1 \\ 0.1 & 0.1 & 0.2 \end{bmatrix}, D_1 = 0.5$$

Subsystem 2

$$A_2 = \begin{bmatrix} 0.01 & 0.01 & 0.02 \\ -0.1 & -0.01 & -0.02 \\ 0.01 & 0.01 & 0.03 \end{bmatrix}, A_{d2} = \begin{bmatrix} -0.01 & 0.01 & 0.02 \\ 0.01 & -0.02 & 0.01 \\ 0 & 0.01 & -0.1 \end{bmatrix}, B_2 = \begin{bmatrix} 0.9 \\ 0.3 \\ -0.6 \end{bmatrix},$$

$$A_{d21} = \begin{bmatrix} 0.2 & 0.1 & 0.1 \\ -0.01 & 0.1 & 0.2 \\ 0.2 & 0.09 & 0.1 \end{bmatrix}, C_2 = \begin{bmatrix} 0.5 \\ 0.2 \\ 0.1 \end{bmatrix}^T, A_{d23} = \begin{bmatrix} -0.2 & 0.1 & 0.1 \\ -0.01 & 0.1 & 0.2 \\ 0.2 & 0.09 & 0.2 \end{bmatrix}, D_2 = -0.6.$$

Subsystem 3

$$A_3 = \begin{bmatrix} 0.03 & -0.02 & -0.01 \\ -0.01 & 0 & -0.01 \\ 0.01 & -0.02 & -0.01 \end{bmatrix}, A_{d3} = \begin{bmatrix} 0.1 & 0.1 & -0.2 \\ 0.1 & 0.2 & -0.1 \\ 0.01 & -0.1 & 0.2 \end{bmatrix}, B_3 = \begin{bmatrix} -0.6 \\ -0.7 \\ 0.9 \end{bmatrix},$$

$$A_{d31} = \begin{bmatrix} 0.2 & -0.1 & -0.2 \\ 0.1 & 0.2 & -0.1 \\ 0.1 & -0.1 & 0.2 \end{bmatrix}, C_3 = \begin{bmatrix} 0.1 \\ -0.2 \\ 0.1 \end{bmatrix}^T, A_{d32} = \begin{bmatrix} -0.1 & 0.1 & -0.2 \\ 0.1 & 0.1 & -0.1 \\ 0.1 & -0.1 & -0.2 \end{bmatrix}, D_3 = -0.1.$$

We choose the switching signal with $\alpha = 0.01$ and the output signal of the weight matrix are given by

$$M_1 = \begin{bmatrix} 0.5 & -0.5 & 0.7 \end{bmatrix}, M_2 = \begin{bmatrix} -0.6 & -0.6 & -0.7 \end{bmatrix}, M_3 = \begin{bmatrix} 0.6 & -0.6 & 0.5 \end{bmatrix},$$

the lower bound of delay is $\tau_m = 2$ and the upper bound of delay is $\tau_M = 4$. The LMIs in 1 are solved with the minimum H_∞ performance, then we obtained the parameters of designed filter as:

$$A_{f1} = \begin{bmatrix} 0.0149 & -0.0352 & 0.0100 \\ 0.0072 & 0.0082 & -0.0471 \\ -0.0263 & 0.0152 & 0.0360 \end{bmatrix}, B_{f1} = \begin{bmatrix} 0.0176 \\ -0.0645 \\ 0.0131 \end{bmatrix}, M_{f1} = \begin{bmatrix} -0.9131 \\ 1.1701 \\ -1.6202 \end{bmatrix}^T,$$

$$A_{f2} = \begin{bmatrix} -0.0170 & -0.0026 & 0.0537 \\ -0.0228 & 0.0306 & 0.0353 \\ -0.0390 & -0.0150 & 0.0314 \end{bmatrix}, B_{f2} = \begin{bmatrix} -0.0667 \\ 0.1708 \\ 0.0320 \end{bmatrix}, M_{f2} = \begin{bmatrix} 2.8102 \\ 3.5217 \\ 3.5705 \end{bmatrix}^T,$$

$$A_{f3} = \begin{bmatrix} 0.0838 & -0.0018 & -0.0408 \\ -0.0263 & -0.0429 & -0.0622 \\ 0.0217 & -0.0437 & -0.0568 \end{bmatrix}, B_{f3} = \begin{bmatrix} -0.0557 \\ -0.0659 \\ -0.1549 \end{bmatrix}, M_{f3} = \begin{bmatrix} -1.0776 \\ 2.9478 \\ -2.3523 \end{bmatrix}^T.$$

Further the error system is exponentially stable with H_∞ performance level $\gamma = 1.9940$. Also, we illustrate the effectiveness of the developed filter by obtaining feasible solution. The disturbance signal of subsystems are: $d_1(k) = 0.1\sin(0.05k)$, $d_2(k) = 0.5\sin(0.02k)$ and $d_3(k) = 0.2\sin(0.03k)$.

Figure 5 indicates the switching signal between the three subsystem in given time intervals. Figures 1, 2 and 3 shows the response of states vector $x_i(k)$ and the filter states $x_{fi}(k)$ where $i = 1, 2, 3$, respectively. It is easy to see that the switching and the coupled term are handled simultaneously and the convergence shows that the designed system is stable. Figure 4 denotes performance of the error state. It is easily illustrated that the developed filter design reduces the disturbance in the system along with the time delays.

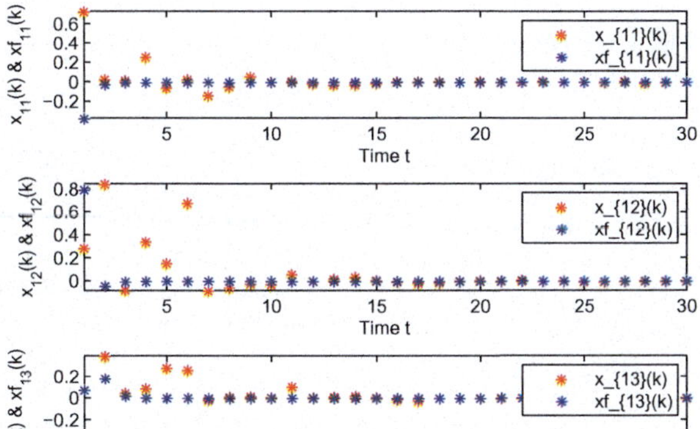

Fig. 1 Response of states x_1(k) and $x_{f1}(k)$

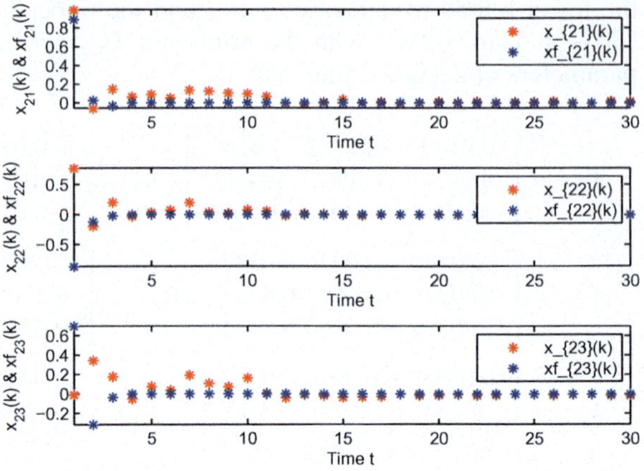

Fig. 2 Response of states x_2(k) and $x_{f2}(k)$

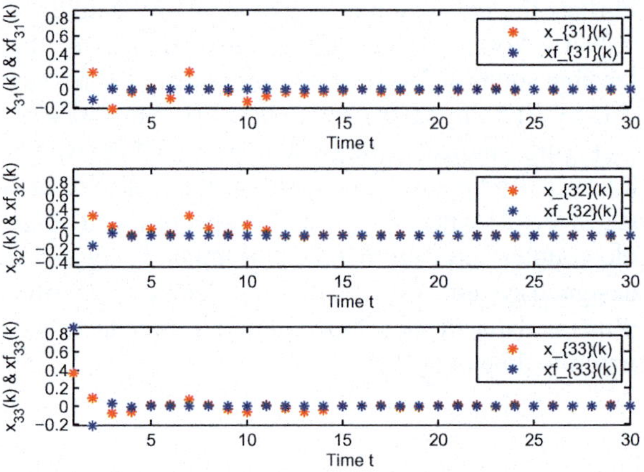

Fig. 3 Response of states x_3(k) and $x_{f3}(k)$

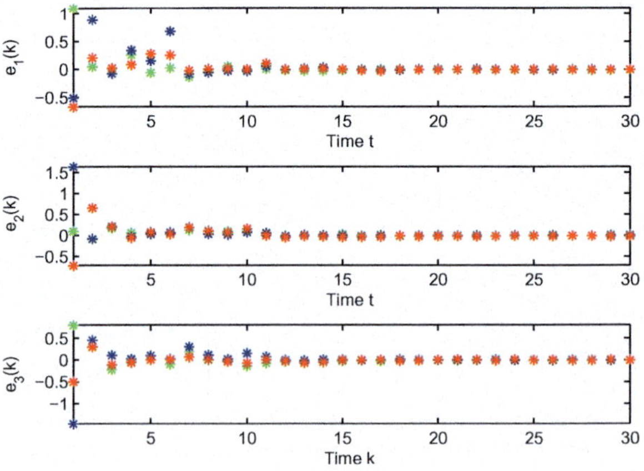

Fig. 4 Response of error state e(k)

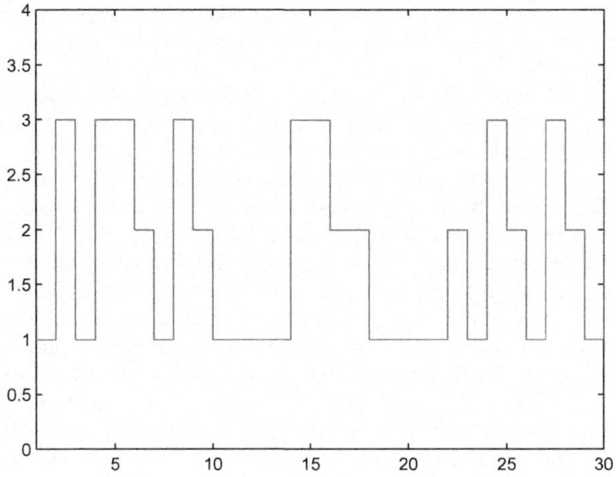

Fig. 5 Switching signal

5 Conclusion

In this paper, we have investigated the exponential mean square stability along with the H_∞ filter design for the discrete time interconnected system with time varying delays. The main contribution is to design a dependent filter for the prescribed interconnected system. The set of LMI constraints are presented to establish the designed system is exponentially mean square stable and H_∞ disturbance rejection level $\gamma > 0$. The set of LMI constraints are solved by standard packages. Finally, examples are illustrated for systems with and without uncertainty to provide the advantage of the proposed results.

References

1. Boyd, S., Ghoui, L.E., Feron, E., Balakrishnan, V.: Linear Matrix Inequalities in System and Control Theory. SIAM (1994)
2. Liu, Y., Wang, Z., Liu, X.: Robust stability of discrete-time stochastic neural networks with time-varying delays. Neurocomputing. **71**(4–6), 823–833 (2008)
3. Liu, Y., Lee, S.M., Kwon, O.M., Park, J.H.: Delay-dependent exponential stability criteria for neutral systems with interval time-varying delays and non linear perturbations. J. Frankl. Inst. **350**(10), 3313–3327 (2013)
4. Sakthivel, R., Selvaraj, P., Kim, Y.J., Lee, D.H., Kwon, O.M., Sakthivel, R.: Robust H_∞ resilient event-triggered control design for T-S fuzzy systems. Discret. & Contin. Dyn. Syst. Ser. S. **15**(11), 3297–3312 (2022)
5. Mathiyalagan, K., Park, J.H., Sakthivel, R.: Robust reliable dissipative filtering for networked control systems with sensor failure. IET Signal Process. **8**(8), 809–822 (2014)
6. Mathiyalagan, K., Balasubramani, M., Chang, X.H., Sangeetha, G.: Finite-time dissipativity-based filter design for networked control systems. J. Adapt. Control. Signal Process. **33**, 1706–1721 (2019)

7. Mathiyalagan, K., Park, J.H., Sakthivel, R., Anthoni, S.M.: Robust mixed H_∞ and passive filtering for networked Markov jump systems with impulses. Signal Process. **101**, 162–173 (2014)
8. Mahmoud, M.S., Al-Sunni, F.M., Yuanqing, X.: Interconnected switched discrete-time systems: robust stability and stabilization. IMA J. Math. Control. Inf. **28**(1), 41–73 (2011)
9. Michiels, W., Niculescu, S.I.: Stability and Stabilization of Time-Delay Systems. Advances in Design and Control. Society for Industrial and Applied Mathematics (2007)
10. Nagpal, K.M., Khargonekar, P.P.: Filtering and smoothing in an H/sup infinity/setting. IEEE Trans. Autom. Control. **36**(2), 152–166 (1991)
11. Regaieg, M.A., Kchaou, M., Bosche, J., El-Hajjaji, A., Chaabane, M.: Robust dissipative observer-based control design for discrete-time switched systems with time-varying delay. IET Control. Theory & Appl. **13**(18), 3026–3039 (2019)
12. Shen, B., Wang, Z., Ding, D., Shu, H.: H_∞ state estimation for complex networks with uncertain inner coupling and incomplete measurements. IEEE Trans. Neural Netw. Learn. Syst. **24**(12), 2027–2037 (2013)
13. Tharanidharana, V., Sakthivel, R., Yong, R., Anthoni, S.M.: Robust finite-time PID control for discrete-time large-scale interconnected uncertain system with discrete-delay. Math. Comput. Simul. **192**, 370–383 (2022)
14. Zhang, L., Yang, X.: On pole assignment of high-order discrete-time linear systems with multiple state and input delays. Discret. & Contin. Dyn. Syst. Ser. S. **15**(11), 3351–3368 (2022)
15. Wei, G., Wang, Z., He, X., Shu, H.: Filtering for networked stochastic time-delay systems with sector nonlinearity. IEEE Trans. Circuits Syst. II Express Brief. **56**(1), 71–75 (2009)
16. Wu, L., Feng, Z., Zheng, W.X.: Exponential stability analysis for delayed neural networks with switching parameters: average dwell time approach. IEEE Trans. Neural Netw. **21**(9), 1396–1407 (2010)
17. Xie, L.: Output feedback H_∞ control of systems with parameter uncertainty. Int. J. Control. **63**(4), 741–750 (1996)
18. Yao, Y., Liang, J., Cao, J.: Stability analysis for switched genetic regulatory networks: an average dwell time approach. J. Frankl. Inst. **348**(10), 2718–2733 (2011)
19. Yang, R., Shi, P., Liu, G.: Filtering for discrete-time networked non-linear systems with mixed random delays and packet dropouts. IEEE Trans. Autom. Control. **56**(11), 2655–2660 (2011)
20. Zhang, B., Zheng, W.H.: H_∞ filter design for nonlinear networked control systems with uncertain packet-loss probability. Signal Process. **92**(6), 1499–1507 (2012)
21. Zhang, Z., Zhang, Z.X., Zhang, H., Shi, P., Karimi, H.R.: Finite-time H_∞ filtering for T-S fuzzy discrete-time systems with time-varying delay and norm-bounded uncertainties. IEEE Trans. Fuzzy Syst. **23**(6), 2427–2434 (2015)
22. Zhang, L., Wang, Y., Huang, Y., Chen, X.: Delay-dependent synchronization for non-diffusively coupled time-varying complex dynamical networks. Appl. Math. Comput. **259**, 510–522 (2015)
23. Zhang, B., Zheng, W.X.: H_∞ filter design for nonlinear networked control systems with uncertain packet-loss probability. Signal Process. **92**, 1499–1507 (2012)

Stability and Observability Analysis of Uncertain Neutral Time-Delay Systems

Nidhal Khorchani, Wiem Jebri, Rafika El Harabi, and Hassen Dahman

Abstract This work discusses the analysis of uncertain neutral time-delay systems, These systems include delays in both the state and derivatives. In actuality, the study of stability and observability in an open-loop is proposed using the Lyapunov-Krasovskii functional. Then, by resolving linear matrix inequalities (LMI) and taking into account optimization theory, new requirements of stability and observability are obtained, In order to support theory development, simulation results will be provided.

Keywords Neutral time-delay · Stability · Parameter uncertainties · Observability · Lyapunov-Krasovskii · LMI

1 Introduction

The diagnostic process for any system should include a step called stability analysis. The behavior of a dynamic system's trajectory around equilibrium points is what a system's stability concept refers to. Therefore, when the starting state is close to an equilibrium point, it is possible to study the evolution of a dynamic system's state trajectory through stability analysis. The earliest studies on neutral time delay systems originate from 1970 and were conducted by [1–3]. In fact, a large number of dynamic systems depend not only on current and past states but also on delayed derivatives. Neutral systems are this kind of system. Numerous studies have been done on the robust stability analysis for linear systems with uncertain delays as well as parameter uncertainties [4, 5]. Furthermore, delays and uncertainty have an impact on the dynamic responses of many industrial processes, including chemical engineering processes, electrical circuits, chemical reactors, and neutral networks. In

N. Khorchani (✉) · W. Jebri · R. El Harabi · H. Dahman
MACS Laboratory, National Engineering School of Gabes, Gabes University, Rue Omar Ibn Elkhatab, Gabes, 6029 Gabes, Tunisia
e-mail: khorchani.nidhal@gmail.com

H. Dahman
LaPhyMNE Laboratory (LR05ES14), FSG, Gabes University, Gabes, Tunisia

A. Ben Makhlouf et al. (eds.), *State Estimation and Stabilization of Nonlinear Systems*, Studies in Systems, Decision and Control 491,
https://doi.org/10.1007/978-3-031-37970-3_7

reality, many dynamic systems depend on more than only the present and past states and their derivatives with delays. Neutral systems are this kind of system. Also, the presence of delay phenomena can lead to instability and divergences.

For a linear system in a stochastic environment, Kalman first introduced the theory of the state observer. Then Luenberger introduced the concepts of a reduced observer and a minimal observer as part of a general theory of observers for deterministic linear systems. Recently, a number of research projects for observer synthesis have been developed. Examples include unidentified entry observers, unidentified detector filters, unidentified structure placement. Therefore, one of the core issues with process control problems is the reconstruction of a system's state variables. Thus, the observation problem is the focus of a sizable portion of automatic research activities. This is motivated by the estimation of the state, which is a critical step required for the synthesis of command laws and for the monitoring or diagnosis of industrial systems [6]. Some research has been done on the observer study for neutral time-delay systems [7–10]. In these searches, we distinguish between adaptable observatories, unidentified observatories, and the Luenberger observatory that we previously used. This paper is organized as follows. Section 2 describes the problem formulation. Section 3 gives stability analysis of uncertain neutral time-delay Systems. Section 4 presents the observability analysis. Section 5 presents simulation results of a numerical example, Finally, is a conclusion in Sect. 6.

2 Problem Formulation

In this section, the uncertain neutral time-delay system described by the following equation:

$$\begin{cases} \dot{x}(t) - \bar{D}\dot{x}(t-\varepsilon_2) = \bar{J}x(t) + \bar{R}x(t-\varepsilon_1) + \bar{B}u(t) \\ y(t) = Cx(t) \\ x(t) = \chi(t) \quad ; \ \forall t \in [-\phi, 0] \end{cases} \tag{1}$$

where $x(t) \in \mathbb{R}^n$ is the state vector, $u(t) \in \mathbb{R}^m$ is the input vector, $y(t) \in \mathbb{R}^p$ indicates the measurement output vector. J, R, D, B and C are known constant matrices with appropriate dimensions.$\varepsilon_1 > 0, \varepsilon_2 > 0$ denotes the state and its derivative time-delay. $\phi = \max\{\varepsilon_1, \varepsilon_2\}$ and $\chi(t)$ is indicates the continuous vector of initial condition.

The parameter uncertainties will be presented at the following forms:

$$\bar{J} = J + \Delta J, \ \bar{R} = R + \Delta R, \ \bar{D} = D + \Delta D \text{ and } \bar{B} = B + \Delta B.$$

Thus, the argument Δ of uncertain matrices depends on the time t with ΔJ, ΔR, ΔD and ΔB. As well, the norm bounded parameter uncertainties terms are given as:

$$\Delta J = E_1 \sum\nolimits_1 F_1$$
$$\Delta R = E_2 \sum\nolimits_2 F_2$$
$$\Delta D = E_3 \sum\nolimits_3 F_3$$
$$\Delta B = E_4 \sum\nolimits_4 F_4$$

where E_1, E_2, E_3, E_4, F_1, F_2, F_3 and F_4 are known constant matrices. After that, $\sum_1$, $\sum_2$, $\sum_3$ and $\sum_4$ denote parameter uncertainties and are added to the nominal matrices (J, R, D and B).

Using the Newton-Lebuniz formula,

$$x(t-\varepsilon_1) = x(t) - \int_{-\varepsilon_1}^{0} \dot{x}(t+\alpha)\, d\alpha$$

Substituting this formula into (1) we can infer

$$\begin{cases} \dot{x}(t) - \bar{D}\dot{x}(t-\varepsilon_2) = (\bar{J} + \bar{R})x(t) - \bar{R} \int_{-\varepsilon_2}^{0} \dot{x}(t+\alpha)d\alpha + \bar{B}u(t) \\ y(t) = Cx(t) \\ x(t) = \chi(t) \quad ; \; \forall t \in [-\phi, 0] \end{cases} \tag{2}$$

3 Stability Analysis of Uncertain Neutral Time-Delay Systems

This section introduces the issue of stability analysis for a class of linear uncertain systems with time-delay of a particular sort. The goal is to obtain constrained LMI criteria for the class of systems whose dynamics depend on the derivation of past states. The development of a system stability criterion (1) that takes into account linear matrix inequalities. The above-mentioned theory determines the condition.

Theorem 1 *The linear uncertain neutral time-delay system (1) with a given constant* $\delta > 0$ *for any* $0 < \phi \leq \delta$*, if there exists a positive symmetric definite matrices* Ξ_1*,* Ξ_2 *and* Ξ_3 *satisfy the the following LMI:*

$$\begin{bmatrix} G(\Xi_1,\overline{J},\overline{R}) & 0 & \bar{D}Y & \overline{B} & \Xi_1\overline{J}^T & \Xi_1 C^T & \Xi_1 & \bar{R}\Xi_3 \\ * & -\Xi_2 & 0 & 0 & \Xi_2\bar{R}^T & 0 & 0 & 0 \\ * & * & -\Xi_3 & 0 & \Xi_3\bar{D}^T & 0 & 0 & 0 \\ * & * & * & I & \overline{B}^T & 0 & 0 & 0 \\ * & * & * & * & -\frac{1}{1+\delta}\Xi_3 & 0 & 0 & 0 \\ * & * & * & * & * & -I & 0 & 0 \\ * & * & * & * & * & * & -\Xi_2 & 0 \\ * & * & * & * & * & * & * & -\frac{1}{\delta}\Xi_3 \end{bmatrix} < 0 \qquad (3)$$

wherein
$G(\Xi_1,\overline{J},\overline{R}) = (\bar{J}+\bar{R})\Xi_1 + \Xi_1(\bar{J}+\bar{R})^T$
I, $(*)$ denote respectively identity matrix and the transposed elements in the symmetric.

Proof We introduce the quadratic Lyapunov Krasovskii functional $V(x,t)$ with a simple integral chosen according to the states and the derivatives of the represented states of this form.

$$\begin{aligned} V(x,t) &= x^T(s)P\,x(s) + \int_{t-\varepsilon_1}^{t} x^T(s)Q_1\,x(s)ds + \int_{t-\varepsilon_2}^{t} \dot{x}^T(s)Q_2\,\dot{x}(s)ds \\ &\quad + \int_{-\varepsilon_1}^{0}\left(\int_{t+\alpha}^{t} \dot{x}^T(s)\;Q_2\dot{x}(s)ds\right)d\alpha \end{aligned} \qquad (4)$$

where P, Q_1 and Q_2 are symmetric positive definite matrices.

Lyapunov-Krasovskii function can be constructed as follows:

$$\begin{aligned} &\dot{V}(x,t) = 2\dot{x}^T(t)Px(t) + x^T(t)Q_1x(t) - x^T(t-\varepsilon_1)Q_1x(t-\varepsilon_1) + \dot{x}^T(t)Q_2\dot{x}(t) \\ &-\dot{x}^T(t-\varepsilon_2)Q_2\dot{x}(t-\varepsilon_2) + \int_{-\varepsilon_1}^{0}\left[\dot{x}^T(t)Q_2\dot{x}(t) - \dot{x}^T(t+\alpha)Q_2\dot{x}(t+\alpha)\right]d\alpha \\ &= 2\dot{x}^T(t)Px(t) + x^T(t)Q_1x(t) - x^T(t-\varepsilon_1)Q_1x^T(t-\varepsilon_1) + (1+\varepsilon_1)\dot{x}^T(t)Q_2\dot{x}(t) \\ &-\dot{x}^T(t-\varepsilon_2)Q_2\dot{x}(t-\varepsilon_2) - \int_{-\varepsilon_1}^{0}\dot{x}^T(t+\alpha)Q_2\dot{x}(t+\alpha)d\alpha \\ &= x^T(t)\left(P(\overline{J}+\overline{R}) + (\overline{J}+\overline{R})^T P + Q_1\right)x(t) - 2x^T(t)P\int_{-\varepsilon_1}^{0}\overline{R}\dot{x}(t+\alpha)d\alpha \\ &+2x^T(t)P\overline{D}\dot{x}(t-\varepsilon_2) + 2x^T(t)P\overline{B}u(t) - x^T(t-\varepsilon_1)Q_1x^T(t-\varepsilon_1) \\ &+(1+\varepsilon_1)\dot{x}^T - \dot{x}^T(t-\varepsilon_2)Q_2\dot{x}(t-\varepsilon_2) - \int_{-\varepsilon_1}^{0}\dot{x}^T(t+\alpha)Q_2\dot{x}(t+\alpha)d\alpha \\ &-2x^T(t)P\int_{-\varepsilon_1}^{0}\overline{R}\dot{x}(t+\alpha)d\alpha = -\int_{-\varepsilon_1}^{0}2x^T(t)P\overline{R}\dot{x}(t+\alpha)d\alpha \\ &\leq \int_{-h}^{0} x^T(t)P\overline{R}S^{-1}\overline{R}^T Px(t) + \dot{x}^T(t+\alpha)Q_2\dot{x}(t+\alpha)d\alpha \\ &\leq \varepsilon_1 x^T(t)P\overline{R}{Q_2}^{-1}\overline{R}^T Px(t) + \int_{-h}^{0}\dot{x}^T(t+\alpha)Q_2\dot{x}(t+\alpha)d\alpha \end{aligned} \qquad (5)$$

The Lyapunov function is given as follows:

$$\begin{aligned}
&= x^T(t)\left(P(\overline{J}+\bar{R})+(\overline{J}+\bar{R})^T P+Q_1+\varepsilon_1 P\bar{R}Q_2{}^{-1}\bar{R}^T P\right)x(t)\\
&+2x^T(t)P\overline{D}\dot{x}(t-\varepsilon_2)+2x^T(t)P\overline{B}u(t)-x^T(t-\varepsilon_1)Q_1x(t-\varepsilon_1)\\
&+x^T(t)\overline{J}^T(1+\varepsilon_1)Q_2\overline{J}x(t)+2x^T(t)\overline{J}^T(1+\varepsilon_1)Q_2\bar{R}^T x(t-\varepsilon_1)\\
&+2x^T(t)\overline{J}^T(1+\varepsilon_1)Q_2\overline{D}\dot{x}(t-\varepsilon_2)+2x^T(t)\overline{J}^T(1+\varepsilon_1)Q_2\overline{B}u(t)\\
&+2x^T(t-\varepsilon_1)\bar{R}^T(1+\varepsilon_1)Q_2\overline{D}\dot{x}(t-\varepsilon_2)+2x^T(t-\varepsilon_1)\bar{R}^T(1+\varepsilon_1)Q_2\overline{B}u(t)\\
&+x^T(t-\varepsilon_1)\bar{R}^T(1+\varepsilon_1)Q_2\bar{R}x(t-\varepsilon_1)+2\dot{x}^T(t-\varepsilon_2)\overline{D}^T(1+\varepsilon_1)Q_2\overline{B}u(t)\\
&+\dot{x}^T(t-\varepsilon_2)\overline{D}^T(1+\varepsilon_1)Q_2\overline{D}\dot{x}(t-\varepsilon_2)+u^T(t)\overline{B}^T(1+\varepsilon_1)Q_2\overline{B}u(t)\\
&-\dot{x}^T(t-\varepsilon_2)Q_2\dot{x}(t-\varepsilon_2)
\end{aligned} \tag{6}$$

In addition, we have $y(t)y(t) < 0$ then,

$$y(t)y(t) \le \vartheta^T \lambda(\varepsilon_1)\,\vartheta < 0 \tag{7}$$

where $\vartheta = \left[x^T(t)\ x^T(t-\varepsilon_1)\ \dot{x}^T(t-\varepsilon_2)\ u^T(t)\right]$
$\lambda(\varepsilon_1) =$

$$\begin{bmatrix}
\alpha(\varepsilon_1) & (1+\varepsilon_1)\bar{J}Q_2\bar{R} & P\bar{D}+(1+\varepsilon_1)\bar{J}^T Q_2\bar{D} & P\bar{B}+(1+\varepsilon_1)\bar{J}^T Q_2\bar{B}\\
* & (1+\varepsilon_1)\bar{R}^T Q_2\bar{J}-Q_1 & (1+\varepsilon_1)\bar{R}^T Q_2\bar{D} & (1+\varepsilon_1)\bar{R}^T Q_2\bar{B}\\
* & * & (1+\varepsilon_1)\bar{D}^T Q_2\bar{D}-Q_2 & (1+\varepsilon_1)\bar{D}^T Q_2\bar{B}\\
* & * & * & (1+\varepsilon_1)\bar{B}^T Q_2\bar{B}-I
\end{bmatrix} \tag{8}$$

where;
$\alpha(h) = P(\bar{J}+\bar{R})+(\bar{J}+\bar{R})^T P+Q_1+\varepsilon_1 P\bar{R}Q_2{}^{-1}\bar{R}^T P+(1+\varepsilon_1)\bar{J}^T Q_2\bar{J}+C^T C$ Certainly, the inequality (7) implies $\dot{V} < 0$ and $\lambda(\varepsilon_1) < 0$. Remember that the matrix $\lambda(\varepsilon_1) < 0$ is monotonically increasing with respect to $\phi < 0$ in the positive definition, therefore $\lambda(\varepsilon_1) < 0$ is equivalent to $0 < \phi \le \delta$ if $\lambda(\delta) < 0$ holds. By Schur complement, $\lambda(\delta) < 0$ holds if and only if:

$$\begin{bmatrix}
G(P,Q_1) & 0 & P\bar{D} & P\bar{B} & \bar{J} & P\bar{R}\\
* & -Q_1 & 0 & 0 & \bar{R}^T & 0\\
* & * & -Q_2 & 0 & \bar{D}^T & 0\\
* & * & * & -I & \bar{B}^T & 0\\
* & * & * & * & -\frac{1}{1+\delta}Q_2{}^{-1} & 0\\
* & * & * & * & * & -\frac{1}{\delta}Q_2
\end{bmatrix} \tag{9}$$

where;
$G(P,Q_1) = P(\bar{J}+\bar{R})+(\bar{J}+\bar{R})^T P+Q_1+C^T C$ Let us consider $diag(\Xi_1, \Xi_2, \Xi_3, I, \Xi_3)$ and multiply equation (9) on both sides through the last matrix. After that, we apply Schur complement in order to get these LMIs:

$$\begin{bmatrix} G(\Xi_1,\overline{J},\overline{R}) & 0 & \bar{D}Y & \overline{B} & \Xi_1\overline{J}^T & \Xi_1 C^T & \Xi_1 & \bar{R}\Xi_3 \\ * & -\Xi_2 & 0 & 0 & \Xi_2\bar{R}^T & 0 & 0 & 0 \\ * & * & -\Xi_3 & 0 & \Xi_3\bar{D}^T & 0 & 0 & 0 \\ * & * & * & I & \overline{B}^T & 0 & 0 & 0 \\ * & * & * & * & -\frac{1}{1+\delta}\Xi_3 & 0 & 0 & 0 \\ * & * & * & * & * & -I & 0 & 0 \\ * & * & * & * & * & * & -\Xi_2 & 0 \\ * & * & * & * & * & * & * & -\frac{1}{\delta}\Xi_3 \end{bmatrix} < 0 \quad (10)$$

where $G(\Xi_1,\overline{J},\overline{R}) = (\bar{J}+\bar{R})\Xi_1 + \Xi_1(\bar{J}+\bar{R})^T$ **End proof**

4 Observability Analysis for Uncertain Neutral Time Delay

A dynamic system's observability analysis involves being able to say if the system's state can be determined based on its entrants and egresses. Additionally, the systems (1) and (2) are described as being observable on $[t_0, t_f]$ $(t_f > t_0)$. If it is possible to estimate their states at a given time t_0 given $x(t_0) = x_0 \in \mathbb{R}^n$ using information from their entrances and exit observations.

Remember that this observability study phase is essential to creating the right conditions for the synthesis of the associated observability for neutral time delay systems in both uncertain. The theory presented below enables to determine the observability requirement for a specific neutral time-delay system.

Theorem 2 *Consider the system (1) of order n with uncertainties, the observability matrix* $obsrv_1$ *as follows:*

$$obsrv_1 = \Big(C\tfrac{1}{\varepsilon_2}(D+\Delta D)^{-1}\tfrac{1}{\omega_1}\; C\tfrac{1}{\varepsilon_2}(D+\Delta D)^{-1}\tfrac{1}{\omega_1}(-\sigma_1)\; C\tfrac{1}{\varepsilon_2}(D+\Delta D)^{-1}\tfrac{1}{\omega_1}(-\sigma_1)^2 \\ \ldots.\; C\tfrac{1}{\varepsilon_2}(D+\Delta D)^{-1}\tfrac{1}{\omega_1}(-\sigma_1)^{n-1} \Big) \quad (11)$$

where $\omega_1 = (-\frac{1}{\varepsilon_2}(D+\Delta D)(J+\Delta J+R+\Delta R) - \frac{1}{4\varepsilon_2^2}D^{-2}(\varepsilon_1(R+\Delta R)+I-(D+\Delta D)^2)^{\frac{1}{2}}\sigma_1 = \frac{1}{2\varepsilon_2}(D+\Delta D)(\varepsilon_1(R+\Delta R)+I-(D+\Delta D))$ *if* $rang(obsrv_1) = n$ *and* $\omega_1 \neq 0$ *then (1) is observable.*

Preuve
The Laplace transform of the system (1) gives:

$$pX(p) - X(0) = (J+\Delta J)X(p) + (R+\Delta R)e^{-\varepsilon_1 p}X(p) \\ + (D+\Delta D)(pe^{-\varepsilon_2 p}X(p) - X(-\varepsilon_2)) \quad (12)$$

therefore

$$X(p) = \frac{I}{p - (J + \Delta J) - (R + \Delta R)e^{-\varepsilon_1 p} - p(D + \Delta D)e^{-\varepsilon_2 p}}(X(p) - (D + \Delta D)X(-\varepsilon_2)) \tag{13}$$

We could not calculate the inverse of this function by the usual formula of the inverse Laplace transform because of the presence of exponential term. So, to deduce the state $x(t)$, we resorted to the approximation of exponential functions. Consider, then, the Taylor approximation of order 1 in the neighborhood of zero for the exponential function, we have:

$$\begin{aligned} X(p) &= \frac{I}{p-(J+\Delta J)-(R+\Delta R)(1-\varepsilon_1 p)-p(D+\Delta D)(1-\varepsilon_2 p)}(X(0) - (D + \Delta D)X(-\varepsilon_2)) \\ &= \frac{I}{-(J+\Delta J+R+\Delta R)+(\varepsilon_1(R+\Delta R)+I+(D+\Delta D)p+\varepsilon_2(D+\Delta D)p^2}(X(0) - (D + \Delta D)X(-\varepsilon_2)) \\ &= \frac{1}{\varepsilon_2}\frac{(D+\Delta D)^{-1}}{-\frac{1}{\varepsilon_2}(D+\Delta D)^{-1}(J+R)+\frac{1}{\varepsilon_2}(D+\Delta D)^{-1}(\varepsilon_1(J+R)+I+(D+\Delta D))p+p^2}(X(0) - (D + \Delta D)X(-\varepsilon_2)) \end{aligned} \tag{14}$$

This expression has the same shape as

$$\frac{1}{\varepsilon_2}F(p)(D + \Delta D)^{-1}(X(0) - (D + \Delta D)X(-\varepsilon_2)) \tag{15}$$

then $F(p) = \frac{1}{p^2+2\sigma_1 p+p^2}$ and $\sigma_1 = \frac{1}{2\varepsilon_2}D^{-1}(\varepsilon_1(R + \Delta R) + I + (D + \Delta D)$

$$\rho^2 = -\frac{1}{\varepsilon_2}(D + \Delta D)(J + \Delta J + R + \Delta R)$$

The state $x(t)$ is obtained as follows:

$$x(t) = \frac{1}{\varepsilon_2}(D + \Delta D) + \frac{1}{\omega_1}e^{-\sigma_1 t}\sin(\omega_1 t)(x(0) - (D + \Delta D)x(-\varepsilon_2)) \tag{16}$$

with

$$\begin{aligned} \rho^2 &= {\omega_1}^2 + {\sigma_1}^2 \\ \omega_1 &= (-\frac{1}{\varepsilon_2}(D + \Delta D)^{-1}(J + \Delta J + R + \Delta R) - \frac{1}{4{\varepsilon_2}^2}(D + \Delta D)^{-2}(\varepsilon_1(R + \Delta R) \\ &\qquad +I + (D + \Delta D))^{\frac{1}{2}} \end{aligned} \tag{17}$$

In order to obtain the state $x(t)$, ω_1 must not be null. Next, the release $y(t)$ will be:

$$y(t) = Cx(t) = C\frac{1}{\varepsilon_2}(D + \Delta D)^{-1}\frac{1}{\omega_1}e^{-\sigma_1 t}\sin(\omega_1 t)(x(0) - (D + \Delta D)x(-\varepsilon_2)) \tag{18}$$

As a result, knowing the output $y(t)$ allows one to obtain$(x(0) - (D + \Delta D)x(-\varepsilon_2))$ provided that $C\frac{1}{\varepsilon_2}(D + \Delta D)^{-1}\frac{1}{\omega_1}e^{-\sigma_1 t}\sin(\omega_1 t)$ is not singular. On the basis of the terms $\sin(\omega_1 t) \neq 0$.
Or

$C\frac{1}{\varepsilon_2}(D + \Delta D)^{-1}\frac{1}{\omega_1}e^{-\sigma_1 t} = \sum_{k=0}^{n-1}\alpha_k(t)C\frac{1}{\varepsilon_2}(D + \Delta D)^{-1}\frac{1}{\omega_1}(-\sigma_1)^k$ et $C\frac{1}{\varepsilon_2}(D + \Delta D)^{-1}$ $\frac{1}{\omega_1}e^{-\sigma_1 t}\sin(\omega_1 t)$ is singular:

$$obsrv_1 = \Big(C\tfrac{1}{\varepsilon_2}(D+\Delta D)^{-1}\tfrac{1}{\omega_1} \; C\tfrac{1}{\varepsilon_2}(D+\Delta D)^{-1}\tfrac{1}{\omega_1}(-\sigma_1) \; C\tfrac{1}{\varepsilon_2}(D+\Delta D)^{-1}\tfrac{1}{\omega_1}(-\sigma_1)^2 \\ .... \; C\tfrac{1}{\varepsilon_2}(D+\Delta D)^{-1}\tfrac{1}{\omega_1}(-\sigma_1)^{n-1} \Big) \tag{19}$$

is of $rank = n$

End proof

5 Numerical Example

In order to demonstrate the stability and observability analysis of uncertain neutral time-delay systems proposed in this research, a numerical example is provided in this section.

$$J = \begin{bmatrix} -1 & 1 \\ 0 & -4 \end{bmatrix}, R = \begin{bmatrix} 0.15 & 0.05 \\ 0 & 0.1 \end{bmatrix}, D = \begin{bmatrix} -0.05 & 0.02 \\ 0.01 & 0 \end{bmatrix}$$

$$B = \begin{bmatrix} 1 \\ 0 \end{bmatrix}, C = \begin{bmatrix} 1 & 1 \end{bmatrix}$$

$$E_1 = \begin{bmatrix} 0.1 \\ 0.1 \end{bmatrix}, E_2 = \begin{bmatrix} 0.1 \\ -0.1 \end{bmatrix}, E_3 = \begin{bmatrix} 0.1 \\ -0.1 \end{bmatrix} \text{ and } E_4 = \begin{bmatrix} -0.1 \\ 0 \end{bmatrix}$$

$$F_1 = \begin{bmatrix} 0.001 & 0.3 \end{bmatrix}, F_2 = \begin{bmatrix} 0.01 & 0.2 \end{bmatrix}, F_3 = \begin{bmatrix} 0.01 & 0.2 \end{bmatrix} \text{ and } F_4 = 0.02$$

where $\sum_1 = \sum_2 = \sum_3 = \sum_4 = 0.2$, $\varepsilon_1 = 0.5$,$\varepsilon_2 = 1$.

- Stability analysis:

Referring to Theorem 1, the resolution of the LMI (3) allows to show that the system (1) is stable. The obtained matrices are as follows:

$$\Xi_1 = \begin{bmatrix} 3.0070 & -4.7706 \\ -4.7706 & 12.3877 \end{bmatrix} > 0 \; \Xi_2 = \begin{bmatrix} 30.1651 & -38.9022 \\ -38.9022 & 76.0571 \end{bmatrix} > 0$$

$$\Xi_3 = \begin{bmatrix} 2.9330 & -3.5685 \\ -3.5685 & 5.1717 \end{bmatrix} > 0$$

These are all positive. Therefore, the system is asymptotically stable.

- Observability analysis:

In this illustration, we'll create the observer from one component if the particular system (1) is observable. The observability matrix denoted by (11), also known as $obsrv_1$, is given by:

$$obsrv_1 = \begin{pmatrix} 0.3398 & -0.2420 \\ 1.0188 & 1.51002 \end{pmatrix} \tag{20}$$

is of $rank = 2$, so according to Theorem 2, the system is observable.

6 Conclusion

In this work, we study the open-loop stability and observability of uncertain neutral delay systems. This analysis is based on the use of the Lyapunov-Krasovskii functional and the Linear Matrix Inequality (LMI) method. To illustrate the approaches suggested to verify theoretical and simulation results.

References

1. Alexandrova, I.V., Zhabko, A.P.: Stability of neutral type delay systems: a joint Lyapunov–Krasovskii and Razumikhin approach. Automatica **106**, 83–90 (2019)
2. Wu, M., He, Y., She, J.H.: Stability Analysis and Robust Control of Time-Delay Systems, vol. 22. Springer, Berlin (2010)
3. Wu, Q., Song, Q., Hu, B., Zhao, Z., Liu, Y., Alsaadi, F.E.: Robust stability of uncertain fractional order singular systems with neutral and time-varying delays. Neurocomputing J. **401**, 145–152. Elsevier (2020)
4. Gomez, M.A., Egorov, A.V., Mondié, S.: Necessary stability conditions for neutral-type systems with multiple commensurate delays. Int. J. Control **92**(5), 1155–1166 (2019)
5. Wang, Y.E., Sun, X.M., Wu, B.: Lyapunov–Krasovskii functionals for input-to-state stability of switched non-linear systems with time-varying input delay. IET **9**(11), 1717–1722 (2015)
6. Vu, V.P., Wang, W.J.: Observer synthesis for uncertain Takagi–Sugeno fuzzy systems with multiple output matrices. IET Control Theory Appl. (2015)
7. Sheng, Z., Lin, C., Chen, B., Wang, Q.-G.: Asymmetric Lyapunov-Krasovskii functional method on stability of time-delay systems, Wiley online library. Int. J. Robust Nonlinear Control **31**(7), 2847–2854 (2021)
8. Hu, G.D.: An observer-based stabilizing controller for linear neutral delay systems. Sib. Math. J. **63**(4), 789–800 (2022)
9. Zhang, Q., Li, Z.Y.: Delay-dependent exponential stability of linear stochastic neutral systems with general delays described by Stieltjes integrals. Int. J. Control, Autom. Syst. 1–10 (2023)
10. Gao, S., Ma, G., Guo, Y.: Robust sliding-mode observer-based multiple-fault diagnosis scheme **25**(2), 1555–1576 (2023)

Zonotopic State Estimation for Uncertain Discrete-Time Switched Linear Systems

Leila Dadi, Haifa Ethabet, and Mohamed Aoun

Abstract This chapter studies the problem of state estimation for a class of discrete-time switched systems. The considered system is subject to disturbances and measurement noise which are supposed to be unknown but bounded in predefined zonotopes. The proposed method consists of two steps. First, a switched L_∞-based observer attenuating the effect of uncertainties is designed to obtain point estimate of the system state. The robust observer is designed using firstly the Luenberger structure and secondly the T-N-L structure. Then, interval state estimation is achieved by integrating robust point estimation with zonotopic analysis techniques. The observers gains are calculated by solving Linear Matrix Inequality (LMI) derived using a Multiple Lyapunov functions (MLF) under an Average Dwell Dime (ADT) switching signal. A numerical example is performed to illustrate the effectiveness of the obtained results.

Keywords Interval estimation · Zonotope · Discrete-time · Switched systems · LMIs · L_∞ term

1 Introduction

Switched systems represent one of the most important classes of hybrid dynamical systems. They appear in several fields and can model a lot of complex systems. They consist of a collection of discrete or continuous-time subsystems and a switching rule that operates switching between them [14]. Due to their powerful modeling capability, this class has witnessed an increasing interest among researchers where the state estimation, stability and control problems dealing with switched systems have been widely introduced during the past years, see for instance ([10, 14, 19]).

L. Dadi (✉) · H. Ethabet · M. Aoun
Research Laboratory Modeling, Analysis and Control of Systems (MACS), University of Gabes, National Engineering School of Gabes, LR16ES22, Zrig Eddakhlania, Tunisia
e-mail: dr.leila.dadi@gmail.com

A. Ben Makhlouf et al. (eds.), *State Estimation and Stabilization of Nonlinear Systems*, Studies in Systems, Decision and Control 491,
https://doi.org/10.1007/978-3-031-37970-3_8

During the past decades, the problem of state estimation and control synthesis for switched systems has been widely investigated that still forms an interesting research domain. State estimation of switched systems has been tackled in many studies such as ([1, 2]). For the stability problem of switched systems, common and multiple Lyapunov functions were introduced in [10, 14]. Most recently, several contributions are developed to improve the theory and application of switched system control. For example, in [20] a distributed filtering problem is considered for switched systems over sensor networks. A robust sliding mode observer is developed in [13] to estimate the state and detect the presence of faults. In [22], stability and stabilization problems for switched systems with mode-dependent average dwell time are introduced.

In practice, industrial systems are usually affected by uncertainties, disturbances and noises. Conventional observers used for state estimation may not be efficient to solve the estimation problems. In this case, interval observers have been introduced to solve this problem. They compute the set of all admissible values and provide two bounds which enclose the real state [8]. Interval observers can provide the upper and lower bounds of the system state by two sub-observers such that the estimation errors dynamics are cooperative and stable. For instance, under the assumption that disturbances are unknown but bounded, [6, 15] have investigated the state estimation based interval observer for continuous- and discrete-time switched linear systems. In [5], interval estimation method is proposed for continuous-time switched linear systems in presence of unknown inputs. Authors in [4] propose an optimal interval observer based H_∞ performances for discrete-time linear switched systems. Recently, in [16], an H_∞ interval observer based state estimation approach is proposed for discrete-time linear switched systems. The most restrictive assumption in aforementioned works is to design a cooperative estimation error dynamics. However, it is not easy and even impossible to design a stable and cooperative error system. To remove the above restrictions, set-membership approach aims to provide compact sets containing the possible state values using predefined geometrical like ellipsoid, parallelotopes and zonotopes. The zonotope based interval estimation methods can achieve a good trade-off between estimation accuracy and computation complexity and have attracted attention of many researchers ([3, 18, 23]). Compared to interval estimation using interval observers, zonotopic techniques are less restrictive since they overcome the cooperativity problem. For instance, zonotopic set-membership state estimation approach for switched systems is proposed in [7]. A state estimation technique based on zonotopes for discrete-time switched systems was introduced in [11]. In [23], zonotope-based interval estimation is addressed for switched systems using Common Lyapunov approach. Unfortunately, such study has not yet been considered in the literature for switched systems, which motivates our research.

Motivated by the above observation, the present work proposes interval estimation for discrete-time switched systems using set-membership approach.The considered switched linear systems is subject to unknown but bounded uncertainties. The methodology proposed in this paper consists in using zonotopic technique to provide interval estimation. The proposed method integrates zonotopic analysis with robust observer design. The disturbances and the measurement noise are assumed to be unknown but bounded by known zonotopes. The proposed observer is designed using firstly the Luenberger structure and secondly the so-called T-N-L structure to help obtain point estimation. First, based on the Luenberger structure, our design extends the preliminary work developed in [23] where the interval estimation for discrete-time switched systems is introduced. The L_∞ observer in [23] is designed by adopting a Common Lyapunov Function (CLF). The most important enhancement with respect to [23] is the idea of using Multiple Lyapunov functions under an ADT condition to compute the observer gains, which reduce such conservatism caused by the CLF. Second, the observer is constructed based on the TNL structure (where T, N and L denote the weighting matrices and gain used in this strategy) adopting MLF under an ADT condition. For both observers, zonotopic analysis is used to help obtain the interval estimation of the state.

This chapter is organized as follows. Some preliminaries are introduced in Sect. 2. In Sect. 3, the problem statement is provided. Robust interval state estimation using both Luenberger and TNL structure are investigated in Sects. 4 and 5. In Sect. 6, simulation results are shown to illustrate the efficiency of the proposed method. Finally, the chapter is concluded in Sect. 7.

2 Preliminaries

In the sequel, the following standard notations and definitions are used. $\mathbb{N}$, $\mathbb{Z}$ and $\mathbb{R}$ denote respectively the sets of natural numbers, integers and real numbers. The set of positive real numbers and positive integers are denoted by $\mathbb{R}_+ = \{\tau \in \mathbb{R} : \tau \geq 0\}$ and $\mathbb{Z}_+ = \mathbb{Z} \cap \mathbb{R}_+$ respectively. $|x|$ denotes the elementwise absolute value of a vector $x \in R^n$. $\mathcal{L}_\infty$ denotes the set of all input u with the property $\|u\| < \infty$. $\overline{1, N}$ denotes the sequence of integers 1, ..., N. I_n denotes the identity matrix with dimension of $(n \times n)$. 0 represents a zero number, vector or matrix with appropriate dimension. $\underline{x}$ and $\overline{x}$ are the lower and upper bounds of a variable x such that $\underline{x} \leq x \leq \overline{x}$. The relation $P \prec 0$ ($P \succ 0$) indicates that $P = P^T$ is negative (positive) definite. The comparison operators $\leq$ and $\geq$ should be understood elementwise for vectors and matrices. The operators $\oplus$ and $\odot$ represent the Minkowski sum and the linear image operators, respectively. The asterisk $*$ denotes the symmetric term in a symmetric matrix. For a signal x, $\|x\|$ represents its Euclidean norm. The L_∞ norm of x is defined as $\|x\|_\infty = \sup_{k\geq 0} \|x_k\|$. Since zonotope techniques represent an important way to estimate the state variables of the system, it is important to recall some properties and definition of zonotope herein.

Definition 1 ([3]) An s-order zonotope $\mathrm{Z} \subset \mathbb{R}^n$ is an affine transformation of a hypercube $\mathrm{B}^s = \left[-1,\ 1\right]^s$ in $\mathbb{R}^n$ which can be expressed as follows:

$$\mathrm{Z} = \langle p, H \rangle = p + H\mathrm{B}^s = \left\{ z \in \mathbb{R}^n : z = p + Hb \right\}$$

where $p \in \mathbb{R}^n$ denotes the center of Z and $H \in \mathbb{R}^{n \times s}$ is called the generator matrix of Z. □

Definition 2 ([23]) For a set $\mathrm{Z} \subset \mathbb{R}^n$, its interval hull $\mathrm{Box}(Z)$ is defined as the smallest interval vector containing it, which is denoted as follows:

$$\mathrm{Z} \subseteq \mathrm{Box}(\mathrm{Z}) = \left[\underline{z}, \overline{z}\right] \tag{1}$$

where $\left[\underline{z}, \overline{z}\right] = \left\{z \in \mathrm{Z},\ \underline{z} < z < \overline{z}\right\}$ is an interval vector, $\underline{z}$ and $\overline{z}$ denote the lower and upper bounds of z. □

Property 1 ([3]) *Consider two sets* $\mathrm{Z}_1 = \langle p_1, H_1 \rangle$ *and* $\mathrm{Z}_2 = \langle p_2, H_2 \rangle$. *The Minkowski sum of* Z_1 *and* Z_2 *is given by:*

$$\mathrm{Z}_1 \oplus \mathrm{Z}_2 = \langle p_1 + p_2\,,\ [H_1 \quad H_2] \rangle\,, \tag{2}$$

where $p_1, p_2 \in \mathbb{R}^n$ *are known vectors and* $H_1, H_2 \in \mathbb{R}^{n \times s}$ *are determined matrices.* □

Property 2 ([3]) *The linear product of a zonotope* $\mathrm{Z} = \langle p, H \rangle$ *and a matrix* $K \in \mathbb{R}^{m \times n}$ *is denoted as* $\odot$ *and defined as:*

$$K \odot \mathrm{Z} = \langle Kp\,,\ KH \rangle\,. \tag{3}$$

□

Property 3 ([23]) *For a zonotope* $\mathrm{Z} = \langle p, H \rangle \subset \mathbb{R}^n$, *its interval hull* $\mathrm{Box}(\mathrm{Z}) = [\underline{z}, \overline{z}]$ *can be obtained by*

$$\begin{cases} \underline{z}_i = p_i - \sum_{j=1}^{s} \left|H_{i,j}\right|\,, & i = 1, \ldots, n \\ \overline{z}_i = p_i + \sum_{j=1}^{s} \left|H_{i,j}\right|\,, & i = 1, \ldots, n \end{cases}\,, \tag{4}$$

with s *is the column number of* H. *According to Definitions 1 and 2, the interval hull of* $\mathrm{Z} = \langle p, H \rangle$ *can also be computed by*

$$\mathrm{Z} = p \oplus H\mathrm{B}^s \subseteq \mathrm{Box}(\mathrm{Z}) = p \oplus \overline{H}\mathrm{B}^n, \tag{5}$$

where $\overline{H} \in \mathbb{R}^{n \times n}$ *is a diagonal matrix as follows:*

$$\overline{H} = \mathrm{Diag}\left(\left[\sum_{j=1}^{s} \left|H_{1,j}\right| \ldots \sum_{j=1}^{s} \left|H_{n,j}\right|\right]\right).$$

□

In the application of zonotopic analysis, the order of the zonotope linearly increases without the reduction operator. Therefore, a useful reduction method has been introduced in [3] in order to bound the order of the zonotope. The following property is proposed to describe this reduction operator.

Property 4 ([3]) *A high-dimensional zonotope* $Z \subset \mathbb{R}^n$ *can be bounded by a lower dimensional one via a reduction operator*

$$Z = \langle p, H \rangle \subseteq \langle p, \downarrow_l (H) \rangle \subseteq \text{Box}(Z), \quad n < l < s \tag{6}$$

where $p \in \mathbb{R}^n$ *and* $H \in \mathbb{R}^{n \times s}$ *are the center and the generation matrix of Z. l is the maximum number of columns of the generated matrix after reduction.* $\downarrow_l (H)$ *represents the complexity reduction operator with* $n < l < s$. □

Definition 3 ([9]) For a switching signal σ and any $0 \le k_a \le k_b$, $N_\sigma(k_a, k_b)$ denotes the number of discontinuities of the switching signal σ on the interval (k_a, k_b). If there exist a positive integer $N_0 \ge 0$ and a scalar $\tau_a > 0$ such that:

$$N_\sigma(k_a, k_b) \le N_0 + \frac{k_b - k_a}{\tau_a} \tag{7}$$

is satisfied $\forall\ 0 \le k_a \le k_b$, then τ_a is called an average dwell time for the switching signal σ. □

Lemma 1 ([24]) *Consider the discrete-time switched system*

$$\begin{cases} x_{k+1} = A_{\sigma_k} x_k + B_{\sigma_k} u_k + w_k \\ y_k = C_{\sigma_k} x_k + v_k \end{cases}, \sigma \in \overline{1, N} \tag{8}$$

Let $0 < \tau < 1$ *and* $\mu > 1$. *Suppose that there exist* $V_{\sigma(k)} : \mathbb{R}^n \to \mathbb{R}_+$ *and two* $\mathcal{K}_\infty$ *functions* c_1 *and* c_2

$$c_1(\|x_k\|) \le V_q(x_k) \le c_2(\|x_k\|) \tag{9}$$

$$\Delta V_q(x_k) \le -\lambda V_q(x_k) \tag{10}$$

$$V_q(x_k) \le \mu V_l(x_k) \tag{11}$$

then, the system (8) *is Input to State Stable (ISS) for any switching signal with an ADT satisfying*

$$\tau_a \ge \tau_a^* = -\frac{\ln(\mu)}{\ln(1 - \lambda)} \tag{12}$$

□

Lemma 1 is common in the literature when the ADT switching signal is considered.

Lemma 2 (Schur complement) *Given the matrices* $R = R^T$, $Q = Q^T$ *and S with appropriate dimensions. The following LMIs are equivalent:*

$$(i)\qquad \begin{bmatrix} Q & S \\ S^T & R \end{bmatrix} \succ 0.$$
$$(ii)\qquad R \succ 0;\ Q - SR^{-1}S^T \succ 0.$$
$$(iii)\qquad Q \succ 0;\ R - S^T Q^{-1} S \succ 0.$$ □

Lemma 3 ([17]) *Given three matrices* $A \in \mathbb{R}^{a\times b}$*,* $B \in \mathbb{R}^{b\times c}$ *and* $C \in \mathbb{R}^{a\times c}$*. If* $\mathrm{rank}(B) = c$*, then the general solution of the equation* $AB = C$ *is defined by:*

$$A = CB^{+} + S(I - BB^{+}) \tag{13}$$

where $S \in \mathbb{R}^{a\times b}$ *is an arbitrary matrix.* □

3 Problem Statement

Consider the following discrete-time switched linear system

$$\begin{cases} x_{k+1} = A_q x_k + B_q u_k + w_k \\ y_k = C x_k + v_k \end{cases} ,q \in \overline{1, N}, N \in \mathbb{N} \tag{14}$$

where $x \in \mathbb{R}^{n_x}$, $u \in \mathbb{R}^{n_u}$ and $y \in \mathbb{R}^{n_y}$ are the state vector, the input and the output, respectively. $w \in \mathbb{R}^{n_w}$ and $v \in \mathbb{R}^{n_v}$ are the unknown disturbances and the measurement noises. A_q, B_q and C are known constant matrices of appropriate dimensions. q is the index of the active subsystem and N is the number of subsystems. The switching signal is assumed to be known. The following assumption is considered.

Assumption 1 The initial state x_0, the disturbances w_k and the measurement noise v_k are supposed to be unknown but bounded by known zonotopes:

$$\begin{aligned} x_0 &\in \mathcal{X}_0 = \langle p_0 \,,\, H_0 \rangle \\ w_k &\in \mathcal{W} = \langle 0 \,,\, H_w \rangle \\ v_k &\in \mathcal{V} = \langle 0 \,,\, H_v \rangle \end{aligned} \tag{15}$$

where p_0 is a known vector, H_0, H_w and H_v are known diagonal matrices. □

The objective of this work is to calculate an interval vector $[\underline{x}_k, \overline{x}_k]$ containing the real state x_k such that

$$\underline{x}_k \leq x_k \leq \overline{x}_k\,, \quad k \in \mathbb{Z}_+$$

holds despite the disturbance.

In this chapter, we investigate the problem of interval estimation for the switched system (14) in presence of unknown but bounded disturbances. Zonotopic method is used to calculate zonotopes enclosing all the possible state trajectories. Firstly, the interval estimation of the state is achieved by integrating the Luenberger structure with zonotopic analysis. After, the so called TNL structure is combined with a zonotopic analysis to calculate the two bounds of the state. Compared to the first method, the proposed TNL structure allows providing more degree of freedom by introducing weighted matrices T_q and N_q. For both approaches, an L_∞ formalism is introduced in order to compute the observers gains taking into account the effect of state disturbances and measurement noise.

4 Robust Interval Estimation Based Luenberger Structure

In this section, interval state estimation of (14) is developed by integrating the robust observer design based Luenberger structure with the zonotopic analysis.

4.1 Robust Observer Design Based Luenberger Structure

This part proposes a robust observer design based on the Luenberger structure for the system (14). This idea has been developed in [23] based on L_∞ formalism to attenuate the effect of uncertainties where the observer gains are computed by adopting a common Lyapunov functions. However, in most cases, the existence of such a common Lyapunov function shared by all modes is not always guaranteed. Accordingly, to reduce this conservatism and improve the estimation accuracy, the observer gains are calculated in the following by adopting a multiple Lyapunov functions under an ADT concept.
Consider the following observer structure for (14):

$$\hat{x}_{k+1} = A_q \hat{x}_k + B_q u_k + L_q(y_k - C\hat{x}_k) \tag{16}$$

where $\hat{x}$ is the state estimation vector and $L_q \in \mathbb{R}^{n_x \times n_y}$ are the observer gains to be designed.
Define the following estimation error:

$$e_k = x_k - \hat{x}_k \tag{17}$$

Then the error dynamic system is given by:

$$e_{k+1} = (A_q - L_q C)e_k + w_k - L_q v_k \tag{18}$$

which can be rewritten as

$$e_{k+1} = \mathcal{A}_q e_k + \mathcal{D}_q d_k, \tag{19}$$

with

$$\mathcal{A}_q = A_q - L_q C, \ \ \mathcal{D}_q = \begin{bmatrix} I & -L_q \end{bmatrix}, \tag{20}$$

$$d_k = \left[w_k \ v_k \right]^T. \tag{21}$$

Next, the main goal is to compute the gains L_q that improve the observer performance and ensure robustness against disturbances. For this purpose, L_∞ technique is used to tune the observer matrices L_q ensuring uncertainties attenuation on the estimation error.

The following theorem is proposed to determine L_q such that the estimation error(19) is stable and satisfies the following L_∞ performance

$$\|e\| \leq \sqrt{\gamma \left(\lambda(1-\lambda)^k V_q(e_0) + \gamma \|d\|_\infty^2\right)} \tag{22}$$

Theorem 1 *Consider the system given by* (14), *assume that there exists a piecewise Lyapunov function* $V_q(e_k) = e_k^T P_q e_k$. *If there exist positive definite diagonal matrices* $P_q \in \mathbb{R}^{n_x \times n_x}$, *constant matrices* $M_l \in \mathbb{R}^{n_x \times n_x}$ *and* $W_q \in \mathbb{R}^{n_x \times n_y}$, $0 < c_1 < c_2$, $\rho > 0$, *for given* $\gamma > 0$, $0 < \beta < 1$ *and* $0 < \lambda < 1$ *then the error dynamics system in* (19) *is stable, satisfies the* L_∞ *performance and the following conditions*

$$\min_{P_q, M_l} \beta\mu + (1-\beta)\rho \quad q \in \overline{1..N} \tag{23}$$

$$c_1 I \leq P_q \leq c_2 I \tag{24}$$

$$\begin{bmatrix} (\lambda-1)P_q & * & * & * \\ 0 & -\rho I & * & * \\ 0 & 0 & -\rho I & * \\ P_q A_q - W_q C & P_q & -W_q & -P_q \end{bmatrix} \prec 0 \tag{25}$$

$$\begin{bmatrix} \lambda P_q & * & * \\ 0 & (\gamma-\rho)I & * \\ I_{n_x} & 0 & \gamma I \end{bmatrix} \succ 0 \tag{26}$$

$$\begin{bmatrix} M_l & P_q \\ P_q & P_q \end{bmatrix} \succeq 0 \tag{27}$$

are satisfied for all $q, l \in \overline{1..N}$, $q \neq l$ *where* $\mu = \frac{c_2}{c_1}$, $M_l = \mu P_l$ *and the average dwell time*

$$\tau_a \geq \tau_a^* = -\frac{\ln(\mu)}{\ln(1-\lambda)} \tag{28}$$

The observer gains L_q *are given by*

$$L_q = P_q^{-1} W_q. \tag{29}$$

In addition, the error (19) *satisfies:*

$$\lim_{k\to\infty} \|e_k\| < \frac{\rho}{c_1\lambda} \|d\|_\infty^2 \tag{30}$$

□

Proof 1 To prove the stability of the system (16) and the L_∞ performances, sufficient conditions are developed in terms of LMIs based on the multiple quadratic Lyapunov functions defined as

$$V_q(e_k) = e_k^T P_q e_k, \quad P_q = P_q^T > 0 \tag{31}$$

The time difference of V_k is given by

$$\begin{aligned}\Delta V_q(e_k) &= V_q(e_{k+1}) - V_q(e_k) \\ &= \begin{bmatrix} e_k \\ d_k \end{bmatrix}^T \begin{bmatrix} \mathcal{A}_q^T P_q \mathcal{A}_q - P_q & \mathcal{A}_q^T P_q \mathcal{D}_q \\ * & \mathcal{D}_q^T P_q \mathcal{D}_q \end{bmatrix} \begin{bmatrix} e_k \\ d_k \end{bmatrix}\end{aligned} \tag{32}$$

Bearing in mind that $P_q \mathcal{D}_q = \begin{bmatrix} P_q & -W_q \end{bmatrix}$ and $P_q A_q - W_q C_q = P_q \mathcal{A}_q$, the LMI (25) is equivalent to

$$\begin{bmatrix} (\lambda - 1)P_q & * & * \\ 0 & -\rho I & * \\ P_q\mathcal{A}_q & P_q\mathcal{D}_q & -P_q \end{bmatrix} \prec 0 \tag{33}$$

Pre-multiplying and post- multiplying (33) with

$$\begin{bmatrix} I & 0 & \mathcal{A}_q^T \\ 0 & I & \mathcal{D}_q^T \end{bmatrix} \tag{34}$$

and its transpose, we obtain:

$$\begin{bmatrix} \mathcal{A}_q^T P_q \mathcal{A}_q - P_q & \mathcal{A}_q^T P_q \mathcal{D}_q \\ * & \mathcal{D}_q^T P_q \mathcal{D}_q \end{bmatrix} + \begin{bmatrix} \lambda P_q & 0 \\ 0 & -\rho I \end{bmatrix} \prec 0 \tag{35}$$

Next, by pre- and post- multiplying (35) with $\begin{bmatrix} e_k^T & d_k^T \end{bmatrix}$ and its transpose, it follows that:

$$V_q(e_{k+1}) - V_q(e_k) < -\lambda V_q(e_k) + \rho d_k^T d_k \tag{36}$$

when $w_k = 0$ and $v_k = 0$, it implies that $d_k = 0$ and (36) is equivalent to

$$V_q(e_{k+1}) - V_q(e_k) < -\lambda V_q(e_k) < 0 \tag{37}$$

Then, the error system in (19) is stable when no uncertainties.

Moreover, inequality (36) can be rewritten as

$$V_q(e_{k+1}) < (1-\lambda)V_q(e_k) + \rho \left\|d\right\|_\infty^2 \tag{38}$$

which implies that

$$\begin{aligned} V_q(e_k) &\leq (1-\lambda)^k V_q(e_0) + \rho \sum_{\tau=0}^{k-1} (1-\lambda)^\tau \left\|d\right\|_\infty^2 \\ &\leq (1-\lambda)^k V_q(e_0) + \rho \frac{(1-\lambda^k)}{\lambda} \left\|d\right\|_\infty^2 \\ &\leq (1-\lambda)^k V_q(e_0) + \frac{\rho}{\lambda} \left\|d\right\|_\infty^2 \end{aligned} \tag{39}$$

Using the Schur complement in Lemma 2, (26) is equivalent to

$$\begin{bmatrix} \lambda P_q & * \\ 0 & (\gamma-\rho)I \end{bmatrix} - \frac{1}{\gamma}\begin{bmatrix} I \\ 0 \end{bmatrix}\begin{bmatrix} I & 0 \end{bmatrix} \succ 0 \tag{40}$$

Pre- and post- multiplying inequality (40) with $\begin{bmatrix} e_k^T & d_k^T \end{bmatrix}$ and its transpose, we obtain

$$e_k^T e_k \leq \gamma \left(\lambda V_q(e_k) + (\gamma-\rho) \left\|d\right\|_\infty^2\right) \tag{41}$$

Substituting (39) into (41) yields

$$\begin{aligned} e_k^T e_k &\leq \gamma \left(\lambda \left((1-\lambda)^k V_q(e_0) + \frac{\rho}{\lambda} \left\|d\right\|_\infty^2\right)\right) \\ &\quad + \gamma \left((\gamma-\rho) \left\|d\right\|_\infty^2\right) \\ &\leq \gamma \left(\lambda(1-\lambda)^k V_q(e_0) + \gamma \left\|d\right\|_\infty^2\right) \end{aligned} \tag{42}$$

Consequently, the L_∞ criterion (22) holds.

Bearing in mind (39) and based on (9) in Lemma 1, the following inequality

$$c_1 \left\|e_k\right\| \leq V_q(e_k) \tag{43}$$

yields

$$\left\|e_k\right\| \leq \frac{1}{c_1}((1-\lambda)^k V_q(e_0) + \frac{\rho}{\lambda} \left\|d_k\right\|_\infty^2) \tag{44}$$

when $k \to \infty$, one can deduce that

$$\lim_{k\to\infty} \left\|e_k\right\| < \frac{\rho}{c_1\lambda} \left\|d_k\right\|_\infty^2 \tag{45}$$

Inequality (45) shows that the estimation error is bounded by a known constant term $\frac{\rho}{c_1\lambda}\left\|d_k\right\|_\infty^2$. For a given c_1 and λ, this bound depends only on ρ which should be optimized as possible to enhance the estimation accuracy. Consequently, optimizing ρ improves the estimation accuracy. On the other hand, we aim to look for an optimum dwell time, this optimum can be obtained by minimizing the variable μ. for a solution of both problems, we add the following objective function to the LMIs conditions:

$$\beta\mu + (1-\beta)\rho, \quad \beta \in \begin{bmatrix} 0 & 1 \end{bmatrix}$$

In addition, the stability at the switching instants is guaranteed based on (11), which yields

$$\mu P_l - P_q \succeq 0 \tag{46}$$

where $q, l \in \overline{1..N}$, $q \neq l$, q is the current mode.
Based on the Schur complement, we obtain

$$\begin{bmatrix} \mu P_l & I \\ I & P_q^{-1} \end{bmatrix} \succeq 0 \tag{47}$$

Multiplying the both sides of (47) by $\begin{bmatrix} I & 0 \\ 0 & P_q \end{bmatrix}$, one can deduce that (27) holds with $M_l = \mu P_l$.

To conclude, all conditions in Theorem 1 are satisfied, which prove that the estimation error dynamics system in (19) is stable and satisfies the L_∞ performance. □

4.2 Zonotopic State Estimation

Note that $\hat{x}$ is a single point estimation of x_k and cannot give any information of the state bounds. Therefore, in this part, we aim to achieve the interval estimation of the state by using zonotopic techniques. We begin by computing an approximation set of the error e_k given in (18) using known zonotopic bounds of state and output disturbances provided in Assumption 1, and then we simply deduce the bounds of the state x_k. From (17), we have

$$x_k = \hat{x}_k + e_k \tag{48}$$

Note that $\hat{x}_k$ can be directly calculated by the observer (16), then, if one can find $\overline{e}_k$ and $\underline{e}_k$ such that $\underline{e}_k \leq e_k \leq \overline{e}_k$, from (48), the interval estimation of x_k can be obtained as follows:

$$\begin{cases} \overline{x}_k = \hat{x}_k + \overline{e}_k \\ \underline{x}_k = \hat{x}_k + \underline{e}_k \end{cases} \tag{49}$$

where $\overline{x}$ and $\underline{x}$ are the upper and the lower bounds of the state x.

Accordingly, we begin by computing the interval estimation of the error e_k given in (18), and then we deduce the two bounds of the state vector x_k. The interval estimation of e_k can be calculated by using the following theorem.

Theorem 2 *For the system* (19)*, given* $x_0 \in \mathcal{X}_0 = \langle \hat{x}_0, H_0 \rangle$*. Assume that* $e_k \in Z_k^e = \langle 0, H_k^e \rangle$*. If Assumption 1 is satisfied,then there exists a zonotope* Z_{k+1}^e *such that* $e_{k+1} \in Z_{k+1}^e = \langle 0, H_{k+1}^e \rangle$ *where*

$$H_{k+1}^e = \left[(A_q - L_q C_q) \downarrow_q (H_k^e) \;\; H_w \;\; - L_q H_v \right] \tag{50}$$

□

Proof 2 Since at the step k, $e_k \in Z_k^e$, the disturbances and the measurement noise satisfy $w_k \in \mathcal{W}$, $v_k \in \mathcal{V}$ and $v_{k+1} \in \mathcal{V}$, then, by using (18) and (15), $e_{k+1} \in Z_{k+1}^e = \langle 0, H_{k+1}^e \rangle$ with

$$Z_{k+1}^e = (A_q - L_q C_q) \odot Z_k^e \oplus \mathcal{W} \oplus (-L_q) \odot \mathcal{V}. \tag{51}$$

Thus,

$$\langle 0, H_{k+1}^e \rangle = (A_q - L_q C_q) \odot \langle 0, H_k^e \rangle \oplus \langle 0, H_w \rangle \oplus (-L_q) \odot \langle 0, H_v \rangle. \tag{52}$$

Using the properties of zonotopes, one can get

$$H_{k+1}^e = \left[(A_q - L_q C_q)(H_k^e) \;\; H_w \;\; - L_q H_v \right] \tag{53}$$

By using the reduction operator, we have $\langle 0, H_k^e \rangle \subseteq \langle 0, \downarrow q(H_k^e) \rangle$. Thus, by replacing the matrix H_k^e by the limit $\downarrow q(H_k^e)$ in (53), we obtain (50).

As detailed in Sect. 2, the column number of the matrix H_k^e should be limited to a fixed one by using the reduction operator $\downarrow_q(.)$. By using this operator, we have $\langle 0, H_k^e \rangle \subseteq \langle 0, \downarrow q(H_k^e) \rangle$. Thus, by replacing the matrix H_k^e by the limit $\downarrow q(H_k^e)$ in (53), we obtain (50).

By using Property 3, the interval estimation of x_k can be computed as follows:

$$\begin{cases} \underline{x}_k(i) = \hat{x}_k(i) - \sum_{j=1}^{n_h} \left| H_k^e(i, j) \right| , i = 1...n_x \\ \overline{x}_k(i) = \hat{x}_k(i) + \sum_{j=1}^{n_h} \left| H_k^e(i, j) \right| , i = 1...n_x \end{cases}, \tag{54}$$

with n_h is the column number of H_k^e. □

5 Robust Interval Estimation Based TNL Structure

In this section, interval state estimation of (14) is introduced by integrating the TNL observer design with the zonotopic analysis. The proposed method can reduce the conservatism of gain matrices L_q and provide more degree of freedom by introducing weighted matrices T_q and N_q.

5.1 Robust Observer Design Based TNL Structure

This part proposes a robust observer design based on the TNL structure introduced in the preliminary work [21] for the system (14). Consider the following observer structure for the system (14):

$$\begin{cases} \xi_{k+1} = T_q A_q \hat{x}_k + T_q B_q u_k + L_q(y_k - C\hat{x}_k) \\ \hat{x}_k = \xi_k + N_q y_k, \end{cases} \tag{55}$$

where ξ_k is an intermediate variable and $\hat{x}_k$ denotes the point state estimation. The matrices $L_q \in \mathbb{R}^{n_x \times n_y}$ are the observer gains to be computed later. The weighted matrices T_q and N_q are constant matrices to be computed satisfying the following condition:

$$T_q + N_q C = I. \tag{56}$$

The relation (56) can be rewritten as follows

$$\begin{bmatrix} T_q & N_q \end{bmatrix} \begin{bmatrix} I \\ C \end{bmatrix} = I. \tag{57}$$

Based on Lemma 3, the general solution of (56) is given by

$$\begin{bmatrix} T_q & N_q \end{bmatrix} = \begin{bmatrix} I \\ C \end{bmatrix}^+ + S_q \left(I - \begin{bmatrix} I \\ C \end{bmatrix} \begin{bmatrix} I \\ C \end{bmatrix}^+ \right). \tag{58}$$

Remark The major difference between the observer (55) and the observer (16) is the presence of additional matrices T_q and N_q. Note that if we choose $T_q = I$ and $N_q = 0$, observer (55) reduces to the observer in (16). In fact, by introducing these matrices, the proposed method provides more design degrees of freedom. □

Define the estimation error as

$$e_k = x_k - \hat{x}_k. \tag{59}$$

Using equation (56), from (14), the state x_k satisfies

$$x_{k+1} = T_q A_q x_k + T_q B_q u_k + T_q w_k + N_q C x_{k+1} \tag{60}$$

It follows that the error dynamics are then

$$e_{k+1} = (T_q A_q - L_q C)e_k + T_q w_k - L_q v_k - N_q v_{k+1} \tag{61}$$

To facilitate the observer design, (61) is rewritten as

$$e_{k+1} = \mathcal{F}_q e_k + \mathcal{G}_q d_k, \tag{62}$$

with

$$\mathcal{F}_q = T_q A_q - L_q C, \ \ \mathcal{G}_q = \begin{bmatrix} I \ L_q \ N_q \end{bmatrix}, \tag{63}$$

$$d_k = \begin{bmatrix} T_q w_k & -v_k & -v_{k+1} \end{bmatrix}^T. \tag{64}$$

Then, an L_∞ approach is used to attenuate the effect of disturbances and measurement noise on the estimation error. The following theorem provides a sufficient condition to determine the weighting matrices T_q and N_q and the gains L_q of the observer (55) such that the estimation error is robust against the disturbances.

Theorem 3 *Consider the system given by* (14)*, assume that there exists a piecewise Lyapunov function* $V_q(e_k) = e_k^T P_q e_k$. *If there exist positive definite diagonal matrices* $P_q \in \mathbb{R}^{n_x \times n_x}$, *constant matrices* $M_l \in \mathbb{R}^{n_x \times n_x}$, $W_q \in \mathbb{R}^{n_x \times n_y}$ *and* $Y_q \in \mathbb{R}^{n_x \times (n_x + n_y)}$, $0 < c_1 < c_2$, $\rho > 0$, $\gamma > 0$ *for given* $0 < \lambda < 1$ *and* $0 < \beta < 1$, *then the error dynamics system in* (62) *is stable, satisfies the* L_∞ *performance* (22) *and the following conditions*

$$\min_{P_q, M_l} \beta\mu + (1-\beta)\rho \quad q \in \overline{1..N} \tag{65}$$

$$c_1 I \leq P_q \leq c_2 I \tag{66}$$

$$\begin{bmatrix} (\lambda-1)P_q & * & * & * & * \\ 0 & -\rho I & * & * & * \\ 0 & 0 & -\rho I & * & * \\ 0 & 0 & 0 & -\rho I & * \\ \Omega_{1q} & P & W_q & \Omega_{2q} & -P_q \end{bmatrix} \prec 0 \tag{67}$$

$$\begin{bmatrix} \lambda P_q & * & * \\ 0 & (\gamma-\rho)I & * \\ I & 0 & \gamma I \end{bmatrix} \succ 0 \tag{68}$$

$$\begin{bmatrix} M_l & P_q \\ P_q & P_q \end{bmatrix} \succeq 0 \tag{69}$$

are satisfied for all $q, l \in \overline{1..N}$, $q \neq l$, *where*
$\Omega_{1q} = P_q\Theta^+\alpha_1 A_q + Y_q\psi\alpha_1 A_q - W_q C^T$
$\Omega_{2q} = P_q\Theta^+\alpha_2 + Y_q\psi\alpha_2$
$\alpha_1 = \begin{bmatrix} I \\ 0 \end{bmatrix}$, $\alpha_2 = \begin{bmatrix} 0 \\ I \end{bmatrix}$, $\Theta = \begin{bmatrix} I \\ C \end{bmatrix}$, $\psi = I - \Theta\Theta^+$.
In addition, if the state x *is bounded, then* $\hat{x}$ *is also bounded under a switching signal with an average dwell time*

$$\tau_a \geq \tau_a^* = -\frac{\ln(\mu)}{\ln(1-\lambda)} \tag{70}$$

where $\mu = \frac{c_2}{c_1}$ *and the observer gains* L_q, T_q *and* N_q *are given by*

$$\begin{cases} L_q = P_q{}^{-1} W_q \\ T_q = \Theta^+\alpha_1 + P_q{}^{-1} Y_q\psi\alpha_1 \\ N_q = \Theta^+\alpha_2 + P_q{}^{-1} Y_q\psi\alpha_2 \end{cases} \tag{71}$$

Moreover, the error (62) *satisfies:*

$$\lim_{k\to\infty} \|e_k\| < \frac{\rho}{c_1\lambda} \|d\|_\infty^2 \tag{72}$$

□

Proof 3 To prove the stability of the observation error (62) and the L_∞ performance, we consider the multiple Lyapunov functions given by $V_q(e_k) = e_k^T P_q e_k$. The increment of V_k is given by

$$\begin{aligned} \Delta V_q(e_k) &= V_q(e_{k+1}) - V_q(e_k) \\ &= \begin{bmatrix} e_k \\ d_k \end{bmatrix}^T \begin{bmatrix} \mathcal{F}_q^T P_q \mathcal{F}_q - P_q & \mathcal{F}_q^T P_q \mathcal{G}_q \\ * & \mathcal{G}_q^T P_q \mathcal{G}_q \end{bmatrix} \begin{bmatrix} e_k \\ d_k \end{bmatrix} \end{aligned} \tag{73}$$

If inequality (67) holds, so it can be rewritten

$$\begin{bmatrix} (\lambda-1)P_q & * & * \\ 0 & -\rho I & * \\ P_q\mathcal{F}_q & P_q\mathcal{G}_q & -P_q \end{bmatrix} \prec 0 \tag{74}$$

Pre- and post- multiplying (74) with $\begin{bmatrix} I & 0 & \mathcal{F}_q^T \\ 0 & I & \mathcal{G}_q^T \end{bmatrix}$ and its transpose, we can get:

$$\begin{bmatrix} \mathcal{F}_q^T P_q \mathcal{F}_q - P_q & \mathcal{F}_q^T P_q \mathcal{G}_q \\ * & \mathcal{G}_q^T P_q \mathcal{G}_q \end{bmatrix} + \begin{bmatrix} \lambda P_q & 0 \\ 0 & -\rho I \end{bmatrix} \prec 0 \tag{75}$$

In addition, by pre- and post- multiplying (75) with $\begin{bmatrix} e_k^T & d_k^T \end{bmatrix}$ and its transpose, it follows that:

$$V_q(e_{k+1}) - V_q(e_k) < -\lambda V_q(e_k) + \rho d_k^T d_k \tag{76}$$

when $w_k = 0$ and $v_k = 0$, it implies that $d_k = 0$ and (76) is equivalent to

$$V_q(e_{k+1}) - V_q(e_k) < -\lambda V_q(e_k) < 0 \tag{77}$$

Thus, the error system in (62) is stable when no uncertainties.

Furthermore, inequality (76) can be rewritten as

$$V_q(e_{k+1}) < (1-\lambda)V_q(e_k) + \rho \|d\|_\infty^2 \tag{78}$$

which implies that

$$\begin{aligned} V_q(e_k) &\leq (1-\lambda)^k V_q(e_0) + \rho \sum_{\tau=0}^{k-1} (1-\lambda)^\tau \|d\|_\infty^2 \\ &\leq (1-\lambda)^k V_q(e_0) + \rho \frac{(1-\lambda^k)}{\lambda} \|d\|_\infty^2 \\ &\leq (1-\lambda)^k V_q(e_0) + \frac{\rho}{\lambda} \|d\|_\infty^2 \end{aligned} \tag{79}$$

LMIs condition in (65), (66), (68) and (69) have been proven in Proof 1. Thus, all conditions in Theorem 3 are satisfied, which prove that the estimation error dynamics system in (62) is stable and satisfies the L_∞ performance. □

5.2 *Zonotopic State Estimation*

Based on the designed observer (55), interval estimation of the state x_k can be achieved based on the zonotopic techniques. We begin by computing the lower and upper bounds of the error e_k given in (61), next, we deduce the lower and the upper bounds of x_k. From (59), we have

$$x_k = \hat{x}_k + e_k \tag{80}$$

Note that $\hat{x}_k$ can be directly calculated by the observer (55), then, if one can find $\overline{e}_k$ and $\underline{e}_k$ such that $\underline{e}_k \leq e_k \leq \overline{e}_k$, from (80), the interval estimation of x_k can be obtained as follows:

$$\begin{cases} \overline{x}_k = \hat{x}_k + \overline{e}_k \\ \underline{x}_k = \hat{x}_k + \underline{e}_k \end{cases} \tag{81}$$

Let us introduce the following theorem in order to compute the interval estimation of the error (61).

Theorem 4 *For the system* (62)*, given* $x_0 \in \mathcal{X}_0 = \langle \hat{x}_0,\ H_0 \rangle$*. Assume that* $e_k \in Z_k^e = \langle 0,\ H_k^e 0,\ H_{k+1}^e \rangle$ *where*

$$H_{k+1}^e = \left[(T_q A_q - L_q C) \downarrow_q (H_k^e)\ T_q H_w\ -L_q H_v\ -N_q H_v \right] \tag{82}$$

□

Proof 4 Considering the same reasoning in Proof 2. Since at the step k, $e_k \in Z_k^e$, the disturbances and the measurement noise satisfy $w_k \in \mathcal{W}$, $v_k \in \mathcal{V}$ and $v_{k+1} \in \mathcal{V}$, using (61) and (15), $e_{k+1} \in Z_{k+1}^e = \langle 0,\ H_{k+1}^e \rangle$ with

$$Z_{k+1}^e = (T_q A_q - L_q C) \odot Z_k^e \oplus T_q \odot \mathcal{W} \oplus (-L_q) \odot \mathcal{V} \oplus (-N_q) \odot \mathcal{V}.$$

Thus,

$$\begin{aligned} \langle 0,\ H_{k+1}^e \rangle = (T_q A_q - L_q C) \odot \langle 0,\ H_k^e \rangle \oplus T_q \odot \langle 0,\ H_w \rangle \oplus (-L_q) \odot \langle 0,\ H_v \rangle \\ \oplus (-N_q) \odot \langle 0,\ H_v \rangle . \end{aligned} \tag{83}$$

Using the properties of zonotopes, one can get

$$H_{k+1}^e = \left[(T_q A_q - L_q C)(H_k^e)\ \ T_q H_w\ \ -L_q H_v\ \ -N_q H_v \right]. \tag{84}$$

By using the reduction operator, we have $\langle 0,\ H_k^e \rangle \subseteq \langle 0, \downarrow q(H_k^e) \rangle$. Thus, by replacing the matrix H_k^e by the limit $\downarrow q(H_k^e)$ in (84), we obtain (82).

By using Property 3, the interval estimation of x_k can be calculated as

$$\begin{cases} \underline{x}_k(i) = \hat{x}_k(i) - \sum_{j=1}^{n_h} \left| H_k^e(i,j) \right| ,\ i = 1...n_x \\ \overline{x}_k(i) = \hat{x}_k(i) + \sum_{j=1}^{n_h} \left| H_k^e(i,j) \right| ,\ i = 1...n_x \end{cases}, \tag{85}$$

with n_h is the column number of H_k^e. □

6 Simulation Results

To illustrate the proposed approach, we consider the discrete-time linear switched system with two subsystems, $N = 2$, described by:

$$\begin{cases} x_{k+1} = A_q x_k + B_q u_k + w_k \\ y_k = C_q x_k + v_k \end{cases} \quad q \in \overline{1, N},\ N \in \mathbb{N} \tag{86}$$

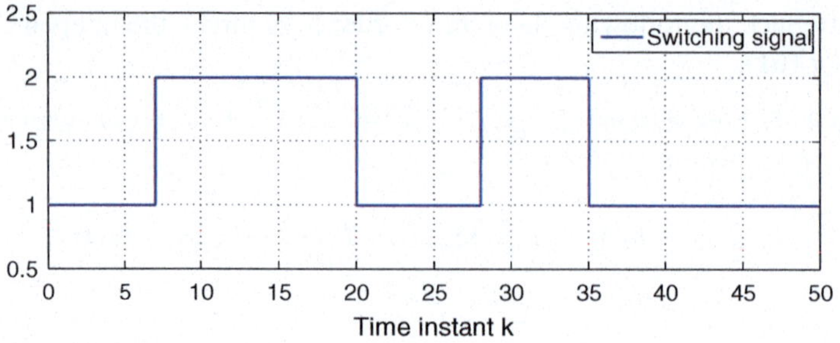

Fig. 1 Switching signal

where

$$A_1 = \begin{bmatrix} 0.6 & 0.5 & 1 \\ 0.2 & 0 & 1 \\ 0 & 0.2 & 0 \end{bmatrix}, A_2 = \begin{bmatrix} 0.9 & -0.8 & 0 \\ 0.1 & 0 & 1 \\ 0 & 0.5 & 0.1 \end{bmatrix}$$

$$B_1 = \begin{bmatrix} 1 \\ 0.1 \\ 1.3 \end{bmatrix}, \quad B_2 = \begin{bmatrix} 0.1 \\ 1 \\ 1 \end{bmatrix}, \quad C = \begin{bmatrix} 1.2 & 0.01 & 0 \\ 0.1 & 1.1 & 0.1 \end{bmatrix}.$$

In order to verify the performance of the proposed results, w_k and v_k are uniformly distributed bounded signals such that $H_w = 0.1$ and $H_v = 0.1$. According to Assumption 1, we select $p_0 = \begin{bmatrix} 0 & 0 & 0 \end{bmatrix}^T$ and $H_0 = I_3$. In addition, to limit the column number of the generator matrix, we set the reduction order of $\downarrow_q (H_k^e)$ as $q = 5$. The evolution of the switching signal is plotted in Fig. 1. It indicates at each instant the active subsystem of the discrete-time switched system.

In the following, the problem of state estimation is considered. Simulation results was developed by using Matlab optimization tools (Yalmip/Sedumi).

6.1 State Estimation Based Luenberger Structure

In this part, interval estimation based on the zonotopic techniques is considered. The optimization problem in (23) can be solved via the software Yalmip/Sedumi (MATLAB). Note that the existence of a feasible solution for the conditions fulfilled in Theorem 1 allows us to improve the accuracy of interval estimation and to provide an optimum dwell time. By solving the LMIs in Theorem 1, the gains L_q are given as:

$$\mathrm{P}_1 = \begin{bmatrix} 0.1 & 0 & 0 \\ 0 & 0.1 & 0 \\ 0 & 0 & 0.4683 \end{bmatrix}, \mathrm{P}_2 = \begin{bmatrix} 0.3283 & 0 & 0 \\ 0 & 0.1354 & 0 \\ 0 & 0 & 0.4306 \end{bmatrix}$$

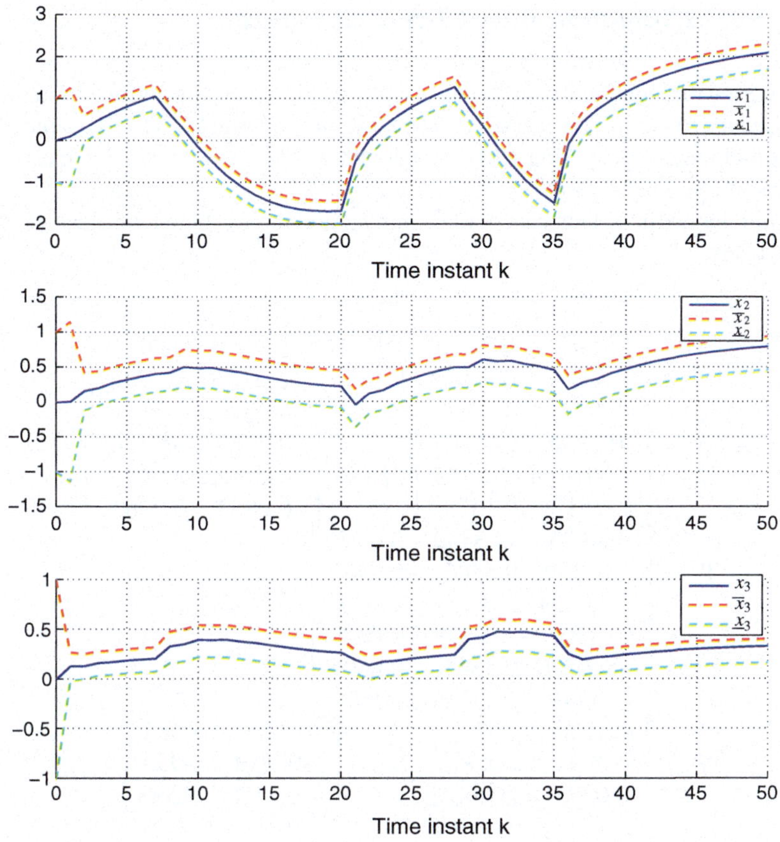

Fig. 2 States x_k and their estimated bounds

$$L_1 = \begin{bmatrix} 0.4533 & 0.4583 \\ 0.1607 & 0.0140 \\ -0.0145 & 0.1776 \end{bmatrix}, L_2 = \begin{bmatrix} 0.7574 & -0.7077 \\ 0.0744 & 0.0249 \\ -0.0346 & 0.4427 \end{bmatrix}$$

In the simulation, $\mu = 4.6827$ which leads to an ADT $\tau_a > 2.2273$. The switching signal plotted in Fig. 1 verifies the average dwell time condition.
Next, interval estimation can be calculated using the zonotopic analysis presented in Theorem 2. The simulation results are presented in Fig. 2. Solid and dotted lines correspond respectively to the real state and their estimated bounds. From these figures, it can be seen that the state vector belongs into an interval vector described by its upper and lower bounds. It follows that, the proposed zonotopic based interval estimation has good performance to estimate the system state. Also, the L_∞ performance applied in the observer design allows ensuring estimation accuracy.

6.2 State Estimation Based T-N-L Structure

Considering the same switching signal used in the first case, the optimization problem proposed in (65) is solved via the software Yalmip/Sedumi(MATLAB). Note that the solution obtained by solving LMIs developed in Theorem 3 provides more design degrees of freedom.

Using the LMIs in Theorem 3, the gains L_q and the weighted matrices T_q and N_q are given as:

$$\mathrm{P}_1 = \begin{bmatrix} 0.102 & 0 & 0 \\ 0 & 0.102 & 0 \\ 0 & 0 & 0.1511 \end{bmatrix}, \mathrm{P}_2 = \begin{bmatrix} 0.1 & 0 & 0 \\ 0 & 0.1 & 0 \\ 0 & 0 & 0.1233 \end{bmatrix}$$

$$\mathrm{L}_1 = \begin{bmatrix} 0.1332 & 0.1707 \\ -0.0803 & -0.1455 \\ -0.0179 & 0.1328 \end{bmatrix}, L_2 = \begin{bmatrix} 0.2765 & -0.2445 \\ -0.0312 & 0.0292 \\ 0.0852 & 0.2660 \end{bmatrix},$$

$$\mathrm{T}_1 = \begin{bmatrix} 0.4633 & -0.3429 & -0.0308 \\ -0.3998 & 0.5506 & -0.0406 \\ -0.0324 & 0.0325 & 1.0030 \end{bmatrix},$$

$$\mathrm{T}_2 = \begin{bmatrix} 0.4202 & -0.0326 & -0.0025 \\ -0.0756 & 0.2345 & -0.0696 \\ 0.1749 & -0.0278 & 0.9973 \end{bmatrix},$$

$$\mathrm{N}_1 = \begin{bmatrix} 0.4216 & 0.3079 \\ 0.2994 & 0.4058 \\ 0.0295 & -0.0298 \end{bmatrix}, \mathrm{N}_2 = \begin{bmatrix} 0.8410 & 0.0253 \\ -0.0050 & 0.6959 \\ -0.1480 & 0.0266 \end{bmatrix}$$

In the simulation, $\mu = 2$ which leads to an ADT $\tau_a > 0.1637$. The switching signal plotted in Fig. 1 verifies the average dwell time condition.

Simulation results are depicted in Fig. 3. From these figures, the TNL observer combined with a zonotopic technique ensure the interval estimation of the state.

Remark It should be noted that, the two proposed techniques are efficient to solve the interval state estimation problem. Meanwhile, compared to the Luenberger structure, the method based on the T-N-L structure provide more degree of freedom to design the observer gains. This can be explained by the introduction of the weighted matrices T_q and N_q. □

7 Conclusion

In this chapter, we have proposed a zonotope based interval estimation for discrete-time switched linear systems subject to unknown but bounded disturbances and measurement noise in predefined zonotopes. A novel interval estimation approach is proposed via the robust observer design and zonotopic analysis techniques. Based firstly

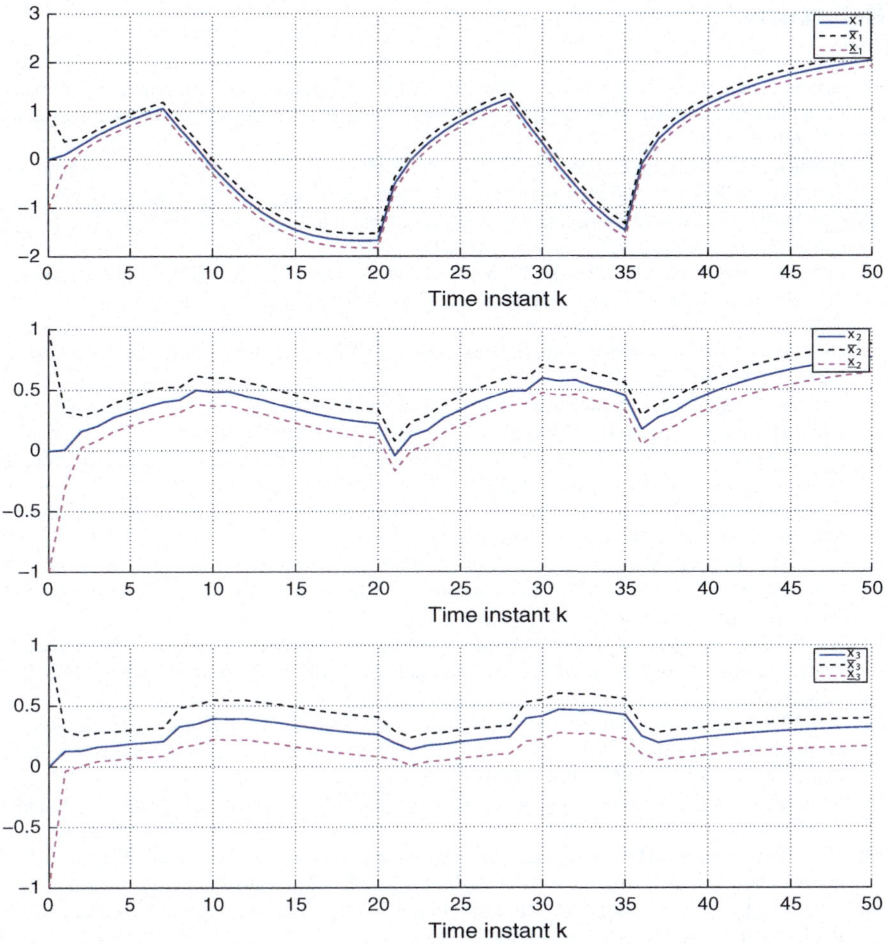

Fig. 3 States x_k and their estimated bounds

on the Luenberger structure and secondly on the TNL structure, a robust observer is designed for state point estimation. Compared to the classical Luenberger structure, the TNL approach allows providing more degree of freedom by introducing weighted matrices. Stability conditions of the proposed observers are expressed in terms of LMIs using multiple quadratic Lyapunov functions. L_∞ formalism is introduced to attenuate the effect of uncertainties. Following that, for both observers, interval state estimation is obtained using zonotopic analysis. Finally, simulation results are presented to show the validity of the proposed results. Extending to more general classes of discrete-time switched systems and considering unknown switching are perspectives for future work.

References

1. Arichi, F., Djemai, M., Cherki, B., Manamanni, N.: Continuous and discrete state estimation for a class of nonlinear switched systems. IEEE Trans. Circuits Syst. II Express Briefs **62**(7), 691–695 (2015)
2. Birouche, A., Daafouz, J., Iung, C.: Observer design for a class of discrete time piecewise-linear systems. In: Analysis and Design of Hybrid Systems 2006, pp. 12–17. Elsevier (2006)
3. Combastel, C.: A state bounding observer based on zonotopes. In: 2003 European Control Conference (ECC), pp. 2589–2594. IEEE (2003)
4. Dinh, T.N., Marouani, G., Raïssi, T., Wang, Z., Messaoud, H.: Optimal interval observers for discrete-time linear switched systems. Int. J. Control **93**(11), 2613–2621 (2020)
5. Ethabet, H., Raïssi, T., Amairi, M., Combastel, C., Aoun, M.: Interval observer design for continuous-time switched systems under known switching and unknown inputs. Int. J. Control **93**(5), 1088–1101 (2020)
6. Ethabet, H., Raissi, T., Amairi, M., Aoun, M.: Interval observers design for continuous-time linear switched systems. IFAC-PapersOnLine **50**(1), 6259–6264 (2017)
7. Fei, Z., Yang, L., Sun, X-M., Ren, S.: Zonotopic set-membership state estimation for switched systems with restricted switching. IEEE Trans. Autom. Control (2021)
8. Gouzé, J-L., Rapaport, A., Hadj-Sadok, M.Z.: Interval observers for uncertain biological systems. Ecol. Model. **133**(1–2), 45–56 (2000)
9. Hespanha, J.P., Stephen Morse, A.: Stability of switched systems with average dwell-time. In: Proceedings of the 38th IEEE Conference on Decision and Control (Cat. No. 99CH36304), vol. 3, pp. 2655–2660. IEEE (1999)
10. Hu, B., Zhai, G., Michel, A.N.: Common quadratic lyapunov-like functions with associated switching regions for two unstable second-order lti systems. Int. J. Control **75**(14), 1127–1135 (2002)
11. Ifqir, S., Puig, V., Ichalal, D., Ait-Oufroukh, N., Mammar, S.: Zonotopic set-membership estimation for switched systems based on wi-radius minimization: vehicle application. IFAC-PapersOnLine **53**(2), 7446–7451 (2020)
12. Jiang, Z.-P., Wang, Y.: Input-to-state stability for discrete-time nonlinear systems. Automatica **37**(6), 857–869 (2001)
13. Li, J., Pan, K., Zhang, D., Qingyu, S.: Robust fault detection and estimation observer design for switched systems. Nonlinear Anal. Hybrid Syst **34**, 30–42 (2019)
14. Liberzon, D., Stephen Morse, A.: Basic problems in stability and design of switched systems. IEEE Control Systems Mag. **19**(5), 59–70 (1999)
15. Marouani, G., Dinh, T.N., Rassï, T., Messaoud, H.: Interval observers design for discrete-time linear switched systems. In: 2018 European Control Conference (ECC), pp. 801–806. IEEE (2018)
16. Marouani, G., Dinh, T.N., Wang, Z., Raïssi, T., Messaoud, H.: Interval observers design for discrete-time linear switched systems. In: Complex Systems: Spanning Control and Computational Cybernetics: Foundations, pp. 245–268. Springer (2022)
17. Rao, C.R., Mitra, S.K. et al.: Generalized inverse of a matrix and its applications. In: Proceedings of the Sixth Berkeley Symposium on Mathematical Statistics and Probability, vol. 1, pp. 601–620. University of California Press Oakland, CA, USA (1972)
18. Sehli, N., Wang, Z., Taarit, K.I., Raïssi, T., Ksouri, M.: Interval estimation for linear discrete-time delay systems. IFAC-PapersOnLine **53**(2), 4798–4803 (2020)
19. Serres, U., Vivalda, J.-C., Riedinger, P.: On the convergence of linears witched systems. IEEE Trans. Autom. Control **56**(2), 320–332 (2010)
20. Wang, D., Wang, Z., Li, G., Wang, W.: Distributed filtering for switched nonlinear positive systems with missing measurements over sensor networks. IEEE Sens. J. **16**(12), 4940–4948 (2016)
21. Wang, Z., Lim, C.-C., Shen, Y.: Interval observer design for uncertain discrete-time linear systems. Syst. Control Lett. **116**, 41–46 (2018)

22. Yin, Y., Zhao, X., Zheng, X.: New stability and stabilization conditions of switched systems with mode-dependent average dwell time. Circuits Syst. Signal Process. **36**(1), 82–98 (2017)
23. Zhang, W., Wang, Z., Raïssi, T., Dinh, T.N., Dimirovski, G.: Zonotope-based interval estimation for discrete-time linear switched systems. In: 21st IFAC World Congress (2020)
24. Zhu, K., Song, Y., Ding, D., Wei, G., Liu, H.: Robust mpc under event-triggered mechanism and round-robin protocol: an average dwell-time approach. Inf. Sci. **457**, 126–140 (2018)

Stability and Stabilisation of Nonlinear Incommensurate Fractional Order Difference Systems

Noureddine Djenina and Adel Ouannas

Abstract In this work, we will discuss the stability of discrete incommensurate fractional order discrete systems, and then we will apply it in the stabilization of two chaotic systems, the predator-prey system and the cancer growth system.

Keywords Stability · Stabilisation · Chaos · Control

1 Introduction

Fractional calculus is a significant tool for mathematical modeling of numerous issues in physics and mathematics [1–3]. Discrete fractional calculus has drawn the interest of a great number of researchers during the last several years, and they have been increasingly interested in its potential applications in neural networks, secure communication, biology, and other domains [4–29]. Recently, numerous different dynamics, including chaos, hyperchaos and coexisting attractors in fractional-order systems, have been explored.

Stability theory addresses the stability of solutions of difference (differential) equations and of trajectories of dynamical systems under small perturbations of initial conditions [30]. Its importance appears in many areas, and among these areas is synchronization and control [31–110]. Many theorems have been developed to test the stability and the asymptotic stability of fractional order discrete systems in the case of commensurate order and the case of incommensurate order [111, 112].

Recently, the behavior of many fractional order discrete systems has been studied, and these systems have shown that they often behave chaotically [113–155].

N. Djenina (✉) · A. Ouannas
Department of Mathematics and Computer Science, University of Larbi Ben M'hidi, Oum El Bouaghi 04000, Algeria
e-mail: noureddine.djeina@univ-oeb.dz

A. Ouannas
e-mail: ouannas.adel@univ-oeb.dz

A. Ben Makhlouf et al. (eds.), *State Estimation and Stabilization of Nonlinear Systems*, Studies in Systems, Decision and Control 491,
https://doi.org/10.1007/978-3-031-37970-3_9

However, this paper is organized in the following order. Section 2 introduces some primary preliminaries associated with discrete fractional calculus, while Sect. 3 discusses some recently established results that have handled the stability of fractional order discrete systems. Section 4 includes the stabilisation of two chaotic systems, followed by the last section that summarizes the achievements of the whole work.

2 Discrete Fractional Calculus

In this section, we introduce the basic discrete fractional calculus that will be useful for our later studies.

2.1 *Basic Concepts*

2.1.1 Gamma Function

Definition 1 ([156]) The gamma function is defined by:

$$\Gamma(z) := \int_0^{\infty} e^{-t} t^{z-1} dt, \tag{1}$$

for those complex numbers z for which the real part of z is positive (it can be shown that the above improper integral converges for all such z).

We have

$$\begin{aligned} \Gamma(1) &= \textstyle\int_0^{\infty} e^{-t} dt \\ &= 1, \end{aligned}$$

and by integration by parts for $x > 0 : \Gamma(x+1) = x\Gamma(x)$, thats give for any positive integer k

$$\Gamma(k) = (k-1)!.$$

We also have the important estimation:

$$\lim_{k \to \infty} \frac{\Gamma(k+\gamma)}{\Gamma(k)k^{\gamma}} = 1, \quad k \in \mathbb{N} \text{ and } \gamma \in \mathbb{C}.$$

2.1.2 Falling Function

Definition 2 ([156]) For $r \in \mathbb{R}$, we define the *falling function*, $t^{(r)}$, read t to the r falling, by:

$$t^{(r)} := \frac{\Gamma(t+1)}{\Gamma(t-r+1)} = t(t-1)(t-2)\cdots(t-r+1), \tag{2}$$

for those values of t and r such that the right-hand side of this last equation make sense. We then extend this definition by making the common convention that $t^{(r)} := 0$ when $t - r + 1$ is a nonpositive integer.

2.2 Fractional Discrete Operators

Now we define fractional sum and **Riemann Liouville** and **Caputo** difference operators, give some of their properties and the relation between them. Frequently, the functions we consider will be defined on a set of the form:

$$\mathbb{N}_a := \{a, a+1, a+2, \ldots\},$$

where $a \in \mathbb{R}$.

2.2.1 Fractional Sum Operator

First we begine with the fractional sum.

Definition 3 ([157]) Let $\alpha > 0$. Then, the $\alpha - th$ fractional sum of $f : \mathbb{N}_a \to \mathbb{R}$ is defined by:

$$\Delta_a^{-\alpha} f(t) := \frac{1}{\Gamma(\alpha)} \sum_{s=a}^{t-\alpha} (t-s-1)^{(\alpha-1)} f(s), \tag{3}$$

for any $t \in \mathbb{N}_{a+\alpha}$.

Remark 1 Note that $\Delta_a^{-\alpha}$ maps functions defined on $\mathbb{N}_a$ to functions defined on $\mathbb{N}_{a+\alpha}$.

Theorem 1 ([157]) *Let $f : \mathbb{N}_a \to \mathbb{R}$ be given, and let $\mu, \nu > 0$. Then:*

$$\Delta_{a+\mu}^{-\nu}[\Delta_a^{-\mu} f(t)] = \Delta_a^{-(\mu+\nu)} f(t) = \Delta_{a+\nu}^{-\mu}[\Delta_a^{-\nu} f(t)], \tag{4}$$

for any $t \in \mathbb{N}_{a+\mu+\nu}$.

Theorem 2 ([157]) *Let $f : \mathbb{N}_a \to \mathbb{R}$ be given, for any $\alpha > 0$ and any positive integer p, we have:*

$$\Delta_a^{-\alpha} \Delta^p f(t) = \Delta^p \Delta_a^{-\alpha} f(t) - \sum_{r=0}^{p-1} \frac{(t-a)^{(\alpha-p+r)}}{\Gamma(\alpha+r-p+1)} \Delta^r f(a), \tag{5}$$

for any $t \in \mathbb{N}_{a+\alpha}$.

2.2.2 Rimann Liouville Operator

Definition 4 ([157]) Let $\alpha > 0$. Then, the $\alpha-$order **Rimann Liouville** fractional difference of a function f defined on $\mathbb{N}_a$ are defined by:

$$\Delta_a^{\alpha} f(t) := \Delta^p \Delta_a^{-(p-\alpha)} f(t) = \frac{1}{\Gamma(p-\alpha)} \Delta^p \sum_{s=a}^{t-(p-\alpha)} (t-s-1)^{(p-\alpha-1)} f(s), \quad \forall t \in \mathbb{N}_{a+p-\alpha}, \tag{6}$$

where $p = [\alpha] + 1$.

Remark 2 If $\alpha \longrightarrow p \in \mathbb{N}$, then

$$\begin{aligned}\lim_{\alpha \to p} \Delta_a^{\alpha} f(t) &= \Delta^p \Delta^{(0)} f(t) \\ &= \Delta^p f(t),\end{aligned}$$

and

$$\begin{aligned}\lim_{\alpha \to p-1} \Delta_a^{\alpha} f(t) &= \Delta^p \Delta^{-1} f(t) \\ &= \Delta^{p-1} f(t).\end{aligned}$$

We can say that the Δ_a^{α} operator is an interpolation of Δ^p operators. Also, it is clear that Δ_a^{α} maps functions defined on $\mathbb{N}_a$ to functions defined on $\mathbb{N}_{a+(p-\alpha)}$.

Theorem 3 ([157]) *Assume $\alpha > 0$ and f is defined on $\mathbb{N}_a$. Then:*

$$\Delta_{a+\alpha}^{\alpha} \Delta_a^{-\alpha} f(t) = f(t), \quad \forall t \in \mathbb{N}_{a+p}, \tag{7}$$

where $p = [\alpha] + 1$.

Remark 3 We see that $\Delta_{a+\alpha}^{\alpha}$ have a right inverse ($\Delta_a^{-\alpha}$), or $\Delta_{a+\alpha}^{-\alpha}$ have a left inverse (Δ_a^{α}).

Theorem 4 ([157]) *Assume $\alpha > 0$ and f is defined on $\mathbb{N}_a$. Then:*

$$\Delta_{a+p-\alpha}^{-\alpha} \Delta_a^{\alpha} f(t) = f(t) - \sum_{r=0}^{p-1} \frac{(t-a)^{(\alpha-p+r)}}{\Gamma(\alpha+r-p+1)} \Delta_a^{r-(p-\alpha)} f(a), \; \forall t \in \mathbb{N}_{a+p}, \tag{8}$$

where $p = [\alpha] + 1$.

Remark 4 We see that

$$\Delta_{a+\alpha}^{-\alpha} \Delta_a^{\alpha} f(t) \neq f(t)$$

generally.

2.2.3 Caputo Operator

Definition 5 [157] Let $\alpha > 0$, $\alpha \notin \mathbb{N}$. Then, the $\alpha-$order **Caputo** fractional difference of a function f defined on $\mathbb{N}_a$, are defined by:

$$ {}^C\Delta_a^\alpha f(t) := \Delta_a^{-(p-\alpha)} \Delta^p f(t) = \frac{1}{\Gamma(p-\alpha)} \sum_{s=a}^{t-(p-\alpha)} (t-s-1)^{(p-\alpha-1)} \Delta^p f(s), \ \ \forall t \in \mathbb{N}_{a+p-\alpha}, \tag{9} $$

where $p = [\alpha] + 1$.

Remark 5 When $\alpha \to p-1 \in \mathbb{N}$, we get

$$ \begin{aligned} \lim_{\alpha \to p-1} {}^C\Delta_a^\alpha f(t) &= \lim_{\alpha \to p-1} \Delta_a^{-(p-\alpha)} \Delta^p f(t) \\ &= \Delta_a^{-1} \Delta^p f(t) \\ &= \Delta^{p-1} f(t) - \Delta^{p-1} f(a). \end{aligned} $$

So ${}^C\Delta_a^\alpha$ is not an interpolation of Δ^p. However, it is used in modeling because it is consistent with the classical initial and boundary conditions. To not be confused we defined ${}^C\Delta_a^p$ by ${}^C\Delta_a^p f(t) := \Delta^p f(t)$, where $p \in \mathbb{N}$. Also, it is clear that ${}^C\Delta_a^\alpha$ maps functions defined on $\mathbb{N}_a$ to functions defined on $\mathbb{N}_{a+(p-\alpha)}$.

Theorem 5 ([157]) *Assume $\alpha > 0$ and f is defined on $\mathbb{N}_a$. Then:*

$$ \Delta_{a+(p-\alpha)}^{-\alpha} {}^C\Delta_a^\alpha f(t) = f(t) - \sum_{r=0}^{p-1} \frac{(t-a)^{(r)}}{r!} \Delta^r f(a), \ \ \forall t \in \mathbb{N}_a, \tag{10} $$

where $p = [\alpha] + 1$.

In particular, if $0 < \alpha \leq 1$ then

$$ \Delta_{a+(1-\alpha)}^{-\alpha} {}^C\Delta_a^\alpha f(t) = f(t) - f(a), \ \ \forall t \in \mathbb{N}_a. \tag{11} $$

Theorem 6 ([157]) *Let $\alpha > 0$, and f is defined on $\mathbb{N}_a$. Then:*

$$ {}^C\Delta_a^\alpha f(t) = \Delta_a^\alpha f(t) - \sum_{r=0}^{p-1} \frac{(t-a)^{(r-\alpha)}}{\Gamma(r-\alpha+1)} \Delta^r f(a), \ \ \forall t \in \mathbb{N}_{a+p-\alpha}, \tag{12} $$

where $p = [\alpha] + 1$.

In particular, when $0 < \alpha < 1$, we have

$$ {}^C\Delta_a^\alpha f(t) = \Delta_a^\alpha f(t) - \frac{(t-a)^{(-\alpha)}}{\Gamma(1-\alpha)} f(a), \ \ \forall t \in \mathbb{N}_{a-\alpha+1}. \tag{13} $$

3 Stability of Incommensurate Fractional Order Difference Systems

In this section, we will present some stability theorems of incommensurate fractional order difference systems. First of all, consider the following linear incommensurate FoDS:

$$\begin{cases} {}^{C}\Delta_0^{\alpha_1} x_1(k+1-\alpha_1) = \sum_{j=1}^{n} a_{1j}x_j(k), \\ {}^{C}\Delta_0^{\alpha_2} x_2(k+1-\alpha_2) = \sum_{j=1}^{n} a_{2j}x_j(k), \qquad k = 0, 1, \ldots, \\ \vdots \\ {}^{C}\Delta_0^{\alpha_n} x_n(k+1-\alpha_n) = \sum_{j=1}^{n} a_{nj}x_j(k), \end{cases} \tag{14}$$

where ${}^{C}\Delta_0^{\alpha_i}$ is the Caputo fractional difference of order α_i, where $0 < \alpha_i \leq 1$, for $i = 1, 2, \ldots, n$, and $A = (a_{ij})_{1\leq i,j\leq n} \in \mathbb{R}^{n\times n}$.

Theorem 7 ([111]) *Consider system (14) subject to the initial vector condition $X(0) = X_0 \in \mathbb{R}^n$. Then,*
• *If all roots of the following characteristic equation:*

$$\det\left(diag\left(z(1-z^{-1})^{\alpha_1}, z(1-z^{-1})^{\alpha_2}, \ldots, z(1-z^{-1})^{\alpha_n}\right) - A\right) = 0, \tag{15}$$

lie inside the unit disk, then the zero solution of system (14) is asymptotically stable.
• *If there exists a zero, say z^* of (15) such that $|z^*| > 1$, then the zero solution of system (15) is not stable.*

Corollary 1 ([111]) *Suppose that A is a triangular matrix with diagonal elements $\lambda_i, i = 1, \ldots, n$. If $-2^{\alpha_i} < \lambda_i < 0, \forall i$, then the zero solution of system (14) is asymptotically stable. Furthermore, such solution is not stable if either $\lambda_i > 0$ or $\lambda_i < -2^{\alpha_i}$, for some i.*

As a matter of fact, the above result, represented by Theorem 8 is deemed one of the main significant results in this section. It can be easily implemented for exploring the stability of some linear incommensurate FoDSs which involve just a triangular matrix A in their forms. On the contrary, one can find it extremely hard to verify condition (15) in the proposed result represented by Theorem 7. Actually, this condition concerns with the full matrix that might be found in the linear incommensurate FoDSs. To deal with this problem, we present below another more practical result which is equivalent to Theorem 7.

Theorem 8 ([111]) *Consider $0 < \alpha_i < 1$, for $i = 1, 2, \ldots, n$, and M is the lowest common multiple of the denominators u_i of α_i's, in which $\alpha_i = \frac{v_i}{u_i}$ and $(u_i, v_i) = 1$, where $u_i, v_i \in \mathbb{Z}_+, \forall i$. Then, the zero solution of system (14), subject to the initial vector condition $X(0) = X_0 \in \mathbb{R}^n$, is:*

• *Asymptotically stable if any zero solution of the polynomial:*

$$\det\left(diag\left(\lambda^{M\alpha_1}, \lambda^{M\alpha_2}, \ldots, \lambda^{M\alpha_n}\right) - (1-\lambda^M)A\right), \tag{16}$$

lie inside the set

$$\mathbb{C}\backslash K^{\gamma},$$

where $\gamma = \frac{1}{M}$ *and where*

$$K^{\gamma} = \left\{z \in \mathbb{C} : |z| \le \left(2\cos\frac{|\arg z|}{\gamma}\right)^{\gamma} \text{ and } |\arg z| \le \frac{\gamma\pi}{2}\right\}. \tag{17}$$

• *Not stable, furthermore, if there is a zero, say* λ, *of (16) such that* $\lambda \in Int K^{\gamma}$.

4 Control of Chaos in Incommensurate Fractional Order Discrete System

4.1 Linear Control

In [158], the authors introduced the following predator–prey discrete system of Leslie type:

$$\begin{aligned} x_{n+1} &= x_n + \delta x_n \left((1-x_n) - \tfrac{y_n}{x_n^2+c}\right), \\ y_{n+1} &= y_n + \delta y_n \left(a - \tfrac{by_n}{x_n}\right), \end{aligned} \tag{18}$$

where a, b, c and δ are non-negative parameters. The incommensurate fractional version of the proposed discrete model (18) can be given as follows:

$$\begin{aligned} \Delta^{\alpha_1} x(t+1-\alpha_1) &= \delta x(t)\left((1-x(t)) - \tfrac{y(t)}{x^2(t)+c}\right), \\ \Delta^{\alpha_2} y(t+1-\alpha_2) &= \delta y(t)\left(a - \tfrac{by(t)}{x(t)}\right), \end{aligned} \tag{19}$$

where $0 < \alpha_1, \alpha_2 < 1, t \in \mathbb{N}$.

For $\delta = 1.05, a = 2.5, b = 5, c = 5.5$ and $(\alpha_1, \alpha_2) = (0.9, 0.8)$, the system 19 has chaotic behaviour [159]. Note that system (19) has $(0, 0)$ as a fixed point.

The controlled system is as follows:

$$\begin{cases} \Delta^{\alpha_1} x(t+1-\alpha_1) = \delta x(t)\left(1 - x(t) - \tfrac{y(t)}{x^2(t)+c}\right) + u_1, \\ \Delta^{\alpha_2} y(t+1-\alpha_2) = \delta y(t)\left(a - \tfrac{by(t)}{x(t)}\right) + u_2, \end{cases} \quad t \in \mathbb{N}_{1-\alpha}. \tag{20}$$

where u_1, u_2, represents the control functions.

Theorem 9 *Under the control law*

$$\begin{aligned} u_1 &= \delta x^2(t) + \tfrac{\delta x(t)y(t)}{x^2(t)+c}, \\ u_2 &= \tfrac{\delta b y^2(t)}{x(t)}, \end{aligned} \tag{21}$$

the equilibrium point (0, 0) *of (20) is asymptotically stable.*

Proof Under the control law 21 the system (20) becomes:

$$\begin{cases} \Delta^{\alpha_1} x(t+1-\alpha_1) = (\delta - 1.5)x(t), \\ \Delta^{\alpha_2} y(t+1-\alpha_2) = (\delta a - 3)y(t), \end{cases}$$

Thus, the Jacobite matrix becomes:

$$J = \begin{pmatrix} \delta - 1.5 & 0 \\ 0 & \delta a - 3 \end{pmatrix}.$$

Note that $\lambda_1 = \delta - 1.5 = -0.45$ and $\lambda_2 = \delta a - 3 = -0.375$. So

$$\begin{aligned} -2^{0.9} &< \lambda_1 < 0, \\ -2^{0.8} &< \lambda_2 < 0. \end{aligned}$$

According to the Theorem 8 the fixed point (0, 0) is asymptotically stable. □

4.2 Nonlinear Control

In recent years, great research efforts and economic resources have been directed to winning the battle against cancer. A major issue is the activation and control of the immune system in its competition against tumor cells. For this purpose, the study of tumor immune dynamics can play a role of paramount importance, given that mathematical modeling of cancer growth is considered as one of the useful tools for developing effective medical therapies, and in light of this, we will be interested in a three-dimensional cancer growth model that was presented and in which is described in [160] as follow:

$$\begin{cases} \frac{dx(t)}{dt} = ax(t)(1-y(t))(1+z(t)) - x^2(t)y(t), \\ \frac{dy(t)}{dt} = by(t)(1-z(t))(1+x(t)) - y^2(t)z(t), \quad t \in \mathbb{R}^+. \\ \frac{dz(t)}{dt} = cz(t)(1-x(t))(1+y(t)) - z^2(t)x(t). \end{cases} \tag{22}$$

The proposed model's flowchart and parameters description are well explained in (23).

Variable	Description	Value
$x(t)$	The number of tumor cells at time t	
$y(t)$	The number of healthy host cells at time t	
$z(t)$	The number of effector immune cells at time t	
a	The growth rate of the tumor cells	0.7455
b	The growth rate of the healthy host cells	0.7367
c	The growth rate of the effector immune cells	0.5619

(23)

In addition, the following initial conditions are taken in consideration:

$$x(0) = 0.4,\ y(0) = 0.5,\ z(0) = 0.5. \tag{24}$$

The fractional order model attached to this system was studied in [164] and it was proved that the system behaves chaotically.

Moreover, biological systems are usually characterized by hereditary properties and nonlocal distributed behaviors. As a consequence, the modeling of these systems by fractional-order differential (or difference) equations has more advantages than integer-order modeling [161–163]. In fact, the statistics are discrete, so the discrete system is closer to modeling and in light of this we will use the following approximation:

$$\frac{dX(t)}{dt} \simeq \frac{X(t+h) - X(t)}{h}, \tag{25}$$

system (22) becomes:

$$\begin{cases} \frac{x(t+h)-x(t)}{h} = ax(t)(1-y(t))(1+z(t)) - x^2(t)y(t), \\ \frac{y(t+h)-y(t)}{h} = by(t)(1-z(t))(1+x(t)) - y^2(t)z(t), \quad t \in \mathbb{R}^+. \\ \frac{z(t+h)-z(t)}{h} = cz(t)(1-x(t))(1+y(t)) - z^2(t)x(t). \end{cases} \tag{26}$$

We multiply both sides by h and put

$$X(t) = X(n),$$

and

$$X(t+h) = X(n+1),$$

where $n \in \mathbb{N}$, $X(t) = (x(t), y(t), z(t))^t$. We finally get the following system

$$\begin{cases} \Delta x(n) = ahx(n)(1-y(n))(1+z(n)) - x^2(n)y(n), \\ \Delta y(n) = bhy(n)(1-z(n))(1+x(n)) - y^2(n)z(n), \quad n \in \mathbb{N}. \\ \Delta z(n) = chz(n)(1-x(n))(1+y(n)) - z^2(n)x(n). \end{cases} \tag{27}$$

where Δ is the forward difference operator ($\Delta X(n) = X(n+1) - X(n)$). Note that system (27) has five equilibrium points [160], four of them are obtained analytically and can be described as follows:

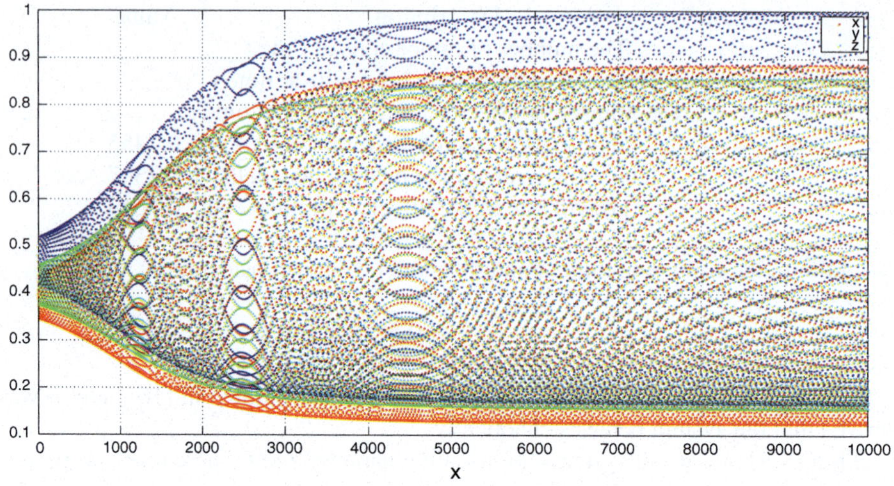

Fig. 1 A numerical simulation of system (27) for system parameters a = 0; 7455; b = 0; 7367; c = 0;5619 and h = 0:326 and initial conditions 20

$$
\begin{aligned}
E_0 &= (0, 0, 0),\\
E_1 &= (0, -1, -2.7979),\\
E_2 &= (-1.28, 0, -1),\\
E_3 &= (-1, -2.9293, 0),\\
E_4 &= (0.5961, 06718, 0.6364).
\end{aligned} \tag{28}
$$

Note that the fixed points E_1, E_2, and E_3 have negative coordinates, indicating that the dynamics cannot take place since it is not possible to define negative populations in system (27).

Discrete system (27) also behaves chaotically for some values of h. Using Matlab, we can get the following simulations that show the chaos of the system (Figs. 1 and 2).

Based on the previous definitions, the incommensurate fractional order version of the system (27) is given as follows:

$$
\begin{cases}
{}^C\Delta^{\alpha_1}x(t) = ahx(t)(1-y(t))(1+z(t)) - x^2(t)y(t),\\
{}^C\Delta^{\alpha_2}y(t) = bhy(t)(1-z(t))(1+x(t)) - y^2(t)z(t), \quad t \in \mathbb{N}_{1-\alpha}.\\
{}^C\Delta^{\alpha_3}z(t) = chz(t)(1-x(t))(1+y(t)) - z^2(t)x(t).
\end{cases} \tag{29}
$$

where $0 < \alpha_1, \alpha_2, \alpha_3 < 1$.

The control theory aims to influence the system to reach the desired result. In this section, we will study the stability of the fixed point E_4, we will prove that this point is unstable, and then we will develop a control law that makes this point stable. The controlled system is as follows:

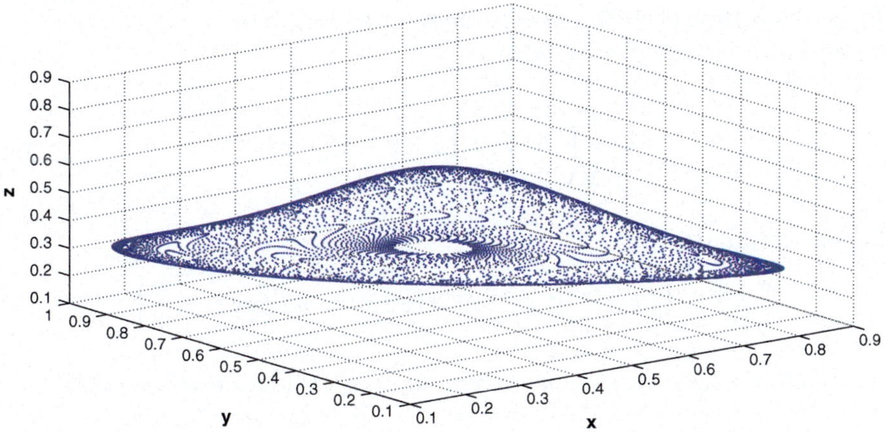

Fig. 2 Chaotic attractor of system (27) for system parameters a = 0; 7455; b = 0; 7367; c = 0; 5619 and h = 0:326 and initial conditions 3

$$\begin{cases} {}^C\Delta^{\alpha_1}x(t) = ahx(t)(1-y(t))(1+z(t)) - x^2(t)y(t) + u, \\ {}^C\Delta^{\alpha_2}y(t) = bhy(t)(1-z(t))(1+x(t)) - y^2(t)z(t), \qquad t \in \mathbb{N}_{1-\alpha}. \\ {}^C\Delta^{\alpha_3}z(t) = chz(t)(1-x(t))(1+y(t)) - z^2(t)x(t). \end{cases} \tag{30}$$

where u represents the control function (can be a chemical compound that eliminates cancer cells).

The linear approximation of the system 6 at E_4 is given by:

$$\begin{pmatrix} {}^C\Delta^{\alpha_1}x(t) \\ {}^C\Delta^{\alpha_2}y(t) \\ {}^C\Delta^{\alpha_3}z(t) \end{pmatrix} = J \begin{pmatrix} x \\ y \\ z \end{pmatrix}, \tag{31}$$

where

$$J = \begin{pmatrix} -0.670\,40 & -0.592\,4 & 4.754\,6 \times 10^{-2} \\ 5.866\,3 \times 10^{-2} & -0.715\,69 & -0.708\,83 \\ -0.599\,90 & 4.708\,5 \times 10^{-2} & -0.635\,03 \end{pmatrix},$$

is the jacobian matrix of $\begin{pmatrix} ahx(t)(1-y(t))(1+z(t)) - x^2(t)y(t) \\ bhy(t)(1-z(t))(1+x(t)) - y^2(t)z(t) \\ chz(t)(1-x(t))(1+y(t)) - z^2(t)x(t) \end{pmatrix}$ at E_4.

Take $\alpha_1 = \frac{3}{4}, \alpha_2 = \frac{2}{4}, \alpha_3 = \frac{1}{4}$.

The asymptotic stability of this type of system has been studied in [111] Then it has been proven that the asymptotic stability of the linear approximation leads to the local asymptotic stability of the nonlinear system in [112].

Proposition 10 ([111]) *The equilibrium point E_4 of (27) is:*
Asymptotically stable if any solution of

$$\det\left(\begin{pmatrix}\lambda^3 & 0 & 0\\ 0 & \lambda^2 & 0\\ 0 & 0 & \lambda\end{pmatrix} - (1-\lambda^4)J\right) = 0, \tag{32}$$

lies outside the set

$$K^{\frac{1}{4}} = \left\{z \in \mathbb{C} : |z| \leq (2\cos 4\,|\arg z|)^{\frac{1}{4}} \text{ and } |\arg z| \leq \frac{\pi}{8}\right\}.$$

Not stable, furthermore, if there is a zero, say λ, of (14) such that $\lambda \in Int K^{\frac{1}{4}}$.

So the solutions to the Eq. 14 are

$$\begin{pmatrix} 1.1803 \\ -0.78303 \\ 1.2109 - 0.23849i \\ 1.2109 + 0.23849i \\ -0.23402 + 0.65078i \\ -0.23402 - 0.65078i \\ -0.83433 + 0.3805i \\ -0.83433 - 0.3805i \\ -0.14254 + 1.2697i \\ -0.14254 - 1.2697i \\ -0.1986 + 1.0211i \\ -0.1986 - 1.0211i \end{pmatrix}.$$

We note that $1.1803 \in Int K^{\frac{1}{4}}$. Hence, the fixed point E_4 is unstable.

To make the system stable we suggest the following result:

Theorem 11 *Under the control law*

$$u = 0.6703993296x + 0.5924y - 4.7546 \times 10^{-2}z, \tag{33}$$

the equilibrium point E_4 of (13) is locally asymptotically stable.

Proof Under the control law 15 the system (12) becomes:

$$\begin{cases} {}^{C}\Delta^{\alpha_1}x(t) = ax(t+\alpha_1-1), \\ {}^{C}\Delta^{\alpha_2}y(t) = by(t+\alpha_2-1)(1-z(t+\alpha_2-1))(1+x(t+\alpha_2-1)) - y^2(t+\alpha_2-1)z(t+\alpha_2-1), \\ {}^{C}\Delta^{\alpha_3}z(t) = cz(t+\alpha_2-1)(1-x(t+\alpha_2-1))(1+y(t+\alpha_2-1)) - z^2(t+\alpha_2-1)x(t+\alpha_2-1). \end{cases}$$

Thus, the Jacobite matrix becomes:

$$J = \begin{pmatrix} -0.000000670\,40 & 0 & 0 \\ 5.866\,3 \times 10^{-2} & -0.715\,69 & -0.708\,83 \\ -0.599\,90 & 4.708\,5 \times 10^{-2} & -0.635\,03 \end{pmatrix}.$$

Note that the solutions of

$$\det \left(\begin{pmatrix} \lambda & 0 & 0 \\ 0 & \lambda & 0 \\ 0 & 0 & \lambda \end{pmatrix} - (1 - \lambda^4) J \right) = 0,$$

are

$$\begin{pmatrix} -6.704\,0 \times 10^{-7} \\ 114.26 \\ -57.129 - 98.951i \\ -57.129 + 98.951i \\ -0.379\,43 - 0.981\,33i \\ -0.379\,43 + 0.981\,33i \\ 1.294\,9 + 6.499\,3 \times 10^{-2}i \\ 1.294\,9 - 6.499\,3 \times 10^{-2}i \\ -0.304\,23 - 1.144\,5i \\ -0.304\,23 + 1.144\,5i \\ -0.611\,2 - 9.818\,5 \times 10^{-2}i \\ -0.611\,2 + 9.818\,5 \times 10^{-2}i \end{pmatrix},$$

All solutions are lies outside the set $K^{\frac{1}{4}}$. According to the Theorem 1 the fixed point E_4 is locally asymptotically stable (Fig. 3). □

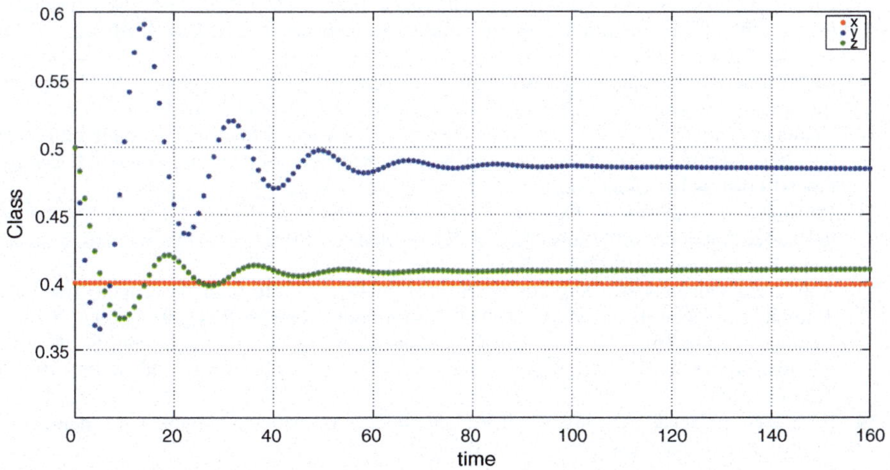

Fig. 3 Numerical simulation showing the stability of the fixed point E_4 under the control law u

5 Conclusions

In this chapter, we discussed the stability of incommensurate fractional order discrete systems. A theoretical result and two applicable results were mentioned, and then we applied them in the stabilisation of tow chaotic systems, predator–prey system and cancer growth system, to see the effectiveness of the mentioned results.

References

1. Djenina, N., Ouannas, A., Oussaeif, T.E., Grassi, G., Batiha, I.M., Momani, S., Albadarneh, R.B.: On the stability of incommensurate h-nabla fractional-order difference systems. Fractal Fract. **6**(3), 158 (2022)
2. Ramadoss, J., Ouannas, A., Tamba, V.K., Grassi, G., Momani, S., Pham, V.T.: Constructing non-fixed-point maps with memristors. Eur. Phys. J. **137**(2), 211 (2022)
3. Shatnawi, M.T., Ouannas, A., Bahia, G., Batiha, I.M., Grassi, G.: The optimal homotopy asymptotic method for solving two strongly fractional-order nonlinear benchmark oscillatory problems. Mathematics **9**(18), 2218 (2021)
4. Hioual, A., Ouannas, A., Grassi, G., Oussaeif, T.E.: Nonlinear nabla variable-order fractional discrete systems: asymptotic stability and application to neural networks. J. Comput. Appl. Math. **423**, 114939 (2023)
5. Almatroud, O.A., Hioual, A., Ouannas, A., Sawalha, M.M., Alshammari, S.: On variable-order fractional discrete neural networks: existence, uniqueness and stability. Fractal Fract. **7**(2), 118 (2023)
6. Al-Shbeil, I., Djenina, N., Jaradat, A., Al-Husban, A., Ouannas, A., Grassi, G.: A new COVID-19 pandemic model including the compartment of vaccinated individuals: global stability of the disease-free fixed point I. Mathematics **11**(3), 576 (2023)
7. Al-Husban, A., Djenina, N., Saadeh, R., Ouannas, A., Grassi, G.: A new incommensurate fractional-order COVID 19: modelling and dynamical analysis. Mathematics **11**(3), 555 (2023)
8. Hioual, A., Oussaeif, T.E., Ouannas, A., Grassi, G., Batiha, I.M., Momani, S.: New results for the stability of fractional-order discrete-time neural networks. Alex. Eng. J. **61**(12), 10359–10369 (2022)
9. Khennaoui, A.A., Ouannas, A., Momani, S., Almatroud, O.A., Al-Sawalha, M.M.: Special fractional-order map and its realization. Mathematics **10**(23), 4474 (2022)
10. Ouannas, A., Horani, M.A., Grassi, G.: The effect of Caputo fractional variable difference operator on a discrete-time hopfield neural network with non-commensurate order RC Karoun. Fractal Fract. **6**(10), 575 (2022)
11. Dababneh, A., Djenina, N., Ouannas, A., Grassi, G., Batiha, I.M., Jebril, I.H.: A new incommensurate fractional-order discrete COVID-19 model with vaccinated individuals compartment. Fractal Fract. **6**(8), 456 (2022)
12. Abbes, A., Ouannas, A., Shawagfeh, N., Grassi, G.: The effect of the Caputo fractional difference operator on a new discrete COVID-19 model. Results Phys. **39**, 105797 (2022)
13. Djenina, N., Ouannas, A., Batiha, I.M., Grassi, G., Oussaeif, T.E., Momani, S.: A novel fractional-order discrete SIR model for predicting COVID-19 behavior. Mathematics **10**(13), 2224 (2022)
14. Hioual, A., Ouannas, A., Oussaeif, T.E., Laouar, Z.: On fractional variable-order neural networks based on the Caputo derivative. Book of Abstracts, vol. 103 (2022)
15. Gasri, A., Khennaoui, A.A., Ouannas, A., Grassi, G., Iatropoulos, A., Moysis, L.: A new fractional-order map with infinite number of equilibria and its encryption application. Complexity **1–18**(4), 2022 (2022)

16. Hioual, A., Ouannas, A., Oussaeif, T.E., Grassi, G., Batiha, I.M., Momani, S.: On variable-order fractional discrete neural networks: solvability and stability. Fractal Fract. **6**(2), 119 (2022)
17. Batiha, I.M., Momani, S., Ouannas, A., Momani, Z., Hadid, S.B.: Fractional-order COVID-19 pandemic outbreak: modeling and stability analysis. Int. J. Biomath. **15**(01), 2150090 (2022)
18. Almatroud, A.O., Djenina, N., Ouannas, A., Grassi, G., Al-Sawalha, M.M.: A novel discrete-time COVID-19 epidemic model including the compartment of vaccinated individuals. Math. Biosci. Eng. **19**, 12387–12404 (2022)
19. Batiha, I.M., Al-Nana, A.A., Albadarneh, R.B., Ouannas, A., Al-Khasawneh, A.: Fractional-order coronavirus models with vaccination strategies impacted on Saudi Arabia's infections. AIMS Math. **7**, 12842–12858 (2022)
20. Khennaoui, A.A., Almatroud, A.O., Ouannas, A., Al-sawalha, M.M., Grassi, G.: The effect of Caputo fractional difference operator on a novel game theory model. Discret. Contin. Dyn. Syst.-Ser. B **26**(8), 5 (2021)
21. Batiha, I.M., Oudetallah, J., Ouannas, A., Al-Nana, A.A., Jebril, I.H.: Tuning the fractional-order PID-controller for blood glucose level of diabetic patients. Int. J. Adv. Soft Comput. Appl. **13**(2), 1–10 (2021)
22. Bahia, G., Ouannas, A., Batiha, I.M., Odibat, Z.: The optimal homotopy analysis method applied on nonlinear time-fractional hyperbolic partial differential equations. Numer. Methods Part. Differ. Equ. **37**(3), 2008–2022 (2021)
23. Ouannas, A., Karouma, A., Grassi, G., Pham, V.T.: A novel secure communications scheme based on chaotic modulation, recursive encryption and chaotic masking. Alex. Eng. J. **60**(1), 1873–1884 (2021)
24. Mellah, M., Ouannas, A., Khennaoui, A.A., Grassi, G.: Fractional discrete neural networks with different dimensions: coexistence of complete synchronization, antiphase synchronization and full state hybrid projective synchronization. Nonlinear Dyn. Syst. Theory **21**(4), 410–419 (2021)
25. Batiha, I.M., Ouannas, A., Momani, S.: Modeling COVID-19 pandemic outbreak using fractional-order systems RB Albadarneh. Int. J. Math. Comput. Sci **16**, 1405–1421 (2021)
26. Khennaoui, A.A., Ouannas, A., Momani, S., Batiha, I.M., Dibi, Z., Grassi, G.: On dynamics of a fractional-order discrete system with only one nonlinear term and without fixed points. Electronics **9**(12), 2179 (2020)
27. Ouannas, A., Khennaoui, A.A., Bendoukha, S., Grassi, G.: On the dynamics and control of a fractional form of the discrete double scroll. Int. J. Bifurc. Chaos **29**(06), 1950078 (2019)
28. Wang, X., Ouannas, A., Pham, V.T., Abdolmohammadi, H.R.: A fractional-order form of a system with stable equilibria and its synchronization. Adv. Differ. Equ. **2018**(1), 1–13 (2018)
29. Ouannas, A., Debbouche, N., Wang, X., Pham, V.T., Zehrour, O.: Secure multiple-input multiple-output communications based on F-M synchronization of fractional-order chaotic systems with non-identical dimensions and orders. Appl. Sci. **8**(10), 1746 (2018)
30. Ahmad, I., Ouannas, A., Shafiq, M., Pham, V.T., Baleanu, D.: Finite-time stabilization of a perturbed chaotic finance model. J. Adv. Res. **32**, 1–14 (2021)
31. Mesdoui, F., Shawagfeh, N., Ouannas, A.: Synchronization methods for chaotic systems involving fractional derivative with a non-singular kernel. J. Appl. Nonlinear Dyn. **11**(02), 375–386 (2022)
32. Wang, B., Ouannas, A., Karaca, Y., Xia, W.F., Jahanshahi, H., Alkhateeb, A.F.: A hybrid approach for synchronizing between two reaction diffusion systems of integer-and fractional-order applied on certain chemical models. Fractals (fractals) **30**(05), 1–11 (2022)
33. Ouannasz, M.M., Batiha, I.M.: A general method for stabilizing the fractional-order discrete neural networks via linear control law. Nonlinear Stud. **29**, 1 (2022)
34. Batiha, I.M., Ouannas, A., Emwas, J.A.: A stabilization approach for a novel chaotic fractional-order discrete neural network. J. Math. Comput. Sci. **11**(5), 5514–5524 (2021)
35. Mesdoui, F., Shawagfeh, N., Ouannas, A.: Global synchronization of fractional-order and integer-order N component reaction diffusion systems: application to biochemical models. Math. Methods Appl. Sci. **44**(1), 1003–1012 (2021)

36. Almatroud, A.O., Ouannas, A., Grassi, G., Batiha, I.M., Gasri, A., Al-Sawalha, M.M.: Different linear control laws for fractional chaotic maps using Lyapunov functional. Arch. Control Sci. **31**(4), 5 (2021)
37. Ouannas, A., Mesdoui, F., Momani, S., Batiha, I., Grassi, G.: Synchronization of FitzHugh-Nagumo reaction-diffusion systems via one-dimensional linear control law. Arch. Control Sci. **31**, 8 (2021)
38. Ouannas, A., Grassi, G., Azar, A.T., Khennaoui, A.A.: Synchronization control in fractional discrete-time systems with chaotic hidden attractors. In: Advanced Machine Learning Technologies and Applications: Proceedings of AMLTA, vol. 6 (2021)
39. Ouannas, A., Wang, X., Pham, V.T., Ziar, T.: $\Lambda - \varphi$ generalized synchronization: application to fractional hyperchaotic systems with arbitrary dimensions and orders. Automatika **61**(4), 554–563 (2020)
40. Ouannas, A., Azar, A.T., Ziar, T.: On inverse full state hybrid function projective synchronization for continuous-time chaotic dynamical systems with arbitrary dimensions. Differ. Equ. Dyn. Syst. **28**(4), 1045–1058 (2020)
41. Zoulikha, C., Ouannas, A.: A New generalized synchronization scheme to control fractional chaotic dynamical systems with different dimensions and orders. Nonlinear Stud. **27**(3), 845–854 (2020)
42. Chougui, Z., Ouannas, A.: A new generalized synchronization scheme to control fractional chaotic dynamical systems with different dimensions and orders. Nonlinear Stud. **27**(3), 4 (2020)
43. Mesdoui, F., Ouannas, A., Shawagfeh, N., Grassi, G., Pham, V.T.: Synchronization methods for the Degn-Harrison reaction-diffusion systems. IEEE Access **8**, 91829–91836 (2020)
44. Khennaoui, A.A., Ouannas, A., Bendoukha, S., Grassi, G., Wang, X., Pham, V.T.: Chaos, control, and synchronization in some fractional-order difference equations. Adv. Differ. Equ. **2019**(1), 1–23 (2019)
45. Ouannas, A., Wang, X., Pham, V.T., Grassi, G., Huynh, V.V.: Synchronization results for a class of fractional-order spatiotemporal partial differential systems based on fractional Lyapunov approach. Bound. Value Probl. **2019**(1), 1–12 (2019)
46. Ouannas, A., Odibat, Z., Shawagfeh, N.: A new Q-S synchronization results for discrete chaotic systems. Differ. Equ. Dyn. Syst. **27**, 413–422 (2019)
47. Ouannas, A., Bendoukha, S., Volos, C., Boumaza, N., Karouma, A.: Synchronization of fractional hyperchaotic Rabinovich systems via linear and nonlinear control with an application to secure communications. Int. J. Control, Autom. Syst. **17**, 2211–2219 (2019)
48. Ouannas, A., Jouini, L., Zehrour, O.: On new generalized hybrid synchronization in chaotic and hyperchaotic discrete-time dynamical systems. J. Appl. Nonlinear Dyn. **8**(3), 435–445 (2019)
49. Ouannas, A., Khennaoui, A.A., Odibat, Z., Pham, V.T., Grassi, G.: On the dynamics, control and synchronization of fractional-order Ikeda map. Chaos, Solitons Fractals **123**, 108–115 (2019)
50. Ouannas, A., Bendoukha, S., Karouma, A., Abdelmalek, S.: A general method to study the co-existence of different hybrid synchronizations in fractional-order chaotic systems. Int. J. Nonlinear Sci. Numer. Simul. **20**(3–4) (2019)
51. Ouannas, A., Khennaoui, A.A., Zehrour, O., Bendoukha, S., Grassi, G., Pham, V.T.: Synchronisation of integer-order and fractional-order discrete-time chaotic systems. Pramana **92**, 1–9 (2019)
52. Ouannas, A., Abdelli, M., Odibat, Z., Wang, X., Pham, V.T., Grassi, G., Alsaedi, A.: Synchronization control in reaction-diffusion systems: application to Lengyel-Epstein system. Complexity **2019**, 15 (2019)
53. Ouannas, A., Abdelli, M., Odibat, Z., Wang, X., Pham, V.T., Grassi, G., Alsaedi, A.: Synchronization control in reaction-diffusion systems: application to Lengyel-Epstein system (2019)
54. Jouini, L., Ouannas, A.: Increased and reduced synchronization between discrete-time chaotic and hyperchaotic systems. Nonlinear Dyn. Syst. Theory **19**(2), 313–318 (2019)

55. Ouannas, A., Bendoukha, S., Khennaoui, A.A., Grassi, G., Wang, X., Pham, V.T.: Chaos synchronization of fractional-order discrete-time systems with different dimensions using two scaling matrices. Open Phys. **17**(1), 942–949 (2019)
56. Ouannas, A., Azar, A.T., Ziar, T.: Control of continuous-time chaotic (hyperchaotic) systems: F-M synchronisation. Int. J. Autom. Control **13**(2), 226–242 (2019)
57. Bendoukha, S., Abdelmalek, S., Ouannas, A.: Secure communication systems based on the synchronization of chaotic systems. Math. Appl. Eng., Model., Soc. Issues **281–311**, 15 (2019)
58. Khennaoui, A.A., Ouannas, A., Bendoukha, S., Grassi, G., Wang, X., Pham, V.T.: Generalized and inverse generalized synchronization of fractional-order discrete-time chaotic systems with non-identical dimensions. Adv. Differ. Equ. **2018**(1), 1–14 (2018)
59. Ouannas, A., Wang, X., Pham, V.T., Grassi, G., Ziar, T.: Coexistence of identical synchronization, antiphase synchronization and inverse full state hybrid projective synchronization in different dimensional fractional-order chaotic. Adv. Differ. Equ. **2018**(1), 1–16 (2018)
60. Ouannas, A., Grassi, G., Wang, X., Ziar, T., Pham, V.T.: Function-based hybrid synchronization types and their coexistence in non-identical fractional-order chaotic systems. Adv. Differ. Equ. **1–12**(11), 2018 (2018)
61. Ouannas, A., Odibat, Z., Alsaedi, A., Hobiny, A., Hayat, T.: Investigation of QS synchronization in coupled chaotic incommensurate fractional order systems. Chin. J. Phys. **56**(5), 1940–1948 (2018)
62. Ouannas, A., Wang, X., Khennaoui, A.A., Pham, V.T., Grassi, G., Bendoukha, S.: The coexistence of different synchronization types in fractional-order discrete-time chaotic systems with non-identical dimensions and orders. Entropy **20**(9), 710 (2018)
63. Gasri, A., Ouannas, A., Ojo, K.S., Pham, V.T.: Coexistence of generalized synchronization and inverse generalized synchronization between chaotic and hyperchaotic systems. Nonlinear Anal.: Model. Control **23**(4), 583–598 (2018)
64. Boudiar, S., Ouannas, A., Bendoukha, S., Zara, A.: Coexistence of different types of chaos synchronization between non-identical and different dimensional dynamical systems. Nonlinear Dyn. Syst. Theory **18**(3), 253–258 (2018)
65. Ouannas, A., Wang, X., Pham, V.T., Grassi, G., Ziar, T.: Co-existence of some synchronization types between non-identical commensurate and incommensurate fractional-order chaotic systems with different dimensions. Adv. Differ. Equ. **21**(10), 12 (2018)
66. Ouannas, A., Khennaoui, A.A., Grassi, G., Bendoukha, S.: On the-chaos synchronization of fractional-order discrete-time systems: general method and examples. Discret. Dyn. Nat. Soc. **2018**, 18 (2018)
67. Grassi, G., Ouannas, A., Pham, V.T.: A general unified approach to chaos synchronization in continuous-time systems (with or without equilibrium points) as well as in discrete-time systems. Arch. Control Sci. **28**(1), 12 (2018)
68. Ouannas, A., Karouma, A.: Different generalized synchronization schemes between integer-order and fractional-order chaotic systems with different dimensions. Differ. Equ. Dyn. Syst. **26**(1–3), 125–137 (2018)
69. Ouannas, A., Zehrour, O., Laadjal, Z.: Nonlinear methods to control synchronization between fractional-order and integer-order chaotic systems. Nonlinear Stud. **25**(1), 12 (2018)
70. Singh, S., Azar, A.T., Bhat, M.A., Vaidyanathan, S., Ouannas, A.: Active control for multi-switching combination synchronization of non-identical chaotic systems. Adv. Syst. Dyn. Control **129–162**, 10 (2018)
71. Ouannas, A., Grassi, G., Karouma, A., Ziar, T., Wang, X., Pham, V.T.: New type of chaos synchronization in discrete-time systems: the FM synchronization. Open Phys. **16**(1), 174–182 (2018)
72. Ouannas, A., Odibat, Z., Hayat, T.: Fractional analysis of co-existence of some types of chaos synchronization. Chaos, Solitons Fractals **105**, 215–223 (2017)
73. Ouannas, A., Azar, A.T., Abu-Saris, R.: A new type of hybrid synchronization between arbitrary hyperchaotic maps. Int. J. Mach. Learn. Cybern. **8**(6), 1887–1894 (2017)
74. Ouannas, A., Wang, X., Pham, V.T., Ziar, T.: Dynamic analysis of complex synchronization schemes between integer order and fractional order chaotic systems with different dimensions. Complexity **2017**, 32 (2017)

75. Ouannas, A., Al-sawalha, M.M.: On inverse full state hybrid projective synchronization of chaotic dynamical systems in discrete-time. Int. J. Dyn. Control **5**, 252–258 (2017)
76. Ouannas, A., Grassi, G., Ziar, T., Odibat, Z.: On a function projective synchronization scheme for non-identical fractional-order chaotic (hyperchaotic) systems with different dimensions and orders. Optik **136**, 513–523 (2017)
77. Ouannas, A., Odibat, Z., Shawagfeh, N., Alsaedi, A., Ahmad, B.: Universal chaos synchronization control laws for general quadratic discrete systems. Appl. Math. Model. **45**, 636–641 (2017)
78. Ouannas, A., Azar, A.T., Vaidyanathan, S.: A robust method for new fractional hybrid chaos synchronization. Math. Methods Appl. Sci. **40**(5), 1804–1812 (2017)
79. Azar, A.T., Vaidyanathan, S., Ouannas, A.: Fractional Order Control and Synchronization of Chaotic Systems. Springer, Berlin (2017)
80. Azar, A.T., Volos, C., Gerodimos, N.A., Tombras, G.S., Pham, V.T., Radwan, A.G.: A novel chaotic system without equilibrium: dynamics, synchronization, and circuit realization. Complexity **2017**, 76 (2017)
81. Ouannas, A., Azar, A.T., Vaidyanathan, S.: A new fractional hybrid chaos synchronisation. Int. J. Model., Identif. Control **27**(4), 314–322 (2017)
82. Ouannas, A., Grassi, G.: A new approach to study the coexistence of some synchronization types between chaotic maps with different dimensions. Nonlinear Dyn. **86**(2), 1319–1328 (2016)
83. Ouannas, A., Al-Sawalha, M.M., Ziar, T.: Fractional chaos synchronization schemes for different dimensional systems with non-identical fractional-orders via two scaling matrices. Optik **127**(20), 8410–8418 (2016)
84. Ouannas, A., Grassi, G.: Inverse full state hybrid projective synchronization for chaotic maps with different dimensions. Chin. Phys. B **25**(9), 090503 (2016)
85. Ouannas, A., Odibat, Z.: On inverse generalized synchronization of continuous chaotic dynamical systems. Int. J. Appl. Comput. Math. **2**, 1–11 (2016)
86. Ouannas, A., Al-sawalha, M.M.: On $\Lambda - \varphi$ generalized synchronization of chaotic dynamical systems in continuous-time. Eur. Phys. J. Spec. Top. **225**(1), 187–196 (2016)
87. Ouannas, A., Abu-Saris, R.: On matrix projective synchronization and inverse matrix projective synchronization for different and identical dimensional discrete-time chaotic systems. J. Chaos **2016**, 29 (2016)
88. Ouannas, A., Al-sawalha, M.M.: Synchronization of chaotic dynamical systems in discrete-time. Adv. Chaos Theory Intell. Control **101–132**, 5 (2016)
89. Ouannas, A., Al-sawalha, M.M.: Synchronization between different dimensional chaotic systems using two scaling matrices. Optik **127**(2), 959–963 (2016)
90. Ouannas, A.: Co-existence of various types of synchronization between hyperchaotic maps. Nonlinear Dyn. Syst. Theory **16**, 312–321 (2016)
91. Ouannas, A.: A new generalized-type of synchronization for discrete-time chaotic dynamical systems. J. Comput. Nonlinear Dyn. **10**(6), 42 (2015)
92. Ouannas, A., Odibat, Z.: Generalized synchronization of different dimensional chaotic dynamical systems in discrete time. Nonlinear Dyn. **81**, 765–771 (2015)
93. Ouannas, A., Abu-Saris, R.: A robust control method for Q-S synchronization between different dimensional integer-order and fractional-order chaotic systems. J. Control Sci. Eng. **55–55**(20), 2015 (2015)
94. Ahlem, G., Ouannas, A.: A general control method for inverse hybrid function projective synchronization of a class of chaotic systems. Int. J. Math. Anal. **9**(9–12), 429–436 (2015)
95. Ouannas, A., Ahlem, G.: General synchronization approach for general chaotic systems in continuous-time. Far East J. Math. Sci. **96**(2), 231 (2015)
96. Ouannas, A., Al-Sawalha, M.M.: A new approach to synchronize different dimensional chaotic maps using two scaling matrices. Nonlinear Dyn. Syst. Theory **15**(4), 400–408 (2015)
97. Ouannas, A.: A new synchronization scheme for general 3D quadratic chaotic systems in discrete-time. Nonlinear Dyn. Syst. Theory **15**(2), 163–170 (2015)

98. Ahlem, G., Ouannas, A.: Synchronization analysis of quadratic chaotic systems in nD. Far East J. Math. Sci. **96**(2), 241 (2015)
99. Ouannas, A.: On full-state hybrid projective synchronization of general discrete chaotic systems. J. Nonlinear Dyn. **2014**, 35 (2014)
100. Ouannas, A., Mahmoud, E.E.: Inverse matrix projective synchronization for discrete chaotic systems with different dimensions. J. Comput. Intell. Electron. Syst. **3**(3), 188–192 (2014)
101. Ouannas, A., Aljazaery, I.: A new method to generate a discrete chaotic dynamical systems using synchronization technique. Far East J. Dyn. Syst **24**(1–2), 15–24 (2014)
102. Ouannas, A.: Chaos synchronization approach for coupled of arbitrary 3-D quadratic dynamical systems in discrete-time. Far East J. Appl. Math. **86**(3), 225 (2014)
103. Ouannas, A.: A synchronization criterion for a class of sinusoidal chaotic maps via linear controller. Int. J. Contemp. Math. Sci. **9**(14), 677–683 (2014)
104. Ouannas, A.: Chaos synchronization approach based on new criterion of stability. Nonlinear Dyn. Syst. Theory **14**(4), 395–401 (2014)
105. Ouannas, A.: A new chaos synchronization criterion for discrete dynamical systems. Appl. Math. Sci. **8**(41), 2025–2034 (2014)
106. Ouannas, A.: Synchronization and inverse synchronization of some different dimensional discrete-time chaotic dynamical systems via scaling matrices. Int. J. Chaos Control Model. Simul. **3**, 1–17 (2014)
107. Ouannas, A.: Nonlinear control method of chaos synchronization for arbitrary 2D quadratic dynamical systems in discrete-time. Int. J. Math. Anal. **8**(53), 2611–2617 (2014)
108. Ouannas, A.: Synchronization of discrete chaotic systems via double scaling matrix in different dimensions. J. Appl. Math. Comput. **3**(7), 1–5 (2014)
109. Ouannas, A.: A new QS synchronization scheme for discrete chaotic systems. Far East J. Appl. Math. **84**(2), 89 (2013)
110. Ouannas, A.: Co-existence of complete synchronization and anti-synchronization in a class of discrete rational chaotic systems. Far East J. Dyn. Syst. **23**(1/2), 41 (2013)
111. Djenina, N., Ouannas, A., Batiha, I.M., Grassi, G., Pham, V.-T.: On the stability of linear incommensurate fractional-order difference systems. Mathematics **8**, 1754 (2020)
112. Shatnawi, M.T., Djenina, N., Ouannas, A., Batiha, I.M., Grassi, G.: Novel convenient conditions for the stability of nonlinear incommensurate fractional-order difference systems. Alex. Eng. J
113. Debbouche, N., Ouannas, A., Batiha, I.M., Grassi, G., Kaabar, M.K.A.: Chaotic behavior analysis of a new incommensurate fractional-order hopfield neural network system. Complexity **1–11**(14), 2021 (2021)
114. Debbouche, N., Ouannas, A., Momani, S., Cafagna, D., Pham, V.T.: Fractional-order biological system: chaos, multistability and coexisting attractors. Eur. Phys. J. Spec. Top. **1–10**, 4 (2021)
115. Ouannas, A., Khennaoui, A.A., Oussaeif, T.E., Pham, V.T., Grassi, G., Dibi, Z.: Hyperchaotic fractional Grassi–Miller map and its hardware implementation. Integration **80**, 13–19 (2021); Ouannas, A., Debbouche, N., Pham, V.T., Kingston, S.L., Kapitaniak, T.: Chaos in fractional system with extreme events. Eur. Phys. J. Spec. Top. **230**(7–8), 2021–2033 (2021)
116. Al-Hussein, A.B.A., Tahir, F.R., Ouannas, A., Sun, T.C., Jahanshahi, H., Aly, A.A.: Chaos suppressing in a three-buses power system using an adaptive synergetic control method. Electronics **10**(13), 1532 (2021)
117. Almatroud, A.O., Khennaoui, A.A., Ouannas, A., Pham, V.T.: Infinite line of equilibriums in a novel fractional map with coexisting infinitely many attractors and initial offset boosting. Int. J. Nonlinear Sci. Numer. Simul. **11** (2021)
118. Debbouche, N., Ouannas, A., Batiha, I.M., Grassi, G.: Chaotic dynamics in a novel COVID-19 pandemic model described by commensurate and incommensurate fractional-order derivatives. Nonlinear Dyn. **1–13**, 20 (2021)
119. Debbouche, N., Momani, S., Ouannas, A., Shatnawi, M.T., Grassi, G., Dibi, Z., Batiha, I.M.: Generating multidirectional variable hidden attractors via newly commensurate and incommensurate non-equilibrium fractional-order chaotic systems. Entropy **23**(3), 261 (2021)

120. Debbouche, N., Almatroud, A.O., Ouannas, A., Batiha, I.M.: Chaos and coexisting attractors in glucose-insulin regulatory system with incommensurate fractional-order derivatives. Chaos, Solitons Fractals **143**, 110575 (2021)
121. Khennaoui, A.A., Almatroud, A.O., Ouannas, A., Al-sawalha, M.M., Grassi, G.: An unprecedented 2-dimensional discrete-time fractional-order system and its hidden chaotic attractors. Math. Probl. Eng. **1–10**(17), 2021 (2021)
122. Talbi, I., Ouannas, A., Khennaoui, A.A., Berkane, A., Batiha, I.M., Grassi, G.: Different dimensional fractional-order discrete chaotic systems based on the Caputo h-difference discrete operator: dynamics, control, and synchronization. Adv. Differ. Equ. **1–15**(13), 2020 (2020)
123. Almatroud, A.O., Khennaoui, A.A., Ouannas, A., Grassi, G., Al-Sawalha, M.M.: Dynamical analysis of a new chaotic fractional discrete-time system and its control. Entropy **22**(12), 1344 (2020)
124. Talbi, I., Ouannas, A., Grassi, G., Khennaoui, A.A., Pham, V.T., Baleanu, D.: Fractional Grassi-Miller map based on the Caputo h-difference operator: linear methods for chaos control and synchronization. Discret. Dyn. Nat. Soc. **1–10**(11), 2020 (2020)
125. Gasri, A., Ouannas, A., Khennaoui, A.A., Bendoukha, S., Pham, V.T.: On the dynamics and control of fractional chaotic maps with sine terms. Int. J. Nonlinear Sci. Numer. Simul. **21**(6) (2020)
126. Khennaoui, A.A., Ouannas, A., Odibat, Z., Pham, V.T., Grassi, G.: On the three-dimensional fractional-order Hénon map with Lorenz-like attractors. Int. J. Bifurc. Chaos **30**(11), 2050217 (2020)
127. Ouannas, A., Khennaoui, A.A., Wang, X., Pham, V.T., Boulaaras, S., Momani, S.: Bifurcation and chaos in the fractional form of Hénon-Lozi type map. Eur. Phys. J. Spec. Top. **229**, 2261–2273 (2020)
128. Ouannas, A., Khennaoui, A.A., Bendoukha, S., Pham, V.T.: The discrete fractional duffing system: chaos, 0–1 test, C0 complexity, entropy, and control. Chaos: Interdiscip. J. Nonlinear Sci. **30**(8), 083131 (2020)
129. Ouannas, A., Khennaoui, A.A., Bendoukha, S., Wang, Z., Pham, V.T.: The dynamics and control of the fractional forms of some rational chaotic maps. J. Syst. Sci. Complex. **33**(3), 584–603 (2020)
130. Ouannas, A., Almatroud, O.A., Khennaoui, A.A., Alsawalha, M.M., Baleanu, D.: Bifurcations, hidden chaos and control in fractional maps. Symmetry **12**(6), 879 (2020)
131. Thoai, V.P., Kahkeshi, M.S., Huynh, V.V., Ouannas, A., Pham, V.T.: A nonlinear five-term system: symmetry, chaos, and prediction. Symmetry **12**(5), 865 (2020)
132. Hadjabi, F., Ouannas, A., Shawagfeh, N., Khennaoui, A.A., Grassi, G.: On two-dimensional fractional chaotic maps with symmetries. Symmetry **12**(5), 756 (2020)
133. Ouannas, A., Khennaoui, A.A., Momani, S., Pham, V.T., El-Khazali, R.: Hidden attractors in a new fractional-order discrete system: chaos, complexity, entropy, and control. Chin. Phys. B **29**(5), 050504 (2020)
134. Ouannas, A., Khennaoui, A.A., Momani, S., Grassi, G., Pham, V.T., El-Khazali, R.: A quadratic fractional map without equilibria: bifurcation, 0–1 test, complexity, entropy, and control. Electronics **9**(5), 748 (2020)
135. Ouannas, A., Khennaoui, A.A., Momani, S., Grassi, G., Pham, V.T.: Chaos and control of a three-dimensional fractional order discrete-time system with no equilibrium and its synchronization. AIP Adv. **10**(4), 045310 (2020)
136. Ouannas, A., Boulaaras, S., Pham, V.T., Taher Azar, A., Khennaoui, A.A.: A fractional map with hidden attractors: chaos and control. Eur. Phys. J. Spec. Top. **229**, 1083–1093 (2020)
137. Talbi, I., Ouannas, A.: Linear Methods for Chaos Control and Synchronization. Fractional Grassi-Miller map based on the Caputo h-difference operator (2020)
138. Ouannas, A., Azar, A.T., Ziar, T.: Fractional inverse full state hybrid projective synchronisation. Int. J. Adv. Intell. Parad. **17**(3–4), 279–298 (2020)
139. Jouini, L., Ouannas, A., Khennaoui, A.A., Wang, X., Grassi, G., Pham, V.T.: The fractional form of a new three-dimensional generalized Hénon map. Adv. Differ. Equ. **2019**(1), 1–12 (2019)

140. Ouannas, A., Khennaoui, A.A., Grassi, G., Bendoukha, S.: On chaos in the fractional-order Grassi-Miller map and its control. J. Comput. Appl. Math. **358**, 293–305 (2019)
141. Huynh, V.V., Ouannas, A., Wang, X., Pham, V.T., Nguyen, X.Q., Alsaadi, F.E.: Chaotic map with no fixed points: entropy, implementation and control. Entropy **21**(3), 279 (2019)
142. Khennaoui, A.A., Ouannas, A., Bendoukha, S., Grassi, G., Lozi, R.P., Pham, V.T.: On fractional-order discrete-time systems: chaos, stabilization and synchronization. Chaos, Solitons Fractals **119**, 150–162 (2019)
143. Ouannas, A., Khennaoui, A.A., Bendoukha, S., Vo, T.P., Pham, V.T., Huynh, V.V.: The fractional form of the Tinkerbell map is chaotic. Appl. Sci. **8**(12), 2640 (2018)
144. Ouannas, A., Wang, X., Khennaoui, A.A., Bendoukha, S., Pham, V.T., Alsaadi, F.E.: Fractional form of a chaotic map without fixed points: chaos, entropy and control. Entropy **20**(10), 720 (2018)
145. Khennaoui, A.A., Ouannas, A., Bendoukha, S., Wang, X., Pham, V.T.: On chaos in the fractional-order discrete-time unified system and its control synchronization. Entropy **20**(7), 530 (2018)
146. Khennaoui, A.A., Ouannas, A., Bekiros, S., Aly, A.A., Alotaibi, A., Jahanshahi, H.: Hidden homogeneous extreme multistability of a fractional-order hyperchaotic discrete-time system: chaos, initial offset boosting, amplitude control, control, and synchronization. Symmetry **15**(1), 139 (2023)
147. Abbes, A., Ouannas, A., Shawagfeh, N., Jahanshahi, H.: The fractional-order discrete COVID-19 pandemic model: stability and chaos. Nonlinear Dyn. **111**(1), 965–983 (2023)
148. Shatnawi, M.T., Abbes, A., Ouannas, A., Batiha, I.M.: A new two-dimensional fractional discrete rational map: chaos and complexity. Phys. Scr. **98**(1), 015208 (2022)
149. Bensid Ahmed, S., Ouannas, A., Al Horani, M., Grassi, G.: The discrete fractional variable-order tinkerbell map: chaos, 0–1 test, and entropy. Mathematics **10**(17), 3173 2022
150. Gasri, A., Ouannas, A., Khennaoui, A.A., Grassi, G., Oussaeif, T.E., Pham, V.T.: Chaotic fractional discrete neural networks based on the Caputo h-difference operator: stabilization and linear control laws for synchronization. Eur. Phys. J. Spec. Top. **231**(10), 1815–1829 (2022)
151. Abbes, A., Ouannas, A., Shawagfeh, N., Khennaoui, A.A.: Incommensurate fractional discrete neural network: chaos and complexity. Eur. Phys. J. Plus **137**(2), 235 (2022)
152. Yousfi, H., Gasri, A., Ouannas, A.: Stabilization of Chaotic h-difference systems with fractional order. Nonlinear Dyn. Syst. Theory, 468
153. Abbes, A., Ouannas, A., Shawagfeh, N.: The incommensurate fractional discrete macroeconomic system: bifurcation, chaos and complexity. Chin. Phys. B **4** (2022)
154. Khennaoui, A.A., Ouannas, A., Momani, S., Dibi, Z., Grassi, G., Baleanu, D.: Hyperchaotic dynamics of a new fractional discrete-time system. Fractals **29**(08), 2140034 (2021)
155. Pham, V.T., Ouannas, A., Volos, C., Kapitaniak, T.: A simple fractional-order chaotic system without equilibrium and its synchronization. AEU-Int. J. Electron. Commun. **86**, 69–76 (2018)
156. Christopher, G., Allan, C.P.: Discrete Fractional Calculus. Springer, Berlin (2010). 978-3-319-25560-6
157. Abdeljawad, T.: On Riemann and Caputo fractional differences. Comput. Math. Appl. **62**, 1602–1611 (2011)
158. Rana, S.M., Kulsum, U.: Bifurcation analysis and chaos control in a discrete-time predator-prey system of Leslie type with simplified Holling type IV functional response. Discret. Dyn. Nat. Soc. **2017**, 9705985 (2017)
159. Saadeh, R., Abbes, A., Al-Husban, A., Ouannas, A., Grassi, G.: The fractional discrete predator–prey model: chaos, control and synchronization. Fractal Fract. **7**, 120 (2023). https://doi.org/10.3390/fractalfract7020120
160. Maddali, R.K., Ahluwalia, D., Chaudhuri, A., Hassan, S.S.: Dynamics of a three dimensional chaotic cancer model. Int. J. Math. Trends Technol. (IJMTT) **53**(5), 353–368 (2018)
161. Debbouche, N., Ouannas, A., Batiha, I.M., Grassi, G.: Chaotic dynamics in a novel COVID-19 pandemic model described by commensurate and incommensurate fractional-order derivatives. Nonlinear Dyn. 1–13

162. Batiha, I.M., Oudetallah, J., Ouannas, A., Al-Nana, A.A., Jebril, I.H.: Tuning the fractional-order PID-controller for blood glucose level of diabetic patients. J. Adv. Soft Comput. Appl. **13**
163. Batiha, I.M., Momani, S., Ouannas, A., Momani, Z., Hadid, S.B.: Fractional-order COVID-19 pandemic outbreak: Modeling and stability analysis. Int. J. Biomath. 2150090
164. Debbouche, N., Ouannas, A., Grassi, G., Al-Hussein, A.B.A., Tahir, F.R., Saad, K.M., Jahanshahi, H., Aly, A.A.: Chaos in cancer tumor growth model with commensurate and incommensurate fractional-order derivatives. Comput. Math. Methods Med. (2022)

Nonlinear Fractional Discrete Neural Networks: Stability, Stabilization and Synchronization

Amel Hioual, Adel Ouannas, and Taki Eddine Oussaeif

Abstract Fractional-order discrete-time neural networks are a type of discrete system characterized by non-integer order difference operators. Because the dynamics of these networks are required for their practical uses, this chapter concentrates on the dynamics of fractional discrete-time neural networks. First, a class of discrete-time fractional-order neural networks with multiple orders is described, and a novel theorem is established to examine the stability issue of the investigated neural networks. Two innovative theorems are shown in specifically, one addressing the existence of the solution for the suggested fractional-order network and the other about its Ulam-Hyers stability. In addition two several numerical examples with simulations that illustrate the previous results. Secondly, an unique chaotic h-fractional discrete neural network model is introduced in this chapter. The chaotic dynamics of the network are stabilized at the origin by an appropriate controller. Furthermore, simulation findings are used to show the viability of the theoretical results obtained herein. Finally, a novel synchronization technique is developed in which a new discrete fractional-order chaotic neural network is regarded the master system and the h-fractional discrete neural network is considered the responder system. Experimentation and numerical simulations are also offered to support the study's key results.

Keywords Fractional-order Caputo h-difference operator · Discrete-time fractional-order neural network · Ulam-Hyers stability · Chaotic discrete fractional-order neural networks · Stabilization · Linear control low · Synchronization

---

A. Hioual (✉)
Laboratory of Dynamical Systems and Control, University of Oum EL-Bouaghi, Oum El Bouaghi, Algeria
e-mail: amel.hioual@univ-oeb.dz

A. Ouannas · T. E. Oussaeif
Department of Mathematics and Computer Science, University of Oum EL-Bouaghi, Oum El Bouaghi, Algeria
e-mail: ouannas.adel@univ-oeb.dz

T. E. Oussaeif
e-mail: taki_maths@live.fr

A. Ben Makhlouf et al. (eds.), *State Estimation and Stabilization of Nonlinear Systems*, Studies in Systems, Decision and Control 491,
https://doi.org/10.1007/978-3-031-37970-3_10

1 Introduction

Considering the long and rich history of fractional calculus, discrete fractional calculus has quickly drawn mathematicians and researchers into its relatively young research field. During this period of time, the theory of discrete fractional calculus has progressed throughout many directions parallel to the theory of fractional calculus, including fractional difference equations, discrete Mittag-Leffler functions, and inequalities with discrete fractional operators, as can be seen in [1–4] and the references therein.

The problem of stability of fractional order systems has recently had extensive consideration. Different types of stability, including asymptotic stability, Mittag-Leffler stability, Ulam Hyers stability, and finite time stability, have been examined in the context of continuous-time systems. In the case of fractional discrete-time systems presenting conditions is essentially sufficient. This results from the fact that the stability of a fractional discrete time system is equivalent to the stability of a fractional continuous system. Several articles in the literature have looked into the stability characteristics of fractional discrete systems. For example: Conditions for incommensurate fractional linear discrete systems stability using the Z transform was presented in [5]. The asymptotic stability of fractional nonlinear discrete systems was also presented in [6] along with some novel practical conditions to guarantee it. In [7], the stability analysis of linear and nonlinear incommensurate fractional-order difference systems was investigated. Moreover, One can highlight a few of the numerous applications of the stability results that included several research areas and topics, for instance, A new theorem that emphasizes that the pandemic ends when an inequality involving the proportion of the population in quarantine is met was reported in [8] along with a novel fractional discrete-time COVID-19 model that takes the number of vaccinated people into account. To analyze the COVID-19 epidemic's dynamics, a straightforward, original fractional-order discrete model was put out in [9]. In [10], a unique discrete-time COVID-19 model was published that makes the number of immunized people an extra state variable in the system equations. More works can be found here [11–13].

When seen macroscopically, many natural occurrences appear to be regular phenomena, but when viewed microscopically, irregularities become apparent. There are actually numerous applications that might make use of the chaotic behavior of dynamic systems. Thus, the synchronization and stabilization of fractional-order chaotic systems is an intriguing topic that has drawn the interest of several researchers in a variety of scientific disciplines. Over the past several years, there has been a lot of study and development on the subject of synchronization in chaotic fractional systems (see [14–65] for example, and references therein). There are several synchronization strategies for fractional differential systems, including linear control synchronization, adaptive sliding mode synchronization, projective synchronization using single sinusoidal coupling, and hybrid chaos synchronization using a resilient mechanism. In fact, the majority of the synchronization strategies for fractional differential systems may also be employed for fractional discrete dynamical systems.

For example, In [66], the dynamics and synchronization of The Fractional Discrete Predator-Prey Model was explored. In [67], chaos management and synchronization of a class of a 2D fractional hyperchaotic map was addressed. Also, a nonlinear control rule was used to create synchronization across two-dimensional fractional discrete-time systems with chaotic hidden attractors In [68]. In [69], the control and synchronization of a two-dimensional fractional-order Lorenz Discrete System and a three-dimensional fractional-order Wang Discrete System was examined. In addition, a unique form of the fractional Grassi-Miller map, as well as better strategies for regulating and synchronizing its dynamics was investigated in [70]. In [71], a novel fractional order discrete-time system with no equilibrium is introduced, and its complicated dynamical behaviors are numerically discussed and a synchronization control technique for two separate fractional order discrete-time systems with hidden attractors is also given. Furthermore, a three fractional chaotic maps based on the well-known 3D Stefanski, Rössler, and Wang maps, and the dynamics of the suggested fractional maps were examined, as well as a combined synchronization method was presented in [72]. The Q-S synchronization for different dimensions chaotic dynamical systems in discrete time was investigated in [73]. In addition, an extension of the notion of hybrid chaotic synchronization in discrete-time by merging full state hybrid projective synchronization and inverse full state hybrid projective synchronization was explored in [74]. The dynamics and control of the fractional form of the discrete double scroll hyperchaotic map, including synchronization, were reported in [75]. In [76], the full-state hybrid projective synchronisation of integer- and fractional-order discrete-time chaotic systems with various dimensions was investigated. Furthermore, the synchronization of fractional-order discrete-time chaotic systems using two scaling matrices was explored in [77]. In, [78] two methods for generalized synchronized synchronization and inverse generalized synchronization of fractional discrete-time chaotic systems with non-identical dimensions were presented. More research papers on synchronization in chaotic discrete fractional systems may be found in the references [79–104].

Stabilization, on the other hand, is a kind of chaotic synchronization control that makes use of the asymptotic stability approach. It is one of the most well-liked and effective control approaches. Numerous studies on this issue have been undertaken, For instance, in [105] it was explained how the h-fractional difference operator may stabilize chaotic dynamics in a fractional order discrete neural network. A study on fractional discrete neural networks' chaotic dynamics, synchronization, and stabilization may be found in [106]. In [107], a Lyapunov-based linear control rule was developed to stabilize the fractional discrete-time Ushio system, which is chaotic. In [108], the stabilization of financial systems utilizing a controller that creates a strong, stable financial closed-loop system was investigated.

Research on fractional neural networks' theory and applications have recently received a lot of interest. Academics have paid a lot of attention to the study of complicated dynamic behavior in neural networks. Many intriguing nonlinear phenomena, including multiple chaotic attractors and hidden attractors, have been identified. Numerous modified fractional neural network models, particularly fractional order neural networks, have been developed because dynamical behaviors are directly tied to the usage of neural networks. Chaos is frequently seen in fractional-order fractional neural network models as a characteristic non-linear behavior. For example, Based on the Caputo derivative, a fractional variable-order neural network was developed in [109]. The chaotic incommensurate fractional-order Hopfield neural network model's various dynamics were investigated in [110]. All of the aforementioned fractional-order models are continuous-time neural networks that are described by fractional order differential equations. There is limited research on the examination of the chaotic behavior and stability for discrete-time fractional-order neural networks. For instance, The chaotic behavior of a discrete-time Hopfield neural network with non-commensurate fractional variable-order was studied in [111] utilizing phase-plot portraiture, bifurcation diagrams, and Lyapunov exponent diagrams. A unique discrete-time fractional Hopfield neural network with incommensurate order was presented in [112], and the dynamical behavior was assessed using the 0–1 test, phase attractors, bifurcation diagrams, and Lyapunov exponents. The chaotic dynamics of the network were stabilized and achieved synchronized dynamics in [113], which presents a unique chaotic h-fractional discrete neural network model. In [114], it was researched how to stabilize chaotic dynamics in discrete fractional-order neural networks. A network model based on the nabla Caputo h-discrete operator was described in [115], along with details on its Mittag-Leffler stability. An introduction to a variable-order fractional discrete neural network model and an analysis of its Ulam-Hyers stability may be found in [116]. In [117], in addition to its uniqueness and uniform stability, a sufficient condition for the existence of the solution of a novel class of discrete-time fractional-order neural networks utilizing discrete nabla operator of variable-order was addressed. New theorems on the asymptotic stability of nonlinear nabla variable-order systems and nonlinear nabla variable-order neural networks were illustrated in [118].

The goal of this chapter is to offer a deeper understanding of the various dynamics of a novel class of discrete fractional neural networks that generalize the models already studied in the literature. Moving on from explaining the model in hand and providing the required and sufficient conditions for its Ulam Hyers stability, we will investigate the chaos synchronization and stabilization of chaotic discrete fractional order neural networks generated by particular system characteristics. Furthermore, the primary developments of the chapter might be recapped in the following order:

- Novel generalized discrete fractional neural networks.
- Rigid method including fixed point method, namely the Krasnoselskii Fixed Point Theorem and Arzela-Ascoli's Theorem, as well as the Ulam Hyers type method.
- New chaotic fractional discrete neural network model.
- Stabilization of chaos using relatively basic controls.
- Chaos synchronize via Linear control laws.
- Thorough method based on innovative Lyapunov-based theorems.

The following is how the paper is structured: Sect. 2 provides a mathematical foundation for discrete fractional calculus. In Sect. 3, a novel discrete fractional neural network is introduced, a specific system solution is provided to aid in the investigation of the solution's existence and Ulam Hyers stability, and several examples with simulations are provided to validate the theoretical results. Section 4 describes a chaotic discrete fractional neural network and its chaotic attractors, as well as bifurcation diagrams and the computation of Lyapunov exponents. The chaotic dynamics of the 3D network are subsequently stabilized at the origin using an appropriate controller. Section 5 uses relatively basic linear controllers to synchronize two chaotic fractional discrete neural networks using Lyapunov-based method.

2 Some Basic Mathematical Background on Fractional Discrete Calculus

For clarification, an overview of the mathematical background linked to the fractional h-difference operators is presented in this section.

Definition 1 (*Fractional sum* [119]) Let $\varkappa : (\hbar\mathbb{N})_{t_0} \to \mathbb{R}$ and $0 < \alpha$ be given. t_0 is a starting point. The $\alpha-$th order $\hbar-$sum is given by

$$_{\hbar}\Delta_{t_0}^{-\alpha}\varkappa(t) = \frac{\hbar}{\Gamma(\alpha)}\sum_{s=\frac{t_0}{\hbar}}^{\frac{t}{\hbar}-\alpha}(t-\sigma(s\hbar))_{\hbar}^{\alpha-1}\varkappa(s\hbar), \quad \sigma(s\hbar) = (s+1)\hbar, \quad t \in (\hbar\mathbb{N})_{t_0+\alpha\hbar}$$

where the h-falling factorial function is defined as

$$t_{\hbar}^{(\alpha)} = \hbar^{\alpha}\frac{\Gamma(\frac{t}{\hbar}+1)}{\Gamma(\frac{t}{\hbar}+1-\alpha)}, \quad t, \alpha \in \mathbb{R}$$

while $(\hbar\mathbb{N})_{t_0+(1-\alpha)\hbar} = \{t_0 + (1-\alpha)\hbar, t_0 + (2-\alpha)\hbar, \ldots\}$.

Definition 2 (*Caputo delta difference* [119]) For $\varkappa(t)$ defined on $(\hbar\mathbb{N})_a$ and $0 < \alpha$. The Caputo like difference is defined by

$${}_{\hbar}^{C}\Delta_{t_0}^{\alpha}\varkappa(t) = {}_{\hbar}\Delta_{t_0}^{-(m-\alpha)}\Delta_{\hbar}^{m}\varkappa(t), \quad t \in (\hbar\mathbb{N})_{t_0+(m-\alpha)\hbar}$$

where $\Delta_{\hbar}\varkappa(t) = \dfrac{\varkappa(t+\hbar)-\varkappa(t)}{\hbar}$ and $m = [\alpha]+1$

If the fractional order α is a positive integer m, then we have the following definition

$${}_{\hbar}^{C}\Delta_{t_0}^{\alpha}\varkappa(t) = \Delta_{\hbar}^{m}\varkappa(t), \quad t \in (\hbar\mathbb{N})_{t_0}.$$

Lemma 1 ([119]) *Here is some useful proprieties used in this paper*

- *Discrete Leibniz integral law*

$${}_{\hbar}\Delta_{t_0+(1-\alpha)\hbar}^{-\alpha}\,{}_{\hbar}^{C}\Delta_{t_0}^{\alpha}\varkappa(t) = \varkappa(t) - \varkappa(t_0), \quad 0 < \alpha \leq 1, \quad t \in (\hbar\mathbb{N})_{t_0+\hbar}$$

- *Fractional Caputo difference of a constant* $\varkappa$

$$ {}^{C}_{\hbar}\Delta^{\alpha}_{t_0}\varkappa = 0, \quad 0 < \alpha \leq 1 $$

- *Delta difference of the h?falling factorial function*

$$ \Delta_s (t - s\hbar)^{(\alpha)}_{\hbar} = -\alpha (t - \sigma(s\hbar))^{(\alpha-1)}_{\hbar}. $$

Lemma 2 (Arzela-Ascoli's Theorem [120]) *A bounded, uniformly Cauchy subset* Ω *of Banach space* E *is relatively compact.*

Lemma 3 (Krasnoselskii Fixed Point Theorem [120]) *Let* Ω *be a nonempty, closed, convex and bounded subset of Banach space* E*. Suppose that* S, T *are two operators, such that*

(i) S *is a contraction;*
(ii) *for any* $\varkappa, y \in \Omega$, $Sx + Ty \in \Omega$*;*
(iii) T *is a continuous and* $B(\Omega)$ *is relatively compact.*

Then for $\varkappa \in \Omega$*, the operator equation* $S\varkappa + T\varkappa = \varkappa$ *has a solution.*

In order to determine some practical stability criteria of the zero equilibrium point for the following nonlinear fractional-order difference system, we basically display a key result stated in [121].

$$ {}^{C}_{\hbar}\Delta^{\alpha}_{t}\varkappa(t) = f(t + \alpha\hbar, \varkappa(t + \alpha\hbar)), \quad t \in (\hbar\mathbb{N})_{t_0+(1-\alpha)\hbar}. \tag{1} $$

Theorem 1 ([121]) *Let* $\varkappa = 0$ *be an equilibrium point of the nonlinear fractional-order discrete system (3). If there exists a positive definite and decrescent scalar function* $V(t, \varkappa(t))$*, such that* ${}^{C}_{\hbar}\Delta^{\alpha}_{t} V(t, \varkappa(t)) \leq 0$*, then the equilibrium point is asymptotically stable*

Lemma 4 ([121]) *The following inequality is accurate for any* $t \in (\hbar\mathbb{N})_{t_0+(1-\alpha)\hbar}$

$$ {}^{C}_{\hbar}\Delta^{\alpha}_{t}\varkappa^2(t) \leq 2\varkappa(t + \alpha\hbar)\, {}^{C}_{\hbar}\Delta^{\alpha}_{t}\varkappa(t), \quad 0 < \alpha \leq 1. \tag{2} $$

3 Stability of Nonlinear Discrete Fractional Neural Networks

In this section, we discuss the existence and stability of a new class of fractional-order discrete neural network. Compared to the literature which already exists regarding the approach used in our work, we state [122] that provided Ulam-Hyers stability results for Caputo nabla fractional difference equations across the linear and non-linear scenarios using generalized Gronwall inequality, while on the other hand, [123] explored the Ulam-Hyers stability of a linear neural networks with fractional order using the discrete version of the Laplace transforme. Additionally, In [124] the existence of solutions for a discrete initial value fractional Duffing equation with forcing term was demonstrated utilizing the Krasnoselskii fixed point theorem and the Hyers-Ulam stability was also discussed.

Based on the studied literature, the model discussed in this section is significantly more involved than those previously researched, and this form of discrete neural network with fractional-order Caputo h-difference operator has not before been examined. As a result, all of the outcomes and comments in this section are original and improved when compared to [122–124].

3.1 Novel Fractional-Order Discrete Neural Network

Discrete-time fractional-order neural networks are quite well-suited to describing the behavior of fractional-order discrete nonlinear systems. In actuality, due to their modeling skills to a sufficient accuracy, they are regarded a candidate for a generic, parametric, non-linear model of a vast category of discrete non-linear systems with fractional-order. In this section, we investigate a sort of fractional-order discrete neural network represented by

$$ {}_{\hbar}^{C}\Delta_{t_{ql}}^{\alpha_q} \varkappa(t) = -A\varkappa(t+\alpha_q\hbar) + B\mathfrak{f}(t+\alpha_q\hbar, \varkappa(t+\alpha_q\hbar)), \tag{3} $$

where
(3) can be reduced as

$$ \begin{cases} {}_{\hbar}^{C}\Delta_{t_0}^{\alpha_0} \varkappa(t) = -A\varkappa(t+\alpha_0\hbar) + B\mathfrak{f}(t+\alpha_q\hbar, \varkappa(t+\alpha_0\hbar)), & t \in \{t_0 + (1-\alpha_0)\hbar, \dots, t_l - \alpha_0\hbar\} \\ {}_{\hbar}^{C}\Delta_{t_l}^{\alpha_1} \varkappa(t) = -A\varkappa(t+\alpha_1\hbar) + B\mathfrak{f}(t+\alpha_q\hbar, \varkappa(t+\alpha_1\hbar)), & t \in \{t_l + (1-\upsilon_1)\hbar, \dots, t_{2l} - \alpha_1\hbar\} \\ \dots \\ {}_{\hbar}^{C}\Delta_{t_{ql}}^{\alpha_q} \varkappa(t) = -A\varkappa(t+\alpha_q\hbar) + B\mathfrak{f}(t+\alpha_q\hbar, x(t+\alpha_q\hbar)), & t \in \{t_{ql} + (1-\alpha_q)\hbar, \dots, t_{(q+1)l} - \alpha_q\hbar\} \end{cases} \tag{4} $$

$t \in (\hbar\mathbb{N})_{t_{ql}+(1-\alpha_q)\hbar}$, $0 < \alpha_q \leq 1$, $k = 0, \dots, m-1$, m is the number of the intervals. ${}_{\hbar}^{C}\Delta_{t_{ql}}^{\alpha_q}$ denotes the fractional-order Caputo $\hbar$-difference operator of order α_q, $\varkappa(t) = (\varkappa_1(t), \dots, \varkappa_p(t))^T \in \mathbb{R}^p$ is the state of the unit at time t and p is the dimension, $A = diag(-a_1, \dots, -a_p)$ where $a_i > 0$ represents the matrix with which the neurons will reset its potential to the resting state when disconnected from the network, $B \in \mathbb{R}_{p*p}$ corresponds to the connection weights and finally $\mathfrak{f}(t, \varkappa(t)) \in C((\hbar\mathbb{N})_{t_{ql}+(1-\alpha_q)\hbar}, \mathbb{R}^p)$ is the activation function.

Lemma 5 *The solution $\varkappa(t)$ of system (3) satisfies*

$$\varkappa(t)=\begin{cases}\varkappa(t_0)+\frac{\hbar}{\Gamma(\alpha_0)}\sum_{s=\frac{t_0}{\hbar}+1-\alpha_0}^{\frac{t}{\hbar}-\alpha_0}(t-\sigma(s\hbar))_{\hbar}^{\alpha_0-1}[-A\varkappa(s\hbar)+B\mathfrak{f}(s\hbar,\varkappa(s\hbar))], \quad t\in\{t_0+\hbar,\ldots,t_l\}\\ \varkappa(t_0)+\sum_{n=1}^{m-1}\frac{\hbar}{\Gamma(\alpha_{n-1})}\sum_{s=\frac{t_{(n-1)l}}{\hbar}+1-\alpha_{n-1}}^{\frac{t_{nl}}{\hbar}-\alpha_{n-1}}(t_{nl}-\sigma(s\hbar))_{\hbar}^{\alpha_{n-1}-1}[-A\varkappa(s\hbar)+B\mathfrak{f}(s\hbar,\varkappa(s\hbar))]\\ +\frac{\hbar}{\Gamma(\alpha_{m-1})}\sum_{s=\frac{t_{(m-1)l}}{\hbar}+1-\alpha_{m-1}}^{\frac{t}{\hbar}-\alpha_{m-1}}(t-\sigma(s\hbar))_{\hbar}^{\alpha_{m-1}-1}[-A\varkappa(s\hbar)+B\mathfrak{f}(s\hbar,\varkappa(s\hbar))], \quad t\in\{t_{(m-1)l}+\hbar,\ldots,t_{ml}\}\end{cases} \tag{5}$$

Proof System (4) is identical to

$$\begin{cases}\varkappa(t)=\varkappa(t_0)+{}_{\hbar}\Delta_{t_0+(1-\alpha_0)\hbar}^{-\alpha_0}[-A\varkappa(t)+B\mathfrak{f}(t,\varkappa(t))], \quad t\in\{t_0+\hbar,\ldots,t_l\}\\ \varkappa(t)=\varkappa(t_l)+{}_{\hbar}\Delta_{t_l+(1-\alpha_1)\hbar}^{-\alpha_1}[-A\varkappa(t)+B\mathfrak{f}(t,\varkappa(t))], \quad t\in\{t_l+\hbar,\ldots,t_{2l}\}\\ \ldots\\ \varkappa(t)=\varkappa(t_{(m-1)l})+{}_{\hbar}\Delta_{t_{(m-1)l}+(1-\alpha_{m-1})\hbar}^{-\alpha(m-1)}[-A\varkappa(t)+B\mathfrak{f}(t,\varkappa(t))], \quad t\in\{t_{(m-1)l}+\hbar,\ldots,t_{ml}\}\end{cases}$$

As for $k=0,\quad t\in\{t_0+(1-\upsilon_0)\hbar,\ldots,t_l-\upsilon_0\hbar\}$, We notice that the solution of (3) may be given by

$$\varkappa(t)=\varkappa(t_0)+{}_{\hbar}\Delta_{t_0+(1-\alpha_0)\hbar}^{-\alpha_0}[-A\varkappa(t)+B\mathfrak{f}(t,\varkappa(t))]$$

We can get even more

$$\varkappa(t)=\varkappa(t_0)+\frac{\hbar}{\Gamma(\alpha_0)}\sum_{s=\frac{t_0}{\hbar}+1-\alpha_0}^{\frac{t}{\hbar}-\alpha_0}(t-\sigma(s\hbar))_{\hbar}^{\alpha_0-1}[-A\varkappa(s\hbar)+B\mathfrak{f}(s\hbar,\varkappa(s\hbar))]$$

and we can have $\varkappa(t_l)=\varkappa(t_0)+\frac{\hbar}{\Gamma(\alpha_0)}\sum_{s=\frac{t_0}{\hbar}+1-\alpha_0}^{\frac{t_l}{\hbar}-\alpha_0}(t_l-\sigma(s\hbar))_{\hbar}^{\alpha_0-1}[-A\varkappa(s\hbar)+B\mathfrak{f}(s\hbar,\varkappa(s\hbar))]$ as an initial condition of

$${}_{\hbar}^{C}\Delta_{t_l}^{\alpha_1}\varkappa(t)=-A\varkappa(t+\alpha_1\hbar)+B\mathfrak{f}(t+\alpha_1\hbar,\varkappa(t+\alpha_1\hbar)), \quad t\in\{t_l+(1-\upsilon_1)\hbar,\ldots,t_{2l}-\alpha_1\hbar\}$$

By employing the same method, we can obtain the expression $x(t)$ for any $t\in\{t_l+(1-\upsilon_1)\hbar,\ldots,t_{2l}-\alpha_1\hbar\}$

$$\begin{aligned}\varkappa(t)&=\varkappa(t_l)+\frac{\hbar}{\Gamma(\alpha_1)}\sum_{s=\frac{t_0}{\hbar}+1-\alpha_1}^{\frac{t}{\hbar}-\alpha_1}(t-\sigma(s\hbar))_{\hbar}^{\alpha_1-1}[-A\varkappa(s\hbar)+B\mathfrak{f}(s\hbar,\varkappa(s\hbar))]\\ &=\varkappa(t_0)+\frac{\hbar}{\Gamma(\alpha_0)}\sum_{s=\frac{t_0}{\hbar}+1-\alpha_0}^{\frac{t_l}{\hbar}-\alpha_0}(t_l-\sigma(s\hbar))_{\hbar}^{\alpha_0-1}[-A\varkappa(s\hbar)+B\mathfrak{f}(s\hbar,\varkappa(s\hbar))]\\ &\quad+\frac{\hbar}{\Gamma(\alpha_1)}\sum_{s=\frac{t_l}{\hbar}+1-\alpha_1}^{\frac{t}{\hbar}-\alpha_1}(t-\sigma(s\hbar))_{\hbar}^{\alpha_1-1}[-A\varkappa(s\hbar)+B\mathfrak{f}(s\hbar,\varkappa(s\hbar))]\end{aligned}$$

Now, for any $t \in \{t_{ql} + (1-\alpha_q)\hbar, \ldots, t_{(q+1)l-\alpha_q\hbar}\}$, $q = 2, 3, \ldots, m-1$. We may obtain the solution $\varkappa(t)$ indicated in the Lemma 5 and the proof is completed. □

Here we lay out the assumptions that will be required in the investigations that will follow in the next sections.

(H_1) for all $t \in (\hbar\mathbb{N})_{t_{(m-1)l}+(1-\alpha_{m-1})\hbar}$, $\mathfrak{f}(t,\varkappa)$ represents a continuous function with respect to $\varkappa$, and there exists a constant $L \in \mathbb{R}^+$ such that

$$\|\mathfrak{f}(t,\varkappa) - \mathfrak{f}(t,\mathfrak{y})\| \leq L\|\varkappa - \mathfrak{y}\|$$

(H_2) There exists a constant $M \in \mathbb{R}^+$ such that

$$M < 1$$

Where

$$\mathfrak{M} = [\mathfrak{M}_A + L\mathfrak{M}_B]\Bigg(\sum_{n=1}^{m-1}\frac{(t_{nl} + (\alpha_{n-1}-1)\hbar + t_{(n-1)l})_{\hbar}^{(\alpha_{n-1})}}{\Gamma(1+\alpha_{n-1})} + \sup_{t\in\{t_{(m-1)l}+\hbar,\ldots,t_{ml}\}}\frac{(t + (\alpha_{m-1}-1)\hbar + t_{(m-1)l})_{\hbar}^{(\alpha_{m-1})}}{\Gamma(1+\alpha_{m-1})}\Bigg)$$

$$\mathfrak{M}_A = \|A\|_\infty, \quad \mathfrak{M}_B = \|B\|_\infty$$

3.2 *Existence of the Solution of Fractional-Order Discrete Neural Network*

This part establishes specific fixed point outcomes that will be required later in the section. By assuming that the starting state is bounded and constructing necessary conditions that rely on the assumptions (H_1) and (H_2) to ensure that at least one solution exists. To do so, we use Kransnoselkii's fixed point theorem 3. For that, we define the following operator:

$$\varphi\varkappa(t) = \varkappa(t_0) + \sum_{n=1}^{m-1}\frac{\hbar}{\Gamma(\alpha_{n-1})}\sum_{s=\frac{t_{(n-1)l}}{\hbar}+1-\alpha_{n-1}}^{\frac{t_{nl}}{\hbar}-\alpha_{n-1}}(t_{nl}-\sigma(s\hbar))_{\hbar}^{\alpha_{n-1}-1}[-A\varkappa(s\hbar) + B\mathfrak{f}(s\hbar,\varkappa(s\hbar))]$$
$$+ \frac{\hbar}{\Gamma(\alpha_{m-1})}\sum_{s=\frac{t_{(m-1)l}}{\hbar}+1-\alpha_{m-1}}^{\frac{t}{\hbar}-\alpha_{m-1}}(t-\sigma(s\hbar))_{\hbar}^{\alpha_{m-1}-1}[-A\varkappa(s\hbar) + B\mathfrak{f}(\varkappa(s\hbar), s\hbar))], \quad t \in (\hbar\mathbb{N})_{t_0+\hbar}$$

It is obvious that $\varkappa$ is a solution of (3). if and only if $\varkappa$ is a fixed point of the operator φ.

For any $t \in (\hbar\mathbb{N})_{t_0+\hbar}$, we define the operators

$$\psi\varkappa(t) = \sum_{n=1}^{m-1} \frac{\hbar}{\Gamma(\alpha_{n-1})} \sum_{s=\frac{t_{(n-1)l}}{\hbar}+1-\alpha_{n-1}}^{\frac{t_{nl}}{\hbar}-\alpha_{n-1}} (t_{nl} - \sigma(s\hbar))_{\hbar}^{\alpha_{n-1}-1} [-A\varkappa(s\hbar) + B\mathfrak{f}(s\hbar, \varkappa(s\hbar))]$$

and

$$\Phi\varkappa(t) = \varkappa(t_0) + \frac{\hbar}{\Gamma(\alpha_{m-1})} \sum_{s=\frac{t_{(m-1)l}}{\hbar}+1-\alpha_{m-1}}^{\frac{t}{\hbar}-\alpha_{m-1}} (t - \sigma(s\hbar))_{\hbar}^{\alpha_{m-1}-1} [-A\varkappa(s\hbar) + B\mathfrak{f}(s\hbar, \varkappa(s\hbar))]$$

Next, a new Theorem is given, which shows the existence of the solution of the fractional-order discrete neural network (3).

Theorem 2 *Suppose that* (H_1) *and* (H_2) *hold and there exists positive* σ *such that* $\|\varkappa(t_0)\| \leq \kappa$. *if the following condition hold*

$$\mathfrak{r} > \frac{\kappa}{1 - \mathfrak{M}}$$

Then, system (3) has at least one bounded solution in $\Omega = \{\varkappa \in \mathbb{R}^p : \|\varkappa\| \leq \mathfrak{r}\}$.

Proof Let $\varkappa : (\hbar\mathbb{N})_{t_0} \to \mathbb{R}^P$, $|\varkappa(t)|$ denotes the norm l_∞ of the vector $\varkappa(t)$, $|\varkappa(t)| = \max_{i=1,\ldots,p} |\varkappa_i(t)|$ and let be the supremum norm $\|\varkappa(t)\| = \sup_{t\in(\hbar\mathbb{N})_{t_0}} \|\varkappa(t)\|$ defined on the set Ω. The matrix norm is used as $\|B\|_\infty = \max_{i=1,\ldots,p} \sum_{j=1}^{p} |b_{ij}|$. it is obvious that Ω is a nonempty, closed, bounded and convex subset of $\mathbb{R}^p$. The proof includes 4 different steps.

1. We demonstrate that $^{\Phi}$ maps Ω into Ω, for any $x \in \Omega$ we have

$$\|\Phi\varkappa(t)\| \leq \|\varkappa(t_0)\| + \frac{\hbar}{\Gamma(\alpha_{m-1})} \sum_{s=\frac{t_{(m-1)l}}{\hbar}+1-\alpha_{m-1}}^{\frac{t}{\hbar}-\alpha_{m-1}} (t - \sigma(s\hbar))_{\hbar}^{\alpha_{m-1}-1} \|$$

$$- A\varkappa(s\hbar) + B\mathfrak{f}(s\hbar, x(s\hbar))\| \leq \kappa + [\mathfrak{M}_A + L\mathfrak{M}_B] \sup_{t\in\{t_{(m-1)l}+\hbar,\ldots,t_{ml}\}}$$

$$\frac{(t + (\alpha_{m-1} - 1)\hbar + t_{(m-1)l})_{\hbar}^{(\alpha_{m-1})}}{\Gamma(1 + \alpha_{m-1})} \|\varkappa\| \leq \kappa + [\mathfrak{M}_A + L\mathfrak{M}_B]r$$

$$\sup_{t\in\{t_{(m-1)l}+\hbar,\ldots,t_{ml}\}} \frac{(t + (\alpha_{m-1} - 1)\hbar + t_{(m-1)l})_{\hbar}^{(\alpha_{m-1})}}{\Gamma(1 + \alpha_{m-1})} \leq \mathfrak{r}$$

This means $\Phi\Omega \subset \Omega$.

2. We must demonstrate that Φ is continuous. Suppose $\varkappa_n$ to be a sequence of Ω fulfilling $\varkappa_n \to \varkappa,\ ;\ (n \to +\infty).;$ then we get

$$\|\Phi\varkappa_n(t) - \Phi\varkappa(t)\| \le \|\varkappa_n(t_0) - \varkappa(t_0)\|$$
$$+ \left\| \frac{\hbar}{\Gamma(\alpha_{m-1})} \sum_{s=\frac{t_{(m-1)l}}{\hbar}+1-\alpha_{m-1}}^{\frac{t}{\hbar}-\alpha_{m-1}} (t-\sigma(s\hbar))_{\hbar}^{\alpha_{m-1}-1}[-A(\varkappa_n(s\hbar) - \varkappa(s\hbar))\right.$$
$$\left. + B(\mathfrak{f}(s\hbar, \varkappa_n(s\hbar)) - \mathfrak{f}(s\hbar, \varkappa(s\hbar)))] \right\| \le \|\varkappa_n(t_0) - \varkappa(t_0)\|$$
$$+ [\mathfrak{M}_A + L\mathfrak{M}_B] \sup_{t\in\{t_{(m-1)l}+\hbar,\ldots,t_{ml}\}} \frac{(t + (\alpha_{m-1}-1)\hbar + t_{(m-1)l})_{\hbar}^{(\alpha_{m-1})}}{\Gamma(1+\alpha_{m-1})}$$
$$\|\varkappa_n(t) - \varkappa(t)\|$$

We may therefore deduce that $\|\Phi\varkappa_n(t) - \Phi\varkappa(t)\| \to 0$ when $n \to +\infty$, , implying that Φ is continuous.

3. We illustrate that Φ is rather compact. We select t', t'' from $\{tql + \hbar, \ldots, t(q+1)l\}$, $q = 1, 2, \ldots, m-1$ and $t' < t''$ and we obtain

$$\|\Phi\varkappa(t') - \Phi\varkappa(t'')\| \le \frac{\hbar}{\Gamma(\alpha_{m-1})} \left\| \sum_{s=\frac{t_{(m-1)l}}{\hbar}+1-\alpha_{m-1}}^{\frac{t'}{\hbar}-\alpha_{m-1}} (t'-\sigma(s\hbar))_{\hbar}^{\alpha_{m-1}-1}[-A\varkappa(s\hbar) + B\mathfrak{f}(s\hbar, \varkappa(s\hbar))]\right.$$
$$\left. - \sum_{s=\frac{t_{(m-1)l}}{\hbar}+1-\alpha_{m-1}}^{\frac{t''}{\hbar}-\alpha_{m-1}} (t''-\sigma(s\hbar))_{\hbar}^{\alpha_{m-1}-1}[-A\varkappa(s\hbar) + B\mathfrak{f}(s\hbar, \varkappa(s\hbar))] \right\|$$
$$\le \frac{\hbar}{\Gamma(\alpha_{m-1})} \left\| \sum_{s=\frac{t_{(m-1)l}}{\hbar}+1-\alpha_{m-1}}^{\frac{t'}{\hbar}-\alpha_{m-1}} (t'-\sigma(s\hbar))_{\hbar}^{\alpha_{m-1}-1}[-A\varkappa(s\hbar) + B\mathfrak{f}(s\hbar, \varkappa(s\hbar))]\right.$$
$$- \sum_{s=\frac{t_{(m-1)l}}{\hbar}+1-\alpha_{m-1}}^{\frac{t_1}{\hbar}-\alpha_{m-1}} (t''-\sigma(s\hbar))_{\hbar}^{\alpha_{m-1}-1}[-A\varkappa(s\hbar) + B\mathfrak{f}(s\hbar, \varkappa(s\hbar))]$$
$$\left. - \sum_{s=\frac{t'}{\hbar}+1-\alpha_{m-1}}^{\frac{t''}{\hbar}-\alpha_{m-1}} (t''-\sigma(s\hbar))_{\hbar}^{\alpha_{m-1}-1}[-A\varkappa(s\hbar) + B\mathfrak{f}(s\hbar, \varkappa(s\hbar))] \right\|$$
$$\le \frac{\hbar}{\Gamma(\alpha_{m-1})} \left\| \sum_{s=\frac{t_{(m-1)l}}{\hbar}+1-\alpha_{m-1}}^{\frac{t'}{\hbar}-\alpha_{m-1}} \left((t'-\sigma(s\hbar))_{\hbar}^{\alpha_{m-1}-1} - (t''-\sigma(s\hbar))_{\hbar}^{\alpha_q-1}\right)[-A\varkappa(s\hbar) + B\mathfrak{f}(s\hbar, \varkappa(s\hbar))]\right.$$
$$\left. - \sum_{s=\frac{t'}{\hbar}+1-\alpha_{m-1}}^{\frac{t''}{\hbar}-\alpha_{m-1}} (t''-\sigma(s\hbar))_{\hbar}^{\alpha_m\ 1-1}[-A\varkappa(s\hbar) + B\mathfrak{f}(s\hbar, \varkappa(s\hbar))] \right\|$$
$$\le \frac{\hbar}{\Gamma(\alpha_{m-1})}[\mathfrak{M}_A + L\mathfrak{M}_B]\Big(\sum_{s=\frac{t_{(m-1)l}}{\hbar}+1-\alpha_{m-1}}^{\frac{t'}{\hbar}-\alpha_{m-1}} |(t'-\sigma(s\hbar))_{\hbar}^{\alpha_{m-1}-1} - (t''-\sigma(s\hbar))_{\hbar}^{\alpha_q-1}|$$
$$+ \sum_{s=\frac{t'}{\hbar}+1-\alpha_{m-1}}^{\frac{t''}{\hbar}-\alpha_{m-1}} (t''-\sigma(s\hbar))_{\hbar}^{\alpha_{m-1}-1}\Big)\mathfrak{r} \quad \to 0, \quad \text{as } t' \to t''.$$

This means that $\{\Phi\varkappa : \varkappa \in \Omega\}$ is a bounded and uniformly Cauchy subset E, and when combined with Azela-Lemma Ascoli'sa Lemma 2, $\Phi\Omega$ is reasonably compact.

4. We select a fixed $\mathfrak{y} \in \Omega$, $\varkappa = \psi\varkappa + \Phi\mathfrak{y}$ for all $q = 0, 1, 2, \ldots, m-1$ and we have

$$
\begin{aligned}
\|\varkappa\| &\leq \|\psi\varkappa\| + \|\Phi\mathfrak{y}\| \\
&\leq \sum_{n=1}^{m-1} \frac{\hbar}{\Gamma(\alpha_{n-1})} \sum_{s=\frac{t_{(n-1)l}}{\hbar}+1-\alpha_{n-1}}^{\frac{t_{nl}}{\hbar}-\alpha_{n-1}} (t_{nl} - \sigma(s\hbar))_{\hbar}^{\alpha_{n-1}-1} \| - A\varkappa(s\hbar) + B\mathfrak{f}(s\hbar, \varkappa(s\hbar))\| \\
&\quad + \|\mathfrak{y}(t_0)\| + \frac{\hbar}{\Gamma(\alpha_{m-1})} \sum_{s=\frac{t_{(m-1)l}}{\hbar}+1-\alpha_{m-1}}^{\frac{t}{\hbar}-\alpha_{m-1}} (t - \sigma(s\hbar))_{\hbar}^{\alpha_{m-1}-1} \| - Ay(s\hbar) + B\mathfrak{f}(s\hbar, y(s\hbar))\| \\
&\leq [\mathfrak{M}_A + L\mathfrak{M}_B]\mathfrak{r} \sum_{n=1}^{m-1} \frac{\hbar}{\Gamma(\alpha_{n-1})} \sum_{s=\frac{t_{(n-1)l}}{\hbar}+1-\alpha_{n-1}}^{\frac{t_{nl}}{\hbar}-\alpha_{n-1}} (t_{nl} - \sigma(s\hbar))_{\hbar}^{\alpha_{n-1}-1} \\
&\quad + \kappa + [\mathfrak{M}_A + L\mathfrak{M}_B]r \frac{\hbar}{\Gamma(\alpha_{m-1})} \sup_{t\in\{t_{(m-1)l}+\hbar,\ldots,t_{ml}\}} \frac{(t + (\alpha_{m-1} - 1)\hbar + t_{(m-1)l})_{\hbar}^{(\alpha_{m-1})}}{\Gamma(1 + \alpha_{m-1})} \\
&\leq \kappa + [\mathfrak{M}_A + L\mathfrak{M}_B]\Big(\sum_{n=1}^{m-1} \frac{(t_{nl} + (\alpha_{n-1} - 1)\hbar + t_{(n-1)l})_{\hbar}^{(\alpha_{n-1})}}{\Gamma(1 + \alpha_{n-1})} \\
&\quad + \sup_{t\in\{t_{(m-1)l}+\hbar,\ldots,t_{ml}\}} \frac{(t + (\alpha_{m-1} - 1)\hbar + t_{(m-1)l})_{\hbar}^{(\alpha_{m-1})}}{\Gamma(1 + \alpha_{m-1})}\Big)\mathfrak{r} \leq \mathfrak{r}
\end{aligned}
$$

Therefore $\varkappa \in \Omega$.

Lastly, we demonstrate that the operator T represents a contraction mapping. In the case of $\varkappa(t), \mathfrak{y}(t)\ in\Omega$, If we take the norm of $\psi\varkappa(t) - \psi\mathfrak{y}(t)$, we get

$$
\begin{aligned}
\|\psi\varkappa(t) - \psi\mathfrak{y}(t)\| &= \Bigg\| \sum_{n=1}^{m-1} \frac{\hbar}{\Gamma(\alpha_{n-1})} \sum_{s=\frac{t_{(n-1)l}}{\hbar}+1-\alpha_{q-n-1}}^{\frac{t_{nl}}{\hbar}-\alpha_{n-1}} (t_{nl} - \sigma(s\hbar))_{\hbar}^{\alpha_{n-1}-1} [-A(\varkappa(s\hbar) - \mathfrak{y}(s\hbar)) \\
&\quad + B(\mathfrak{f}(s\hbar, \varkappa(s\hbar)) - \mathfrak{f}(s\hbar, \mathfrak{y}(s\hbar)))] \Bigg\| \\
&\leq [\mathfrak{M}_A + L\mathfrak{M}_B]\|\varkappa - \mathfrak{y}\| \sum_{n=1}^{m-1} \frac{(t_{nl} + (\alpha_{n-1} - 1)\hbar + t_{(n-1)l})_{\hbar}^{(\alpha_{n-1})}}{\Gamma(1 + \alpha_{n-1})} < \|\varkappa - \mathfrak{y}\|
\end{aligned}
$$

It may be deduced from (H_2) that the operator ψ is a contraction mapping. According to Lemma 3, $\varphi = \psi + \Phi$ has a fixed point in Ω that is a solution of (3). □

3.3 Ulam-Hyers Stability of Fractional-Order Discrete Neural Network

Definition 3 (*Ulam-Hyers stability* [122]) We say (3) is Ulam-Hyers stable if there exists $c > 0$ such that for arbitrary $\epsilon > 0$, if $x \in \mathbb{R}^p$ satisfies

$$\|{}_{\hbar}^{C}\Delta_{t_{ql}}^{\alpha_q} \varkappa(t) + A\varkappa(t+\alpha_q\hbar) - B\mathfrak{f}(t+\alpha_q\hbar, \varkappa(t+\alpha_q\hbar)\| \leq \epsilon, \tag{6}$$

for $t \in (\hbar\mathbb{N})_{t_{ql}+(1-\alpha_q)\hbar}, 0 < \alpha_k \leq 1, q = 0, \ldots, m-1$
then there exists a solution $\mathfrak{y}$ of (3) satisfying

$$\|\varkappa(t) - \mathfrak{y}(t)\| \leq c\epsilon$$

At first, the following Lemma is proved.

Lemma 6 *if $\varkappa(t)$ solves (6), then*

$$\Bigg\|\varkappa(t)-\varkappa(t_0) + \sum_{n=1}^{m-1}\frac{\hbar}{\Gamma(\alpha_{n-1})}\sum_{s=\frac{t_{(n-1)l}}{\hbar}+1-\alpha_{n-1}}^{\frac{t_{nl}}{\hbar}-\alpha_{n-1}} (t_{nl}-\sigma(s\hbar))_{\hbar}^{\alpha_{n-1}-1}[A\varkappa(s\hbar) - B\mathfrak{f}(s\hbar, \varkappa(s\hbar))]$$

$$+\frac{\hbar}{\Gamma(\alpha_{m-1})}\sum_{s=\frac{t_{(m-1)l}}{\hbar}+1-\alpha_{m-1}}^{\frac{t}{\hbar}-\alpha_{m-1}} (t-\sigma(s\hbar))_{\hbar}^{\alpha_{m-1}-1}[A\varkappa(s\hbar) - B\mathfrak{f}(s\hbar, x(s\hbar))]\Bigg\| \leq d\epsilon$$

where

$$d = \left(\sum_{n=1}^{m-1}\frac{(t_{nl}+(\alpha_{n-1}-1)\hbar + t_{(n-1)l})_{\hbar}^{(\alpha_{n-1})}}{\Gamma(1+\alpha_{n-1})} + \sup_{t\in\{t_{(m-1)l}+\hbar,\ldots,t_{ml}\}}\frac{(t+(\alpha_{m-1}-1)\hbar + t_{(m-1)l})_{\hbar}^{(\alpha_{m-1})}}{\Gamma(1+\alpha_{m-1})}\right)$$

Proof If $\varkappa(t)$ satisfies (6), there exists a function $\mathfrak{g}(t)$ that satisfies. $\|\mathfrak{g}(t)\| \leq \epsilon$ such that

$${}_{\hbar}^{C}\Delta_{t_{ql}}^{\alpha_q}\varkappa(t) + A\varkappa(t+\alpha_q\hbar) - B\mathfrak{f}(t+\alpha_q\hbar, \varkappa(t+\alpha_q\hbar) = \mathfrak{g}(t), \quad t \in \{t_{ql}+\hbar, \ldots, t_{(q+1)l}\}, \quad q = 0, 1, 2, \ldots, m-1$$

this implies

$$\begin{cases} {}_{\hbar}^{C}\Delta_{t_0}^{\alpha_0}\varkappa(t) + A\varkappa(t+\alpha_0\hbar) - B\mathfrak{f}(t+\alpha_0\hbar, \varkappa(t+\alpha_0\hbar) = \mathfrak{g}(t), & t \in \{t_0+\hbar, \ldots, t_l\}, \\ {}_{\hbar}^{C}\Delta_{t_l}^{\alpha_1}\varkappa(t) + A\varkappa(t+\alpha_1\hbar) - B\mathfrak{f}(t+\alpha_1\hbar, \varkappa(t+\alpha_1\hbar) = \mathfrak{g}(t), & t \in \{t_l+\hbar, \ldots, t_{2l}\}, \\ \ldots \\ {}_{\hbar}^{C}\Delta_{t_{(m-1)l}}^{\alpha_{m-1}}\varkappa(t) + A\varkappa(t+\alpha_{m-1}\hbar) - B\mathfrak{f}(t+\alpha_{m-1}\hbar, \varkappa(t+\alpha_{m-1}\hbar) = \mathfrak{g}(t), & t \in \{t_{(m-1)l}+\hbar, \ldots, t_{ml}\}, \end{cases}$$

As a result, we have

$$\begin{cases} \varkappa(t_l) = \varkappa(t_0) + \dfrac{\hbar}{\Gamma(\alpha_0)} \sum_{s=\frac{t_0}{\hbar}+1-\alpha_0}^{\frac{t_l}{\hbar}-\alpha_0} (t_l - \sigma(s\hbar))_{\hbar}^{\alpha_0 - 1}[-A\varkappa(s\hbar) + B\mathfrak{f}(s\hbar, \varkappa(s\hbar))] \\ \quad + \dfrac{h}{\Gamma(\alpha_0)} \sum_{s=\frac{t_0}{\hbar}+1-\alpha_0}^{\frac{t_l}{\hbar}-\alpha_0} (t_l - \sigma(s\hbar))_{\hbar}^{\alpha_0 - 1} \mathfrak{g}(s\hbar) \\ \varkappa(t_{2l}) = \varkappa(t_l) + \dfrac{\hbar}{\Gamma(\alpha_1)} \sum_{s=\frac{t_l}{\hbar}+1-\alpha_1}^{\frac{t_{2l}}{\hbar}-\alpha_1} (t_{2l} - \sigma(s\hbar))_{\hbar}^{\alpha_1 - 1}[A\varkappa(s\hbar) - B\mathfrak{f}(s\hbar, \varkappa(s\hbar))] \\ \quad + \dfrac{\hbar}{\Gamma(\alpha_1)} \sum_{s=\frac{t_l}{\hbar}+1-\alpha_1}^{\frac{t_{2l}}{\hbar}-\alpha_1} (t_{2l} - \sigma(s\hbar))_{\hbar}^{\alpha_1 - 1} \mathfrak{g}(s\hbar) \\ \ldots \\ \varkappa(t) - \varkappa(t_{(m-1)l} + \dfrac{\hbar}{\Gamma(\alpha_{m-1})} \sum_{s=\frac{t_{(m-1)l}}{\hbar}+1-\alpha_{m-1}}^{\frac{t}{\hbar}-\alpha_{m-1}} (t - \sigma(s\hbar))_{\hbar}^{\alpha_{m-1} - 1}[A\varkappa(s\hbar) - B\mathfrak{f}(s\hbar, \varkappa(s\hbar))] \\ \quad = \dfrac{\hbar}{\Gamma(\alpha_{m-1})} \sum_{s=\frac{t_{(m-1)l}}{\hbar}+1-\alpha_{m-1}}^{\frac{t}{\hbar}-\alpha_{m-1}} (t - \sigma(s\hbar))_{\hbar}^{\alpha_{m-1} - 1} \mathfrak{g}(s\hbar) \end{cases}$$

as a result of this we get at the following equality

$$\begin{aligned} &\Bigg\| \varkappa(t) - \varkappa(t_0) + \sum_{n=1}^{m-1} \frac{\hbar}{\Gamma(\alpha_{n-1})} \sum_{s=\frac{t_{(n-1)l}}{\hbar}+1-\alpha_{n-1}}^{\frac{t_{nl}}{\hbar}-\alpha_{n-1}} (t_{nl} - \sigma(s\hbar))_{\hbar}^{\alpha_{n-1}-1}[A\varkappa(s\hbar) - B\mathfrak{f}(s\hbar, \varkappa(s\hbar))] \\ &\quad + \frac{\hbar}{\Gamma(\alpha_{m-1})} \sum_{s=\frac{t_{(m-1)l}}{\hbar}+1-\alpha_{m-1}}^{\frac{t}{\hbar}-\alpha_{m-1}} (t - \sigma(s\hbar))_{\hbar}^{\alpha_{m-1}-1}[A\varkappa(s\hbar) - B\mathfrak{f}(s\hbar, \varkappa(s\hbar))] \Bigg\| \\ &= \Bigg\| \sum_{n=1}^{m-1} \frac{\hbar}{\Gamma(\alpha_{n-1})} \sum_{s=\frac{t_{(n-1)l}}{\hbar}+1-\alpha_{n-1}}^{\frac{t_{nl}}{\hbar}-\alpha_{n-1}} (t_{nl} - \sigma(s\hbar))_{\hbar}^{\alpha_{n-1}-1} \mathfrak{g}(s\hbar) \\ &\quad + \frac{\hbar}{\Gamma(\alpha_{m-1})} \sum_{s=\frac{t_{(m-1)l}}{\hbar}+1-\alpha_{m-1}}^{\frac{t}{\hbar}-\alpha_{m-1}} (t - \sigma(s\hbar))_{\hbar}^{\alpha_{m-1}-1} \mathfrak{g}(s\hbar) \Bigg\| \\ &\leq \left(\sum_{n=1}^{m-1} \frac{(t_{nl} + (\alpha_{n-1} - 1)\hbar + t_{(n-1)l})_{\hbar}^{(\alpha_{n-1})}}{\Gamma(1 + \alpha_{n-1})} + \sup_{t \in \{t_{(m-1)l}+\hbar, \ldots, t_{ml}\}} \frac{(t + (\alpha_{m-1} - 1)\hbar + t_{(m-1)l})_{\hbar}^{(\alpha_{m-1})}}{\Gamma(1 + \alpha_{m-1})} \right) \|\mathfrak{g}\| \\ &\leq \left(\sum_{n=1}^{m-1} \frac{(t_{nl} + (\alpha_{n-1} - 1)\hbar + t_{(n-1)l})_{\hbar}^{(\alpha_{n-1})}}{\Gamma(1 + \alpha_{n-1})} + \sup_{t \in \{t_{(m-1)l}+\hbar, \ldots, t_{ml}\}} \frac{(t + (\alpha_{m-1} - 1)\hbar + t_{(m-1)l})_{\hbar}^{(\alpha_{m-1})}}{\Gamma(1 + \alpha_{m-1})} \right) \epsilon \end{aligned}$$

which completes the proof. □

The Ulam-Hyers stability of the discrete fractional-order neural network (3) is now demonstrated using a novel theorem.

Theorem 3 *Under assumption* (H_1) *and* (H_2) *system (3) is Ulam-Hyers stable.*

Proof It is obvious that the solution $\mathfrak{y}(t)$ of (3) satisfies

$$\mathfrak{y}(t) = \varkappa(t_0) + \sum_{n=1}^{m-1} \frac{\hbar}{\Gamma(\alpha_{n-1})} \sum_{s=\frac{t_{(n-1)l}}{\hbar}+1-\alpha_{n-1}}^{\frac{t_{nl}}{\hbar}-\alpha_{n-1}} (t_{nl} - \sigma(s\hbar))_{\hbar}^{\alpha_{n-1}-1} [A\mathfrak{y}(s\hbar) - B\mathfrak{f}(s\hbar, \mathfrak{y}(s\hbar))]$$
$$+ \frac{\hbar}{\Gamma(\alpha_{m-1})} \sum_{s=\frac{t_{(m-1)l}}{\hbar}+1-\alpha_{m-1}}^{\frac{t}{\hbar}-\alpha_{m-1}} (t - \sigma(s\hbar))_{\hbar}^{\alpha_{m-1}-1} [A\mathfrak{y}(s\hbar) - B\mathfrak{f}(s\hbar, \mathfrak{y}(s\hbar))]$$

Therefore,

$$\|\varkappa(t) - \mathfrak{y}(t)\| = \Bigg\| \varkappa(t) - \varkappa(t_0) - \sum_{n=1}^{m-1} \frac{\hbar}{\Gamma(\alpha_{n-1})} \sum_{s=\frac{t_{(n-1)l}}{\hbar}+1-\alpha_{n-1}}^{\frac{t_{nl}}{\hbar}-\alpha_{n-1}} (t_{nl} - \sigma(s\hbar))_{\hbar}^{\alpha_{n-1}-1} [-A\mathfrak{y}(s\hbar) + B\mathfrak{f}(s\hbar, \mathfrak{y}(s\hbar))]$$
$$- \frac{\hbar}{\Gamma(\alpha_{m-1})} \sum_{s=\frac{t_{(m-1)l}}{\hbar}+1-\alpha_{m-1}}^{\frac{t}{\hbar}-\alpha_{m-1}} (t - \sigma(s\hbar))_{\hbar}^{\alpha_{m-1}-1} [-A\mathfrak{y}(s\hbar) + B\mathfrak{f}(s\hbar, \mathfrak{y}(s\hbar))] \Bigg\|$$
$$= \Bigg\| \varkappa(t) - \varkappa(t_0) - \sum_{n=1}^{m-1} \frac{\hbar}{\Gamma(\alpha_{n-1})} \sum_{s=\frac{t_{(n-1)l}}{\hbar}+1-\alpha_{n-1}}^{\frac{t_{nl}}{\hbar}-\alpha_{n-1}} (t_{nl} - \sigma(s\hbar))_{\hbar}^{\alpha_{n-1}-1} [-A\mathfrak{y}(s\hbar) + B\mathfrak{f}(s\hbar, \mathfrak{y}(s\hbar))]$$
$$- \frac{\hbar}{\Gamma(\alpha_{m-1})} \sum_{s=\frac{t_{(m-1)l}}{\hbar}+1-\alpha_{m-1}}^{\frac{t}{\hbar}-\alpha_{m-1}} (t - \sigma(s\hbar))_{\hbar}^{\alpha_{m-1}-1} [-A\mathfrak{y}(s\hbar) + B\mathfrak{f}(s\hbar, \mathfrak{y}(s\hbar))]$$
$$- \sum_{n=1}^{m-1} \frac{\hbar}{\Gamma(\alpha_{n-1})} \sum_{s=\frac{t_{(n-1)l}}{\hbar}+1-\alpha_{n-1}}^{\frac{t_{nl}}{\hbar}-\alpha_{n-1}} (t_{nl} - \sigma(s\hbar))_{\hbar}^{\alpha_{n-1}-1} [-A\varkappa(s\hbar) + B\mathfrak{f}(s\hbar, \varkappa(s\hbar))]$$
$$- \frac{\hbar}{\Gamma(\alpha_{m-1})} \sum_{s=\frac{t_{(m-1)l}}{\hbar}+1-\alpha_{m-1}}^{\frac{t}{\hbar}-\alpha_{m-1}} (t - \sigma(s\hbar))_{\hbar}^{\alpha_{m-1}-1} [-A\varkappa(s\hbar) + B\mathfrak{f}(s\hbar, \varkappa(s\hbar))]$$
$$+ \sum_{n=1}^{m-1} \frac{\hbar}{\Gamma(\alpha_{n-1})} \sum_{s=\frac{t_{(n-1)l}}{\hbar}+1-\alpha_{n-1}}^{\frac{t_{nl}}{\hbar}-\alpha_{n-1}} (t_{nl} - \sigma(s\hbar))_{\hbar}^{\alpha_{n-1}-1} [-A\varkappa(s\hbar) + B\mathfrak{f}(s\hbar, \varkappa(s\hbar))]$$
$$+ \frac{\hbar}{\Gamma(\alpha_{m-1})} \sum_{s=\frac{t_{(m-1)l}}{\hbar}+1-\alpha_{m-1}}^{\frac{t}{\hbar}-\alpha_{m-1}} (t - \sigma(s\hbar))_{\hbar}^{\alpha_{m-1}-1} [-A\varkappa(s\hbar) + B\mathfrak{f}(s\hbar, \varkappa(s\hbar))] \Bigg\|$$
$$\leq \Bigg\| \varkappa(t) - \varkappa(t_0) - \sum_{n=1}^{m-1} \frac{\hbar}{\Gamma(\alpha_{n-1})} \sum_{s=\frac{t_{(n-1)l}}{\hbar}+1-\alpha_{n-1}}^{\frac{t_{nl}}{\hbar}-\alpha_{n-1}} (t_{nl} - \sigma(s\hbar))_{\hbar}^{\alpha_{n-1}-1} [-A\varkappa(s\hbar) + B\mathfrak{f}(s\hbar, \varkappa(s\hbar))]$$

$$- \frac{\hbar}{\Gamma(\alpha_{m-1})} \sum_{s=\frac{t_{(m-1)l}}{\hbar}+1-\alpha_{m-1}}^{\frac{t}{\hbar}-\alpha_{m-1}} (t - \sigma(s\hbar))_{\hbar}^{\alpha_{m-1}-1} [-A\varkappa(s\hbar) + B\mathfrak{f}(s\hbar, \varkappa(s\hbar))] \Bigg\|$$
$$+ \Bigg\| \sum_{n=1}^{m-1} \frac{\hbar}{\Gamma(\alpha_{n-1})} \sum_{s=\frac{t_{(n-1)l}}{\hbar}+1-\alpha_{n-1}}^{\frac{t_{nl}}{\hbar}-\alpha_{n-1}} (t_{nl} - \sigma(s\hbar))_{\hbar}^{\alpha_{n-1}-1} [-A(\varkappa(s\hbar) - \mathfrak{y}(s\hbar)) + B(\mathfrak{f}(s\hbar, \varkappa(s\hbar)) - \mathfrak{f}(s\hbar, \mathfrak{y}(s\hbar)))] \Bigg\|$$

$$+\left\|\frac{\hbar}{\Gamma(\alpha_{m-1})}\sum_{s=\frac{t_{(m-1)l}}{\hbar}+1-\alpha_{m-1}}^{\frac{t}{\hbar}-\alpha_{m-1}}(t-\sigma(s\hbar))_{\hbar}^{\alpha_{m-1}-1}[-A(\varkappa(s\hbar)-\mathfrak{y}(s\hbar))+B(\mathfrak{f}(s\hbar,\varkappa(s\hbar))-\mathfrak{f}(s\hbar,\mathfrak{y}(s\hbar)))]\right\|$$

$$\leq\left(\sum_{n=1}^{m-1}\frac{(t_{nl}+(\alpha_{n-1}-1)\hbar+t_{(n-1)l})_{\hbar}^{(\alpha_{n-1})}}{\Gamma(1+\alpha_{n-1})}+\sup_{t\in\{t_{(m-1)l}+\hbar,\ldots,t_{ml}\}}\frac{(t+(\alpha_{m-1}-1)\hbar+t_{(m-1)l})_{\hbar}^{(\alpha_{m-1})}}{\Gamma(1+\alpha_{m-1})}\right)\epsilon$$

$$+[\mathfrak{M}_A+L\mathfrak{M}_B]\left\{\sum_{n=1}^{m-1}\frac{(t_{nl}+(\alpha_{n-1}-1)\hbar+t_{(n-1)l})_{\hbar}^{(\alpha_{n-1})}}{\Gamma(1+\alpha_{n-1})}\right.$$

$$\left.+\sup_{t\in\{t_{(m-1)l}+\hbar,\ldots,t_{ml}\}}\frac{(t+(\alpha_{m-1}-1)\hbar+t_{(m-1)l})_{\hbar}^{(\alpha_{m-1})}}{\Gamma(1+\alpha_{m-1})}\right\}\|\varkappa(t)-\mathfrak{y}(t)\|$$

$$\leq\frac{d}{1-\mathfrak{M}}\epsilon\leq c\epsilon$$

Where

$$c=\frac{d}{1-\mathfrak{M}}$$

Which conclude that (3) is Ulam-Hyers stable. □

3.4 Examples and Simulations

Example 1 Considering the following discrete fractional-order neural networks

$$\begin{cases}{}^{C}_{\hbar}\Delta_{t_0}^{\alpha_0}\varkappa(t)=-A\varkappa(t+\alpha_0\hbar)+B\tanh(\varkappa(t+\alpha_0 h)), & t\in\{t_0+(1-\alpha_0)\hbar,\ldots,t_l-\alpha_0\hbar\}\\ {}^{C}_{\hbar}\Delta_{t_l}^{\alpha_1}\varkappa(t)=-A\varkappa(t+\alpha_1\hbar)+B\tanh(\varkappa(t+\alpha_1\hbar)), & t\in\{t_l+(1-\alpha_l)\hbar,\ldots,t_{2l}-\alpha_1\hbar\}\\ {}^{C}_{\hbar}\Delta_{t_{2l}}^{\alpha_2}\varkappa(t)=-A\varkappa(t+\alpha_2\hbar)+B\tanh(\varkappa(t+\alpha_2\hbar)), & t\in\{t_{2l}+(1-\alpha_2)\hbar,\ldots,t_{3l}-\alpha_2\hbar\}\end{cases}\tag{7}$$

where $\hbar=0.5$, $q=0,1,2$, $t\in(\hbar\mathbb{N})_{t_{ql}-(1-\alpha_q)\hbar}$ and $(\alpha_0,\alpha_1,\alpha_2)=(0.05, 0.1, 0.15)$.
$\tanh(\varkappa(t))=(\tanh\varkappa_1(t),\tanh(\varkappa_2(t))^T$,

$$A=\begin{bmatrix}0.2&0&0\\0&0.2&0\\0&0&0.2\end{bmatrix},\quad B=\begin{bmatrix}0.002&-0.004&0.0015\\-0.002&0.001&-0.002\\-0.0025&0.0015&-0.003\end{bmatrix},\quad x(0)=\begin{bmatrix}0.9\\0.6\\0.3\end{bmatrix}$$

We see that for $\mathfrak{M}_A=0.2$ and $\mathfrak{M}_B=0.0075$, the assumptions (H_1) and (H_2) are met. This leads in $M=0.9866<1$;, and since the criteria of Theorem 2 are satisfied for $\mathfrak{r}\approx 67.16$, at least one solution must exist. For $c=\dfrac{d}{1-\mathfrak{M}}$ where $d=4.7547$ and the system (7) is Ulam-Hyers stable, the Theorem 3 is also fulfilled.

The numerical solution in Fig. 1 and the initial condition $\varkappa(0)=(0.9,0.6,0.3)T$ are obtained using the following mathematical formulae, and they demonstrate the stability of the neural network under consideration.

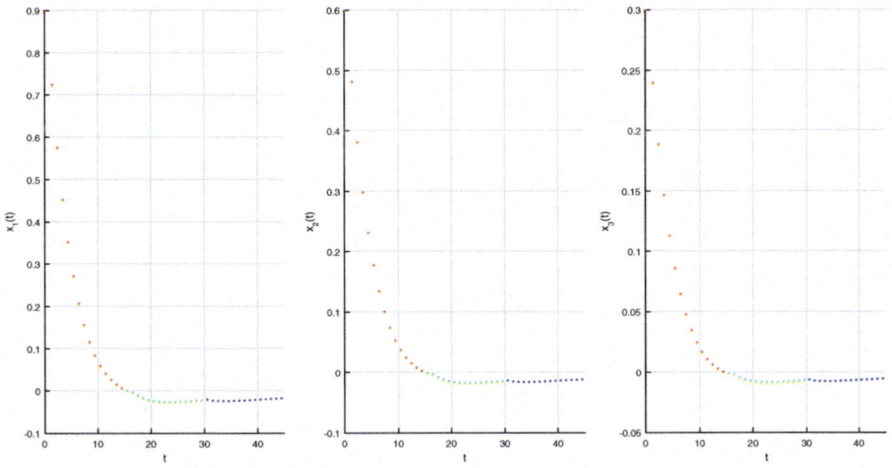

Fig. 1 Numerical solution of the variable-order neural network (7) (red = 0.05, green = 0.1, blue = 0.15)

$$\begin{cases} \varkappa(t) = \varkappa(t_0) + \dfrac{\hbar^{\alpha_0}}{\Gamma(\alpha_0)} \sum_{t_0+1}^{t} \dfrac{\Gamma(t-j+\alpha_0)}{\Gamma(t-j+1)} (-A\varkappa(j) + B \tanh(\varkappa(j))), \\ \varkappa(t) = \varkappa(t_l) + \dfrac{\hbar^{\alpha_1}}{\Gamma(\alpha_1)} \sum_{t_l+1}^{t} \dfrac{\Gamma(t-j+\alpha_0)}{\Gamma(t-j+1)} (-A\varkappa(j) + B \tanh(\varkappa(j))), \\ \varkappa(t) = \varkappa(t_{2l}) + \dfrac{\hbar^{\alpha_2}}{\Gamma(\alpha_2)} \sum_{t_{2l}+1}^{t} \dfrac{\Gamma(t-j+\alpha_0)}{\Gamma(t-j+1)} (-A\varkappa(j) + B \tanh(\varkappa(j))), \end{cases}$$

Example 2 Let be the two-dimensional fractional discrete time neural network shown below.

$$ {}^{C}_{\hbar}\Delta^{\alpha_q}_{t_{ql}} \varkappa(t) = -A\varkappa(t+\alpha_q h) + B\sin(\varkappa(t+\alpha_q h)), \quad 0 < \alpha_q \le 1 \quad t \in (\hbar\mathbb{N})_{ql+(1-\alpha_q)\hbar} \quad q = 0, 1, 2, 3. \tag{8}$$

where $sin(\varkappa(t)) = (sin(\varkappa_1(t)), sin(\varkappa_2(t)))^T$
suppose $\varkappa(0) = (1, -1)^T$, $(\alpha_0, \alpha_1, \alpha_2, \alpha_3) = (0.5, 0.4, 0.3, 0.2)$, $\hbar = 1.3$ and $m = 4$

$$A = \begin{bmatrix} 0.046 & 0 \\ 0 & 0.046 \end{bmatrix}, \quad B = \begin{bmatrix} 0.0003 & 0 \\ -0.0004 & 0.0002 \end{bmatrix},$$

We may discuss the stability over several time domains, because the parameters fulfill (H_1) and (H_2) for $\mathfrak{M}_A = 0.046$, $\mathfrak{M}_B = 0.0006$, and; $M = 0.98581$; and all the requirements in Theorem 2 for the existence of the solution are true with $\mathfrak{r} \approx 70.42$. Theorem 3 holds true for the constants $d = 21.155$ and $c = \dfrac{d}{1-M}$, ensuring that discrete-time neural networks (8) are Ulam-Hyers stable. Figure 2 depicts the numerical solution of the system.

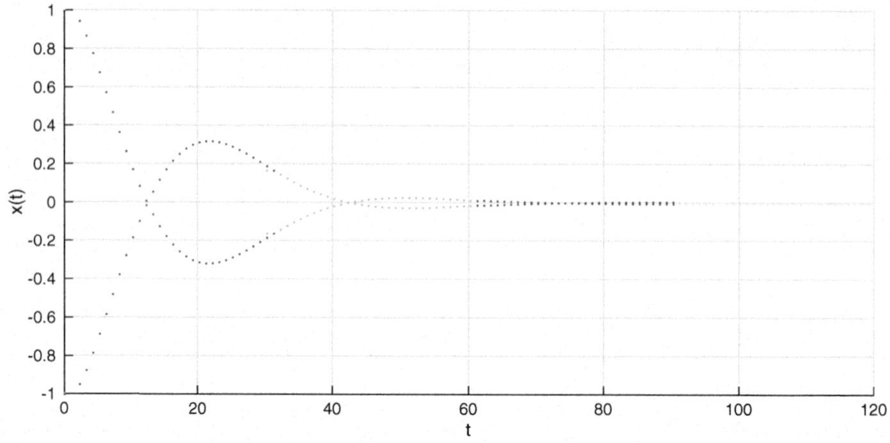

Fig. 2 The numerical solution of neural network (8)

Example 3 Consider the following neural network

$$\begin{cases} {}^{C}_{\hbar}\Delta^{\alpha_q}_{t_{ql}}\varkappa_1(t) = -a_1\varkappa_1(t+\alpha_q\hbar) + b_{11}\sin(\varkappa_1(t+\alpha_q\hbar)) + b_{12}\sin(x_2(t+\alpha_q\hbar)), \\ {}^{C}_{\hbar}\Delta^{\alpha_q}_{t_{ql}}\varkappa_2(t) = -a_2\varkappa_2(t+\alpha_q\hbar) + b_{21}\sin(\varkappa_1(t+\alpha_q\hbar)) + b_{22}\sin(\varkappa_2(t+\alpha_q\hbar)), \end{cases} \tag{9}$$

where $t \in (\hbar\mathbb{N})_{t_{ql}+(1-\alpha_q)\hbar}$, $m = 4$ and $0 < \alpha_q \leq 1$.

Let be $t \in [0, 173]$, $\hbar = 1.9$, $(\alpha_0, \alpha_1, \alpha_2, \alpha_3) = (0.8, 0.6, 0.3, 0.4)$.

and

$$A = \begin{bmatrix} 0.015 & 0 \\ 0 & 0.015 \end{bmatrix}, \quad B = \begin{bmatrix} -0.0003 & 0 \\ -0.0004 & 0.0002 \end{bmatrix}, \quad x(0) = \begin{bmatrix} -0.001 \\ 0.0009 \end{bmatrix}$$

We obtain $\mathfrak{M}_A = 0.015$ and $\mathfrak{M}_B = 0.0006$ from the supplied data. The assumptions (H_1) and (H_2) clearly hold with; $M = 0.921211$;. As a result, all of the criteria of Theorems 2 and 3 are met. As a result, system (9) has at least one Ulam-Hyers stable solution for $\mathfrak{r} \approx 0.0127$, $d = 59.051$, and $c = \dfrac{d}{1-M}$.

The stability is shown in Fig. 3 using the following numerical formulas

$$\begin{cases} \varkappa_1(t) = \varkappa_1(t_{ql}) + \dfrac{\hbar^{\alpha_q}}{\Gamma(\alpha_q)} \sum_{t_q+1}^{i} \dfrac{\Gamma(t-j+\alpha_q)}{\Gamma(t-j+1)} (-a_1\varkappa_1(j) + b_{11}\sin(\varkappa_1(j)) + b_{12}\sin(\varkappa_2(j))) \\ \varkappa_2(t) = \varkappa_2(t_{ql}) + \dfrac{\hbar^{\alpha_q}}{\Gamma(\alpha_q)} \sum_{t_q+1}^{i} \dfrac{\Gamma(t-j+\alpha_q)}{\Gamma(t-j+1)} (-a_2\varkappa_2(j) + b_{21}\sin(\varkappa_1(j)) + b_{22}\sin(\varkappa_2(j)) \end{cases}$$

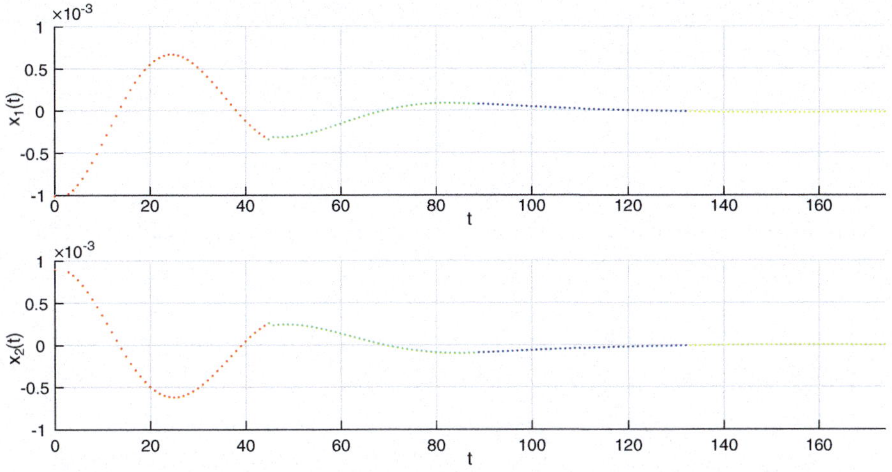

Fig. 3 Numerical solution of neural network (9)

4 Stabilization of Fractional Discrete Chaotic Neural Networks

In this section, an unique theory is demonstrated with the goal of stabilizing the chaotic trajectories of the three-dimensional fractional-order discrete neural network at zero by providing very simple linear control laws.

4.1 Noval Chaotic Discrete Fractional-Order Neural Network

We provide first a chaotic discrete neural networks using the Caputo like difference operator. The authors of [125] used the α Caputo operator to develop a three-dimensional continuous fractional-order Hopfield neural network model, which is given by

$$\begin{cases} D^{\alpha}\varkappa_1(t) = -\varkappa_1(t) + 2\tanh(\varkappa_1(t)) - 1.2\tanh(\varkappa_2(t)) - 0.1\tanh(\varkappa_3(t)), \\ D^{\alpha}\varkappa_2(t) = -\varkappa_2(t) + 2\tanh(\varkappa_1(t)) + 1.71\tanh(\varkappa_2(t)) + 1.15\tanh(\varkappa_3(t)), \\ D^{\alpha}\varkappa_3(t) = -\varkappa_3(t) - 4.75\tanh(\varkappa_1(t)) + 1.1\tanh(\varkappa_3(t)), \end{cases} \tag{10}$$

In this case, we may define the following three-dimensional fractional-order discrete neural network model using the aputo like difference operator:

$$\begin{cases} {}^{C}_{\hbar}\Delta^{\alpha}_{t_0}\varkappa_1(t) = -a_1\varkappa_1(t) + b_{11}\tanh(\varkappa_1(t)) + b_{12}\tanh(\varkappa_2(t)) + b_{13}\tanh(\varkappa_3(t)) + I_1, \\ {}^{C}_{\hbar}\Delta^{\alpha}_{t_0}\varkappa_2(t) = -a_2\varkappa_2(t) + b_{21}\tanh(\varkappa_1(t)) + b_{22}\tanh(\varkappa_2(t)) + b_{23}\tanh(\varkappa_3(t)) + I_2, \\ {}^{C}_{\hbar}\Delta^{\alpha}_{t_0}\varkappa_3(t) = -a_3\varkappa_3(t) + b_{31}\tanh(\varkappa_1(t)) + b_{32}\tanh(\varkappa_2(t)) + b_{33}\tanh(\varkappa_3(t)) + I_3, \end{cases} \tag{11}$$

Let be the following parameters $a_1 = a_2 = a_3 = 1$, and $I_1 = I_2 = I_3 = 0$ also the matrix B is defined as follow

$$B = \begin{bmatrix} 2 & -1.2 & -0.1 \\ 2 & 1.71 & 1.15 \\ -4.75 & 0 & 1.1 \end{bmatrix}$$

The following implicit discrete numerical solution is defined to describe the characteristics of the discrete fractional neural network:

$$\begin{cases} \varkappa_1(k) = \varkappa_1(0) + \frac{1}{\Gamma(\alpha)} \sum_{j=1}^{k} \frac{\Gamma(k-j+\alpha)}{\Gamma(k-j+1)} [-\varkappa_1(j-1) + b_{11}\tanh(\varkappa_1(j-1)) \\ \quad + b_{12}\tanh(\varkappa_2(j-1)) + b_{13}\tanh(\varkappa_3(j-1))], \\ \varkappa_2(k) = \varkappa_2(0) + \frac{1}{\Gamma(\alpha)} \sum_{j=1}^{k} \frac{\Gamma(k-j+\alpha)}{\Gamma(k-j+1)} [-\varkappa_2(j-1) + b_{21}\tanh(\varkappa_1(j-1)) \\ \quad + b_{22}\tanh(\varkappa_2(j-1) + b_{23}\tanh(\varkappa_3(j-1))], \\ \varkappa_3(k) = \varkappa_3(0) + \frac{1}{\Gamma(\alpha)} \sum_{j=1}^{k} \frac{\Gamma(k-j+\alpha)}{\Gamma(k-j+1)} [-\varkappa_3(j-1) + b_{31}\tanh(\varkappa_1(j-1)) \\ \quad + b_{32}\tanh(\varkappa_2(j-1) + b_{33}\tanh(\varkappa_3(j-1))], \end{cases} \tag{12}$$

We will numerically discuss the dynamics of fractional-order discrete neural networks (11). The maximum Lyapunov exponent (LE) of a dynamical model, as is widely known, is an essential measure that describes the rate of dispersion of infinitely near trajectories. It is frequently used to identify dynamic behavior and system chaos. For fractional-order discrete maps, numerous LE calculation techniques are been proposed. To compute the LE of fractional-order discrete neural networks, the Jacobian matrix approach for Lyapunov exponents of discrete fractional maps. The tangent map in matrix J_i is as follows:

$$J_k = \begin{pmatrix} c_1(k) & c_2(k) & c_3(k) \\ d_1(k) & d_2(k) & d_3(k) \\ e_1(k) & e_2(k) & e_3(k) \end{pmatrix} \tag{13}$$

where

$$\begin{cases} c_i(k) = c_i(0) + \frac{1}{\Gamma(\alpha)} \sum_{j=1}^{k} \frac{\Gamma(k-j+\alpha)}{\Gamma(k-j+1)} [-c_i(j-1) + b_{11}c_i(j-1)(1-\tanh^2(\varkappa_1(j-1))) \\ \quad + b_{12}d_i(j-1)(1-\tanh^2(\varkappa_2(j-1))) + b_{13}e_i(j-1)(1-\tanh^2(\varkappa_3(j-1)))] \\ d_i(k) = d_j(0) + \frac{1}{\Gamma(\alpha)} \sum_{j=1}^{k} \frac{\Gamma(k-j+\alpha)}{\Gamma(k-j+1)} (-d_i(j-1) + b_{21}c_i(j-1)(1-\tanh^2(\varkappa_1(j-1))) \\ \quad + b_{22}d_i(j-1)(1-\tanh^2(\varkappa_2(j-1))) + b_{23}e_i(j-1)(1-\tanh^2(\varkappa_3(j-1)))], \\ e_i(k) = e_i(0) + \frac{1}{\Gamma(\alpha)} \sum_{j=1}^{k} \frac{\Gamma(k-j+\alpha)}{\Gamma(k-j+1)} (-e_i(j_1) + b_{31}c_i(j-1)(1-\tanh^2(\varkappa_1(j-1))) \\ \quad + b_{33}e_i(j-1)(1-\tanh^2(\varkappa_3(j-1)))] \end{cases} \tag{14}$$

where $i = 1, 2, 3$ and the initial conditions satisfy:

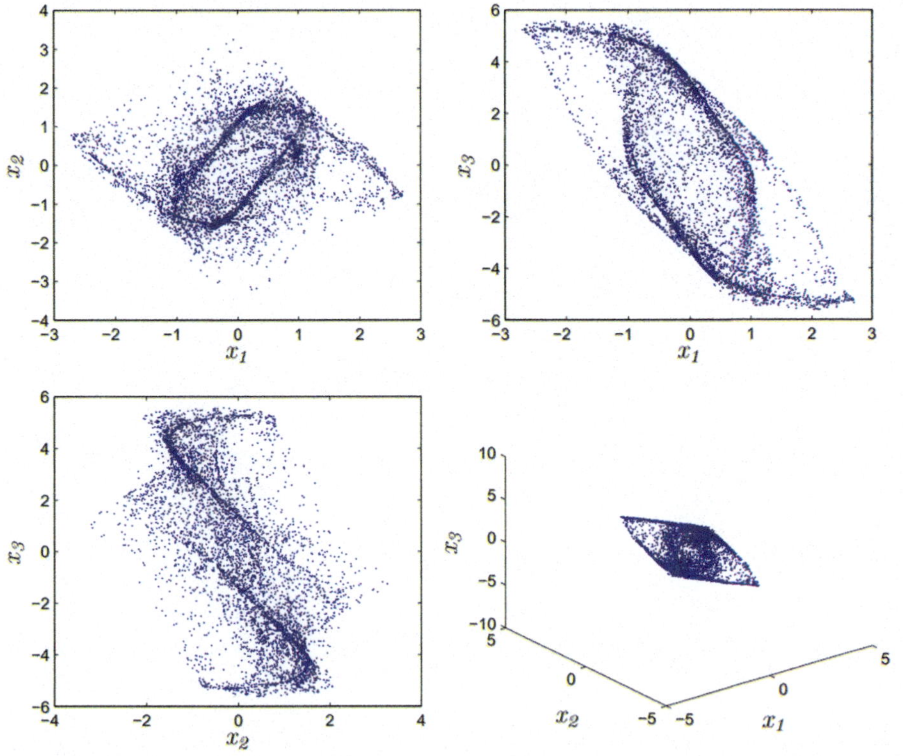

Fig. 4 Chaotic neural networks (11)

$$\begin{pmatrix} c_1(0) & c_2(0) & c_3(0) \\ d_1(0) & d_2(0) & d_3(0) \\ e_1(0) & e_2(0) & e_3(0) \end{pmatrix} = \begin{pmatrix} 1 & 0 & 0 \\ 0 & 1 & 0 \\ 0 & 0 & 1 \end{pmatrix} \tag{15}$$

Then LE of the discrete fractional-order neural network can be derived from the following equation

$$\lambda(\varkappa_0) = \frac{1}{i} \ln |J_i|. \tag{16}$$

The phase portraits of the fractional neural network are shown in Fig. 4 when $\alpha \in [0, 1]$. The system exhibits chaotic behavior, as seen in this illustration. To properly examine the impact of fractional-order difference on system behavior, Exponents of Lyapunov and bifurcation diagrams are shown. In Fig. 5, the Lyapunov exponents for the system with the aforementioned parameters are shown, indicating that validates the system's chaotic behavior In addition, in Fig. 5 The bifurcation diagram is depicted by altering the value of the fractional derivative. According to the bifurcation diagram in Fig. 5 for $\alpha > 0.962$, the system enters chaos via a period-doubling bifurcation.

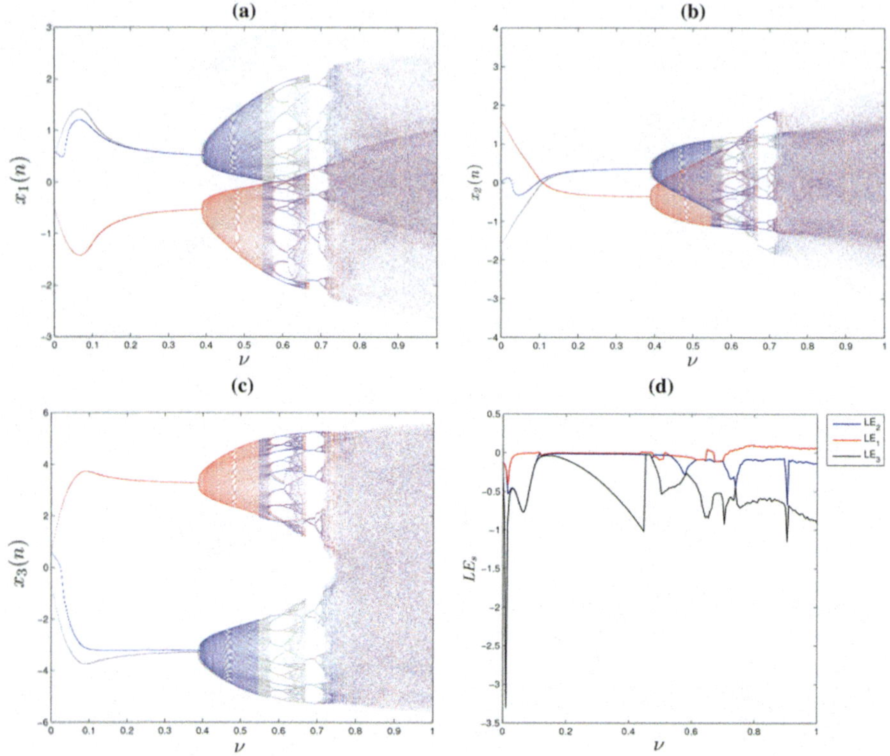

Fig. 5 Coexisting multiple bifurcation diagrams and LEs of the fractional neural network (11) for $\alpha \in [0, 1]$ fore the initial conditions $\varkappa_1(0) = 0.02$, $\varkappa_2(0) = 0.5$, $\varkappa_3(0) = 0.8$(red diagram), $\varkappa_1 = -0.02$, $\varkappa_2 = -0.5$, $\varkappa_3 = -0.8$and $\varkappa_1 = 0.02$, $\varkappa_2 = -0.5$, $\varkappa_3 = 0.8$ (blue diagram): **a** in the $\varkappa_1 - \alpha$ plane, **b** in the $\varkappa_2 - \alpha$ plane, **c** in the $\varkappa_3 - \alpha$ plane, **d** Lyapunov exponents corresponding to **a** (colour figure online)

4.2 *Chaos Control of the Fractional-Order Discrete Neural Network*

Now, a unique stabilization result is presented and obtained applying the Lyapunov direct approach with the goal of obtaining proper control for the chaos that emerged in the fractional-order discrete-time neural network (11). For this, we will try to come up with some appropriate linear control laws that will stabilize all three states of system (11) at the origin. Following that, we will do many simulations to see if this task was completed correctly. Let us begin with the following theorem,

Theorem 4 *The fractional-order neural network (11) is stabilized under the linear control laws* $W = (W_1, W_2, W_3]^T$ *defined as follow*

$$\begin{cases} W_1(t) = -(|b_{11}| + |b_{12}| + |b_{13}|)sgn(\varkappa_1(t)), \\ W_2(t) = -(|b_{21}| + |b_{22}| + |b_{23}|)sgn(\varkappa_2(t)), \\ W_3(t) = -(|b_{31}| + |b_{33}|)sgn(\varkappa_3(t)). \end{cases} \tag{17}$$

Proof The controlled fractional-order neural network, which can be inferred by incorporating the controllers listed in (17) into system (11), is stated as follows:

$$ {}_{\hbar}^{C}\Delta_{t_0}^{\alpha}\varkappa(t) = -A\varkappa(t) + B\tanh(\varkappa(t)) + W \tag{18}$$

Which can then be displayed in the following manner:

$$\begin{cases} {}_{\hbar}^{C}\Delta_{t_0}^{\alpha}\varkappa_1(t) = -\varkappa_1(t) + 2\tanh(\varkappa_1(t)) - 1.2\tanh(\varkappa_2(t)) - 0.1\tanh(\varkappa_3(t)) \\ \qquad - (|b_{11}| + |b_{12}| + |b_{13}|)\, sgn\,(\varkappa_1(t)), \\ {}_{\hbar}^{C}\Delta_{t_0}^{\alpha}\varkappa_2(t) = -\varkappa_2(t) + 2\tanh(\varkappa_1(t)) + 1.71\tanh(\varkappa_2(t)) + 1.15\tanh(\varkappa_3(t)) \\ \qquad - (|b_{21}| + |b_{22}| + |b_{23}|)\, sgn\,(\varkappa_2(t)), \\ {}_{\hbar}^{C}\Delta_{t_0}^{\alpha}\varkappa_3(t) = -\varkappa_3(t) - 4.75\tanh(\varkappa_1(t)) + 1.1\tanh(\varkappa_3(t)) \\ \qquad - (|b_{31}| + |b_{33}|)\, sgn\,(\varkappa_3(t)), \end{cases} \tag{19}$$

First, considering the following Lyapunov function $V(t)$

$$V(t) = \frac{1}{2}\sum_{i=1}^{3}\varkappa_i(t) \tag{20}$$

Combining discrete Caputo fractional-order difference operator with order α to the Lyapunov function (20) with respect to time along with Lemma 4, we derive the following

$$\begin{aligned} {}_{\hbar}^{C}\Delta_{t_0}^{\alpha}V(t) &= \frac{1}{2}\sum_{i=1}^{3}{}_{\hbar}^{C}\Delta_{t_0}^{\alpha}\varkappa_i(t) \\ &\leq \sum_{i=1}^{3}\varkappa_i(t)\,{}_{\hbar}^{C}\Delta_{t_0}^{\alpha}\varkappa_i(t) \\ &= \varkappa_1(t)[-\varkappa_1(t) + b_{11}\tanh(\varkappa_1(t)) + b_{12}\tanh(\varkappa_2(t)) + b_{13}\tanh(\varkappa_3(t)) \\ &\qquad - (|b_{11}| + |b_{12}| + |b_{13}|)\, sgn\,(\varkappa_1(t))] \\ &+ \varkappa_2(t)[-\varkappa_2(t) + b_{21}\tanh(\varkappa_1(t)) + b_{22}\tanh(\varkappa_2(t)) + b_{23}\tanh(\varkappa_3(t)) \\ &\qquad - (|b_{21}| + |b_{22}| + |b_{23}|)\, sgn\,(\varkappa_2(t))] \\ &+ \varkappa_3(t)[-\varkappa_3(t) + b_{31}\tanh(\varkappa_1(t)) + b_{33}\tanh(\varkappa_3(t)) \\ &\qquad - (|b_{31}| + |b_{33}|)\, sgn\,(\varkappa_3(t))] \\ &\leq -\varkappa_1(t)^2 - \varkappa_2^2(t) - \varkappa_3^2(t) \\ &+ |\varkappa_1(t)|[|b_{11}|\tanh(\varkappa_1(t)) + |b_{12}|\tanh(\varkappa_2(t)) + |b_{13}|\tanh(\varkappa_3(t))] \end{aligned}$$

$$
\begin{aligned}
&- \varkappa_1(t)\,(|b_{11}| + |b_{12}| + |b_{13}|)\, sgn\,(\varkappa_1(t))] \\
&+ |\varkappa_2(t)|[|b_{21}| \tanh(\varkappa_1(t)) + |b_{22}| \tanh(\varkappa_2(t)) + |b_{23}| \tanh(\varkappa_3(t))] \\
&- \varkappa_2(t)\,(|b_{21}| + |b_{22}| + |b_{23}|)\, sgn\,(\varkappa_2(t))] \\
&+ |\varkappa_3(t)|[|b_{31}| \tanh(\varkappa_1(t)) + |b_{33}| \tanh(\varkappa_3(t))] \\
&- \varkappa_3(t)\,(|b_{31}| + |b_{33}|)\, sgn\,(\varkappa_3(t))] \\
&\leq -(\varkappa_1(t)^2 + \varkappa_2^2(t) + \varkappa_3^2(t))
\end{aligned}
$$

which indicates ${}^{C}_{\hbar}\Delta^{\alpha}_{t_0} V(t) \leq 0$,. As a result, based on Theorem 1, the dynamics of the suggested fractional-order discrete neural network (11) are stabilized by the linear control law (17). □

The fractional order value in the numerical simulation is $\alpha = 0.8$. Figs. 6 and 7 show, respectively, the development of states and the phase space plot of the controlled fractional neural network (11). These figures clearly illustrate that the chaotic dynamics of the fractional discrete neural network (11) are efficiently converged to the origin as $t = 20$, confirming the efficiency of the control rules established (17)

5 Synchronization of Fractional Discrete Chaotic Neural Networks

This section demonstrates the exceptional effectiveness of the suggested approach for synchronizing fractional discrete neural network systems. Both the master and slave systems are modeled as fractional-order discrete chaotic neural networks. The slave system is stated by

$$
\begin{cases}
{}^{C}_{\hbar}\Delta^{\alpha}_{t_0} \mathfrak{y}_1(t) = -a_1\mathfrak{y}_1(t) + b_{11}\tanh(\mathfrak{y}_1(t)) + b_{12}\tanh(\mathfrak{y}_2(t)) + b_{13}\tanh(\mathfrak{y}_3(t)) + I_1 + D_1(t), \\
{}^{C}_{\hbar}\Delta^{\alpha}_{t_0} \mathfrak{y}_2(t) = -a_2\mathfrak{y}_2(t) + b_{21}\tanh(\mathfrak{y}_1(t)) + b_{22}\tanh(\mathfrak{y}_2(t)) + b_{23}\tanh(\mathfrak{y}_3(t)) + I_2 + D_2(t), \\
{}^{C}_{\hbar}\Delta^{\alpha}_{t_0} \mathfrak{y}_3(t) = -a_3\mathfrak{y}_3(t) + b_{31}\tanh(\mathfrak{y}_1(t)) + b_{32}\tanh(\mathfrak{y}_2(t)) + b_{33}\tanh(\mathfrak{y}_3(t)) + I_3 + D_3(t),
\end{cases}
\tag{21}
$$

where $D_i (i = 1, 2, 3)$ represents control input to be determined and the master system is as follows:

$$
\begin{cases}
{}^{C}_{\hbar}\Delta^{\alpha}_{t_0} \varkappa_1(t) = -a_1\varkappa_1(t) + b_{11}\tanh(\varkappa_1(t)) + b_{12}\tanh(\varkappa_2(t)) + b_{13}\tanh(\varkappa_3(t)) + I_1, \\
{}^{C}_{\hbar}\Delta^{\alpha}_{t_0} \varkappa_2(t) = -a_2\varkappa_2(t) + b_{21}\tanh(\varkappa_1(t)) + b_{22}\tanh(\varkappa_2(t)) + b_{23}\tanh(\varkappa_3(t)) + I_2, \\
{}^{C}_{\hbar}\Delta^{\alpha}_{t_0} \varkappa_3(t) = -a_3\varkappa_3(t) + b_{31}\tanh(\varkappa_1(t)) + b_{32}\tanh(\varkappa_2(t)) + b_{33}\tanh(\varkappa_3(t)) + I_3,
\end{cases}
\tag{22}
$$

The master system's settings are the same as those of the slave system. The initial conditions for salve and master systems are $[\varkappa_1(0), \varkappa_2(0), \varkappa_3(0),] = [3, 4, 3]$ and $[\mathfrak{y}_1(0), \mathfrak{y}_2(0), \mathfrak{y}_3(0),] = [7, 7, 7]$.

The error system is defined by subtracting the master system (22) from the slave system (21)

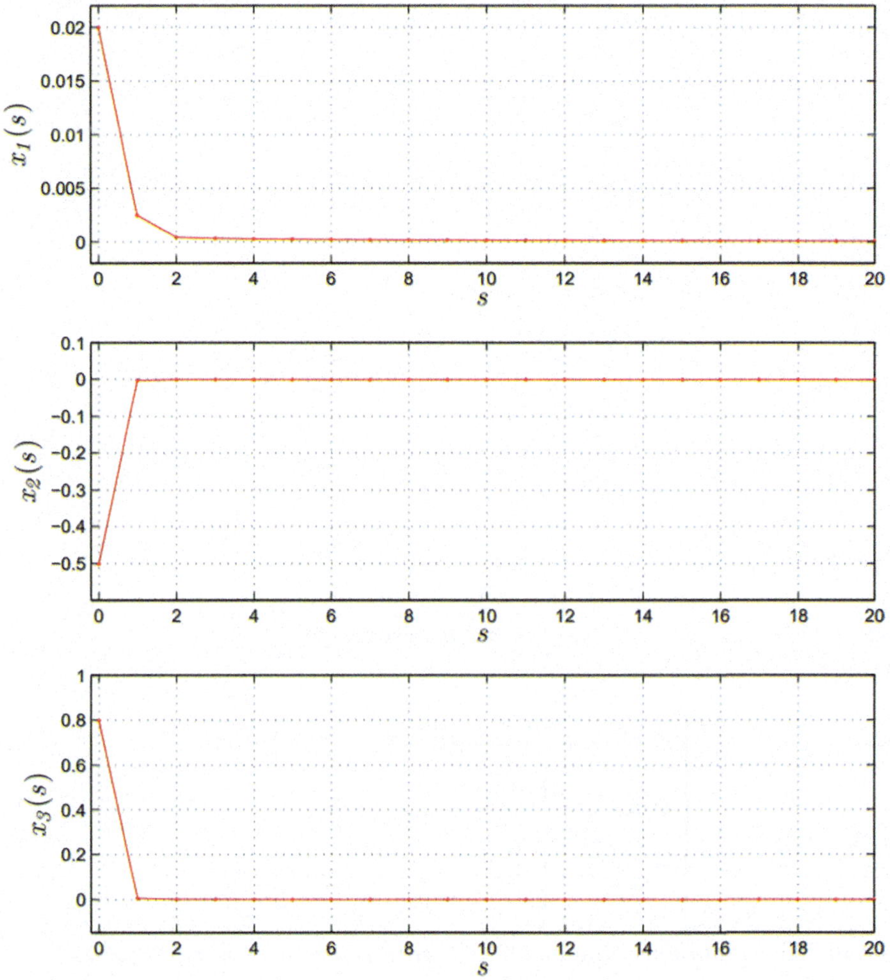

Fig. 6 Stabilization of fractional-order discrete neural networks (11) via the control laws

$$e(t) = (e_1(t), e_2(t), e_3(t))^T = (\eta_1(t) - \varkappa_1(t), \eta_2(t) - \varkappa_2(t), \eta_3(t) - \varkappa_3(t))^T \tag{23}$$

Now , a theorem employing a Lyapunov approach is demonstrated, with the goal of synchronizing the master-slave (21) and (22) through linear controllers D_1, D_2 and D_3.

Theorem 5 *The master system (22) and slave system (21) are synchronized if the controlled $D_1(t)$, $D_2(t)$ and $D_3(t)$ are chosen as follow*

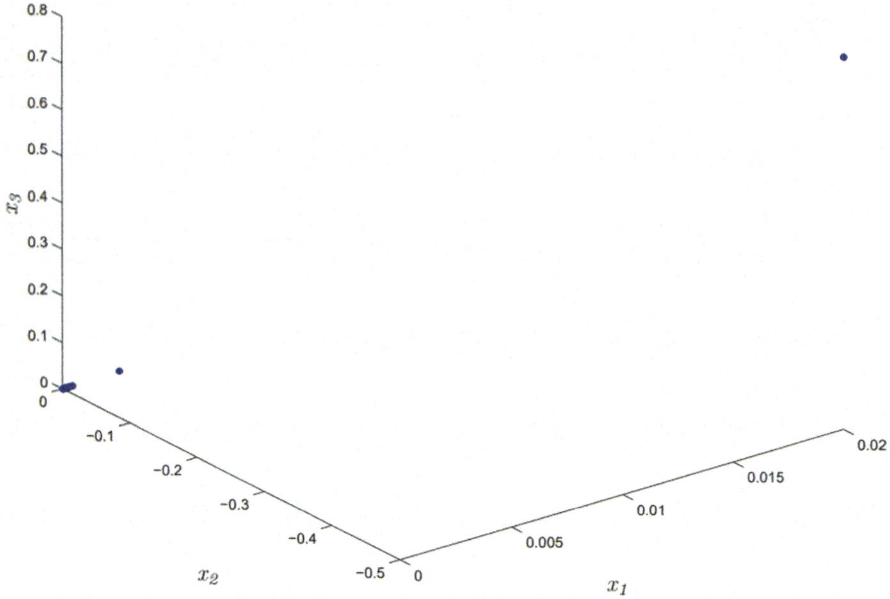

Fig. 7 Stabilitation of the variable-order chaotic neural networks

$$\begin{cases} D_1(t) = -\left[\dfrac{(|b_{12}| + |b_{13}|)^2}{4} + |b_{11}|\right] e_1(t), \\ D_2(t) = -\left[\dfrac{|b_{21}|^2}{4} + |b_{22}|\right] e_2(t), \\ D_3(t) = -\left[\dfrac{(|b_{31}| + |b_{13}|)^2}{4} + |b_{33}|\right] e_3(t). \end{cases} \tag{24}$$

Proof Taking into consideration (23), the error system may be expressed as

$$\begin{cases} {}^{C}_{\hbar}\Delta^{\alpha}_{t_0} e_1(t) = -a_1 e_1(t) + b_{11}(\tanh(\mathfrak{y}_1(t)) - \tanh(\varkappa_1(t))) + b_{12}(\tanh(\mathfrak{y}_2(t)) - \tanh(\varkappa_2(t))) \\ \qquad + b_{13}(\tanh(\mathfrak{y}_3(t)) - \tanh(\varkappa_3(t))) - \left[\dfrac{(|b_{12}| + |b_{13}|)^2}{4} + |b_{11}|\right] e_1(t), \\ {}^{C}_{\hbar}\Delta^{\alpha}_{t_0} e_2(t) = -a_2 e_2(t) + b_{21}(\tanh(\mathfrak{y}_1(t)) - \tanh(\varkappa_1(t))) + b_{22}(\tanh(\mathfrak{y}_2(t)) - \tanh(\varkappa_2(t))) \\ \qquad + b_{23}(\tanh(\mathfrak{y}_3(t)) - \tanh(\varkappa_3(t))) - \left[\dfrac{(|b_{21}| + |b_{23}|)^2}{4} + |b_{22}|\right] e_2(t), \\ {}^{C}_{\hbar}\Delta^{\alpha}_{t_0} e_3(t) = -a_3 e_3(t) + b_{31}(\tanh(\mathfrak{y}_1(t)) - \tanh(\varkappa_1(t))) + b_{32}(\tanh(\mathfrak{y}_2(t)) - \tanh(\varkappa_2(t))) \\ \qquad + b_{33}(\tanh(\mathfrak{y}_3(t)) - \tanh(\varkappa_3(t))) - \left[\dfrac{b_{31}^2}{4} + |b_{33}|\right] e_3(t), \end{cases} \tag{25}$$

Taking the Lyapunov function (26) and applying Lemma 4, we have

$$V(t) = \frac{1}{2}\sum_{i=1}^{3} e_i(t), \tag{26}$$

$$\begin{aligned}
{}^{C}_{\hbar}\Delta^{\alpha}_{t_0} V(t) &= \frac{1}{2}\sum_{i=1}^{3} {}^{C}\nabla^{\delta(t)}_{t_0} e_i(t) \le \sum_{i=1}^{3} e_i(t)\, {}^{C}_{\hbar}\Delta^{\alpha}_{t_0} e_i(t)\\
&= e_1(t)\{-a_1 e_1(t) + b_{11}(\tanh(\eta_1(t)) - \tanh(\varkappa_1(t))) + b_{12}(\tanh(\eta_2(t)) - \tanh(\varkappa_2(t)))\\
&\qquad + b_{13}(\tanh(\eta_3(t)) - \tanh(\varkappa_3(t))) - \left[\frac{(|b_{12}| + |b_{13}|)^2}{4} + |b_{11}|\right] e_1(t)\},\\
&+ e_2(t)\{-a_2 e_2(t) + b_{21}(\tanh(\eta_1(t)) - \tanh(\varkappa_1(t))) + b_{22}(\tanh(\eta_2(t)) - \tanh(\varkappa_2(t)))\\
&\qquad + b_{23}(\tanh(\eta_3(t)) - \tanh(\varkappa_3(t))) - \left[\frac{(|b_{21}| + |b_{23}|)^2}{4} + |b_{22}|\right] e_2(t)\},\\
&+ e_3(t)\{-a_3 e_3(t) + b_{31}(\tanh(\eta_1(t)) - \tanh(\varkappa_1(t))) + b_{32}(\tanh(\eta_2(t)) - \tanh(\varkappa_2(t)))\\
&\qquad + b_{33}(\tanh(\eta_3(t)) - \tanh(\varkappa_3(t))) - \left[\frac{b_{31}^2}{4} + |b_{33}|\right] e_3(t)\},\\
&\le -a_1 e_1(t)^2 - a_2 e_2^2(t) - a_3 e_3^2(t)\\
&+ e_1(t)[|b_{11}|(\tanh(\eta_1(t)) - \tanh(\varkappa_1(t))) + |b_{12}|(\tanh(\eta_2(t)) - \tanh(\varkappa_2(t)))\\
&+ |b_{13}|(\tanh(\eta_3(t)) - \tanh(\varkappa_3(t)))] - \left[\frac{(|b_{12}| + |b_{13}|)^2}{4} + |b_{11}|\right] e_1^2(t)\\
&+ e_2(t)[|b_{21}|(\tanh(\eta_1(t)) - \tanh(\varkappa_1(t))) + |b_{22}|(\tanh(\eta_2(t)) - \tanh(\varkappa_2(t)))\\
&+ |b_{23}|(\tanh(\eta_3(t)) - \tanh(\varkappa_3(t)))] - \left[\frac{(|b_{21}| + |b_{23}|)^2}{4} + |b_{22}|\right] e_2^2(t)\\
&+ e_3(t)[|b_{31}|(\tanh(\eta_1(t)) - \tanh(\varkappa_1(t))) + |b_{32}|(\tanh(\eta_2(t)) - \tanh(\varkappa_2(t)))\\
&+ |b_{33}|(\tanh(\eta_3(t)) - \tanh(\varkappa_3(t)))] - \left[\frac{b_{31}^2}{4} + |b_{33}|\right] e_3^2(t),\\
&\le -a_1 e_1(t)^2 - a_2 e_2^2(t) - a_3 e_3^2(t) + |b_{11}| e_1^2(t) + |b_{12}| e_1(t)e_2(t) + |b_{13}| e_1(t)e_3(t)\\
&- \left[\frac{(|b_{12}| + |b_{13}|)^2}{4} + |b_{11}|\right] e_1^2(t) + |b_{21}| e_2(t)e_1(t) + |b_{22}| e_2^2(t) + |b_{23}| e_2(t)e_3(t)\\
&- \left[\frac{(|b_{21}| + |b_{23}|)^2}{4} + |b_{22}|\right] e_2^2(t) + |b_{31}| e_3(t)e_1(t) + |b_{32}| e_3(t)e_2(t) + |b_{33}| e_3^2(t)\\
&- \left[\frac{b_{31}^2}{4} + |b_{33}|\right] e_3^2(t),\\
&\le (|b_{12}| + |b_{21}|)|e_1(t)||e_2(t)| + |b_{23}||e_2(t)||e_3(t)| + (|b_{13}| + |b_{31}|)|e_1(t)||e_3(t)|\\
&- \left(\frac{b_{12}| + |b_{21}|}{2}|e_1(t) - e_2(t)|\right)^2 - \left(\frac{b_{23}|}{2}|e_2(t) - e_3(t)|\right)^2 - \left(\frac{b_{13}| + |b_{31}|}{2}|e_3(t) - e_1(t)|\right)^2\\
&- \left(\left(\frac{|b_{12}| + |b_{21}|}{2}\right)^2 + 1\right) e_1^2(t) - \left(\left(\frac{|b_{23}|}{2}\right)^2 + 1\right) e_2^2(t) - \left(\left(\frac{|b_{13}| + |b_{31}|}{2}\right)^2 + 1\right) e_3^2(t)\\
&\le 0.
\end{aligned}$$

From Theorem 1, the dynamics of the fractional discrete error system are said to be stabilized at zero. As a result, the synchronization between the discrete fractional

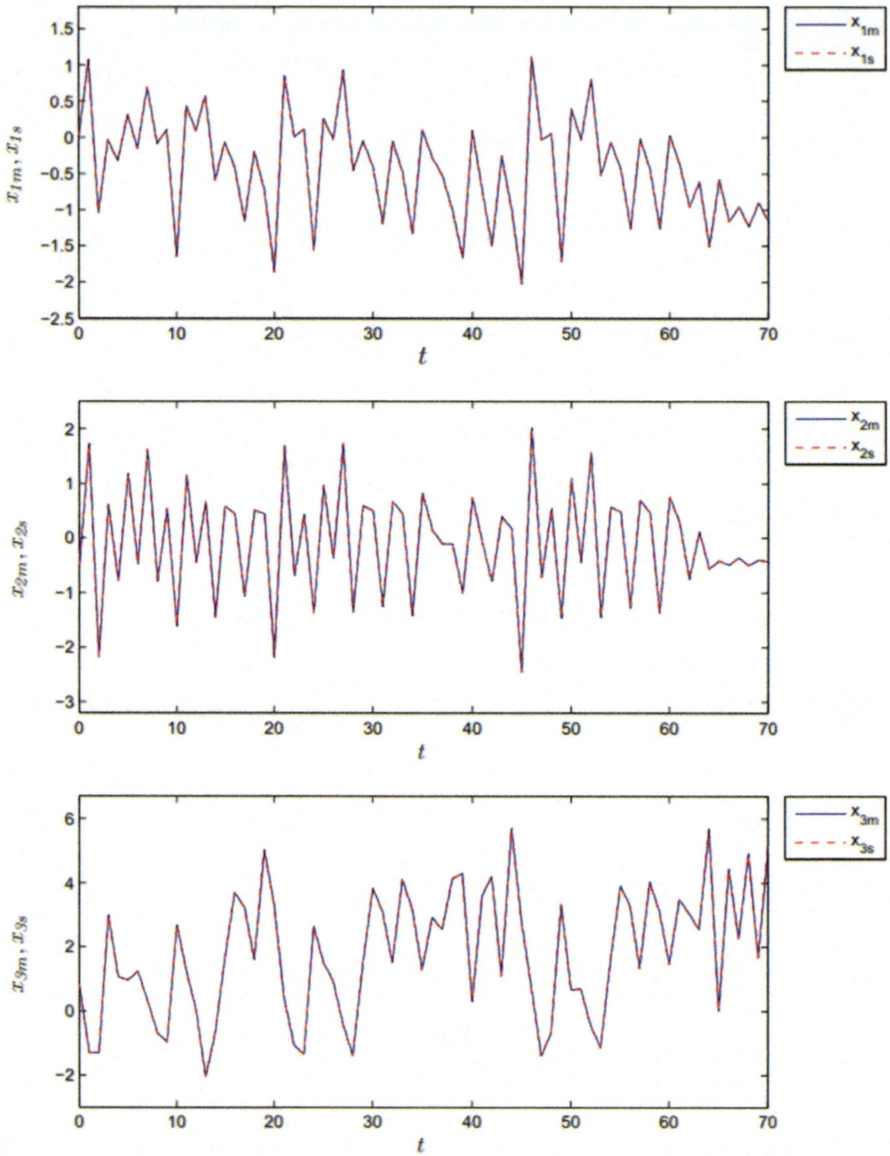

Fig. 8 Chaotic state trajectory of the master-slave fractional-order neural networks (21)–(22)

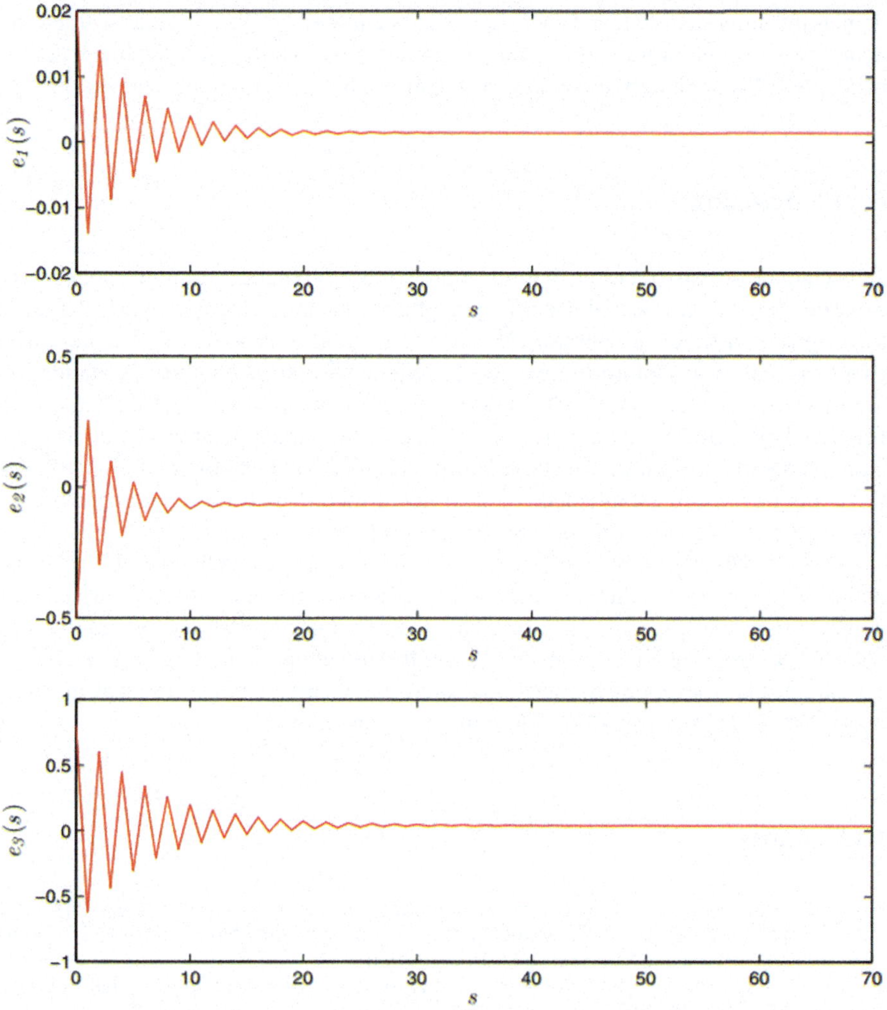

Fig. 9 Time evolution of the synchronization errors

neural network master system (22) and slave system (21) is obtained using linear control laws, which completes the proof. □

To mimic the states of the $\hbar$-fractional error system and synchronized systems (21) and (22), the predictor corrector technique with $\hbar = 1$ is utilized. With $\alpha = 0.8$, the starting values are $e_1(0) = 0.1$, $e_2(0) = 0.05$, and $e_3(0) = 0.02$. Figure 8 depicts the temporal development of the chaotic trajectories of the fractional neural network for both master and slave systems, i.e. systems (22) and (21); meanwhile Fig. 9

depicts the numerical states of the $\hbar$-fractional system (23). The two chaotic fractional neural networks exhibit synchronized dynamics, demonstrating that the linear control principles (24) are an effective and easy instrument for chaos synchronization.

6 Conclusions

This chapter provided a contribution to the study of dynamics in non-integer order discrete systems. We describe a new type of discrete-time neural networks based on the Caputo h-difference operator. The system's stability is addressed by presenting two theorems that explore the existence, uniqueness, and Ulam Hyers stability of the investigated neural networks. Furthermore, the chaotic attractors of a new 3D fractional discrete neural network represented by the same operator, as well as bifurcation diagrams and the greatest Lyapunov exponents, are disclosed. Two additional theorems have been established. First, a novel theory with the goal of stabilizing the chaotic trajectories of the h-fractional neural network at zero was demonstrated. Then, a new theory is introduced that guarantees that two chaotic 3D fractional neural networks may achieve synchronized dynamics using relatively simple linear control principles. In terms of future objectives for our research, we intend to create some hardware implementations of chaotic fractional neural networks. We will also investigate possible applications of fractional chaotic neural networks, such as image encryption [126] and pseudo-random number generators [127].

References

1. Diaz, J.B., Osler, T.J.: Differences of fractional order. Math. Comput. **28**(125), 185–202 (1974)
2. Granger, C.W., Joyeux, R.: An introduction to long-memory time series models and fractional differencing. J. Time Ser. Anal. **1**(1), 15–29 (1980)
3. Gray, H.L., Zhang, N.F.: On a new definition of the fractional difference. Math. Comput. **50**(182), 513–529 (1988)
4. Isaacs, G.L.: Exponential laws for fractional differences. Math. Comput. **35**(151), 933–936 (1980)
5. Djenina, N., Ouannas, A., Batiha, I.M., Grassi, G., Pham, V.T.: On the stability of linear incommensurate fractional-order difference systems. Mathematics **8**(10), 1754 (2020)
6. Shatnawi, M.T., Djenina, N., Ouannas, A., Batiha, I.M., Grassi, G.: Novel convenient conditions for the stability of nonlinear incommensurate fractional-order difference systems. Alex. Eng. J. **61**(2), 1655–1663 (2022)
7. Djenina, N., Ouannas, A., Oussaeif, T.E., Grassi, G., Batiha, I.M., Momani, S., Albadarneh, R.B.: On the stability of incommensurate h-Nabla fractional-order difference systems. Fractal Fract. **6**(3), 158 (2022)
8. Dababneh, A., Djenina, N., Ouannas, A., Grassi, G., Batiha, I.M., Jebril, I.H.: A new incommensurate fractional-order discrete COVID-19 model with vaccinated individuals compartment. Fractal Fract. **6**(8), 456 (2022)
9. Djenina, N., Ouannas, A., Batiha, I.M., Grassi, G., Oussaeif, T.E., Momani, S.: A novel fractional-order discrete SIR model for predicting COVID-19 behavior. Mathematics **10**(13), 2224 (2022)

10. Almatroud, A.O., Djenina, N., Ouannas, A., Grassi, G., Al-Sawalha, M.M.: A novel discrete-time COVID-19 epidemic model including the compartment of vaccinated individuals. Math. Biosci. Eng. **19**, 12387–12404 (2022)
11. Al-Shbeil, I., Djenina, N., Jaradat, A., Al-Husban, A., Ouannas, A., Grassi, G.: A new COVID-19 pandemic model including the compartment of vaccinated individuals: global stability of the disease-free fixed point. Mathematics **11**(3), 576 (2023)
12. Abbes, A., Ouannas, A., Shawagfeh, N., Jahanshahi, H.: The fractional-order discrete COVID-19 pandemic model: stability and chaos. Nonlinear Dyn. **111**(1), 965–983 (2023)
13. Al-Husban, A., Djenina, N., Saadeh, R., Ouannas, A., Grassi, G.: A new incommensurate fractional-order COVID 19: modelling and dynamical analysis. Mathematics **11**(3), 555 (2023)
14. Debbouche, N., Ouannas, A., Grassi, G., Al-Hussein, A.B.A., Tahir, F.R., Saad, K.M., Jahanshahi, H., Aly, A.A.: Chaos in cancer tumor growth model with commensurate and incommensurate fractional-order derivatives. Comput. Math. Methods Med. **2022** (2022)
15. Wang, B., Ouannas, A., Karaca, Y., Xia, W.F., Jahanshahi, H., Alkhateeb, A.F., Nour, M.: A hybrid approach for synchronizing between two reaction-diffusion systems of integer-and fractional-order applied on certain chemical models. Fractals **30**(05), 1–11 (2022)
16. Mesdoui, F., Shawagfeh, N., Ouannas, A.: Global synchronization of fractional-order and integer-order N component reaction diffusion systems: application to biochemical models. Math. Methods Appl. Sci. **44**(1), 1003–1012 (2021)
17. Almatroud, A.O., Ouannas, A., Grassi, G., Batiha, I.M., Gasri, A., Al-Sawalha, M.M.: Different linear control laws for fractional chaotic maps using Lyapunov functional. Arch. Control Sci. **31**(4) (2021)
18. Ouannas, A., Mesdoui, F., Momani, S., Batiha, I., Grassi, G.: Synchronization of FitzHugh-Nagumo reaction-diffusion systems via one-dimensional linear control law. Arch. Control Sci. **31** (2021)
19. Gasri, A., Ouannas, A., Khennaoui, A.A., Bendoukha, S., Pham, V.T.: On the dynamics and control of fractional chaotic maps with sine terms. Int. J. Nonlinear Sci. Numer. Simul. **21**(6), 589–601 (2020)
20. Ouannas, A., Wang, X., Pham, V.T., Ziar, T.: $\Lambda - \phi$ generalized synchronization: application to fractional hyperchaotic systems with arbitrary dimensions and orders. Automatika: Casopis za automatiku, mjerenje, elektroniku, racunarstvo i komunikacije **61**(4), 554–563 (2020)
21. Ouannas, A., Azar, A.T., Ziar, T.: On inverse full state hybrid function projective synchronization for continuous-time chaotic dynamical systems with arbitrary dimensions. Differ. Equ. Dynam. Syst. **28**(4), 1045–1058 (2020)
22. Khennaoui, A.A., Ouannas, A., Odibat, Z., Pham, V.T., Grassi, G.: On the three-dimensional fractional-order Hénon map with Lorenz-like attractors. Int. J. Bifurc. Chaos **30**(11), 2050217 (2020)
23. Zoulikha, C., Ouannas, A.: A New generalized synchronization scheme to control fractional chaotic dynamical systems with different dimensions and orders. Nonlinear Stud. **27**(3), 845–854 (2020)
24. Mesdoui, F., Ouannas, A., Shawagfeh, N., Grassi, G., Pham, V.T.: Synchronization methods for the Degn-Harrison reaction-diffusion systems. IEEE Access **8**, 91829–91836 (2020)
25. Ouannas, A., Azar, A.T., Ziar, T.: Fractional inverse full state hybrid projective synchronisation. Int. J. Adv. Intell. Parad. **17**(3–4), 279–298 (2020)
26. Ouannas, A., Wang, X., Pham, V.T., Grassi, G., Huynh, V.V.: Synchronization results for a class of fractional-order spatiotemporal partial differential systems based on fractional Lyapunov approach. Bound. Value Probl. **2019**(1), 1–12 (2019)
27. Ouannas, A., Khennaoui, A.A., Grassi, G., Bendoukha, S.: On chaos in the fractional-order Grassi-Miller map and its control. J. Comput. Appl. Math. **358**, 293–305 (2019)
28. Ouannas, A., Bendoukha, S., Volos, C., Boumaza, N., Karouma, A.: Synchronization of fractional hyperchaotic Rabinovich systems via linear and nonlinear control with an application to secure communications. Int. J. Control Autom. Syst. **17**, 2211–2219 (2019)

29. Ouannas, A., Khennaoui, A.A., Odibat, Z., Pham, V.T., Grassi, G.: On the dynamics, control and synchronization of fractional-order Ikeda map. Chaos, Solitons Fractals **123**, 108–115 (2019)
30. Ouannas, A., Bendoukha, S., Karouma, A., Abdelmalek, S.: A general method to study the co-existence of different hybrid synchronizations in fractional-order chaotic systems. Int. J. Nonlinear Sci. Numer. Simul. **20**(3–4), 351–359 (2019)
31. Ouannas, A., Abdelli, M., Odibat, Z., Wang, X., Pham, V.T., Grassi, G., Alsaedi, A.: Synchronization control in reaction-diffusion systems: application to Lengyel-Epstein system. Complexity **2019** (2019)
32. Ouannas, A., Azar, A.T., Ziar, T.: Control of continuous-time chaotic (hyperchaotic) systems: F-M synchronisation. Int. J. Autom. Control **13**(2), 226–242 (2019)
33. Bendoukha, S., Abdelmalek, S., Ouannas, A.: Secure communication systems based on the synchronization of chaotic systems. Math. Appl. Eng., Model., Soc. Issues 281–311 (2019)
34. Wang, X., Ouannas, A., Pham, V.T., Abdolmohammadi, H.R.: A fractional-order form of a system with stable equilibria and its synchronization. Adv. Differ. Equ. **2018**(1), 1–13 (2018)
35. Ouannas, A., Wang, X., Pham, V.T., Grassi, G., Ziar, T.: Coexistence of identical synchronization, antiphase synchronization and inverse full state hybrid projective synchronization in different dimensional fractional-order chaotic systems. Adv. Differ. Equ. **2018**(1), 1–16 (2018)
36. Ouannas, A., Grassi, G., Wang, X., Ziar, T., Pham, V.T.: Function-based hybrid synchronization types and their coexistence in non-identical fractional-order chaotic systems. Adv. Differ. Equ. **2018**, 1–12 (2018)
37. Ouannas, A., Odibat, Z., Alsaedi, A., Hobiny, A., Hayat, V.: Investigation of QS synchronization in coupled chaotic incommensurate fractional order systems. Chin. J. Phys. **56**(5), 1940–1948 (2018)
38. Ouannas, A., Debbouche, N., Wang, X., Pham, V.T., Zehrour, O.: Secure multiple-input multiple-output communications based on F-M synchronization of fractional-order chaotic systems with non-identical dimensions and orders. Appl. Sci. **8**(10), 1746 (2018)
39. Gasri, A., Ouannas, A., Ojo, K.S., Pham, V.T.: Coexistence of generalized synchronization and inverse generalized synchronization between chaotic and hyperchaotic systems. Nonlinear Anal.: Model. Control **23**(4), 583–598 (2018)
40. Pham, V.T., Ouannas, A., Volos, C., Kapitaniak, T.: A simple fractional-order chaotic system without equilibrium and its synchronization. AEU-Int. J. Electron. Commun. **86**, 69–76 (2018)
41. Boudiar, S., Ouannas, A., Bendoukha, S., Zara, A.: Coexistence of different types of chaos synchronization between non-identical and different dimensional dynamical systems. Nonlinear Dyn. Syst. Theory **18**(3), 253–258 (2018)
42. Ouannas, A., Wang, X., Pham, V.T., Grassi, G., Ziar, T.: Co-existence of some synchronization types between non-identical commensurate and incommensurate fractional-order chaotic systems with different dimensions. Adv. Differ. Equ. **21**(10) (2018)
43. Grassi, G., Ouannas, A., Pham, V.T.: A general unified approach to chaos synchronization in continuous-time systems (with or without equilibrium points) as well as in discrete-time systems. Arch. Control Sci. **28**(1) (2018)
44. Ouannas, A., Karouma, A.: Different generalized synchronization schemes between integer-order and fractional-order chaotic systems with different dimensions. Differ. Equ. Dyn. Syst. **26**(1–3), 125–137 (2018)
45. Ouannas, A., Zehrour, O., Laadjal, Z.: Nonlinear methods to control synchronization between fractional-order and integer-order chaotic systems. Nonlinear Stud. **25**(1) (2018)
46. Singh, S., Azar, A.T., Bhat, M.A., Vaidyanathan, S., Ouannas, A.: Active control for multi-switching combination synchronization of non-identical chaotic systems. In: Advances in System Dynamics and Control, pp. 129–162. IGI Global (2018)
47. Ouannas, A., Odibat, Z., Hayat, T.: Fractional analysis of co-existence of some types of chaos synchronization. Chaos, Solitons Fractals **105**, 215–223 (2017)
48. Ouannas, A., Wang, X., Pham, V.T., Ziar, T.: Dynamic analysis of complex synchronization schemes between integer order and fractional order chaotic systems with different dimensions. Complexity **2017** (2017)

49. Ouannas, A., Grassi, G., Ziar, T., Odibat, Z.: On a function projective synchronization scheme for non-identical fractional-order chaotic (hyperchaotic) systems with different dimensions and orders. Optik **136**, 513–523 (2017)
50. Ouannas, A., Azar, A.T., Vaidyanathan, S.: A robust method for new fractional hybrid chaos synchronization. Math. Methods Appl. Sci. **40**(5), 1804–1812 (2017)
51. Azar, A.T., Vaidyanathan, S., Ouannas, A.: Fractional Order Control and Synchronization of Chaotic Systems, vol. 688. Springer, Berlin (2017)
52. Azar, A.T., Volos, C., Gerodimos, N.A., Tombras, G.S., Pham, V.T., Radwan, A.G., Vaidyanathan, S., Ouannas, A., Munoz-Pacheco, J.M.: A novel chaotic system without equilibrium: dynamics, synchronization, and circuit realization. Complexity **2017** (2017)
53. Ouannas, A., Azar, A.T., Vaidyanathan, S.: A new fractional hybrid chaos synchronisation. Int. J. Model. Ident. Control **27**(4), 314–322 (2017)
54. Ouannas, A., Grassi, G.: A new approach to study the coexistence of some synchronization types between chaotic maps with different dimensions. Nonlinear Dyn. **86**(2), 1319–1328 (2016)
55. Ouannas, A., Al-Sawalha, M.M., Ziar, T.: Fractional chaos synchronization schemes for different dimensional systems with non-identical fractional-orders via two scaling matrices. Optik **127**(20), 8410–8418 (2016)
56. Ouannas, A., Grassi, G.: Inverse full state hybrid projective synchronization for chaotic maps with different dimensions. Chin. Phys. B **25**(9), 090503 (2016)
57. Ouannas, A., Odibat, Z.: On inverse generalized synchronization of continuous chaotic dynamical systems. Int. J. Appl. Comput. Math. **2**, 1–11 (2016)
58. Ouannas, A., Al-sawalha, M.M.: Synchronization between different dimensional chaotic systems using two scaling matrices. Optik **127**(2), 959–963 (2016)
59. Ouannas, A.: Co-existence of various types of synchronization between hyperchaotic maps. Nonlinear Dyn. Syst. Theory **16**, 312–321 (2016)
60. Ouannas, A., Abu-Saris, R.: A robust control method for QS synchronization between different dimensional integer-order and fractional-order chaotic systems. J. Control Sci. Eng. **2015**, 55–55 (2015)
61. Ahlem, G., Ouannas, A.: A general control method for inverse hybrid function projective synchronization of a class of chaotic systems. Int. J. Math. Anal. **9**(9–12), 429–436 (2015)
62. Ouannas, A., Ahlem, G.: General synchronization approach for general chaotic systems in continuous-time. Far East J. Math. Sci. **96**(2), 231 (2015)
63. Ahlem, G., Ouannas, A.: Synchronization analysis of quadratic chaotic systems in nD. Far East J. Math. Sci. **96**(2), 241 (2015)
64. Ouannas, A.: A synchronization criterion for a class of sinusoidal chaotic maps via linear controller. Int. J. Contemp. Math. Sci. **9**(14), 677–683 (2014)
65. Ouannas, A.: Chaos synchronization approach based on new criterion of stability. Nonlinear Dyn. Syst. Theory **14**(4), 395–401 (2014)
66. Saadeh, R., Abbes, A., Al-Husban, A., Ouannas, A., Grassi, G.: The fractional discrete predator-prey model: chaos control and synchronization. Fractal Fract. **7**(2), 120 (2023)
67. Khennaoui, A.A., Ouannas, A., Bekiros, S., Aly, A.A., Alotaibi, A., Jahanshahi, H., Alsubaie, H.: Hidden homogeneous extreme multistability of a fractional-order hyperchaotic discrete-time system: chaos, initial offset boosting, amplitude control, control, and synchronization. Symmetry **15**(1), 139 (2023)
68. Ouannas, A., Grassi, G., Azar, A.T., Khennaoui, A.A.: Synchronization control in fractional discrete-time systems with chaotic hidden attractors. In: Advanced Machine Learning Technologies and Applications: Proceedings of AMLTA, vol. 2020, pp. 661–669. Springer Singapore (2021)
69. Talbi, I., Ouannas, A., Khennaoui, A.A., Berkane, A., Batiha, I.M., Grassi, G., Pham, V.T.: Different dimensional fractional-order discrete chaotic systems based on the Caputo H-difference discrete operator: dynamics, control, and synchronization. Adv. Differ. Equ. **2020**, 1–15 (2020)
70. Talbi, I., Ouannas, A., Grassi, G., Khennaoui, A.A., Pham, V.T., Baleanu, D.: Fractional Grassi-Miller map based on the Caputo H-difference operator: linear methods for chaos control and synchronization. Discret. Dyn. Nat. Soc. **2020**, 1–10 (2020)

71. Ouannas, A., Khennaoui, A.A., Momani, S., Grassi, G., Pham, V.T.: Chaos and control of a three-dimensional fractional order discrete-time system with no equilibrium and its synchronization. AIP Adv. **10**(4), 045310 (2020)
72. Khennaoui, A.A., Ouannas, A., Bendoukha, S., Grassi, G., Wang, X., Pham, V.T., Alsaadi, F.E.: Chaos, control, and synchronization in some fractional-order difference equations. Adv. Differ. Equ. **2019**(1), 1–23 (2019)
73. Ouannas, A., Odibat, Z., Shawagfeh, N.: A new Q-S synchronization results for discrete chaotic systems. Differ. Equ. Dyn. Syst. **27**, 413–422 (2019)
74. Ouannas, A., Jouini, L., Zehrour, O.: On new generalized hybrid synchronization in chaotic and hyperchaotic discrete-time dynamical systems. J. Appl. Nonlinear Dyn. **8**(3), 435–445 (2019)
75. Ouannas, A., Khennaoui, A.A., Bendoukha, S., Grassi, G.: On the dynamics and control of a fractional form of the discrete double scroll. Int. J. Bifurc. Chaos **29**(06), 1950078 (2019)
76. Ouannas, A., Khennaoui, A.A., Zehrour, O., Bendoukha, S., Grassi, G., Pham, V.T.: Synchronisation of integer-order and fractional-order discrete-time chaotic systems. Pramana **92**, 1–9 (2019)
77. Ouannas, A., Bendoukha, S., Khennaoui, A.A., Grassi, G., Wang, X., Pham, V.T.: Chaos synchronization of fractional-order discrete-time systems with different dimensions using two scaling matrices. Open Phys. **17**(1), 942–949 (2019)
78. Khennaoui, A.A., Ouannas, A., Bendoukha, S., Grassi, G., Wang, X., Pham, V.T.: Generalized and inverse generalized synchronization of fractional-order discrete-time chaotic systems with non-identical dimensions. Adv. Differ. Equ. **2018**(1), 1–14 (2018)
79. Bendoukha, S., Ouannas, A., Wang, X., Khennaoui, A.A., Pham, V.T., Grassi, G., Huynh, V.V.: The co-existence of different synchronization types in fractional-order discrete-time chaotic systems with non-identical dimensions and orders. Entropy **20**(9), 710 (2018)
80. Khennaoui, A.A., Ouannas, A., Bendoukha, S., Wang, X., Pham, V.T.: On chaos in the fractional-order discrete-time unified system and its control synchronization. Entropy **20**(7), 530 (2018)
81. Ouannas, A., Khennaoui, A.A., Grassi, G., Bendoukha, S.: On the-chaos synchronization of fractional-order discrete-time systems: general method and examples. Discret. Dyn. Nat. Soc. **2018** (2018)
82. Ouannas, A., Grassi, G., Karouma, A., Ziar, T., Wang, X., Pham, V.T.: New type of chaos synchronization in discrete-time systems: the FM synchronization. Open Phys. **16**(1), 174–182 (2018)
83. Ouannas, A., Al-sawalha, M.M.: On inverse full state hybrid projective synchronization of chaotic dynamical systems in discrete-time. Int. J. Dyn. Control **5**, 252–258 (2017)
84. Ouannas, A., Abu-Saris, R.: On matrix projective synchronization and inverse matrix projective synchronization for different and identical dimensional discrete-time chaotic systems. J. Chaos **2016** (2016)
85. Ouannas, A., Al-Sawalha, M.M.: Synchronization of chaotic dynamical systems in discrete-time. Adv. Chaos Theory Intell. Control 101–132 (2016)
86. Ouannas, A.: A new generalized-type of synchronization for discrete-time chaotic dynamical systems. J. Comput. Nonlinear Dyn. **10**(6) (2015)
87. Ouannas, A., Odibat, Z.: Generalized synchronization of different dimensional chaotic dynamical systems in discrete time. Nonlinear Dyn. **81**, 765–771 (2015)
88. Ouannas, A.: A new synchronization scheme for general 3D quadratic chaotic systems in discrete-time. Nonlinear Dyn. Syst. Theory **15**(2), 163–170 (2015)
89. Ouannas, A.: On full-state hybrid projective synchronization of general discrete chaotic systems. J. Nonlinear Dyn. **2014** (2014)
90. Ouannas, A., Aljazaery, I.: A new method to generate a discrete chaotic dynamical systems using synchronization technique. Far East J. Dyn. Syst **24**(1–2), 15–24 (2014)
91. Ouannas, A.: Chaos synchronization approach for coupled of arbitrary 3-D quadratic dynamical systems in discrete-time. Far East J. Appl. Math. **86**(3), 225 (2014)

92. Ouannas, A.: A new chaos synchronization criterion for discrete dynamical systems. Appl. Math. Sci. **8**(41), 2025–2034 (2014)
93. Ouannas, A.: Synchronization and inverse synchronization of different dimensional discrete chaotic systems via scaling matrix. Int. J. Chaos, Control, Model. Simul. **3**(4), 1–12 (2014)
94. Ouannas, A.: Nonlinear control method of chaos synchronization for arbitrary 2D quadratic dynamical systems in discrete-time. Int. J. Math. Anal. **8**(53), 2611–2617 (2014)
95. Ouannas, A.: Synchronization of discrete chaotic systems via double scaling matrix in different dimensions. J. Appl. Math. Comput. **3**(7), 1–5 (2014)
96. Ouannas, A.: A new QS synchronization scheme for discrete chaotic systems. Far East J. Appl. Math. **84**(2), 89 (2013)
97. Ouannas, A., Al-Sawalha, M.M.: A new approach to synchronize different dimensional chaotic maps using two scaling matrices. Nonlinear Dyn. Syst. Theory **15**(4), 400–408 (2015)
98. Shatnawi, M.T., Abbes, A., Ouannas, A., Batiha, I.M.: Hidden multistability of fractional discrete non-equilibrium point memristor based map. Phys. Scr. (2023)
99. Khennaoui, A.A., Ouannas, A., Batiha, I.M., Pham, V.T.: Hidden chaotic attractors in fractional-order discrete-time systems. In: Fractional Order Systems and Applications in Engineering, pp. 227–243. Academic Press (2023)
100. Ouannas, A., Batiha, I.M., Pham, V.T.: Fractional Discrete Chaos: Theories, Methods and Applications (2023)
101. Khennaoui, A.A., Almatroud, A.O., Ouannas, A., Al-sawalha, M.M., Grassi, G., Pham, V.T.: The effect of caputo fractional difference operator on a novel game theory model. Discret. Contin. Dyn. Syst.-Ser. B **26**(8) (2021)
102. Khennaoui, A.A., Ouannas, A., Momani, S., Batiha, I.M., Dibi, Z., Grassi, G.: On dynamics of a fractional-order discrete system with only one nonlinear term and without fixed points. Electronics **9**(12), 2179 (2020)
103. Ouannas, A., Mahmoud, E.E.: Inverse matrix projective synchronization for discrete chaotic systems with different dimensions. J. Comput. Intell. Electron. Syst. **3**(3), 188–192 (2014)
104. Ouannas, A.: Co-existence of complete synchronization and anti-synchronization in a class of discrete rational chaotic systems. Far East J. Dyn. Syst. **23**(1/2), 41 (2013)
105. Ouannasz, M.M., Batiha, I.M.: A general method for stabilizing the fractional-order discrete neural networks via linear control law. Nonlinear Stud. **29** (2022)
106. Gasri, A., Ouannas, A., Khennaoui, A.A., Grassi, G., Oussaeif, T.E., Pham, V.T.: Chaotic fractional discrete neural networks based on the Caputo H-difference operator: stabilization and linear control laws for synchronization. Eur. Phys. J. Spec. Top. **231**(10), 1815–1829 (2022)
107. Yousfi, H., Gasri, A., Ouannas, A.: Stabilization of Chaotic H-difference systems with fractional order. Nonlinear Dyn. Syst. Theory **468** (2022)
108. Ahmad, I., Ouannas, A., Shafiq, M., Pham, V.T., Baleanu, D.: Finite-time stabilization of a perturbed chaotic finance model. J. Adv. Res. **32**, 1–14 (2021)
109. Hioual, A., Ouannas, A., Oussaeif, T.E., Laouar, Z.: On fractional variable-order neural networks based on the Caputo derivative. In: Book of Abstracts, p. 103 (2022)
110. Debbouche, N., Ouannas, A., Batiha, I.M., Grassi, G., Kaabar, M.K., Jahanshahi, H., Aly, A., Aljuaid, A.M.: Chaotic behavior analysis of a new incommensurate fractional-order hopfield neural network system. Complexity **2021**, 1–11 (2021)
111. Karoun, R.C., Ouannas, A., Horani, M.A., Grassi, G.: The effect of Caputo fractional variable difference operator on a discrete-time hopfield neural network with non-commensurate order. Fractal Fract. **6**(10), 575 (2022)
112. Abbes, A., Ouannas, A., Shawagfeh, N., Khennaoui, A.A.: Incommensurate fractional discrete neural network: chaos and complexity. Eur. Phys. J. Plus **137**(2), 235 (2022)
113. Gasri, A., Ouannas, A., Khennaoui, A.A., Grassi, G., Oussaeif, T.E., Pham, V.T.: Chaotic fractional discrete neural networks based on the Caputo H-difference operator: stabilization and linear control laws for synchronization. Eur. Phys. J. Spec. Top. **231**(10), 1815–1829 (2022)

114. Batiha, I.M., Ouannas, A., Emwas, J.A.: A stabilization approach for a novel chaotic fractional-order discrete neural network. J. Math. Comput. Sci. **11**(5), 5514–5524 (2021)
115. A. Hioual, T. E., Oussaeif, A. Ouannas, G. Grassi, I. M. Batiha, S. Momani, New results for the stability of fractional-order discrete-time neural networks. Alex. Eng. J. **61**(12), 10359–10369 (2022)
116. Hioual, A., Ouannas, A., Oussaeif, T.E., Grassi, G., Batiha, I.M., Momani, S.: On variable-order fractional discrete neural networks: solvability and stability. Fractal Fract. **6**(2), 119 (2022)
117. Almatroud, O.A., Hioual, A., Ouannas, A., Sawalha, M.M., Alshammari, S., Alshammari, M.: On variable-order fractional discrete neural networks: existence. Uniqueness Stability. Fractal Fract. **7**(2), 118 (2023)
118. Hioual, A., Ouannas, G., Grassi, T., Oussaeif, E.: Nonlinear nabla variable-order fractional discrete systems: asymptotic stability and application to neural networks. J. Comput. Appl. Math. **423**, 114939 (2023)
119. Baleanu, D., Wu, G.C., Bai, Y.R., Chen, F.L.: Stability analysis of Caputo-like discrete fractional systems. Commun. Nonlinear Sci. Numer. Simul. **48**, 520–530 (2017)
120. Burton, T.A., Furumochi, T.: Krasnoselskii's fixed point theorem and stability. Nonlinear Anal. Theory Methods Appl. **49**(4), 445–454 (2002)
121. Baleanu, D., Wu, G.C., Bai, Y.R., Chen, F.L.: Stability analysis of Caputo-like discrete fractional systems. Commun. Nonlinear Sci. Numer. Simul. **48**, 520–530 (2017)
122. Chen, C., Bohner, M., Jia, B.: Ulam-Hyers stability of Caputo fractional difference equations. Math. Methods Appl. Sci. **42**(18), 7461–7470 (2019)
123. Jonnalagadda, J.M.: Hyers-Ulam stability of fractional nabla difference equations. Int. J. Anal **2016**, 1–5 (2016)
124. Selvam, A.G.M., Baleanu, D., Alzabut, J., Vignesh, D., Abbas, S.: On Hyers-Ulam Mittag-Leffler stability of discrete fractional Duffing equation with application on inverted pendulum. Adv. Differ. Equ. **2020**(1), 1–15 (2020)
125. Zhang, R., Qi, D., Wang, Y.: Dynamics analysis of fractional order three-dimensional Hopfield neural network. In: 2010 Sixth International Conference on Natural Computation, vol. 6, pp. 3037–3039. IEEE (2010)
126. Zhang, S., Zheng, J., Wang, X., Zeng, Z., He, S.: Initial offset boosting coexisting attractors in memristive multidouble-scroll Hopfield neural network. Nonlinear Dyn. **102**(4), 2821–2841 (2020)
127. Zhang, S., Zheng, J., Wang, X., Zeng, Z.: Multi-scroll hidden attractor in memristive HR neuron model under electromagnetic radiation and its applications. Chaos: Interdiscip. J. Nonlinear Sci. **31**(1), 011101 (2021)

LMI-Based Designs for Feedback Stabilization of Linear/Nonlinear Discrete-Time Systems in Reciprocal State Space: Synthesis and Experimental Validation

Assem Thabet, Aya Hassine, Noussaiba Gasmi, Ghazi Bel Haj Frej, and Mohamed Boutayeb

Abstract In many studies cases, generating a control laws using the state and/or output variables in Reciprocal State Space (RSS) form can be more easy than in Standard State Space (SSS) one. The formulation of a new stabilization control problem using feedback principle for linear discrete-time systems in RSS is presented in this brief. In contrast to the existing approaches, the considered model is described in RSS framework with an extension to Lipschitz nonlinear discrete-time systems. The asymptotic stability based on Lyapunov functions of the closed-loop system is guaranteed. The control design resolution problem is ensured through the Linear Matrix Inequalities (LMI) technique, the lemmas and the use of slack-variables and Differential Mean Value Theorem (DMVT). The performance of the proposed approach are shown through the experimental results using Real Time Implementation (RTI) with an Arduino MEGA 2560 board.

Keywords Reciprocal State Space (RSS) · Linear discrete-time systems · Lipschitz nonlinear systems · Standard State Space (SSS) · State/output feedback · Arduino MEGA 2560 board

---

A. Thabet (✉) · A. Hassine
MACS Laboratory, National Engineering School of Gabes (ENIG), University of Gabes, Gabes, Tunisia
e-mail: assem.thabet@issig.rnu.tn

N. Gasmi
Laboratoire d'Informatique & Télécommunications, ECAM Rennes - Louis de Broglie, Campus de Ker Lann - Bruz, 35091 Rennes, France
e-mail: noussaiba.gasmi@ecam-rennes.fr

G. B. H. Frej
IMS Laboratory, CNRS UMR 5218, Bordeaux INP, University of Bordeaux, 33405 Talence, France
e-mail: ghazi.bel_haj_frej@bordeaux-inp.fr

M. Boutayeb
Centre de Recherche en Automatique de Nancy, CRAN UMR 7039 CNRS, University of Lorraine, 54400 Cosnes-et-Romain, France
e-mail: mohamed.boutayeb@univ-lorraine.fr

A. Ben Makhlouf et al. (eds.), *State Estimation and Stabilization of Nonlinear Systems*, Studies in Systems, Decision and Control 491,
https://doi.org/10.1007/978-3-031-37970-3_11

1 Introduction

Although using state variables to design control systems in Standard Space State (SSS) is classical method (using feedback approaches [1, 2]) for their feasibility and real time implementation in many applications [3–5], state derivatives are more easily generated in many engineering applications.

In this context and to purpose a solution to this control problem, several stabilization methods have been proposed in SSS [6–12] with particular attention has been given to Lipschitz non-linear systems (using the state or output feedback principle [13, 14]).

Despite there is an important literature on the approaches developed to study in SSS, many limitations exist. Effectively, derivative feedback controllers has recently attracted particular attention and being applied to many engineering problems in the new Reciprocal State Space (RSS). This is due to several operating and engineering systems (vibration control problems [15], photovoltaic systems [16], mobile robot [17], active suspension systems [18, 19], electronic impulse system [20, 21], etc.) are described through information generated from the state or output derivatives. From where, several research works and control law synthesis methods covering different aspects (sliding Mode Control, tracking problems, stabilization, etc.) by state or output derivative feedback in RSS have been developed for linear systems [22–25] as well as nonlinear ones [26–28] in continuous time case.

Despite this important literature dealing with this control problem in the case of continuous time systems, almost no approach has been proposed for the discrete time case in RSS. Moreover, the case of the design of output feedback control in the RSS framework is not well developed [24, 29].

In this context, motivated by the latest results of [22, 23] for linear continuous time systems and also determined to overcome the aforementioned limitations. This chapter propose a novel approach for the design of a stabilization approach with output and state feedback for Linear and Lipschitz nonlinear discrete-time systems. Besides, the main idea of this chapter lies in the use of the Differential Mean Value Theorem(DMVT) which allows to transform the dynamics of the nonlinear function terms as a class of Linear Parameter Varying (LPV) systems based on the works of [3, 30].

The proposed method involves the use of the output and state feedback design in RSS for linear and Lipschitz dicrete-time systems. The closed loop stability of the overall system is ensured through the use of the appropriate Lyapunov functions. The stabilization control design resolution problem is ensured under the use of DMVT principle with certain lemmas and through Linear Matrix Inequalities (LMI) techniques.

Besides, the main contributions of this chapter are:

- The synthesis of a control approach based on state and output feedback principle in RSS contrary to the various standard approaches in SSS.
- The design of an extension to the case of discrete-time nonlinear Lipschitz systems.

- The use of the Lipschitz propriety for control problem design in the new RSS form where the nonlinear system is transformed to LPV system using the principle of DMVT.

The structure of this chapter is organized as follows:

First, problems statement and preliminaries are presented. Second, the synthesis method of feedback control law design (based state and output variables) to linear discrete-time systems in RSS form is presented in details. Third, an extension to the stabilization of Lipschitz nonlinear discrete-time system based on the use of DMVT is displayed. Finally, real-time implementation results are discussed by demonstrating the performance of the proposed approach through an Arduino MEGA 2560 board which used as an emulator (Hardware In the Loop).

Notation: The following notation will be used throughout this chapter:

- In a matrix, the notation $(\star)$ is used for the blocks induced by symmetry.
- $\bar{Q}^T$ is the transposed matrix of $\bar{Q}$; if $\bar{Q}$ is a square matrix then the notation $\bar{Q} > 0$ ($\bar{Q} < 0$) means that $\bar{Q}$ is positive definite (negative definite).
- the set $Co(x, y) = \{\lambda x + (1 - \lambda)y, 0 \leq \lambda \leq 1\}$ is the convex hull of x, y;
- $e_s(j) = \left(\underbrace{0, ..., 0, \overbrace{1}^{j\text{th}}, 0, ..., 0}_{s-components} \right)^T \in \mathbb{R}^s$, $s \geq 1$, is the vector of the canonical basis of $\mathbb{R}^s$.

2 Problem Statement and Preliminaries

The notion of RSS is introduced as a special case of linear singular systems [27]. Two Formulations of linear and nonlinear systems has been treated in [23, 24].

Within this framework, two approaches to modeling this new RSS reform have been adopted: transformation of nonlinear singular system or the use of state derivatives directly in the system representation [24, 31]. Moreover, the mathematical proofs of the different notions (controllability, stability based Lyapunov approach, ...) and the existence of a unique solution for linear and nonlinear systems, are detailed in [16, 20, 21, 27, 32].

This is based on an essential condition [25, 28, 29]: if the matrix $\bar{A}$ is not Hurwitz, as $(\bar{A}, \bar{B})$ is controllable, a preliminary control should be used to make it Hurwitz.

Now, let's consider the standard linear discrete time system representation in SSS form:

$$\begin{cases} x_{k+1} = \overline{A}x_k + \overline{B}u_k \\ y_k = \overline{C}x_k \end{cases} \tag{1}$$

where $x_k \in \mathbb{R}^n$ is the state vector and $u_k \in \mathbb{R}^m$ and $y_k \in \mathbb{R}^p$ is the input vector. Matrices $\overline{A}$,$\overline{B}$ and $\overline{C}$ are constant and have appropriate dimensions.

Then the following lemmas are considered in this chapter.

Lemma 1 ([33]) *Given two matrices X, Y, of appropriate dimensions, then the following inequality holds for any constant* $\overline{\eta} > 0$*:*

$$X^T Y + Y^T X \leq \overline{\eta} X^T X + \overline{\eta}^{-1} Y^T Y \tag{2}$$

Lemma 2 ([30]) *Consider a nonlinear function* $g : \mathbb{R}^n \to \mathbb{R}^n$*, the following two items are equivalent:*

- g is globally Lipschitz with respect to its argument, i.e.,

$$\| g(a) - g(b) \| \leqslant \gamma_g \| a - b \|, \quad \forall\, a,\ b \in \mathbb{R}^n. \tag{3}$$

- there exist constants $\underline{\mathfrak{g}_{ij}}$ and $\overline{\mathfrak{g}_{ij}}$ so that for all $\forall\, a, b \in \mathbb{R}^n$ there exist $z_i \in Co(a, b)$, $z_i \neq a$, $z_i \neq b$ and functions $\mathfrak{g}_{ij} : \mathbb{R}^n \times \mathbb{R}^n \longrightarrow \mathbb{R}$ satisfying the following equality:

$$g(a) - g(b) = \sum_{i,j=1}^{q,n} \mathfrak{g}_{ij}(z_i) \mathrm{H}_{ij}(a - b) \tag{4}$$

and $\underline{\mathfrak{g}_{ij}} \leqslant \mathfrak{g}_{ij}(z_i) \leqslant \overline{\mathfrak{g}_{ij}}$ where $\mathfrak{g}_{ij}(z_i) = \dfrac{\partial g_i}{\partial x_j}(z_i)$ and $\mathrm{H}_{ij} = e_q(i) e_n^T(j)$.

This chapter seeks to design a state and output feedback control for linear/nonlinear discrete-time systems in RSS framework that ensure the asymptotic stability based on the proper Lyapunov functions of the closed-loop system.

Remark 1 The proposed approach involves with an essential condition. The matrix $\overline{A}$ must be non singular [22].

3 Feedback Controllers Design for Linear Discrete-Time Systems

In this section, the control approach is designed based on the choice of Lyapunov stability function to stabilize linear discrete-time systems without disturbances in RSS form.

3.1 State Feedback Controller

The objective is to find a simple derivatives-state feedback control law in the following form:

$$u_k = -\overline{K} x_{k+1} \tag{5}$$

where $\overline{K} \in \mathbb{R}^{m\times n}$. Then, when integrating (5) in (1) the closed-loop system becomes:

$$x_{k+1} = \overline{A}x_k - \overline{BK}x_{k+1} \tag{6}$$

Supposing that the matrix $\overline{A}$ is non singular, (6) can be given in the following RSS form:

$$x_k = \overline{A}^{-1}(I + \overline{BK})x_{k+1} \tag{7}$$

Now, in order to guarantee that (7) is asymptotically stable, the following state feedback control law is proposed:

Theorem 1 *For linear discrete time system* (1) *with state feedback control* (5)*, the closed-loop RSS system* (7) *is asymptotically stable if there exists SPD matrices P ,$\overline{X}$ and matrices $\overline{W}$, $\overline{T}$ of appropriate dimensions with a scalars $\overline{\beta}$, $\overline{\eta_x} > 0$, such that the following LMI are feasible:*

$$\begin{bmatrix} \overline{X} - \overline{T}^T & -\overline{W}^T & -\overline{TA}^{-1}\overline{B} \\ \star & -\overline{A} & 0 & 0 \\ \star & \star & (-\overline{\eta_x}^{-1}I_s)^{-1} & 0 \\ \star & \star & \star & (-\overline{\eta_x}I_s)^{-1} \end{bmatrix} < 0 \tag{8}$$

$$\begin{bmatrix} \overline{\beta}I_n & (\overline{PA}^{-1} - \overline{TP})^T \\ \star & \overline{\beta}I_n \end{bmatrix} > 0 \tag{9}$$

where $\overline{K} = \overline{WX}^{-1}$.

Proof First, the Lyapunov function is chosen as $V_k = x_k^T Px_k$. Then, $V_{k+1} - V_k$ gives:

$$V_{k+1} - V_k = x_{k+1}^T Px_{k+1} - x_k^T Px_k < 0 \tag{10}$$

Replacing x_k by (7) in (10)leads to:

$$x_{k+1}^T \Psi_x x_{k+1} < 0 \tag{11}$$

with $\Psi_x = P - (\overline{A}^{-1})^T P\overline{A}^{-1} - \overline{K}^T\overline{B}^T(\overline{A}^{-1})^T P\overline{A}^{-1} - (\overline{A}^{-1})^T P\overline{A}^{-1}\overline{BK}$.

Thereafter, similarly to [22] considering the following new variable:

$$\overline{PA}^{-1} \triangleq \overline{TP} \tag{12}$$

Ψ_x becomes:

$$P - P^T\overline{A}^{-1} - \overline{K}^T\overline{B}^T(\overline{A}^{-1})^T\overline{T}P - P\overline{T}^T\overline{A}^{-1}\overline{BK} < 0 \tag{13}$$

Now, pre and post multiplying (13) by a new variable $\overline{X} = \overline{X}^T = P^{-1}$ and considering $\overline{W} = \overline{K}\overline{X}$ leads to:

$$\overline{X} - \overline{T}^T\overline{A}^{-1}\overline{X} - \underbrace{\overline{W}^T\overline{B}^T(\overline{A}^{-1})^T\overline{T} - \overline{T}\overline{A}^{-1}\overline{B}\overline{W}}_{\overline{Q}} < 0 \tag{14}$$

Now, by applying Lemma 1 on the term $\overline{Q}$ (where $X^T = \overline{T}\overline{A}^{-1}\overline{B},\ \ Y = \overline{W}$), (14) becomes:

$$\overline{X} - \overline{T}^T\overline{A}^{-1}\overline{X} - \overline{\eta}_x\overline{T}\overline{A}^{-1}\overline{B}\overline{B}^T(\overline{A}^{-1})^T\overline{T} - \overline{\eta}_x^{-1}\overline{W}^T\overline{W} < 0 \tag{15}$$

Second, the choice of new variable $\overline{T}$ must satisfy:

$$(\overline{P}\overline{A}^{-1} - \overline{T}\overline{P})(\overline{P}\overline{A}^{-1} - \overline{T}\overline{P})^T < \overline{\beta}^2 I_n \tag{16}$$

and by applying the the Schur complement [34] in (15) it is easy to find the inequality (8). □

3.2 *Output Feedback Controller*

The second objective is to find a output feedback control scheme in the following form:

$$u_k = -\overline{F}y_{k+1} = -\overline{F}\overline{C}x_{k+1} \tag{17}$$

where $\overline{F} \in \mathbb{R}^{m \times p}$. Similarly to the previous section, when integrating (17) in (1) the closed-loop system becomes:

$$x_k = \overline{A}^{-1}(I + \overline{B}\overline{F}\overline{C})x_{k+1} \tag{18}$$

Now, in order to guarantee that (18) is asymptotically stable, the following output feedback control law is proposed:

Theorem 2 *For linear discrete time system* (1) *with output feedback control* (17), *the closed-loop RSS system* (18) *is asymptotically stable if there exists: SPD matrices* P, $\overline{X}$, *rectangular matrix* $\overline{T}$ *and unstructured matrix* $\overline{S}$ *of appropriate dimensions and scalars* $\overline{\eta}_y$; $\overline{\beta}$; $\overline{\gamma}$, *as such the following LMIs are feasible:*

$$\begin{bmatrix} \overline{X} - \overline{X}\ (\overline{WC})^T & (\overline{B}^T(\overline{A}^{-1})^T\overline{T})^T & -\overline{C}^T\overline{W}^T\overline{B}^T(\overline{A}^{-1})^T \\ \overline{T}\ \overline{A}^T \quad 0 & 0 & 0 \\ \star \quad \star \quad -(\overline{\eta}_y I_s)^{-1} & 0 & 0 \\ \star \quad \star \quad \star & -(\overline{\eta}_y^{-1} I_s)^{-1} & 0 \\ \star \quad \star \quad \star & \star & -\frac{1}{\overline{\gamma}}\overline{X} \end{bmatrix} < 0 \tag{19}$$

$$blk - diag(\overline{V}_1, \overline{V}_2) > 0 \tag{20}$$

with

$$\overline{V}_1 = \begin{bmatrix} \overline{\beta} I_n & (\overline{PA}^{-1} - \overline{TP})^T \\ \star & \overline{\beta} I_n \end{bmatrix} > 0 \tag{21}$$

$$\overline{V}_2 = \begin{bmatrix} \overline{\gamma} I_n & (\overline{CX} - \overline{SC})^T \\ \star & \overline{\gamma} I_n \end{bmatrix} > 0 \tag{22}$$

where $\overline{F} = \overline{WS}^{-1}$.

Proof First, the same Lyapunov function and change of variable given in the previous section is used to synthesize an output feedback controller. This leads to:

$$\begin{aligned} V_{k+1} - V_k = x_{k+1}^T[P - (\overline{A}^{-1})^T P\overline{A}^{-1} - (\overline{A}^{-1})^T P\overline{A}^{-1}\overline{BFC} - \overline{C}^T\overline{F}^T\overline{B}^T(\overline{A}^{-1})^T P\overline{A}^{-1} + \\ + \overline{C}^T\overline{F}^T\overline{B}^T(\overline{A}^{-1})^T P\overline{A}^{-1}\overline{BFC}]x_{k+1} < 0 \end{aligned} \tag{23}$$

which can be rewritten as $x_{k+1}^T \Psi_y x_{k+1} < 0$. Then, with the use of schur complement [34], Ψ_y becomes

$$\Psi_y = \begin{bmatrix} (1) & -\overline{C}^T\overline{F}^T\overline{B}^T(\overline{A}^{-1})^T \\ \star & -P^{-1} \end{bmatrix} < 0 \tag{24}$$

where:
$(1) = P - (\overline{A}^{-1})^T P\overline{A}^{-1} - (\overline{A}^{-1})^T P\overline{A}^{-1}\overline{BFC} - \overline{C}^T\overline{F}^T\overline{B}^T(\overline{A}^{-1})^T P\overline{A}^{-1}$.

Next, with the same steps presented in the previous section and applied respectively to the term Ψ_y:

1. Consider the same new variable given in (12).
2. pre and post multiply by $diag(\overline{X}, I)$ with $\overline{X} = \overline{X}^T = P^{-1}$.

and according (24), Ψ_y becomes:

$$\begin{bmatrix} (2) & -\overline{XC}^T\overline{F}^T\overline{B}^T(\overline{A}^{-1})^T \\ \star & -P^{-1} \end{bmatrix} \tag{25}$$

with:
$(2) = \overline{X} - \overline{X}(\overline{A}^{-1})^T\overline{T} - \overline{T}^T\overline{A}^{-1}\overline{BFCX} - \overline{XC}^T\overline{F}^T\overline{B}^T(\overline{A}^{-1})^T\overline{T}$.

Thereafter, considering the following new variables:

$$\overline{CX} \triangleq \overline{SC} \tag{26}$$

and using $\overline{W} = \overline{FS}$, (25) is can be rewritten as:

$$\begin{bmatrix} (3) & -\overline{C}^T\overline{W}^T\overline{B}^T(\overline{A}^{-1})^T \\ \star & -P^{-1} \end{bmatrix} \tag{27}$$

where:
$(3) = \overline{X} - \overline{X}(\overline{A}^{-1})^T\overline{T} - \overline{T}^T\overline{A}^{-1}\overline{BWC} - \overline{C}^T\overline{W}^T\overline{B}^T(\overline{A}^{-1})^T\overline{T}$.

Now, in (27), the Lemma 1 is applied on term $\overline{T}^T\overline{A}^{-1}\overline{BWC} - \overline{C}^T\overline{W}^T\overline{B}^T(\overline{A}^{-1})^T\overline{T}$ (where $X = \overline{WC},\ Y = \overline{B}^T(\overline{A}^{-1})^T\overline{T}$ with scalar $\overline{\eta}_y$). It is easy to find the inequality (19). Then, the choice of new variables must satisfy:

$$(\overline{CX} - \overline{SC})(\overline{CX} - \overline{SC})^T < \overline{\gamma}^2 I_n \tag{28}$$

This leads to the inequality (22). □

4 Extension to Lipschitz Nonlinear Discrete-Time Systems

In this part, the steps of the synthesis of feedback control laws for Lipschitz discrete-time systems are detailed. The basic idea is to use the DMVT on the nonlinear function to transform the global nonlinear system to LPV system. The following non-linear system class is considered:

$$\begin{cases} x_{k+1} = \overline{A}x_k + \overline{B}u_k + \overline{g}(k, x_k) \\ y_k = \overline{C}x_k \end{cases} \tag{29}$$

where $x_k \in \mathbb{R}^n$, $u_k \in \mathbb{R}^m$ and $y_k \in \mathbb{R}^p$ are the state, the input and the output, respectively. $A \in \mathbb{R}^{n\times n}$, $B \in \mathbb{R}^{n\times m}$ and $C \in \mathbb{R}^{p\times n}$ are constant matrices. $\overline{g} : \mathbb{R}^n \to \mathbb{R}^q$ is a Lipschitz nonlinear vector.

First, the use of Lemma 2 on $\overline{g}(k, x_k)$ and assume that $\overline{g}(k, 0) = 0$ leads to:

$$\overline{g}(k, x(k)) - \overline{g}(k, 0) = \sum_{i,j=1}^{q,n} \pi_{ij} H_{ij} x_k = \overline{\Upsilon}(\vartheta)x_k \tag{30}$$

with $\underline{g}_{ij} \leqslant \pi_{ij} \leqslant \overline{g}_{ij}$ and $\mathrm{H}_{ij} = e_q(i)e_n^T(j)$. The parameter ϑ belongs to the bounded convex set $\mathcal{G}_{qn}$ for which the set of vertices is defined by

$$\mathcal{V}_{\mathcal{G}_{nq}} = \{\pi \in \mathbb{R}^{n\times q} \text{ and } \pi_{ij} \in \{\underline{g}_{ij}, \overline{g}_{ij}\}\}. \tag{31}$$

Then, similarly to the work of [3] and using (30) the affine matrix function is:

$$\mathcal{A}_{RSS}(\vartheta)x_k = \overline{A}x_k + \overline{g}(k, x_k) = (\overline{A} + \Upsilon(\vartheta))x_k \tag{32}$$

Now, the same development given in section 1 leads to:

Corollary 1 *For nonlinear discrete time system* (29) *with the same state feedback control* (5)*, the closed-loop RSS system is asymptotically stable with the same matrices and variables given in Theorem 1, such that the following LMI are feasible:*

$$Block - diag(S(\vartheta^1), ..., S(\vartheta^{2^{nq}})) < 0, \qquad \vartheta^i \in \mathcal{V}_{\mathcal{G}_{nq}} \tag{33}$$

where: $S(\alpha^i)$ *is given by* (8)–(9) *with the substitution of the term* $\overline{A}$ *by* $\mathcal{A}_{RSS}(\vartheta)$ *and* $\overline{K} = \overline{WX}^{-1}$.

Remark 2 With the same analogy, the case where the output of the system is also nonlinear (of the form $y_k = h(k, x_k)$) can be treated with the application of the DMVT which will make it possible to replace the term Cx_k by $\mathcal{C}_{RSS}(\kappa)$ obtained with the same approach given in (31) and (32).

5 Experimental Results

The experimental validation of the performances offered by the proposed approach in this brief was carried out with two examples through RTI by the use of the board ARDUINO MEGA 2560 as a real-time emulator (All technical implementation details are given in [3] and all magnitudes are in Volts [V]). In these examples, a noise was added to the system: sinusoidal signals with variable frequencies (between 85 and 420 Hz) and amplitude (±0.35 V).

5.1 State Feedback Stabilization: Nonlinear Case

For the first example, the considered system is described as:

$$\overline{A} = \begin{pmatrix} 1 & T & 0 & 0 \\ -48.6T & 1-1.25T & 48.6T & 0 \\ 0 & 0 & 1 & T \\ 19.5T & 0 & -19.5T & 1 \end{pmatrix}, \overline{B} = \begin{pmatrix} 0 \\ 21.6T \\ 0 \\ 0 \end{pmatrix}, \overline{C} = \begin{pmatrix} 1\,0\,0\,0 \\ 0\,1\,0\,0 \end{pmatrix},$$

$$\overline{g}(k, x_k) = \begin{bmatrix} 0.025sin(x_1(k)) \\ 0 \\ 0 \\ 0 \end{bmatrix}.$$

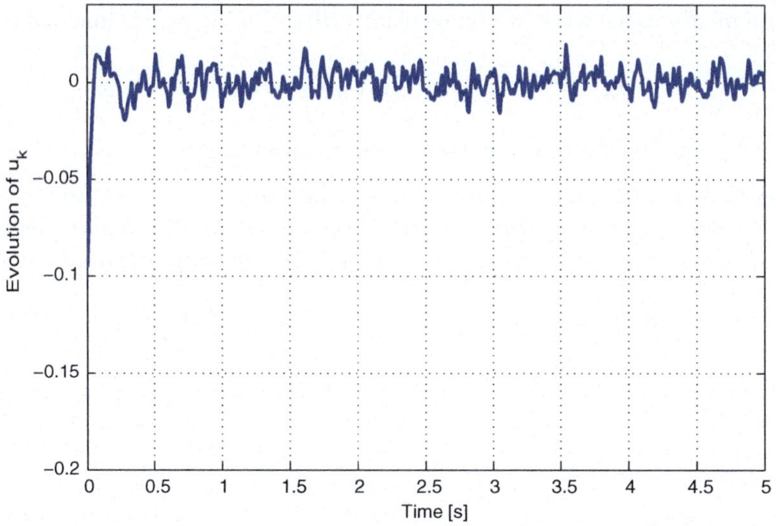

Fig. 1 Variation of control $u(k)$ with state feedback control

Where: $\gamma_g = 0.025$ and $\mathcal{V}_{\mathcal{G}_{21}} = \pm\gamma_g$ and $T = 0.01s$. Subsequently, the resolutions of LMIs (8)–(9) using $YALMIP^{®}$, leads to the following control gain:

$$\overline{X} = \begin{bmatrix} 1.219 & 1.325 & 1.187 & 1.1489 \\ 1.325 & 1.4138 & 1.2984 & 1.2667 \\ 1.187 & 1.2984 & 1.1535 & 1.1138 \\ 1.1489 & 1.2667 & 1.1138 & 1.0724 \end{bmatrix},$$

$$\overline{T} = \begin{bmatrix} 142.4352 & 142.2819 & 141.0694 & 140.2987 \\ 8.8548 & 11.4078 & 8.3966 & 7.8768 \\ -306.6624 & -315.4162 & -302.8622 & -300.0159 \\ 156.5254 & 162.878 & 154.9498 & 152.9941 \end{bmatrix},$$

$$\overline{K} = \begin{bmatrix} -3.1588 & -3.1825 & 2.0674 & -1.9105 \end{bmatrix}.$$

Next, Figs. 1 and 2 show the evolution of the control and the state of the closed-loop system with state feedback control $\left(\text{with } x(0) = \begin{bmatrix} 0.1 & 0.25 & 0.1 & 0.15 \end{bmatrix}^T \right)$.

Figures 1 and 2 show that the proposed control law ensures the stability of the system with reduced amplitudes and perturbations.

5.2 *Output Feedback Stabilization: Linear Case*

For the second example, the dynamic equation for the considered system is described as:

$$\overline{A} = \begin{bmatrix} 0.705 & 0.2 \\ 0.15 & 0.85 \end{bmatrix}, \overline{B} = \begin{bmatrix} 0.1 \\ 0.05 \end{bmatrix}, \overline{C} = \begin{bmatrix} 1 & 0 \end{bmatrix}.$$

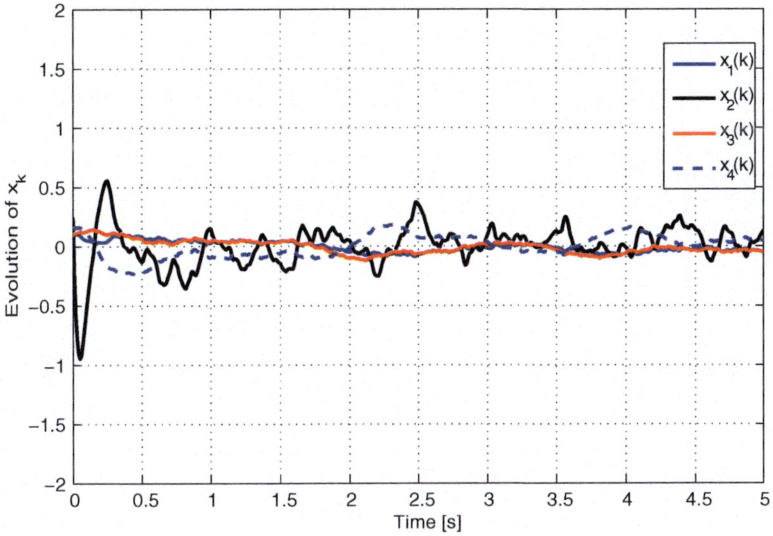

Fig. 2 Evolution of $x(k)$ with state feedback control

Then, similarly, the resolution of the LMIs (19)–(22) provides the following control gain:

$\overline{X} = \begin{bmatrix} 0.4291 & 0.2485 \\ 0.2485 & 1.1865 \end{bmatrix}, \overline{T} = \begin{bmatrix} 1.542 & -0.1737 \\ -0.4411 & 1.865 \end{bmatrix},$

$\overline{W} = -2.4968$ and $\overline{F} = -5.8184$.

Now, Figs. 3, 4 and 5 plot the evolution of the control, the output and the state of the closed-loop system $\left(\text{with } x(0) = \begin{bmatrix} 0.85 & 0.5 \end{bmatrix}^T\right)$.

Based on the findings of Figs. 3, 4 and 5, it is deduced that the states are asymptotically stable with the output feedback law control. All the figures reveal that the proposed method provides good performances in the presence of variable external disturbances.

6 Conclusion

In this chapter, a state and an output feedback design methodology for a linear discrete-time systems in RSS form have been presented with an extension to Lipschitz nonlinear systems. The proposed approach ensure the asymptotic stability of the closed-loop system with unknown disturbances (variable in terms of amplitude and frequency) through the use of Lyapunov's stability theory. The resolution of the proposed control design problem is achieved using the Linear Matrix Inequalities (LMI) technique, the lemmas and the introducing of Slack-variables and the DMVT.

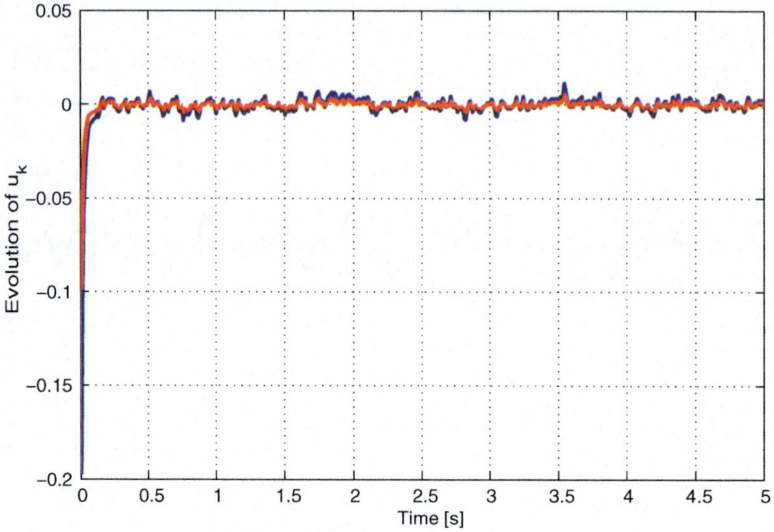

Fig. 3 Variation of $u(k)$ with output feedback control

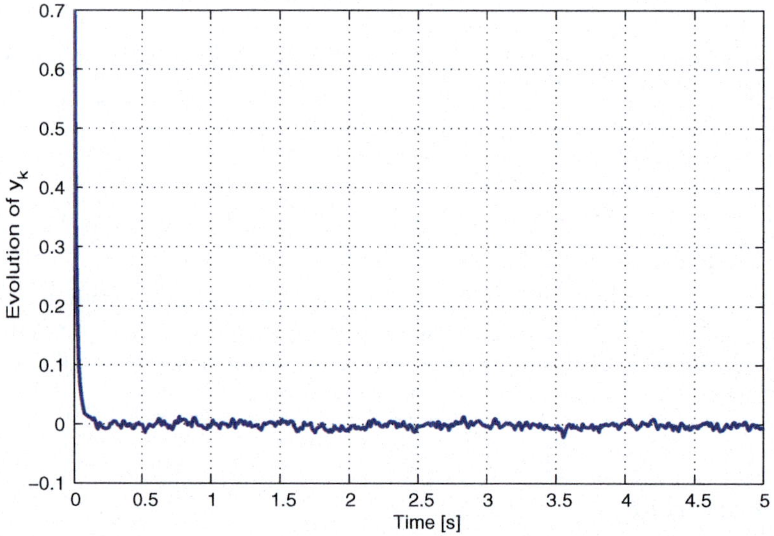

Fig. 4 Evolution of $y(k)$ with output feedback control

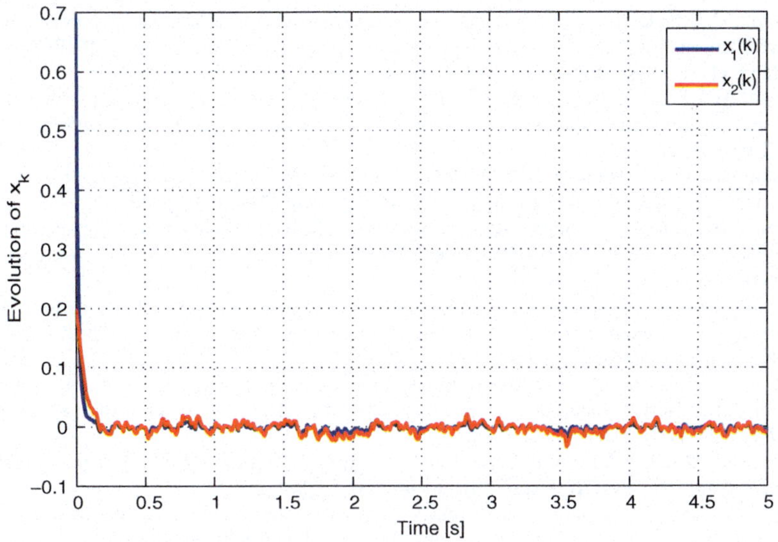

Fig. 5 Evolution of $x(k)$ with output feedback control

The experimental results trough Arduino MEGA 2560 board with RTI (real time emulator) show the performances of stabilization given by the proposed method.

Expanding proposed synthesis methods with nonlinear output and the generalization of the proposed method for robust scheme might be a direction for further studies.

References

1. Min, H., Duan, N., Yu, X., Fei, S.: Tracking-based control for constrained nonlinear systems under parametric uncertainties. IEEE Trans. Cir. Syst. II: Express Briefs 1–1 (2020). https://doi.org/10.1109/TCSII.2020.3010100
2. Su, Y., Zheng, C.: A simple unified control for output feedback finite-time stabilization of uncertain planar systems without controllable/observable linearization. IEEE Trans. Cir. Syst. II: Express Briefs (2020). https://doi.org/10.1109/TCSII.2020.2996208
3. Gasmi, N., Boutayeb, M., Thabet, A., Aoun, M.: Robust control of a class of nonlinear discrete-time systems: design and experimental results on a real-time emulator. Actuators **10**, 303 (2021). https://doi.org/10.3390/act10110303
4. Gasmi, N., Boutayeb, M., Thabet, A., Aoun, M.: Sliding window based nonlinear H_∞ filtering: design and experimental results. IEEE Trans. Cir. Syst. II: Express Briefs **66**, 302–306 (2019)
5. Thabet, A., Frej, G.B.H., Boutayeb, M.: Observer-based feedback stabilization for lipschitz nonlinear systems with extension to H_∞ performance analysis: Design and experimental results. IEEE Trans. Control Syst. Technol. **26**, 321–328 (2018)
6. Hussain, S., Xie, S.Q., Jamwal, P.K.: Robust nonlinear control of an intrinsically compliant robotic gait training orthosis. IEEE Trans. Syst. Man Cybern.: Syst. **43**(3), 655–665 (2013)

7. Wang, D., Liu, D., Li, H., Luo, B., Ma, H.: An approximate optimal control approach for robust stabilization of a class of discrete-time nonlinear systems with uncertainties. IEEE Trans. Syst. Man Cybern.: Syst. **46**(5), 713–717 (2016)
8. Saad, W., Sellami, A., Garcia, G.: Robust stabilization of one-sided Lipschitz nonlinear systems via adaptive sliding mode control. J. Vibr. Control (2019). https://doi.org/10.1177/1077546319889413
9. Badreddine, E.H., Hicham, E.A., Abdelaziz, H.: New approach to robust observer-based control of one-sided Lipschitz non-linear systems. IET Control Theory Appl. **13**, 333–342 (2019)
10. Deniz, N., Murillo, M., Sanchez, G., Giovanini, L.: Robust stability of moving horizon estimation for non-linear systems with bounded disturbances using adaptive arrival cost. IET Control Theory Appl. **14**, 2879–2888 (2020)
11. Aminsafaee, M., Shafiei, M.H.: Delayed state feedback predictive control of uncertain nonlinear discrete-time switched systems with input constraint and state delays. IET Control Theory Appl. **14**, 2937–2947 (2020)
12. Du, J., Hu, X., Sun, Y.: Adaptive robust nonlinear control design for course tracking of ships subject to external disturbances and input saturation. IEEE Trans. Syst. Man Cybern.: Syst. **50**(1), 193–202 (2020)
13. Kim, S., Ahn, C.: Auto-tuner-based controller for quadcopter attitude tracking applications. IEEE Trans. Cir. Syst. II: Express Briefs **66**, 2012–2016 (2019)
14. Guo, C., Xie, X.: Global output feedback control of nonlinear time-delay systems with input matching uncertainty and unknown output function. Int. J. Syst. Sci. **50**, 713–725 (2019)
15. Yazici, H., Sever, M.: L_2 gain state derivative feedback control of uncertain vehicle suspension systems. J. Vib. Control **24**(16), 3779–3794 (2018)
16. Kaewpraek, N., Assawinchaichote, W.: H_∞ fuzzy state-feedback control plus state-derivative-feedback control synthesis for photovoltaic systems. Asian J. Control **18**, 1441–1452 (2016)
17. Boukas, E.: Static output feedback control for linear descriptor systems: LMI approach. In: IEEE International Conference on Mechatronics and Automations, July 20–August 1, Niagara Falls, Ontario, Canada, pp. 1230–1234 (2005)
18. Da Silva, E., Assuncao, E., Teixeira, M.C.M., Cardim, R.: Robust controller implementation via state-derivative feedback in an active suspension system subjected to fault. In: International Conference on Control and Fault-Tolerant Systems (SysTol), Nice, France, pp. 752–757 (2013)
19. Da Silva, E., Assunção, E., Teixeira, M.C.M., Faria, F.A., Buzachero, L.F.S.: Parameter-dependent lyapunov functions for state-derivative feedback control in polytopic linear systems. Int. J. Control **84**, 1377–1386 (2011)
20. Tseng, Y.: Control design for system with lipschitz nonlinearity of state derivative variables in reciprocal state space form. In: International Conference on Image Processing, Electrical and Computer Engineering, pp. 26–32, July 8–9, Singapore (2015)
21. Thabet, A.: Adaptive-state feedback control for lipschitz nonlinear systems in reciprocal-state space: design and experimental results. Proc IMechE Part I: J. Syst. Control Eng. **233**, 144–152 (2019)
22. Sever, M., Yazici, H.: LMI-based designs for robust state and output derivative feedback guaranteed cost controllers in reciprocal state space form. Int. J. Control (2019). https://doi.org/10.1080/00207179.2019.1697465
23. Tseng, Y., Wang, Y.N.: Sliding mode control with state derivative output feedback in reciprocal state space form. Abstr. Appl. Anal. **2013**, 1–12 (2013)
24. Tseng, Y.: Sliding mode control with state derivative feedback in novel reciprocal state space form. Adv. Appl. Nonlinear Control Syst. **635**, 159–184 (2016)
25. Abdelaziz, T.H.S.: Optimal control using derivative feedback for linear systems. J. Syst. Control Eng. **224**, 185–202 (2010)
26. Thabet, A., Gasmi, N., Frej, G.B.H., Boutayeb, M.: Sliding mode control for Lipschitz nonlinear systems in reciprocal state space: synthesis and experimental validation. IEEE Trans. Circuits Syst. II Express Briefs (2020). https://doi.org/10.1109/TCSII.2020.3018016
27. Tseng, Y., Kwak, S., Yedavalli, R.: Stability, controllability and observability criteria for the reciprocal state space framework. In: Proceedings of American Control Conference, 4–6 June, Denver, CO, USA (2003). https://doi.org/10.1109/ACC.2003.1242535

28. Thabet, A., Frej, G.B.H., Gasmi, N., Mansouri, K., Chauveau, E.: State and output feedback control for Lipschitz nonlinear systems in reciprocal state space: synthesis and real-time validation. Bull. Electr. Eng. Inf. (2022). https://doi.org/10.11591/eei.v11i5.3521
29. Tseng, Y., Wang, Y.N.: Sliding mode control with state derivative output feedback in reciprocal state space form. Abstr. Appl. Anal. **2013** (2016). https://doi.org/10.1155/2013/590524
30. Zemouche, A., Rajamani, R., Phanomchoeng, G., Boulkroune, B., Rafaralahy, H., Zasadzinski, M.: Circle criterion-based H_∞ observer design for Lipschitz and monotonic nonlinear systems-Enhanced LMI conditions and constructive discussions. Automatica **85**, 412–425 (2017)
31. Ilchman, A., Reis, T.: Suverys in Differential Algebraic Systems I. Springer (2013). https://doi.org/10.1007/978-3-642-34928-7
32. Kwak, S.K., Washington, G.N., Yedavalli, R.K.: Acceleration based vibration control of distributed parameter systems using the reciprocal state-space framework. J. Sound Vibr. **251**, 543–557 (2002)
33. Yu, L.: Robust Control-the Method of Linear Matrix Inequality Processing, 1st edn. Tsinghua University Press, Beijing, China (2002)
34. Boyd, S., Ghaoui, L.E., Ferron, E., Balakrishnan, V.: Linear Matrix Inequalities in Systems and Control Theory, 15th edn. Studies in Applied Mathematics SIAM, Philadelphia (1994)

Overview on Active Fault-Tolerant Control

Hajer Mlaych and Atef Khedher

Abstract Compared to the fault-diagnosis field, fault-tolerance is still a recent topic that is missing standard definitions and classifications. Indeed, fault-tolerant control (FTC) problem has gained utmost significance in recent decades. The objective in this chapter is to give the necessary background for the understanding of the FTC problem. We define standard terms that are often used in FTC and we provide a review of the different fault types that might occur in automated systems. A classification of the different FTC techniques is given, with an emphasis on active FTC methods. A comparison of the different active strategies is also provided.

Keywords FTC · Nonlinear systems · Active methods · Faults

1 Introduction

The reliability and dependability of an automated process are the major concerns of manufacturers in their various fields. The theory of automated system control has found widespread application in industrial processes. Such control theories must ensure a specific performance of the plant and ensure it's stability, when the system's components are working correctly.

However, increasing levels of automation in the process correspond to a higher potential for faults to happen. These malfunctions can affect sensors, actuators and system components [1]. Thus, the process monitoring, known as fault-detection and Isolation (FDI) process, is necessary for ensuring a safe plant operation.

Nevertheless, for highly automated industrial systems, repair or maintenance cannot be achieved immediately. Minor faults can eventually result in critical failures and a conventional feedback-control may even lead to system instability. Consequently, detecting and identifying theses faults is no longer sufficient. The existence

H. Mlayeh (✉) · A. Khedher
LATIS, Ecole nationale d'Ingénieurs de Sousse, Université de Sousse, Erriadh, 4023 Sousse, Tunisia
e-mail: mlayehhajer@gmail.com

A. Ben Makhlouf et al. (eds.), *State Estimation and Stabilization of Nonlinear Systems*, Studies in Systems, Decision and Control 491,
https://doi.org/10.1007/978-3-031-37970-3_12

of faults must be taken into account during the system control design. Attempts have been undertaken to create control techniques able to instantly correct the respective anomaly while keeping a nominal performance. Such control is known as (FTC) [2].

Studies that explore FTC problem have either employed analytical or hardware redundancy. The latter is necessary in safety-critical systems, but rarely used in other industrial processes because it is financially expansive. Nevertheless, redundant sensors are generally used since they are much easier and less expensive than actuators. Consequently, using analytical redundancy with actuators can significantly reduce the cost of instruments and maintenance.

There are two types of analytical FTC techniques: passive and active. Passive approaches involve designing controllers that can already withstand a predetermined range of failures. For such failures, a fixed-parameters controller can be designed with the ability to control both nominal system and the system affected by these "known" failures [3].

However, the effectiveness of these methods is limited since it is not feasible to predict all faults and their impact on the system beforehand.

On the other hand, by using the FDI process, active techniques respond promptly to faults by adjusting the control law. The controllers are adjusted on-line depending on the fault type, to preserve the nominal system performance [4]. It is apparent that active techniques are more practical and favorable because they can address a broad range of faults.

According to the information delivered by the FDI bloc, if it is possible to accommodate for the fault effect or to sustain the performance of the defective system near the nominal one, then, active techniques are associated with accommodation/reconfiguration. If the nominal performance cannot be preserved, the operation is known as restructuring. The latter case consists of modifying the control objectives or the system structure. For instance, one can reduce the number of controlled outputs. Moreover, if the reference inputs are not reachable anymore, other references must be redefined. Consequently, the system is lead into a "degraded operating mode". For specific faults, however, even with degraded functionality, the system cannot be kept in operation. The objective is then, to safely turn it off.

In this chapter, we present a general overview of the FTC problem. We first define standard terms that are often used in FTC. Then, we provide a review of different fault types that might occur in automated systems. Finally, a classification and a comparison of the different FTC techniques is given, with an emphasis on active FTC methods.

2 Definitions

In the sequel, we give a short summary of terms, commonly employed in both diagnosis and fault-tolerance [5].

Fault: "is typically an unexpected or unusual occurrence that deviates from what is considered normal or acceptable for a given system or situation". The eventual causes can be design or human errors, use, deterioration, ageing, etc. Fault consequences differ from one plant to another. For instance, it can lead to performance degradation, lower quality, energy waste, lower production, environmental and human damages, etc.

Failure: Results from a fault. It induces a persistent impairment of the system's capacity to execute a necessary task under certain operating circumstances.

FDD/FDI module: (Fault Detection and Diagnosis/Identification) unit that tracks the system's health. The FDI system reports to a surveillance module of the seriousness of the failure.

Fault detection: It recognizes the existence of a malfunction in the system.

Fault isolation: Once the fault is detected, it locates the root cause by isolating the system components that are not in nominal operation mode.

Fault identification: Estimates the magnitude and the type of the failure.

FTCS: A Fault-Tolerant Control System is equipped with the necessary hardware architecture and software mechanisms that allow to achieve a satisfactory performance, not only in normal conditions, but also with one or several faults in the plant.

3 Overview of Fault Types

3.1 Classification of Faults

Faults can be classified according to different aspects such as the intensity, the origin and the location of the failure, the way it affects the system behavior, it's temporal evolution, etc. [2].

3.1.1 According to the Fault Localization

The different faults causing malfunctions in the system can either be internal, or external faults.

- ***External faults***: result from the system's environment such as wind effects, perturbations, human-operator errors.
- ***Internal faults***: result from factors that are relatively controllable, such as: ageing, use, bad maintenance, etc. They can cause malfunctions in the different components of the system: actuators, sensors, plant.

Internal faults can then be classified into three types according to their location in the system: actuator, component or sensor faults, as illustrated in Fig. 1.

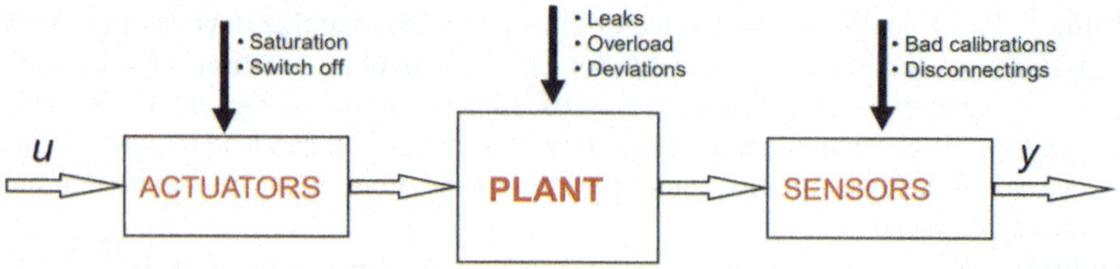

Fig. 1 Different types of internal faults

- ***Actuator faults***: An actuator is a necessary mechanical component, used for converting energy into specific motion. Consequently, an actuator fault acts on the operating part of the system and affects the input signals. It represents a partial/total failure of one or more actuators in the system. More precisely, it is possible to sort actuator faults into three main types: stuck, float and loss of effectiveness of the actuator.
  . A stuck or frozen actuator failure means that the actuator is blocked at certain position (starting from the instant t_f) and has no response to the control input (1). Consequently, the system can no longer be controlled by this "stuck" actuator.

$$u(t) = u(t_f) = const, \qquad \forall t > t_f \tag{1}$$

  . A floating actuator does not provide any moment or force and does not affect the control system anymore.

$$u(t) = 0, \qquad \forall t > t_f. \tag{2}$$

  . A loss of the control effectiveness occurs when the efficiency of one or multiple actuators reduces. It is also known as partial actuator failure. The actuator acts similarly to the nominal case, but with a degraded efficiency (exp: loss of engine power, valve leakage...)

$$u(t) = ku_n(t), \quad 0 < k < 1, \quad \forall t > t_f \tag{3}$$

  where u_n represents the nominal control.

- ***Sensor faults***: A sensor is a component that measures signals from the internal states and converts them into reachable signal outputs that can be used by the operator. Sensor faults can be classified into four categories: bias/offset, drift, freezing/stuck and loss of accuracy.
  . Bias failure is a fixed difference between the measured and real signal.
  . A drift failure corresponds to an increase in the measurement errors over time.
  . A frozen or stuck failure occurs when the output is continuously indicating a constant value, that is different from the real one.
  . The loss of accuracy is when output measurements do not correspond to the true values of the signals.

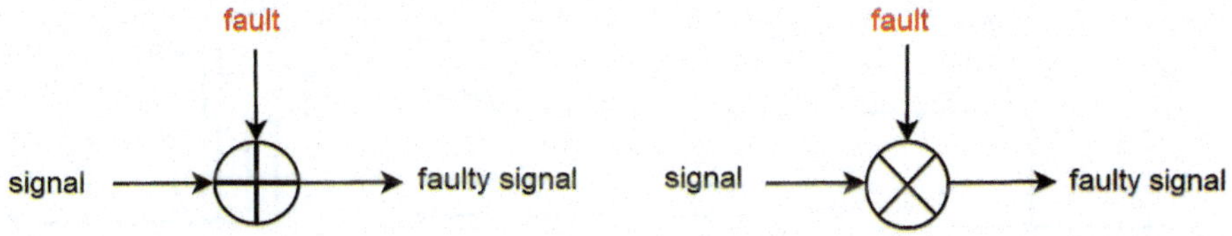

Fig. 2 Multiplicative and additive faults

- ***Component/system faults***: Such failures affect the system components itself. They are characterized by a modification in the parameters, that makes the dynamic system's model invalid. The mathematical representation of component faults is often difficult to obtain, and requires intensive experimental tests. Generally, component faults result in a modification of the states equation. This modification can either be parametrical or structural. It is often considered that, faults that do not affect actuators, nor sensors, are arbitrarily classified within component/system faults.

3.1.2 According to the Fault Modelling

Failures are modeled depending on how they impact the behavior of the system. Two forms of modelling can be distinguished: additive and multiplicative form (Fig. 2).

- ***Additive faults*** where changes in the output are independent of known inputs, and depend of the fault magnitude. This form is generally used to represent actuator and sensor faults. Exp: offsets in actuators and sensors, disturbances, etc.
- ***Multiplicative faults*** represent a gradual or abrupt modification of parameters which causes changes in the output. These changes depend of both inputs and fault magnitude. This form is generally used to represent system and actuator faults. Exp: corrosion, deterioration, loss of energy, erosion, etc.

3.1.3 According to the Temporal Aspect

Faults can be classified depending on their evolution in time. They can either be abrupt, intermittent or gradual (Fig. 3).

- ***Abrupt faults*** are suddenly-occurring faults, characterized by their discontinuous behavior in time.

$$\beta\left(t-t_f\right)=\begin{cases}\delta,\ t\geq t_f\\ 0,\ t<t_f\end{cases} \tag{4}$$

  Where $\beta\left(t-t_f\right)$ represents the temporal fault behaviour t_f the time of fault incidence. Such faults arisen (as a step change), from a sudden breakdown of a component (total or partial) and induce significant changes in the system behavior. Exp: Offset.

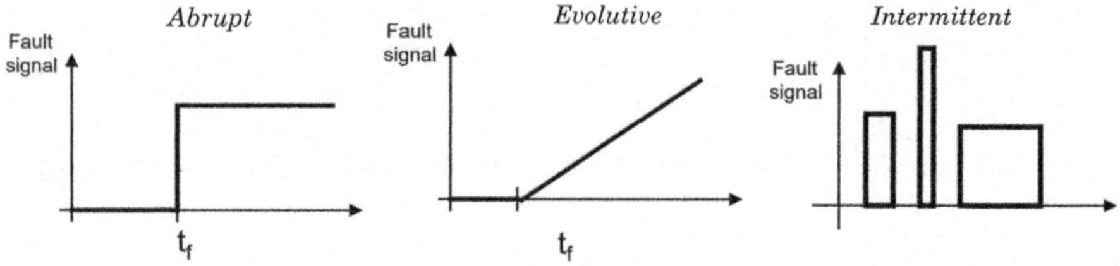

Fig. 3 Temporal evolution of faults

- ***intermittent fault*** is a malfunction of a component or a system, that occurs in discrete intervals, usually irregular. The more the system is complex, the more likely intermittent faults occur. Such faults are attributed to unknown system interactions, component degradation, loose contact, intermittent breakdown of sensors. Intermittent faults are sometimes known as "soft" faults since they are not continuous over time and disappear in an unpredictable way. Thus, their immediate effect can be negligible but its repetition may lead to serious failures.
- ***Gradual/Evolutive faults*** are characterized with a slow temporal evolution.

$$\beta\left(t - t_f\right) = \begin{cases} \delta\left(1 - \exp\left(-\alpha\left(t - t_f\right)\right)\right), & t \geq t_f \\ 0, & t < t_f \end{cases} \tag{5}$$

Where α is a positive constant. Such faults can be considered as a slight but abnormal increasing over time. They usually characterize a gradual component wear. Evolutive faults are difficult to detect, because they are often confused with slowly-varying parameters, representing the non-stationary process.

4 Overview of FTC Methods

Using the FDI process to provide informations about a system's status is an important step [6]. However, it is evident that the detection and isolation of the failure is not enough if no further actions are taken.

A conventional control might cause undesirable system behavior and even instability. On the other hand, FTC methods take into consideration the occurrence of faults during the control design and are able to instantly correct or "tolerate" the respective anomaly while maintaining a nominal performance [7–12]. If the latter cannot be maintained, the system may be driven in a degraded mode, but not at the cost of stability. In addition to maintaining stability, the primary objective of a FTCS is to restore desired performances in the presence of faults, such as:

- accuracy
- speed
- tracking error

- disturbance rejection
- avoid actuators saturation, etc.

The restored performances (post-fault), differ according to the nature of the system, the redundant components, the gravity/type of the failure and the necessary required performances.

In the sequel, we present a classification of FTC methods using analytical redundancy.

4.1 Classification of FTC Methods

The classification of FTC methods is performed according to different specifications: control algorithms, design tools, field of application, type of the system, reconfiguration mechanisms, etc. A classification according to design approaches is given in Fig. 4. FTC approaches can be classified into: heuristic and analytical.

4.1.1 Heuristic Methods

FTC strategy requires high flexibility in order to adapt to the wide range of unexpected situations. Indeed, in the presence of faults, the system may evolve into uncontrollable operating regions where we may not have a mathematical model, describing this

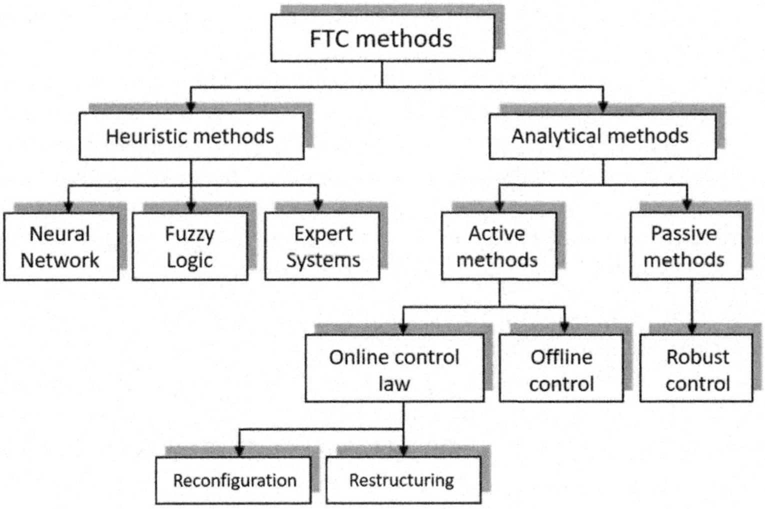

Fig. 4 Classification of FTC methods

evolution. In this case, a heuristic approach can be the solution to the FTC problem, mainly by introducing the concept of deep learning or expert systems [13, 14].

These methods fit into the general framework of "intelligent" control, which generally requires:

- The evaluation of the system performances
- Memorizing the newly calculated parameters with the corresponding operating points
- Re-updating the control law.

Expert Systems

Recently, expert systems have been employed in system monitoring and were proved to be suitable as well for FTC problems [15]. Commonly, the specific problem is given to the expert system by means of specific facts and data. A knowledge base stores the specific required knowledge. Then, an inference tool derives conclusions and proposes possible remedies. The expert system communicates with the computer by introducing problem data and requesting subsequent conclusion. The generally used expert systems are suitable for problems that are not solvable in a straightforward manner, because there is no exact knowledge of the underlying principles. Consequently, heuristic knowledge expressed as facts and rules and based upon human experience, has to be used. When the expert system is related to the process inputs or outputs, we then talk about an on-line expert system.

Literature [16] proposed a FTC strategy based on online expert system. The aim is to reconfigure the control law when actuator/sensor faults or misoperation occur. The online expert system consists mainly of four parts:

- The analytical solution which relies on modifying the coefficients of the system. The latter represent symptoms of the system failures. The controller is then modified according to these failures.
- The process-knowledge base includes analytical and heuristic knowledge, consisting of system performance and failures statistics.
- The knowledge acquisition part consists of compiling the process-specific knowledge such as abnormal behavior, theoretical system model, misoperation and failures.
- The inference mechanism executes the FTC, depending on controller trees, the observable symptoms, the system history and the fault probabilities.

Expert systems offer a useful formalism for FTC thanks to their ability to consider sub-problem abstractions and diverse data sources. The expert system-based FTC can effectively enhance the performance of the system's operational features. Nonetheless, this improvement are constrained with reliable control elements, good knowledge of the system, and a sufficient-post-failures controllability and observability.

Neural Networks

FTC methods based on neural networks learning techniques were introduced by Farell [17]. Other studies were then developed for NL systems and employed to handle actuator and sensor faults in aerospace plants [18, 19].

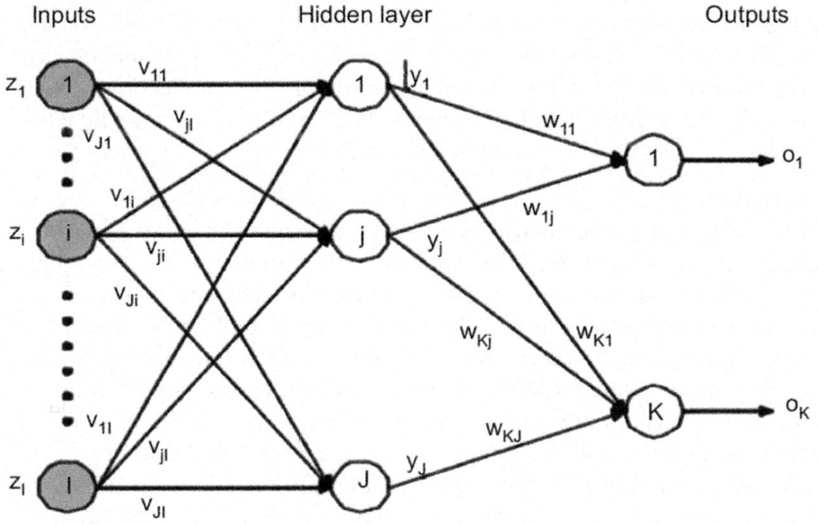

Fig. 5 Neural network model with 3 layers [19]

A neural network (NN) [20, 21] has a parallel structure that possesses several output and input nodes. The neural network also encompasses the inter-neuron connections within its various layers. For instance, a three-layers system is shown in Fig. 5. Data is sent through synapses from the first layer of input neurons $(z_1 \ldots z_I)$, to the following layer of neurons $(y_1 \ldots y_J)$. Then, the third layer of output neurons $(o_1 \ldots o_K)$ receives the data sent via synapses. The more the system is complex, the more it has layers of neurons. The "weights" are parameters, stored in the synapses, and used to modify or alter the data during computations. Two weight matrices are shown in Fig. 5: Matrix $[v]$ and matrix $[w]$. These weights are trained by the neural network.

The NN can be used via two strategies: a model-based strategy (MBNN) and a knowledge-based strategy (KBNN). The latter trains a NN to recognize faults depending on certain features. On the other hand, a MBNN provides analytical redundancy for fault detection process.

In [19], KBNN and MBNN are used for both detection and reconfiguration of sensor failures, occurring in aircraft dynamics. The detection of the sensor failure is done using the KBNN fault detection. Another model (MBNN) is employed for the reconfiguration process of the failing sensor. The MBNN approach uses the generalized regression NN (GRNN), which is a type of the radial-basis-function of the NN and gives efficient results for function approximation.

KBNN and MBNN are proved to form a robust FTC system that works faster in handling sensor failures and accommodating failures. It efficiently accommodates the intermittent failures as well.

Despite the fact that NN is an appealing methodology for diverse fields with numerous benefits, in practice, however, a NN has low capabilities for fault tolerance and, lacking a proper design, it can not be integrally qualified as fault tolerant [22]. Moreover, artificial NN has the disadvantage of being computationally expensive and hardware dependent since it requires a processor with parallel processing power. Furthermore, no specific rule for the determination of the NN structure exists. The proper network structure is determined via trial and error and experience.

Fuzzy Logic

The "intelligent" methods such as NN, fuzzy and expert systems possess the ability of "learning" the system model from input and output data, or "learning" fault-knowledge from experience history. They are employed for the construction of analytical model for residual generation or for making the fault analysis decisions.

Both fuzzy logic and expert systems were used for fault diagnosis and fault accommodation schemes. Reference [23] used an expert system technique for the supervision of an F-16 aircraft affected by actuator failures. Theses failures were compensated for, using a controller based on a fuzzy model reference-learning.

FTC methods based on fuzzy logic were applied to numerous nonlinear systems such as gas furnace, motors, aircrafts, nuclear power plants, etc. However, developing a mathematical model for these NL systems can be a challenging task, especially that the exactitude of linear models is not always convenient for the system objectives. To overcome these issues, "Takagi–Sugeno" fuzzy systems, used in multiple studies [24–26], can serve as an analytical model for nonlinear systems.

In [27], the authors developed an intelligent FTC system for a turbine engine. They started by designing a fuzzy model of Takagi–Sugeno using a hierarchical learning structure. Then, FTC is performed using an adaptive fuzzy control. The online-learning abilities of the fuzzy control are employed to detect the unknown dynamics related to failures.

Literature [28] used Takagi–Sugeno fuzzy model for the description of a space-vehicle dynamics, affected by actuator faults. First, the authors proposed a general fault model, integrating various gain and bias faults. Then, a bank of residuals for FDI process were provided using sliding-mode observer. To compensate for the fault effects, two accommodation schemes were used to deal with both cases where the system state is available and not.

Through simulations, the authors have proved that the designed FDI and FTC algorithms performed well with actuator failures. Nevertheless, the FTC design does not consider the delays resulting from the FDI and FTC algorithms, which might affect seriously the stability of the system.

4.1.2 Analytical Methods

If there is an available mathematical model of the process, analytical FTC methods can be used for the accommodation of fault effects, under the constraint that these effects can be quantified. Referring to Fig. 4, analytical FTC strategies are divided into 2 types: passive and active methods.

Passive Methods

Passive approaches involve designing controllers that can already withstand a predetermined range of failures, without using on-line faults information [3]. For such failures, a fixed-parameters controller can be designed with the ability to control both nominal system and the system affected by these "known" failures. That is, the affected system continues its operation with the same system structure and controller. The efficiency of the approach depends then on the nominal system's robustness.

Passive FTC is also called a reliable control method. Indeed, designing a passive FTCS can be done mathematically using a reliable stabilization problem [29], where several controllers are designed for the stabilization of a single system. i.e., one or a combination of these controllers can stabilize the plant which is useful especially for controller failures. For such control design, the multiple controllers form the analytical redundancies.

There exist numerous literatures based on robust control methods [18, 30] such as H_∞ optimization method, quantitative feedback control, variable structure control, control allocation (CA), the frequency domain approach, Youla (YJBK) parameterization.

The main concept of passive approach is making the plant robust against a restrictive set of failures and how they impact control functions. This technique is simple to implement and suitable for restricted cases of faults. However, its fault tolerance capability is very limited since it does not use diagnosis information such as the fault occurrence, the localization and the seriousness of the fault. Moreover, passive approaches cannot account for a wide range of faults such as unforeseen ones. Consequently, the effectiveness of these methods is limited since it is not feasible to predict all faults and their impact on the system beforehand.

Active Methods

Using the FDI process, active techniques respond promptly to faults by adjusting the control law, instead of using one fixed controller for different situations. The controllers are adjusted depending on the type of the fault and it's magnitude, to preserve the nominal performance and the original desired properties of the plant [4].

Figure 6 shows a typical structure of an active FTCS, which is formed of three main components: an FDD scheme, a reconfiguration algorithm and a reconfigurable controller. The harmony between these modules is the key to an efficient active FTCS. The objective of the FDD unit is detecting and isolating, any occurring fault in the process. It should also estimate, online, the fault parameters, the impaired system model, and the system variables. Based on the aforementioned informations, the reconfiguration mechanism re-designs the reconfigurable controller in order to maintain the desired performance and the stability. Moreover, in order to allow tracking of a command trajectory, a feedforward controller can also be redesigned. If the failure makes the system work with a degraded performance and/or causes actuator saturation, then, a reference governor may also be redesigned so that command inputs can be adjusted according to the new requirements.

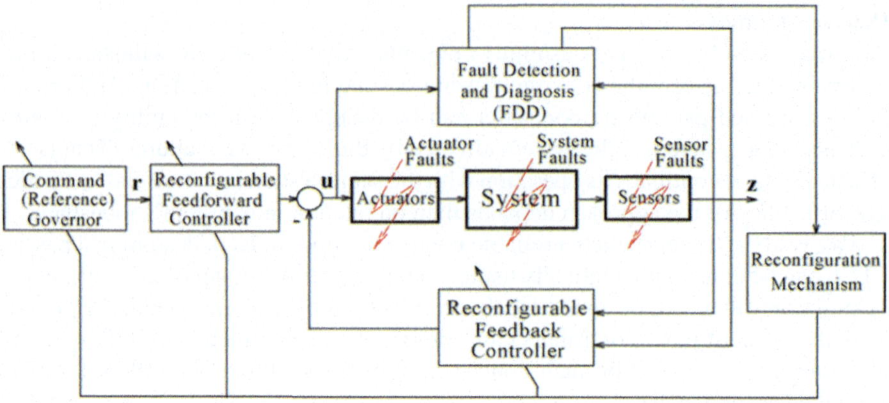

Fig. 6 Basic structure of an active FTCS

Online/Off-line control laws: As regards the controller implementation, two approaches are generally considered: The first is to design a set of controllers off-line and to store them in a database, based on the FDI bloc decisions. The most suitable controller is selected to be used by the reconfigurable controller [30] (such as multiple-model technique). The other strategy is to synthesis new controllers online, in real-time [31] (such as eigenstructure assignment, predictive control, linear quadratic regulator...). Obviously, the first approach is more restrictive regarding the type of failures dealt with. Online controllers, on the other hand, offer a larger spectrum of faults that can be dealt with, but are more difficult with respect to real-time challenges.

Reconfiguration/restructuring: Considering the controller redesign, several literatures classify active FTCS as reconfigurable versus restructurable [18, 30]. If the required objectives as well as the performance of the impaired system are maintained close to the nominal ones, then, active approach correspond to reconfiguration/accommodation.

However, in case of a critical fault (for instance, a complete actuator loss), the nominal performance and the control objectives may not be achievable anymore. Thus, a restructuring strategy is necessary. The latter consists of modifying the control objectives or the system structure such as the controlled outputs number. If the nominal reference inputs are not reachable, new command inputs have to be redefined. Consequently, the system continues operating with the healthy actuators, but in a degraded mode. Moreover, for specific faults, even with degraded functionality, the system cannot be kept in operation. The objective is then, to safely turn it off (Fig. 7 [32]).

Compared to passive FTC approaches, active methods are indeed more complex to implement and present a real-time challenge in decision making. However, it is apparent that active techniques are more practical and favorable because they can address a broad range of faults. Additionaly, due to their real-time capability, these

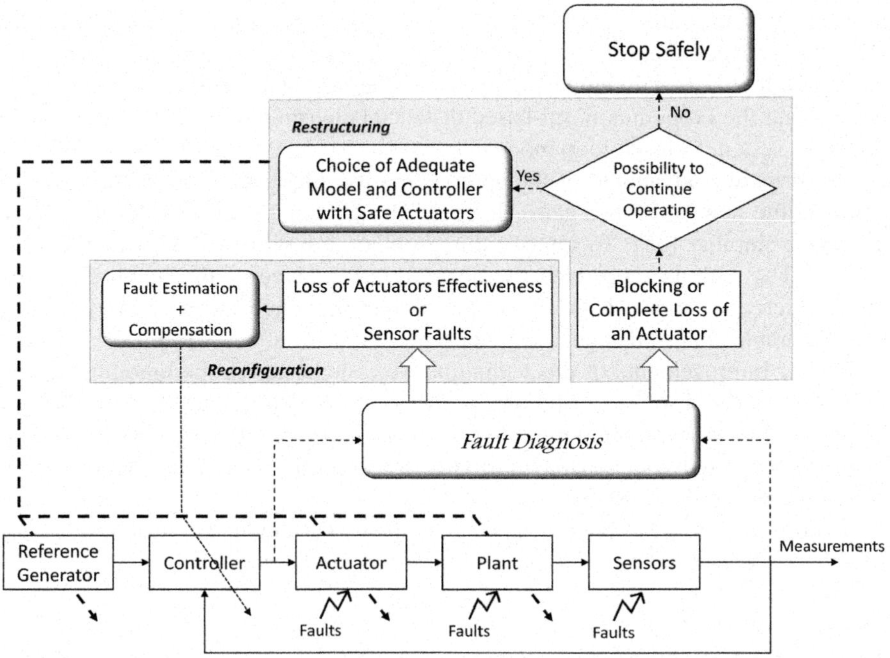

Fig. 7 Active FTCS: reconfiguration versus restructuring

methods are particularly adequate for implementation in critical systems with the possibility to work in degraded mode.

In the sequel, a brief overview of some active reconfiguration methods will be presented.

4.2 *Active Reconfiguration Methods*

In the literature, different methods were investigated in active reconfiguration strategy, such as: pseudo-inverse, eigenstructure assignment, Linear Matrix Inequality (LMI) [33], multiple models, adaptive and predictive controls, etc. A brief overview of most common methods is presented in the sequel.

4.2.1 Pseudo-inverse

The pseudo-inverse (PIM) technique was first studied in aeronautical field [34, 35]. In the latter, the authors used the PIM to recover the original system performance in the presence of component failures. The original scheme was modified so that the impaired system stability could be maintained. The concept of the PIM is to modify the state-feedback control law, such that the closed loop dynamic of the impaired system is approximately equal to the one of the nominal system. This is done by minimizing the Frobenius-norm-based distance between a reference model and the impaired system's closed loop model.

The principle advantage of the approach is its simplicity to implement, and to compute the new controller gain. However, the obtained solution does not always guarantee stability [35]. To remedy this problem, a modified PIM was developed in [36]. The authors embedded the desired performance of the impaired system into a reference model. Then, using the perfect model following concept, a new reconfigurable control design was proposed.

Further improvements of this technique were suggested in the literature such as [37]. The author proposed searching for the solution within a set of admissible reference models, rather than looking for an optimal approximate solution that doesn't assure the impaired system behaviour. This technique allows quantifying the accommodation procedure' robustness.

The limitations of this approach is that the faulty system model must be available. Moreover, the incorporation of uncertainties may not be possible.

4.2.2 Eigenstructure Assignment

The eigenstructure assignment (EA) method was firstly presented in [38] and has been the subject of several studies [39–41]. The advantage of this technique, in comparison with the pseudo-inverse, is that once a solution exists, it guarantees the good stability of the faulty process. The concept of this technique is to align the defected system's eigenstructure with the one of the idealized plant. The first task of the reconfiguration mechanism is to obtain a gain matrix such that the eigenvalues of the impaired and the nominal system are the same in closed loop. The second objective is making the impaired system's eigenvectors very close to those of the nominal process.

Literature [39] used an EA-based technique for the reconfiguration of a control system affected by component failures. Once these component failures occur, or some operating conditions vary, the proposed technique is capable of recovering the nominal system's eigenstructure to a maximum extent by reconfiguring the control system. If the full state feedback is available, then the reconfigured plant's stability is guaranteed. Whereas, if only output feedback is available, the approach can recover the dominant eigenstructure of the healthy system and the stability can be maintained under certain constraints.

The main benefit of the EA is that the performance of the system can be determined in terms of eigenstructure and it is ensured that stability will be maintained. The main downside of this approach is the need to full state-feedback which lessens the potential application to high-order systems. Besides, observers will have to be implemented for the estimation of system states.

4.2.3 Multiple Models

In a given process, if faults can be specified through a finite set of fault modes, then the process can be presented by a definite number of "multiple models" (MM) [42]. The technique of multiple models allow controlling a nonlinear system in a large operating range, split into different linearized regions.

To handle with anticipated component faults, literature [43] proposed an adaptive control scheme using a multiple model representation. A similar approach, based on adaptive control, was developed in [44]. The first model is associated with the nominal operation of the plant, while other models are characterized by the occurrence of a particular failure, leading the process outside the nominal operating range. To handle actuator faults, literature [45] developed an integrated active FTCS based on an Interacting Multiple Model (IMM) strategy. Similarly, literature [46] developed a technique on the basis of the interaction of multiple kalman filters for the disgnosis and the accommodation of actuator/sensor faults. This approach provides a robust system state estimation and can reconfigure the control law using one (or a combination) of appropriate models.

Considering the case where each model represents a particular fault case, the main drawback is that only a predefined, limited set of faults can be considered. Indeed, the majority of multiple model-based approaches assume that one fault at a time can occur in the system, and consequently a unique model can be selected. If the defected system's model doesn't correspond to the set of the available models, the corresponding control method can then lead the system to instability.

4.2.4 Predictive Control

The ability of the predictive control method (PCM) to solve FTC problems was demonstrated in [47]. The proposed approach allows the reconfiguration of the controller in the presence of a failure, such that stability can be preserved and the performance is nearly the same as the nominal case. The predictive control consists of solving, at each time step, an optimal control problem, i.e, minimizing a definite criteria under certain constraints.

As mentioned in [48], the PCM can effectively handle the various control constraints (for instance, the actuators' physical limitations), with a limited number of parameters. The interest of this method is to predict the system's future behavior, by modifying, online, the different control constraints such that an acceptable performance level can be maintained.

In [51], a reconfiguration scheme is proposed for nonlinear systems based on a combination of generalized predictive control and multiple model estimator. The efficiency of the strategy was proved through the application to a joint of a space robot and to an inverted pendulum. Another interesting study was led by [50] where nonlinear model predictive control was used along with neural networks for the reconfiguration of an all-thruster spacecraft affected by thrusters faults.

It is worth noting that predictive control increases computational load as it needs online optimization. Thus, various studies tend to use simplified models to decrease the computational effort such as [50].

4.3 Synthesis

Based on the literature overview of FTC methods discussed in previous sections, we present in Table 1, a summary of the advantages and limitations of each class of methods. It should be noted that the "suitable" technique for FTC depends on the type of system considered, the available model, and the nature/gravity of the fault.

Table 1 Advantages and limitations of FTC methods

Method	Advantages	Limitations	References
Heuristic methods			
• Expert systems	– Ability to consider diverse data sources – Improvement of the operating characteristics of the system	– Controllability and observability conditions must be verified (post-fault)	[15, 49]
• Neural networks	– Allows integrated control schemes – Can be used for both FDI and FTC	– Computationally expensive – No specific rule for determining NN structure	[18, 19]
• Fuzzy logic	– Dealing with unknown/complex model dynamics – Handling nonlinearities	– Requires a lot of data – Manual tuning/time-consuming	[28]
Analytical methods			
• Passive	– Simple to implement – Suitable for restricted cases of faults	– Limited range of faults – FDI is not incorporated	[3, 18]
• Active			
PI	– Simple to implement	– Stability not always guaranteed – System model must be available	[4, 37]
EA	– Stability always guaranteed	– Full state feedback is required – Observers are required	[41]
MM	– Robust for a finite set of failures	– One fault at a time can be handled	[46]
PCM	– Manging different control constraints	– High computational effort	[50]

5 Conclusion

This chapter was dedicated for the state of the art of the FTC problem. We first defined standard terms that are often used in the field of FDI as well as FTC. Then, we provided an overview of the different fault types that might occur in automated systems. Next, a classification of the different FTC techniques was given, with an emphasis on active reconfiguration methods. A synthesis section was also provided to summarize the advantages and drawbacks of each class of methods.

References

1. Severson, K., Chaiwatanodom, P., Braatz, R.D.: Perspectives on process monitoring of industrial systems. IFAC-PapersOnLine **48**(21), 931–939 (2015). https://doi.org/10.1016/j.ifacol.2015.09.646. 9th IFAC Symposium on Fault Detection, Supervision and Safety for Technical Processes SAFEPROCESS 2015
2. Amin, A.A., Hasan, K.M.: A review of fault tolerant control systems: advancements and applications. Measurement **143**, 58–68 (2019)
3. Nasiri, A., Nguang, S.K., Swain, A., Almakhles, D.: Passive actuator fault tolerant control for a class of mimo nonlinear systems with uncertainties. Int. J. Control **92**(3), 693–704 (2019)
4. Abbaspour, A., Mokhtari, S., Sargolzaei, A., Yen, K.K.: A survey on active fault-tolerant control systems. Electronics **9**(9), 1513 (2020)
5. Sadeghzadeh, I., Zhang, Y.: A review on fault-tolerant control for unmanned aerial vehicles (uavs). In: Infotech@ Aerospace 2011, p. 1472 (2011)
6. Etlili, D., Khedher, A., Errachdi, A.: Sliding mode observer for fractional order nonlinear dynamical systems. In: 2022 5th International Conference on Advanced Systems and Emergent Technologies, pp. 103–108 (2022)
7. Mlayeh, H., Ghachem, S., Nasri, O., Benothman, K.: Accommodation of successive actuator failures with optimal control for a class of nonlinear systems. In: 2016 17th International Conference on Sciences and Techniques of Automatic Control and Computer Engineering (STA), pp. 217–222. IEEE (2016)
8. Mlayeh, H., Ghachem, S., Nasri, O., Benothman, K.: Nonlinear fault tolerant systems with multiple actuator failures using progressive accommodation. In: 2017 International Conference on Control, Automation and Diagnosis (ICCAD), pp. 396–401. IEEE (2017)
9. Mlayeh, H., Ghachem, S., Nasri, O., Othman, K.B.: Nonlinear fault recovery of a quadrotor using progressive accommodation. Stud. Inf. Control **29**(1), 67–75 (2020)
10. Mlayeh, H., Ghachem, S., Nasri, O., Othman, K.B.: Stabilization of a quadrotor vehicle using PD and recursive nonlinear control techniques. Int. Rev. Aerosp. Eng. (IREASE) **14**(4) (2021). https://doi.org/10.15866/irease.v14i4.19739
11. Mlayeh, H., Othman, K.B.: Attitude control of a drone subject to disturbance and structural fault. In: 2022 IEEE 21st International Conference on Sciences and Techniques of Automatic Control and Computer Engineering (STA), pp. 232–236. IEEE (2022)
12. Mlayeh, H., Ben Othman, K.: Nonlinear accommodation of a dc-8 aircraft affected by a complete loss of a control surface. Stud. Inf. Control **31**(3), 107–116 (2022). https://doi.org/10.24846/v31i3y202210
13. Yen, G.G., Ho, L.-W.: Intelligent on-line fault tolerant control for unanticipated catastrophic failures. ISA Trans. **43**(4), 549–569 (2004)
14. Witczak, M.: Fault diagnosis and fault-tolerant control strategies for non-linear systems. Lect. Notes Electr. Eng. **266**, 375–392 (2014)

15. Wagner, W.P.: Trends in expert system development: a longitudinal content analysis of over thirty years of expert system case studies. Expert Syst. Appl. **76**, 85–96 (2017)
16. Liu, W.: An on-line expert system-based fault-tolerant control system. Expert Syst. Appl. **11**(1), 59–64 (1996)
17. Farrell, J., Berger, T., Appleby, B.: Using learning techniques to accommodate unanticipated faults. IEEE Control Syst. Mag. **13**(3), 40–49 (1993)
18. Fekih, A.: Fault diagnosis and fault tolerant control design for aerospace systems: a bibliographical review. In: 2014 American Control Conference, pp. 1286–1291 (2014). https://doi.org/10.1109/ACC.2014.6859271
19. Singh, S., Murthy, T.R.: Neural network-based sensor fault accommodation in flight control system. J. Intell. Syst. **22**(3), 317–333 (2013)
20. Miri, D., Khedher, A., BenOthman, K.: State and faults estimation via artificial neural networks. In: 2020 20th International Conference on Sciences and Techniques of Automatic Control and Computer Engineering (STA), pp. 57–62 (2020). https://doi.org/10.1109/STA50679.2020.9329322
21. Miri, D., Khedher, A., BenOthman, K.: Tracking of trajectory and fault estimation of miabot robot using an artificial neural network. In: 2021 18th International Multi-Conference on Systems, Signals and Devices (SSD), pp. 1296–1301 (2021). https://doi.org/10.1109/SSD52085.2021.9429397
22. Torres-Huitzil, C., Girau, B.: Fault and error tolerance in neural networks: a review. IEEE Access **5**, 17322–17341 (2017). https://doi.org/10.1109/ACCESS.2017.2742698
23. Kwong, W.A., Passino, K.M., Laukonen, E.G., Yurkovich, S.: Expert supervision of fuzzy learning systems for fault tolerant aircraft control. Proc. IEEE **83**(3), 466–483 (1995)
24. Khedher, A., Elleuch, I., Othman, K.B.: Observer desing for non linear takagi-sugeno fuzzy systems. application to fault tolerant control. In: Advances in Observer Design and Observation for Nonlinear Systems: Fundamentals and Applications, pp. 101–133. Springer (2022)
25. Youssef, I., Khedher, A., Othman, K.B.: Fault tolerant control for non linear uncertain takagi-sugeno fuzzy systems based on proportional integral sliding mode observer. In: 2021 International Conference on Control, Automation and Diagnosis (ICCAD), pp. 1–5 (2021). https://doi.org/10.1109/ICCAD52417.2021.9638771
26. Khedher, A., Elleuch, I., BenOthman, K.: Adaptive proportional integral observer design for interval type 2 takagi-sugeno fuzzy systems. Math. Probl. Eng. **2021**, 1–13 (2021)
27. Diao, Y., Passino, K.M.: Stable fault-tolerant adaptive fuzzy/neural control for a turbine engine. IEEE Trans. Control Syst. Technol. **9**(3), 494–509 (2001)
28. Shen, Q., Jiang, B., Cocquempot, V.: Fuzzy logic system-based adaptive fault-tolerant control for near-space vehicle attitude dynamics with actuator faults. IEEE Trans. Fuzzy Syst. **21**(2), 289–300 (2013). https://doi.org/10.1109/TFUZZ.2012.2213092
29. Jiang, J.: Fault-tolerant control systems-an introductory overview. Acta Autom. Sinica **31**(1), 161–174 (2005)
30. Zhang, Y., Jiang, J.: Bibliographical review on reconfigurable fault-tolerant control systems. Annu. Rev. Control **32**(2), 229–252 (2008)
31. Cieslak, J., Henry, D., Zolghadri, A., Goupil, P.: Development of an active fault-tolerant flight control strategy. J. Guid. Control Dyn. **31**(1), 135–147 (2008)
32. Noura, H., Theilliol, D., Ponsart, J.-C., Chamseddine, A.: Fault-tolerant Control Systems: Design and Practical Applications. Springer (2009)
33. Andrade, J.P., Campos, V., Potts, A., Garcia, C.: Damping improvement of a f-16 aircraft through linear matrix inequalities. IFAC-PapersOnLine **50**(1), 3947–3952 (2017)
34. Ostroff, A.J.: Techniques for accommodating control effector failures on a mildly statically unstable airplane. In: 1985 American Control Conference, pp. 906–913. IEEE (1985)
35. Gao, Z., Antsaklis, P.J.: Stability of the pseudo-inverse method for reconfigurable control systems. Int. J. Control **53**(3), 717–729 (1991)
36. Gao, Z., Antsaklis, P.J.: Reconfigurable control system design via perfect model following. Int. J. Control **56**(4), 783–798 (1992)

37. Staroswiecki, M.: Fault tolerant control: the pseudo-inverse method revisited. Ifac Proc. Vol. **38**(1), 418–423 (2005)
38. Andry, A., Shapiro, E., Chung, J.: Eigenstructure assignment for linear systems. IEEE Trans. Aerosp. Electron. Syst. **5**, 711–729 (1983)
39. Jiang, J.: Design of reconfigurable control systems using eigenstructure assignments. Int. J. Control **59**(2), 395–410 (1994)
40. Wang, A.-P., Lin, S.-F.: The parametric solutions of eigenstructure assignment for controllable and uncontrollable singular systems. J. Math. Anal. Appl. **248**(2), 549–571 (2000)
41. Brizuela-Mendoza, J.A., Sorcia-Vázquez, F.D., Rumbo-Morales, J.Y., Lozoya-Ponce, R.E., Rodríguez-Cerda, J.C.: Active fault tolerant control based on eigenstructure assignment applied to a 3-dof helicopter. Asian J. Control (2020)
42. Khedher, A., Othman, K.B., Maquin, D., Benrejeb, M.: Active sensor faults tolerant control for takagi-sugeno multiple models. In: Proceedings of the 6th WSEAS International Conference on Dynamical Systems and Control, pp. 68–73 (2010)
43. Maybeck, P.S.: Application of multiple model adaptive algorithms to reconfigurable flight control. In: Control and Dynamic Systems, vol. 52, pp. 291–320. Elsevier (1992)
44. Maybeck, P.S., Stevens, R.D.: Reconfigurable flight control via multiple model adaptive control methods. IEEE Trans. Aerosp. Electron. Syst. **27**(3), 470–480 (1991). https://doi.org/10.1109/7.81428
45. Zhang, Y., Jiang, J.: Integrated active fault-tolerant control using imm approach. IEEE Trans. Aerosp. Electron. Syst. **37**(4), 1221–1235 (2001). https://doi.org/10.1109/7.976961
46. Yang, Z., Izadi-zamanabadi, R., Blanke, M.: On-line multiple-model based adaptive control reconfiguration for a class of nonlinear control systems. IFAC Proc. Vol. **33**(11), 729–734 (2000). 4th IFAC Symposium on Fault Detection, Supervision and Safety for Technical Processes 2000 (SAFEPROCESS 2000), Budapest, Hungary, 14–16 June 2000
47. Maciejowski, J.: Modelling and predictive control: enabling technologies for reconfiguration. Annu. Rev. Control **23**, 13–23 (1999)
48. Aström, K.J., Albertos, P., Blanke, M., Isidori, A., Schaufelberger, W., Sanz, R.: Control of Complex Systems. Springer (2011)
49. Tashakori, A., Ektesabi, M.: A simple fault tolerant control system for hall effect sensors failure of bldc motor. In: 2013 IEEE 8th Conference on Industrial Electronics and Applications (ICIEA), pp. 1011–1016. IEEE (2013)
50. Tavakoli, M., Assadian, N.: Predictive fault-tolerant control of an all-thruster satellite in 6-dof motion via neural network model updating. Adv. Space Res. **61**(6), 1588–1599 (2018)
51. Kanev, S., Verhaegen, M.: Controller reconfiguration for non-linear systems. Control Eng. Pract. **8**(11), 1223–1235 (2000)

The Nonlinear Progressive Accommodation: Design and Methodology

Hajer Mlayeh, Sahbi Ghachem, and Atef Khedher

Abstract In critical plants, the accommodation procedure must be performed with utmost speed to prevent the potential failure(s) damages. In relation to the optimal control issue, the Progressive Accommodation (PA) is an effective solution to overcome this issue. In this chapter, we define the concept and the objectives of the PA and we justify the choice of the iterative algorithm. Then, we present the limitations of the linear PA when it comes to nonlinear (NL) systems and thus, the necessity to use the NL optimal control. The SDRE method is presented as an approximate solution of the NL optimal control problem, as well as the related NL PA. To demonstrate the suggested strategy, an example is provided at the end.

Keywords Nonlinear systems · SDRE · Iterative algorithm · PA

1 Introduction

Real-time applications require a different approach than most FTC methodologies are using. Indeed, applying accommodated control immediately after establishing the diagnosis [1–3] is not suitable and assumes the system be "ideal". Indeed, there is a transient period after the diagnosis stage which is needed to compute the new FTC. During this critical period, the system continues to be controlled by nominal control law which can cause serious performance degradation and even instability. That's why, it is important to reduce this time period especially for safety-critical systems, to prevent the potential failure(s) damages. The theory of the PA was developed for minimizing such risks, based on the linear quadratic (LQR) optimal control.

H. Mlayeh (✉) · A. Khedher
LATIS, Ecole nationale d'Ingénieurs de Sousse, Université de Sousse, Erriadh, 4023 Sousse, Tunisia
e-mail: mlayehhajer@gmail.com

S. Ghachem
LARA Automatic, National Engineering School of Tunis, BP37 Belvedere, 1002 Tunis, Tunisia

A. Ben Makhlouf et al. (eds.), *State Estimation and Stabilization of Nonlinear Systems*, Studies in Systems, Decision and Control 491,
https://doi.org/10.1007/978-3-031-37970-3_13

There exist few literatures dealing with the PA such as: [4–8]. In all the aforementioned literatures, only linear systems were investigated.

For nonlinear systems, recent papers [9, 10] have proved the efficiency of the PA method for a local analysis. The authors of the aformentioned papers considered an input-affine nonlinear system and used a linear approximation around an operating point in order to obtain the linearized form. The latter was used to compute the linear FTC. The linear PA was then applied to a first-order-nonlinear system. Although the linear approximation strategy is simple, it may not be applicable to certain initial conditions and causes system's divergence. A solution to this problem is the use of a Nonlinear Progressive Accommodation scheme (NPA).

In this NL context, the few existing literatures [10–13], used the iterative methods of the Successive-Galerkin-Approximation (SGA) and the State-Dependent-Riccati Equation (SDRE) in order to prove the NPA efficiency. The application didn't include a realistic NL system and was restricted to simple equations. Recently, specific studies [14–19] have recalled the NPA theory and applied it for the recovery of real-critical plants, particularly: a quadrotor drone and a fixed-wing DC-8 aircraft where different control problems and fault scenarios where investigated.

In this chapter, we first define the concept and the objectives of the PA and we justify the choice of the iterative algorithm used in this work. Then, we present the limitations of the linear PA when it comes to nonlinear (NL) systems and thus, the necessity to use the NL optimal control. Thereafter, we present the SDRE method, as an approximate solution of the NL control problem and the related NL PA. Finally, to demonstrate the suggested strategy, a numerical example is provided.

2 The Concept of the Progressive Accommodation

In the follow-up, the progressive accommodation is presented in the general context of linear systems.

2.1 Definition

The PA consists of solving the optimal control problem through an iterative scheme, applied online and converging toward the optimal FTC. Figure 1 illustrates the framework of the system using the PA [5].

2.2 Comparison Between the Classical FTC and the PA

In this section, we do a comparison between the PA and the classical FTC schemes (using the LQR approach). We suppose that we want to control the linear system represented in (1).

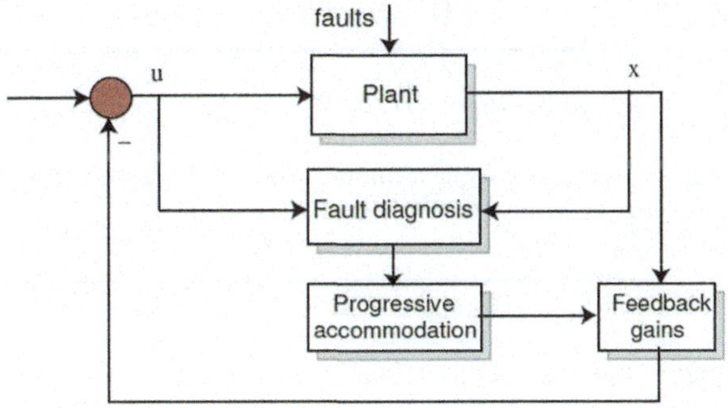

Fig. 1 PA framework

$$\dot{x} = A_n x + B_n u_n \tag{1}$$

Where n denotes the nominal condition, $x \in \Re^n$ is the state vector, $u \in \Re^m$ is the input vector, and A, B are the system and input matrices respectively. The nominal control u_n, to be found, must minimize the performance index defined in (2).

$$J(x) = \int_0^\infty (x^T Qx + u^T Ru)\mathrm{d}t \tag{2}$$

where $Q \in \Re^{n\times m}$ is a symmetric positive semi definite matrix that is used to penalize the state action and $R \in \Re^{m\times m}$ is a positive definite matrix that affects the control action. Once the penalization matrices are selected, the control law is given by:

$$u_n = -F_n x = -R^{-1} B_n^T P_n x \tag{3}$$

where P_n is the solution of the Algebraic–Riccati-Equation (ARE) (4).

$$A_n^T P_n + P_n A_n - P_n B_n R^{-1} B_n^T P_n + Q = 0 \tag{4}$$

We suppose that an actuator or system fault has occurred at time t_f. Then, the representation of the system (5) becomes:

$$\begin{cases} \dot{x} = A_n x + B_n u_n, & t \in [0, t_f[\\ \dot{x} = A_f x + B_f u_n, & t \in [t_f, \infty[\end{cases} \tag{5}$$

The objective of the fault accommodation procedure is to find a new control law u_{fc} (where fc denotes the fault correction), that takes into consideration the impact of the fault and controls the faulty system (5). This control law must ensure the system's stability and maintain an acceptable performance that is close to the one of the nominal system (1). To do so, a direct fault-tolerant control law can be generated using the ARE (6) while considering the "faulty" matrices (A_f, B_f).

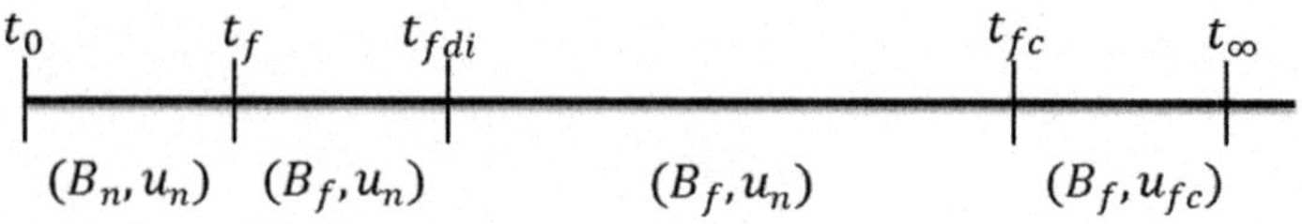

Fig. 2 The control sequence of the classical NL accommodation

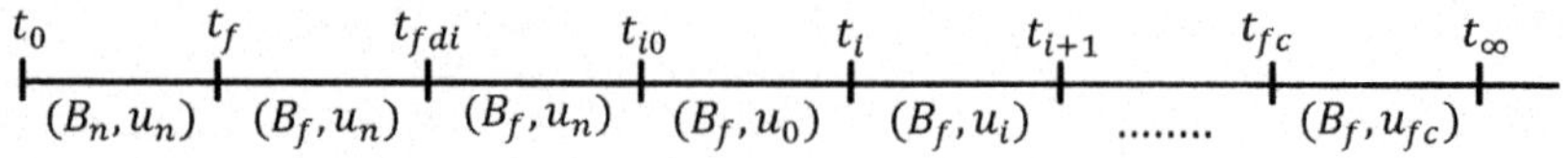

Fig. 3 The control sequence of the accommodated system using the PA

$$A_f^T P_{fc} + P_{fc} A_f - P_{fc} B_f R^{-1} B_f^T P_{fc} + Q = 0 \tag{6}$$

The new FTC law is given by (7) where P_{fc} is the solution of the ARE (6).

$$u_{fc} = -F_{fc} x = -R^{-1} B_f^T P_{fc} x \tag{7}$$

This classical NL accommodation theory can be represented by 4 time intervals as illustrated in Fig. 2 (where t_{fdi} denotes the diagnosis delay).

Let t_{i0} be the initial iteration's application instant using the iterative method of PA. The control sequence of the accommodated system, using the PA, is described in Fig. 3. This control sequence is characterized by the following time periods:

1. $t \in [t_0, t_f[$: The nominal system is under u_n control.
2. $t \in [t_f, t_{fdi}[$: A fault has occurred at time t_f. the diagnosis bloc is running in order to identify the fault and the system affected by the fault is still under the control u_n.
3. $t \in [t_{fdi}, t_{i0}[$: the fault is identified at time t_{fdi} but the PA has not started because the first stabilizing control u_0 is not computed yet. Until so, the affected system is still under u_n.
4. $t \in [t_{i0}, t_1[$: The process of the PA is now launched. The first stabilizing control u_0 is computed and applied to the faulty system at time t_{i0}. The iterative algorithm starts computing the next control u_1, to be applied at time t_1.
5. $t \in [t_i, t_{i+1}[$: The process of the PA is still running. The stabilizing control (number i) u_i is computed and applied to the faulty system at time t_i. The iterative algorithm starts computing the next control u_{i+1}, to be applied at time t_{i+1}.
6. $t \in [t_{fc}, \infty[$: The iterative algorithm has now reached the optimal solution at time t_{fc}. The faulty system is finally under the FTC u_{fc}.

Referring to Fig. 3, we can conduct that the main PA advantage is that the defected system works under u_n for only $\Delta = t_{i0} - t_{fdi}$ which represents the required delay to apply the first control u_0.

Contrarily, in the classic scheme, the defected system operates under u_n for $\Delta = t_{fc} - t_{fdi}$, until the computation of the last iteration (ideal solution) u_{fc}. This proves the effectiveness of the PA scheme regarding convergence speed and stability.

2.3 *The Iterative Algorithm of Newton–Kleinman*

As was mentioned in the previous sections, in order to minimize the delay $(t_{fc} - t_{fdi})$, the progressive accommodation (PA) uses an iterative algorithm also known as an "anytime" algorithm. Specific desired properties can be included in such algorithms to have a "satisfying behavior":

- The quality of the produced result can be measured precisely and it is an increasing function of time.
- The improvement in the result's quality is more significant at the first steps of the algorithm and it decreases over time.
- The algorithm can be interrupted whenever in order to provide a result.

In [8], one finds a comparison between two PA strategies using "anytime" algorithms: The Newton–Kleinman (NK) algorithm and the Matrix-Sign Function (MSF). Both strategies are used in the frame of the linear quadratic FTC. Simulations proved that (NK) algorithm took more time for convergence but performed better in terms of the closed loop admissibility.

In 1968, Kleinman [20] used Newton's technique for solving, iteratively, the ARE which explains the nomination of the "Newton–Kleinman" (or the "Newton–Raphson" method in other literatures [6]).

The objective of the NK algorithm, is to converge to the same optimal control (7), while solving the ARE in an iterative way. The theory of NK can be summarized as follows:

1. Selecting a matrix $P_0 \in \Re^{n \times n}$.
2. Solve Eq. (8) of "Lyapunov" to get the matrix P_{i+1} ($i = 0, 1, 2 \ldots, N$ is the number of iterations).

$$P_{i+1}(A - BF_i) + (A - BF_i)^T P_{i+1} = -Q - F_i^T R F_i \tag{8}$$

3. Compute the corresponding gain:

$$F_i = R^{-1} B^T P_i \tag{9}$$

4. Finally, each time a new iteration F_i is computed, the control u_i is applied:

$$u_i = -F_i x \tag{10}$$

5. The algorithm stops after N iterations, when $P_{i+1} = P_i$.

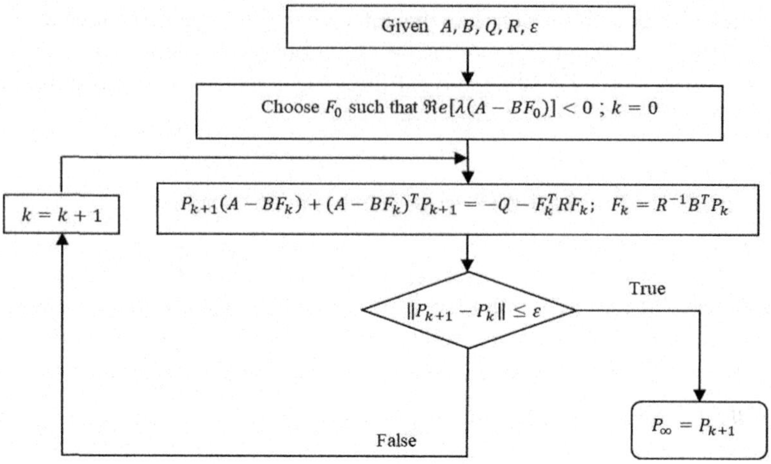

Fig. 4 The Newton–Kleinman algorithm

The steps of the NK algorithm can be summarized in Fig. 4. The consecutive controls u_i, (10) ($i = 0, 1, 2, \ldots N$), are applied on the interval $[t_i, t_{i+1}[$ each time they are available until the algorithm reaches the final FTC. The initialization of the NK algorithm requires 2 parameters:

- t_{i0} is the time needed to compute and apply the control u_0 which explains the interval $[t_{fdi}, t_{i0}]$ presented in Fig. 3.
- The initializing gain matrix F_0.

For the initialization of the algorithm, if the control ($u_n = -F_n x$) stabilizes the defected system, i.e., the matrix $(A - B_f F_n)$ is Hurwitz, then it is appropriate to choose $F_0 = F_n$. Else, another stabilizing gain F_0 must be found. Concerning convergence and stability aspects, it was demonstrated in [20] that the NK algorithm guarantees the convergence if the following assumptions are satisfied:

- There is $P_0 \in \Re^{n \times n}$, satisfying that $(A - B_f F_0)$ is Hurwitz.
- The detectability and stabilizability assumptions must be verified ([21], Definition 2.1).

If the above assumptions hold, then the following results are implied:

- $(A - B_f F_i)$ is Hurwitz for all $i = 1, 2, \ldots$.
- All produced iterations P_i defined in (8), guarantee the stability and lead to the solution P_∞ ($\lim_{i \to \infty} P_i = P_\infty$):

$$0 \leqslant P_\infty \leqslant \cdots \leqslant P_{i+1} \leqslant P_i \leqslant \cdots \leqslant P_0. \tag{11}$$

We can point out that choosing the appropriate gain matrix makes convergence way quicker. We can find several methods for this aim e.g., the "Bass Armstrong" or the "Chandrasekhar"'s methods [22].

3 Limitations of the Linear PA

As linear systems are concerned, there exists various studies based on PA such as [4–8]. For instance, in [6, 7], the linear PA was investigated using the NK algorithm. The efficiency of the iterative method was proved compared to the classical one, where the application was to a Boeing 747. In all the aforementioned literatures, only linear systems were investigated.

Literatures [9, 10] have proved the efficiency of the PA method for a local analysis. The authors of the aformentioned papers considered an input-affine nonlinear system. The main idea was to use a linear approximation around an operating point in order to obtain the linearized form (1). The optimal control problem is then described by an ARE, as defined in Sect. 2.2. The linear PA was finally applied to the nonlinear system and its efficiency, compared to the direct method, was proved.

It was proved, however, [9, 10], that although the linear approximation strategy is simple, it may not be applicable to certain initial conditions and causes system's divergence. Indeed, if we define t_c as the application time of the first corrective control, then: Independently of how close t_c is, there is always an initial state $x(0)$ that makes the NL closed-loop-defected system leave the stability limit and diverge.

A solution to this problem is the use of a NL PA scheme (NPA).

4 Nonlinear Control: Interest and Motivations

Recently, the topic of NL control has been increasingly attracting the attention of researchers and control engineers from different areas: Robotics, aircraft control, biomedical applications.

In addition to the limitations of the linear approximation method, there exist other motivations that justify the interest in nonlinear control techniques [23]:

Improvement of the operation range:Linear control techniques require the assumption of a small operation range so that the linear model becomes valid. However, in case where a large operation range is required, the nonlinearities in the system can't be appropriately compensated for. Consequently, the linear controller is no longer effective and is very likely to cause instability. On the other hand, NL controllers may deal, directly, with the nonlinearities for a very large operation range.

Analyzing hard nonlinearities: Linear control rely also on the condition that the model is undoubtedly linearizable. Nevertheless, the discontinuous nature of some nonlinearities does not permit a linear approximation. These "hard nonlinearities" can involve: saturation, hysteresis, Coloumb friction, etc. and they commonly cause undesired behavior such as instability or specious limit cycles. Unlike linear methods, the nonlinear techniques can predict and compensate for these undesirable effects.

Handling model uncertainties: In linear controllers design, it is generally assumed that the system model parameters are well known. However, these model parameters can be "uncertain" which may be caused by an abrupt change or a slow variation of these parameters (e.g., the pressure of the ambient air during the aircraft flight). Thus, linear controllers that rely on inaccurate parameter values, can easily lead to instability or performance degradation. The solution is to tolerate model uncertainties by introducing nonlinearities in the controller part of the system. For this aim, NL control methods can be used such as adaptive control and robust control.

There exist other reasons to adopt NL control methods, for example: The performance and cost optimality. Indeed, for the linear control, high quality sensors and actuators might be needed to achieve linear behavior in the desired operation range. However, for NL control, less expensive components may be used with NL characteristics.

As regards performance optimality, there exist controllers that produce fast response although being inherently nonlinear (e.g., On-Off type controller).

All the previously-mentioned motivations can explain the important position that the NL control occupies in control engineering, especially in modern technologies such as robots or high-quality aircrafts.

As regards the context of the present chapter, the problem of the progressive accommodation, in its NL counterpart, is related to the NL optimal control which will be presented in the next part.

5 NL Optimal Control Problem

5.1 Definition

The optimal control problem consists of finding a control for a given system while achieving a specific optimality criterion. The cost function, to be optimized, depends on the control and state variables.

Let the minimization problem be:

$$\min_{u(.)} \int_0^T L(x, u)\mathrm{d}t + V(x(T)) \tag{12}$$

subject to:

$$\dot{x} = f(x) + g(x)u, \qquad x \in \Re^n, \ u \in \Re^m \tag{13}$$

where $f(x) : \Re^n \to \Re^n$ and $g(x) : \Re^n \to \Re^{n\times m}$ are continuously differentiable functions. In this optimization problem, the objective is to find the trajectory $(x(t), u(t))$ minimizing the cost function:

$$J(x,u)=\int_0^T L(x,u)\mathrm{d}t+V(x(T)) \tag{14}$$

where T represents the control horizon, $L(x,u)$ is the integral cost and $V(x(T))$ is the terminal cost.

5.2 *Infinite-Horizon Optimal Control*

The optimal control problem includes several variations such as the optimal LQR. For nonlinear systems, we are interested in the infinite-horizon optimal control:

$$\begin{cases} \bullet\, T=\infty \\ \bullet\, V=0 \end{cases} \tag{15}$$

Thus, the optimal control problem is:

$$\begin{cases} \min\limits_{u(.)} \int_0^\infty L(x,u)\mathrm{d}t \\ \dot{x}=f(x)+g(x)u \end{cases} \tag{16}$$

If there exists a solution to this problem, with a finite cost, then the term $L(x,u)$ must converge to 0 as $t\to\infty$. If $T<\infty$, the optimal control is a *finite horizon* problem.

A solution to (16) is obtained using the *Hamilton–Jacobi–Bellman* (HJB) equation [24].

5.3 *Hamilton–Jacobi–Bellman Equation*

In literature [25], Bellman proposes a solution to the optimal control problem by breaking it into separate sub-problems.

The optimality principle of Bellman: An optimal policy has the following criteria: Independently of the initial decision and initial state, each subpolicy must itself constitute an optimal policy with regarding the initial state of the subpolicy.

In other words, and referring to Fig. 5: If C belongs to an optimal path from A to B, then sub-paths A to C and C to B are also optimal. The Dynamic Programming (DP) method is an optimization technique that is used to solve a wide range of problems, having a particular sequential form. This technique is based on Bellman's principle of optimality. As a consequence, the optimal control problem defined in (16), is reduced to the partial-differential-equation of HJB:

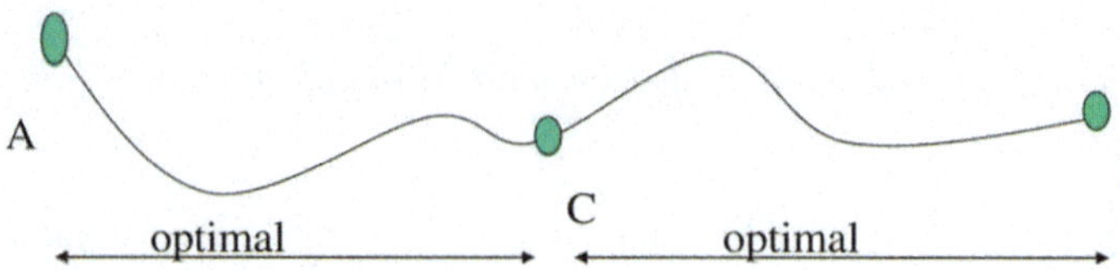

Fig. 5 Bellman's principle of optimality

$$\frac{\partial V^*}{\partial x} f - \frac{1}{4}\left(\frac{\partial V^*}{\partial x} g R^{-1} g^T \frac{\partial V^{*T}}{\partial x}\right) + x^T Q x = 0 \tag{17}$$

where $\frac{\partial V^*}{\partial x}$ denotes the gradient of V^* with respect to x $\left(\frac{\partial V^*}{\partial x} = \left[\frac{\partial V^*}{\partial x_1} \dots \frac{\partial V^*}{\partial x_n}\right]\right)$.

V^* represents the performance index (or the cost function) defined by:

$$V^*(x) = \min_{u(.)} \int_0^\infty (x^T Q(x) x + u^T R(x) u) \mathrm{d}t \tag{18}$$

where $Q(x) \geq 0$ and $R(x) > 0$ for all $x \in \Re$. For the sake of simplicity, and so that $V^*(x)$ is continuously differentiable, the matrices Q and R are commonly constant.

If the HJB equation (17) admits a solution that is positive definite and continuously differentiable, then the NL optimal control is designed as follows:

$$u^* = -\frac{1}{2} R^{-1} g^T \frac{\partial V^{*T}}{\partial x}. \tag{19}$$

An important property of the HJB equation is that it is considered as a "global" approach, since it provides a solution for every initial condition of the optimal control problem.

5.4 Limitations of the HJB Equation

The main difficulty about NL optimal control problem is to find a solution of the HJB differential equation. The latter is extremely hard to solve especially with the complex structure of NL system (13). Consequently, several numerical and approximate solutions were developed such as: polynomial approximation [24], Markov chain approximation [26], the adaptive cross approximation [27], successive Galerkin approximation [13], dicrete-time approximation [28], the Heuristic DP approach [29], SDRE method [30], etc.

In this chapter, our emphasis would be on the SDRE approach which is employed to find the NL optimal control. The latter is used in the NL accommodation procedure.

6 The SDRE Method

6.1 Definition

The majority of literatures mentioned in the previous section do offer an approximate solution to the HJB equation, but they suffer from a heavy computational load.

The SDRE approach, however, doesn't necessitate resolving the HJB equation. It is first proposed in [31] as an extension to the ARE as it enables real-time solutions for NL control problems.

The SDRE method produces nonlinear feedback controls by keeping "nonlinearities" in the states. Concurrently, a design flexibility is offered through the choice of weighting matrices. This strategy consists of factorizing factorize the NL dynamics into the product of a state-dependant matrix function and the state vector itself. This factorization is referred to as the "parameterization" procedure. The latter allows capturing the nonlinearities of the system by transferring the NL model to a linear-form structure having state-dependent coefficient matrices (SDC). Through SDC matrices, and for a given point in state space, we solve the ARE online to obtain the optimal control.

Owing to its aforementioned properties, the SDRE approach offers several advantages:

- It is both simple to implement and effective in resolving NL complicated control problems.
- The SDC parameterization is not unique which creates additional degrees of freedom and so, enhances the controller performance.
- Using weighting matrices, one can act on the control effort and control accuracy, which can prevent some issues like the actuators' saturation.

6.2 The SDRE Concept

The NL system (13) can be represented as follows (where $B(x) = g(x)$):

$$\dot{x} = f(x) + B(x)u \tag{20}$$

The SDC parameterization is the main key of the SDRE approach where the system (20) must be factorized into SDC matrices. Assuming that $f(0) = 0$, and $f(.) \in C^1(\Re^n)$, there is always a matrix $A(x)$ satisfying:

$$f(x) = A(x)x \tag{21}$$

Hence, the SDC parameterization of the NL system (20) becomes:

$$\dot{x} = A(x)x + B(x)u \tag{22}$$

It is worth mentioning that, for $n > 1$, this parameterization is not unique. When designing the control law, the most suitable SDC parameterization must be chosen to satisfy the control objectives. One important criteria for choosing the "right" parameterization is the controllability matrix:

$$C_{trb}(x) = [B(x) \quad A(x)B(x) \ \ldots \ A^{n-1}(x)B(x)]. \tag{23}$$

For linear system theory, the system is Controllable if C_{trb} (a constant), has a full rank. Similarly, in the NL case, the SDC factorization must ensure that $C_{trb}(x)$ has full rank for the whole domain $(rank(C_{trb}(x)) = n)$.

As it was explained in Sect. 5.2, the performance index (24) must be minimized:

$$J(x) = \int_0^{\infty} (x^T Q(x)x + u^T R(x)u)dt \tag{24}$$

The matrices A and B as well as the penalization matrices Q and R are computed for the given state and dealt with as constants, then similarily to the LQR approach, $u(x)$ is given by:

$$u(x) = -R^{-1}(x)B^T(x)P(x)x = -F(x)x \tag{25}$$

$P(x)$ is the solution of the SDRE:

$$A^T(x)P(x) - P(x)B(x)R^{-1}(x)B^T(x)P(x) + P(x)A(x) + Q(x) = 0 \tag{26}$$

The SDRE approach is represented in Fig. 6 which can be summarized in the following steps:

1. Measure the states x. If all states are not available for measurement, an observer must be incorporated.
2. Compute the model states $A(x)$, $B(x)$, $Q(x)$ and $R(x)$.
3. Solve the SDRE to obtain $P(x)$.
4. Evaluate the feedback control law (25).

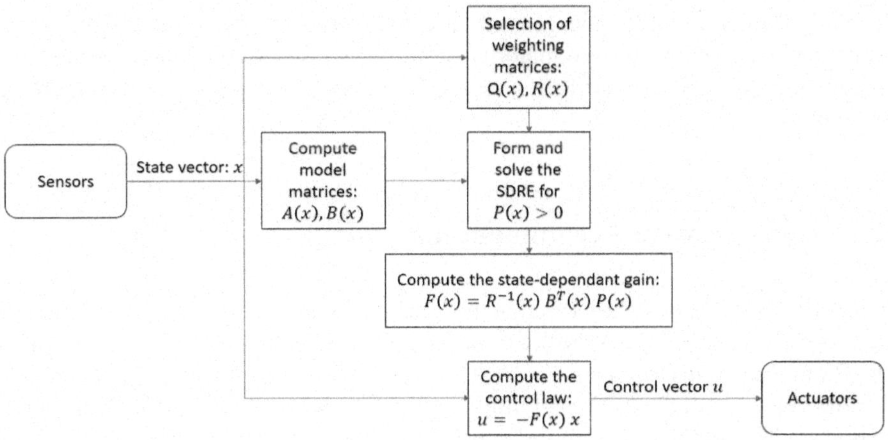

Fig. 6 Principle steps of the SDRE method

6.3 *Stability Properties*

The SDRE technique have the same stability and robustness aspects of the LQR method (with the exception that properties for NL systems hold locally). In [32], it was demonstrated that, to ensure local asymptotic stability, Theorem 1 must be verified.

Theorem 1 *The NL system* (22)*, having the feedback control law* (25)*, is locally asymptotically stable providing that:*

1. $x \in \Re^n$ *(*$n \geqslant 1$*).*
2. $P(x)$ *is the unique, positive-definite, symmetric, pointwise-stabilizing solution of the SDRE* (26)*.*
3. $A(x)$*,* $B(x)$*,* $Q(x)$*, and* $R(x)$ *are* C^1 *matrix-valued functions.*
4. *The pair* $\{(A(x), B(x))\}$ *is a pointwise controllable parameterization of the NLsystem* (20)*, for all* x*.*

The aforementioned theorem presents the necessary conditions that ensure local asymptotic stability of the NL system. Moreover, these conditions are satisfied for a wide range of NL systems (22).

As regards global asymptotic stability, this implies the stability of the plant for any IC which is in general, hard to achieve. This particular case is presented in Theorem 2 [32].

Theorem 2 *If conditions 3 and 4 hold, and the closed-loop SDC matrix* (27)*, is symmetric for all* x*, then, the closed-loop-NL-system is globally asymptotically stable.*

$$A_{cl}(x) = A(x) - B(x)F(x) \tag{27}$$

It can be noticed that, in the general case, it is hard to ensure the conditions of Theorem 2. Therefore, for the majority of NL systems, we can guarantee the local asymptotic stability in an attraction region. A method for estimating this region can be found in [32].

7 NL Progressive Accommodation (NPA)

Similarily to the LQR theory (used in the linear accommodation), the SDRE method admits a direct and an iterative solution. The present section is the NL extension of the PA presented in Sect. 2.

The objective of the NPA is minimizing the interval $[t_{fdi}, t_{fc}]$ through the application of the controls $u_i(x)$ on the interval $[t_i, t_{i+1}[\subset [t_{fdi}, t_{fc}]$ rather than waiting for $u_{fc}(x)$ (Refer to Fig. 3).

The controls $u_i(x)$ are produced using the state-dependent-iterative algorithm of NK. The latter produces, each time, a control $u_i(x)$ that converges to the final FTC after N steps.

In the NL case, the NK consists in starting with a matrix $P_0(x) \in \Re^{n\times n}$ (corresponding to $u_0(x)$) and solving Eq. (28) of Lyapunov to get the matrix $P_{i+1}(x)$.

$$\begin{aligned}&(A(x) - B_f(x)F_i(x))^T P_{i+1}(x) + P_{i+1}(x)(A(x) - B_f(x)F_i(x)) = \\ &- Q(x) - F_i(x)^T R(x)F_i(x)\end{aligned} \tag{28}$$

given:

$$F_i(x) = R^{-1}(x)B_f(x)^T P_i(x) \tag{29}$$

Once $P_i(x)$ is computed, the iterative NL control law is applied online using Eq. (30).

$$u_i(x) = -F_i(x)x \tag{30}$$

We should mention that all stability and convergence properties discussed in the linear case of the NK algorithm (Sect. 2.3), still hold for the NL case. That is, if $(A(x) - B_f(x)F_0(x))$ is poitwize Hurwitz, then, the NPA can stabilize the faulty system and the following results are implied:

- $(A(x) - B_f(x)F_i(x))$ is Hurwitz for all $i = 1, 2, \ldots N$.
- All produced iterations $P_i(x)$ converge monotonically to the final solution $(P_\infty(x) = P_{fc}(x))$: $\lim\limits_{i\to\infty} P_i(x) = P_{fc}(x)$.
- $0 \leqslant P_{fc}(x) \leqslant \cdots \leqslant P_{i+1}(x) \leqslant P_i(x) \leqslant \cdots \leqslant P_0(x)$.

Finally, the system under the NPA can be written as follows:

$$\begin{cases} \dot{x} = A(x)x + B_n(x)u_n, & t \in [t_0, t_f[\\ \dot{x} = A(x)x + B_f(x)u_n, & t \in [t_f, t_{fdi}[\\ \dot{x} = A(x)x + B_f(x)u_n, & t \in [t_{fdi}, t_{i0}[\\ \dot{x} = A(x)x + B_f(x)u_0, & t \in [t_0, t_1[\\ \vdots & \\ \dot{x} = A(x)x + B_f(x)u_i, & t \in [t_i, t_{i+1}[\quad i = 1, 2 \dots N \\ \vdots & \\ \dot{x} = A(x)x + B_f(x)u_{fc}, & t \in [t_{fc}, \infty[\end{cases} \tag{31}$$

8 Illustrative Example

In this section, we apply the NPA procedure for the accommodation of a second-order NL system (32). We adopt the same notations used in previous sections.

$$\dot{x} = \begin{pmatrix} -x_1^3 - x_2 \\ x_1 + x_2 \end{pmatrix} + \begin{pmatrix} 0 \\ 1 \end{pmatrix} u \tag{32}$$

- **Nominal conditions**

We first apply the SDRE strategy to control the system (32) in nominal conditions. To do so, we need to obtain the form (22) by defining the SDC matrices (33).

$$\begin{aligned} A_n(x) &= \begin{pmatrix} -x_1^2 & -1 \\ 1 & 1 \end{pmatrix} \\ B_n(x) &= \begin{pmatrix} 0 \\ 1 \end{pmatrix} \end{aligned} \tag{33}$$

The controllability condition (23) is verified since $rank(C_{trb}(x)) = 2,\ \forall x$. By choosing $Q(x) = I_2$ and $R(x) = 1$, we can solve the SDRE (26) to obtain the nominal controller $u_n(x)$. Strating from the IC $x_0 = (4.3 \quad 4.3)$, we obtain the outputs as shown in Fig. 7. The system is stabilized after $t = 4s$.

- **Faulty conditions**

We suppose an actuator failure occuring at $t = 0.2s$. To see its impact on the system, we apply the control $u_n(x)$ such that:

$$\dot{x} = A_n(x)x + B_f(x)u_n, \qquad t \in [t_f, \infty[\tag{34}$$

where $B_f(x) = \begin{pmatrix} 0 \\ 0.4 \end{pmatrix}$. The system starts oscillating as showcased in Fig. 8.

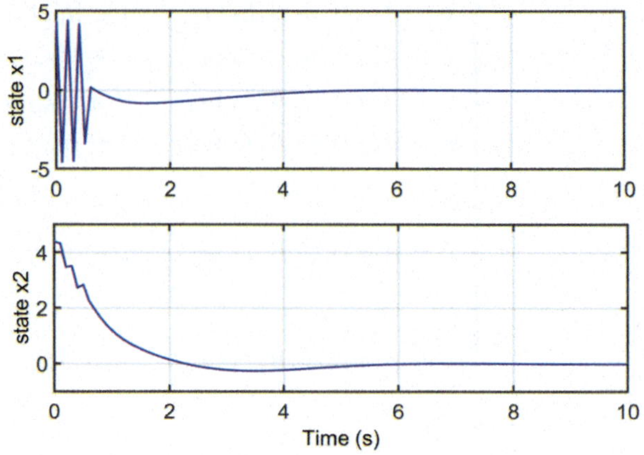

Fig. 7 Outputs in nominal conditions

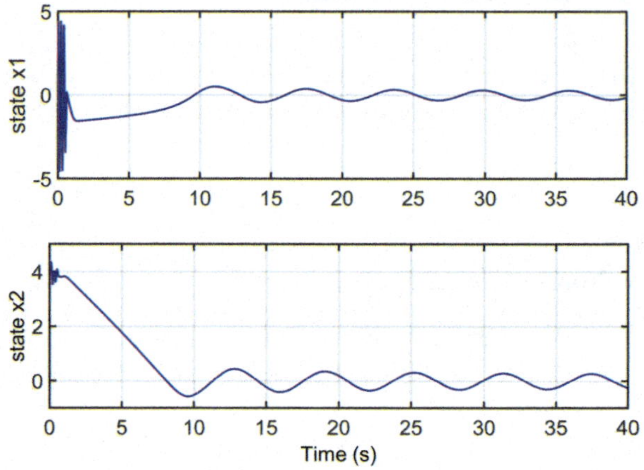

Fig. 8 Outputs in faulty conditions

- **Accommodated system**

The objective of the accommodation procedure is to re-stabilize the system and to acheive a performance that is close to the nominal case. Before applying the NPA, we first try to stabilize the outputs using the direct FTC as described in Sect. 2.2. The optimal control $u_{fc}(x)$ is computed using the SDRE, while considering the matrix $B_f(x)$. As expected, the classical procedure, Fig. 9, could not accommodate the defected system that diverges at $t = 0.9s$.

To apply the NPA, as described in Fig. 3, we start with the initial gain $F_0(x) = F_n(x)$ and we apply it at $t = 0.3s$. Next, Eq. (28) is solved and $u_i(x)$ is applied on $[t_i,\ t_{i+1}]$. When gains $F_i(x)$ (29) change no more, the algorithm stops. The results in Fig. 10 show that the outputs converge to 0 at $t = 8s$ and the system behaviour is close to the nominal case which proves the efficiency of the NPA.

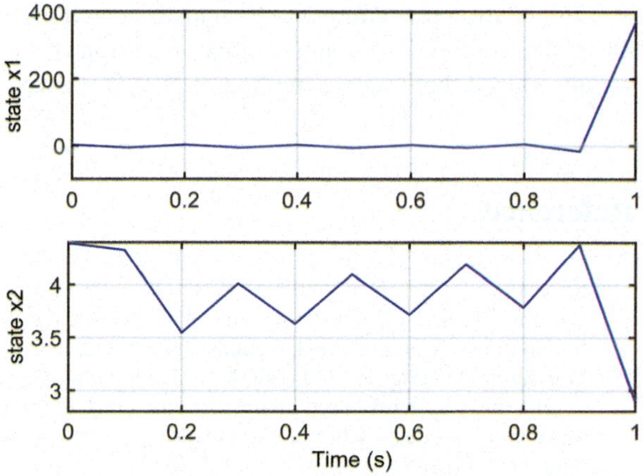

Fig. 9 Outputs under classical FTC

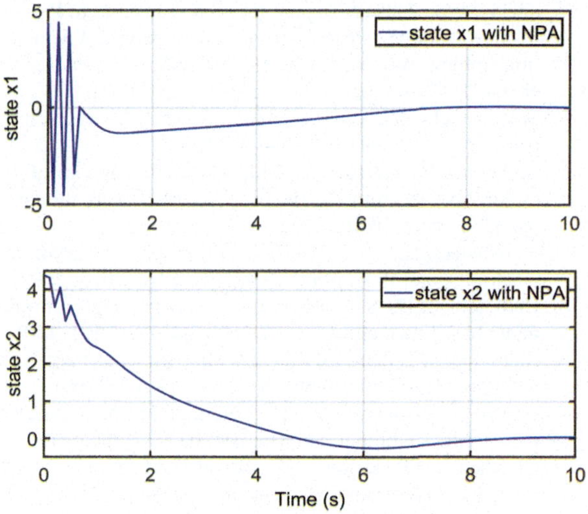

Fig. 10 Outputs under NPA

9 Conclusion

In the fault accommodation procedure, the computation time of the FTC can represent a serious risk of instability to the system. The PA was proven to be an effective solution to this problem.

In this chapter, we first introduced the PA concept and objectives in case of linear models. We compared the classical accommodation scheme (using the direct solution of the optimal LQR), to the PA scheme (using the iterative algorithm of (NK)). Then, we presented the limitations encountered with the linear PA when it comes to using NL system models. The necessity of using NL control methods in the accommodation procedure was then justified and motivated. After that, we presented the SDRE method, as an approximate solution of the NL control problem. Finally, we introduced the iterative method of NPA, followed by an illustrative example.

References

1. Khedher, A., Elleuch, I., Othman, K.B.: Observer desing for non linear takagi-sugeno fuzzy systems. Application to fault tolerant control. In: Advances in Observer Design and Observation for Nonlinear Systems: Fundamentals and Applications, pp. 101–133. Springer (2022)
2. Youssef, I., Khedher, A., Othman, K.B.: Fault tolerant control for non linear uncertain takagi-sugeno fuzzy systems based on proportional integral sliding mode observer. In: 2021 International Conference on Control, Automation and Diagnosis (ICCAD), pp. 1–5 (2021). https://doi.org/10.1109/ICCAD52417.2021.9638771
3. Khedher, A., Elleuch, I., BenOthman, K.: Adaptive proportional integral observer design for interval type 2 takagi-sugeno fuzzy systems. Math. Probl. Eng. **2021**, 1–13 (2021)
4. Staroswiecki, M.: Progressive accommodation of actuator faults in the linear quadratic control problem. In: 2004 43rd IEEE Conference on Decision and Control (CDC) (IEEE Cat. No.04CH37601), vol. 5, pp. 5234–52415 (2004)
5. Staroswiecki, M., Yang, H., Jiang, B.: Progressive accommodation of aircraft actuator faults. IFAC Proc. Vol. **39**(13), 825–830 (2006)
6. Ciubotaru, B., Staroswiecki, M.: Fault tolerant control of the B747 short-period mode using progressive accommodation. In: 2006 IEEE International Conference on Control Applications, pp. 3288–3294. IEEE (2006)
7. Staroswiecki, M., Yang, H., Jiang, B.: Progressive accommodation of parametric faults in linear quadratic control. Automatica **43**(12), 2070–2076 (2007)
8. Ciubotaru, B.D., Staroswiecki, M.: Anytime algorithm for parametric faults accommodation under handling quality constraints. IFAC Proc. Vol. **42**(8), 887–892 (2009)
9. Raharijaona, T., Outbib, R., Ouladsine, M., Ghachem, S.: Comments on actuator fault accommodation. IFAC Proc. Vol. **42**(8), 1611–1616 (2009)
10. Ghachem, S., Benothman, K., Benrejeb, M.: Nonlinear approach of actuator fault accommodation. Int. J. Comput. Sci. Iss. (IJCSI) **8**(4), 290 (2011)
11. Ghachem, S., Benothman, K., Benrejeb, M.: Actuator fault accommodation with optimal control for a class of nonlinear systems. Int. Rev. Autom. Control **5**(2), 274–283 (2012)
12. Ghachem, S., Benothman, K., Benrejeb, M.: Nonlinear progressive accommodation of actuator fault with optimal control. In: 10th International Multi-Conferences on Systems, Signals & Devices 2013 (SSD13), pp. 1–6. IEEE (2013)
13. Ghachem, S., Othman, K., Benrejeb, M.: Nonlinear progressive accommodation of actuator fault. Syst. Autom. Control 1–16 (2017). https://doi.org/10.1515/9783110470468-001
14. Mlayeh, H., Ghachem, S., Nasri, O., Benothman, K.: Accommodation of successive actuator failures with optimal control for a class of nonlinear systems. In: 2016 17th International Conference on Sciences and Techniques of Automatic Control and Computer Engineering (STA), pp. 217–222. IEEE (2016)
15. Mlayeh, H., Ghachem, S., Nasri, O., Benothman, K.: Nonlinear fault tolerant systems with multiple actuator failures using progressive accommodation. In: 2017 International Conference on Control, Automation and Diagnosis (ICCAD), pp. 396–401. IEEE (2017)
16. Mlayeh, H., Ghachem, S., Nasri, O., Othman, K.B.: Nonlinear fault recovery of a quadrotor using progressive accommodation. Stud. Inf. Control **29**(1), 67–75 (2020)
17. Mlayeh, H., Ghachem, S., Nasri, O., Othman, K.B.: Stabilization of a quadrotor vehicle using PD and recursive nonlinear control techniques. Int. Rev. Aerosp. Eng. (IREASE) **14**(4) (2021). https://doi.org/10.15866/irease.v14i4.19739

18. Mlayeh, H., Othman, K.B.: Attitude control of a drone subject to disturbance and structural fault. In: 2022 IEEE 21st International Conference on Sciences and Techniques of Automatic Control and Computer Engineering (STA), pp. 232–236. IEEE (2022)
19. Mlayeh, H., Ben Othman, K.: Nonlinear accommodation of a dc-8 aircraft affected by a complete loss of a control surface. Stud. Inf. Control **31**(3), 107–116 (2022). https://doi.org/10.24846/v31i3y202210
20. Kleinman, D.: On an iterative technique for riccati equation computation. IEEE Trans. Autom. Control **13**(1), 114–115 (1968). https://doi.org/10.1109/TAC.1968.1098829
21. Feitzinger, F., Hylla, T., Sachs, E.W.: Inexact kleinman-newton method for riccati equations. SIAM J. Matrix Anal. Appl. **31**(2), 272–288 (2009)
22. Banks, H., Ito, K.: A numerical algorithm for optimal feedback gains in high dimensional linear quadratic regulator problems. SIAM J. Control Optim. **29**(3), 499–515 (1991)
23. Slotine, J.-J.E., Li, W. et al.: Applied Nonlinear Control, vol. 199. Prentice Hall Englewood Cliffs, NJ (1991)
24. Kalise, D., Kunisch, K.: Polynomial approximation of high-dimensional hamilton-jacobi-bellman equations and applications to feedback control of semilinear parabolic pdes. SIAM J. Sci. Comput. **40**(2), 629–652 (2018)
25. Bellman, R.: Dynamic Programming, 1st edn. Princeton University Press, Princeton, NJ, USA (1957)
26. Xueping Li, Song, Q.S.: Markov chain approximation methods on generalized hjb equation. In: 2007 46th IEEE Conference on Decision and Control, pp. 4069–4074 (2007)
27. Wang, Z., Li, Y.: An adaptive cross approximation method for the hamilton-jacobi-bellman equation. IFAC-PapersOnLine **50**(1), 6289–6294 (2017)
28. Kharroubi, I., Langrené, N., Pham, H.: Discrete time approximation of fully nonlinear hjb equations via bsdes with nonpositive jumps. Ann. Appl. Probab. **25** (2013). https://doi.org/10.1214/14-AAP1049
29. Mu, C., Wang, D., He, H.: Data-driven finite-horizon approximate optimal control for discrete-time nonlinear systems using iterative hdp approach. IEEE Trans. Cybern. **48**(10), 2948–2961 (2017)
30. Korayem, A.H., Nekoo, S.R., Korayem, M.H.: Optimal sliding mode control design based on the state-dependent riccati equation for cooperative manipulators to increase dynamic load carrying capacity. Robotica **37**(2), 321–337 (2019). https://doi.org/10.1017/S0263574718001030
31. Pearson, J.D.: Approximation methods in optimal control i. sub-optimal control†. J. Electr. Control **13**(5), 453–469 (1962). https://doi.org/10.1080/00207216208937454
32. Steinfeldt, B., Tsiotras, P.: A state-dependent riccati equation approach to atmospheric entry guidance (2010). https://doi.org/10.2514/6.2010-8310

Linear Methods for Stabilization and Synchronization h-Fractional Chaotic Maps

Abderrahmane Abbes and Adel Ouannas

Abstract In this chapter, we investigate the stabilization and synchronization of different fractional chaotic maps described by the Caputo h-difference operator using linear control laws. First, we proposed different chaotic maps and presented some simple linear control laws intended to control and synchronize the dynamics of the maps. In addition, using adaptive linear controllers, we introduce different suitable schemes to achieve synchronization between h-fractional chaotic maps. All findings were obtained using theorems based on the Lyapunov approach and the characteristics of the Caputo h-difference operator. Finally, the validity of the developed linear controllers is verified by implementing relevant simulation results.

Keywords Caputo h-difference operator · Linear control laws · Stabilization · Synchronization

1 Introduction

Chaos control and synchronization are important concepts in the field of nonlinear dynamics, which deals with the study of complex systems that exhibit unpredictable and irregular behavior [8, 48, 49, 56, 61–64, 64–73, 95, 104]. In recent years, there has been a growing interest in the study of fractional discrete systems, which are a type of dynamic system that involves fractional calculus. The study of chaos control and synchronization of fractional discrete systems has important applications in various fields, including engineering, physics, and biology. For example, these techniques can be used to control and synchronize the behavior of chaotic circuits, chaotic

A. Abbes (✉)
Department of Mathematics, University Badji mokhtar, 23000 Annaba, Algeria
e-mail: abderrahmane.abbes@univ-annaba.dz

A. Ouannas
Department of Mathematics and Computer Science, University of Larbi Ben M'hidi, 04000 Oum El Bouaghi, Algeria
e-mail: ouannas.adel@univ-oeb.dz

A. Ben Makhlouf et al. (eds.), *State Estimation and Stabilization of Nonlinear Systems*, Studies in Systems, Decision and Control 491,
https://doi.org/10.1007/978-3-031-37970-3_14

oscillators, and chaotic biological systems [15, 28, 37, 42, 42, 45–47, 58, 59, 59, 60, 66, 73–78, 97, 98, 101].

Fractional calculus is a branch of mathematics that deals with the generalization of traditional calculus to include non-integer order derivatives and integrals. Fractional discrete systems are mathematical models that describe the behavior of dynamic systems with discrete-time fractional calculus [2–4, 6, 10, 11, 14, 21–24, 27, 30–34, 36, 38–41, 60, 85, 92, 99, 100].

Chaos control is the process of manipulating the dynamics of a chaotic system to prevent it from exhibiting unpredictable behavior. This is achieved by applying a small perturbation to the system that can stabilize its behavior and prevent it from diverging. In the case of fractional discrete systems, chaos control can involve the application of various control strategies such as feedback control, adaptive control, and sliding mode control [5, 9, 13, 18, 22, 29, 35, 35, 54, 55, 79, 79, 80, 80, 86, 86, 87, 93, 94]. Synchronization is the process of making two or more chaotic systems behave identically or very similarly. This is achieved by using a coupling scheme that connects the systems, which can involve various coupling functions such as linear, nonlinear, and delayed coupling. In the case of fractional discrete systems, synchronization can involve the design of appropriate coupling schemes that can ensure the synchronization of the systems [19, 43, 44, 51, 53, 81, 81–83, 88–91, 104, 105].

Chaos control and synchronization of fractional discrete systems can be achieved using various control strategies, including linear control laws. A linear control law is a type of feedback control that uses a linear relationship between the input and output of the system to stabilize its behavior. In the chaos control of fractional discrete systems using linear control law, the control input is proportional to the difference between the desired and actual state of the system. The effectiveness of linear control law in controlling chaos in fractional discrete systems depends on the system's properties and the control parameters. The control parameters must be selected carefully to ensure the stability of the system while preventing it from exhibiting chaotic behavior [9, 12, 12, 18, 18, 26, 50, 96, 102, 106, 107].

Synchronization of fractional discrete systems using linear control law involves the design of appropriate coupling schemes that can ensure the synchronization of the systems. The coupling scheme can be designed using a linear control law that connects the states of the systems and drives them toward synchronization. The synchronization of fractional discrete systems using linear control law has been successfully applied in various applications, such as secure communication systems, power electronics, and image encryption systems. The use of linear control law for chaos control and synchronization of fractional discrete systems provides a simple and effective approach that can be easily implemented in practice. However, the choice of the appropriate linear control technique and the control parameters remains an active area of research in the field of nonlinear dynamics [1, 26, 52, 57, 84, 103].

In this chapter, we will introduce different linear control laws for controlling and synchronizing the dynamics of certain chaotic maps based on Caputo h-differences operator. This objective is achieved by introducing novel theorems that exploit the Lyapunov-based method. The chapter is organized as follows. Section 2 introduces the definition of the Caputo h-difference operator and useful results related to the Lyapunov stability. In Sect. 3 linear control laws are proposed to stabilize the dynamics of the considered fractional maps at the origin. In Sect. 4 different schemes are proposed, which enables synchronization to be achieved between master and slave system based on identical chaotic maps. Finally, simulation results are reported to highlight the effectiveness of the conceived approach.

2 Fractional h-Difference Operator

In this section, we will review some of the basic concepts and definitions related to the fractional h-difference operators [17].

Definition 1 ([25]) Let $h > 0$ and $(h\mathbb{N})_\theta = \{\theta, \theta + h, \theta + 2h, \ldots\}$ $(\theta \in \mathbb{R})$. The difference operator Δ_h is defined by

$$\Delta_h z(s) = \frac{z(s+h) - z(s)}{h}, \tag{1}$$

where $z : (h\mathbb{N})_\theta \to \mathbb{R}$.

Definition 2 The h-falling fractional function defined as

$$s_h^{(\gamma)} = h^\gamma \frac{\Gamma(\frac{s}{h} + 1)}{\Gamma(\frac{s}{h} + 1 - \gamma)} \tag{2}$$

Definition 3 The γth order fractional h-summation is given by

$$_h\Delta_\theta^{-\gamma} g\,(s) == \frac{h}{\Gamma\,(\gamma)} \sum_{\tau=\frac{\theta}{h}}^{\frac{s}{h}-\gamma} (s - (\tau + 1)h)_h^{(\gamma-1)}\, \Delta g\,(\tau h)\,, \quad s \in (h\mathbb{N})_{0+h\gamma} \tag{3}$$

Next, we define the Caputo-like h-difference operator.

Definition 4 ([7]) The Caputo-like h-difference operator is defined by

$$^C_h\Delta_\theta^\gamma g(s) = {_h\Delta_\theta^{-(m-\gamma)}} \Delta_h^m g(s) \tag{4}$$

for $\gamma \notin \mathbb{N}$ is the fractional order, $s \in (h\mathbb{N})_{\theta+(m-\gamma)h}$ and $m = \lceil\gamma\rceil + 1$.

Now, we will present the following Theorem with the aim of identifying the stability of discrete fractional systems based on the Caputo fractional h-difference operator.

Theorem 1 ([16]) *The zero equilibrium point of the nonlinear h-fractional difference system*

$$ {}_{h}^{C}\Delta_{\theta}^{\gamma} z(s) = g(s + h\gamma, z(s + h\gamma)), \tag{5}$$

where $0 < \gamma \leq 1$ *and* $\forall s \in (h\mathbb{N})_{\theta+(1-\gamma)h}$, *is* ***asymptotically stable*** *if there exists a positive definite and decrescent scalar function* $V(s, z(s))$ *such that*

$$ {}_{h}^{C}\Delta_{\theta}^{\gamma} V(s, z(s)) \leq 0. \tag{6}$$

Lemma 1 *For any discrete time* $s \in (h\mathbb{N})_{\theta+(1-\gamma)h}$, *we have*

$$ {}_{h}^{C}\Delta_{\theta}^{\gamma} z^2(s) \leq 2z(s + h\gamma) {}_{h}^{C}\Delta_{\theta}^{\gamma} z(s), \quad 0 < \upsilon \leq 1. \tag{7}$$

3 Linear Methods for Chaos Stabilisation of h-Fractional Maps

In general, the capacity to control or stabilize chaotic dynamical systems is of particular interest. "Control" refers to adding additional adaptively updated terms to the chaotic system so that its states are close to zero. In this section, Here, we present linear control laws to control chaotic fractional discrete systems defined by the Caputo h-difference operator. In addition, the Lyapunov direct method is used to prove the asymptotic convergence of the controllers.

Consider the following fractional map

$$ {}_{h}^{C}\Delta_{\theta}^{\gamma} z(s) = g\left(z(s + h\gamma)\right). \tag{8}$$

Let $C(s)$ be the adaptive controller. Then the controlled fractional discrete system (8) is formulated as

$$ {}_{h}^{C}\Delta_{\theta}^{\gamma} z(s) = g\left(z(s + h\gamma)\right) + C(s + h\gamma). \tag{9}$$

By using the stability conditions of the h-fractional discrete system reported in Theorem 1, we can investigate that the zero fixed point of the system (9) is asymptotically stable.

3.1 *Stabilization of h-Fractional Ikeda Map Using Scalar Control Law*

Consider the following h-fractional Ikeda map [12]

$$\begin{cases} {}_{h}^{C}\Delta_{\theta}^{\gamma} z_1(s) &= 1 + \Big(z_1(s+h\gamma)\cos f(s+h\gamma) - z_2(s+h\gamma)\sin f(s+h\gamma)\Big)\alpha \\ & \quad -z_1(s+h\gamma), \\ {}_{h}^{C}\Delta_{\theta}^{\gamma} z_2(s) &= \Big(z_1(s+h\gamma)\sin f(s+h\gamma) + z_2(s+h\gamma)\cos f(s+h\gamma)\Big)\alpha \\ & \quad -z_2(s+h\gamma), \end{cases} \tag{10}$$

where α is arbitrary and $f(s) = 0.4 - \frac{6}{1+z_1^2(s)+z_2^2(s)}$. In order to figure out the dynamic the map, the following numerical formulas are formulated:

$$\begin{cases} z_1(m) = z_1(0) + \dfrac{h^\gamma}{\Gamma(\gamma)} \displaystyle\sum_{l=0}^{m-1} \dfrac{\Gamma(m+1-l+\gamma)}{\Gamma(m+1-l-1)} \Big(1 + \big(z_1(l)\cos f(l) - z_2(l)\sin f(l)\big)\alpha \\ \qquad\qquad - z_1(l)\Big), \\ z_2(m) = z_2(0) + \dfrac{h^\gamma}{\Gamma(\gamma)} \displaystyle\sum_{l=0}^{m-1} \dfrac{\Gamma(m+1-l+\gamma)}{\Gamma(m+1-l-1)} \Big(\big(z_1(l)\sin f(l) + z_2(l)\cos f(l)\big)\alpha \\ \qquad\qquad - z_2(l)\Big), \quad m = 1, 2, \ldots. \end{cases} \tag{11}$$

These implicit equations are used to investigate the chaotic behaviors of the fractional Ikeda map (10). When $h = 0.1$, $\alpha = 0.9$ and the initial states $z_1(0) = 0$, $z_2(0) = 0$, then the h-fractional Ikeda map (10) show a chaotic behavior, as can be seen in the bifurcation diagram and the phase portraits presented in Fig. 1.

The linear control law for the fractional Ikeda map (10) is presented in the following theorem.

Theorem 2 *The h-fractional Ikeda map* (10) *is stabilized subject to the following scalar control law*

$$C = -1. \tag{12}$$

Proof The controlled fractional Ikeda map can be described as:

$$\begin{cases} {}_{h}^{C}\Delta_{\theta}^{\gamma} z_1(s) &= 1 + \Big(z_1(s+h\gamma)\cos f(s+h\gamma) - z_2(s+h\gamma)\sin f(s+h\gamma)\Big)\alpha \\ & \quad -z_1(s+h\gamma) + C, \\ {}_{h}^{C}\Delta_{\theta}^{\gamma} z_2(s) &= \Big(z_1(s+h\gamma)\sin f(s+h\gamma) + z_2(s+h\gamma)\cos f(s+h\gamma)\Big)\alpha \\ & \quad -z_2(s+h\gamma), \end{cases} \tag{13}$$

Substituting the Eq. (12) into (13), we obtain

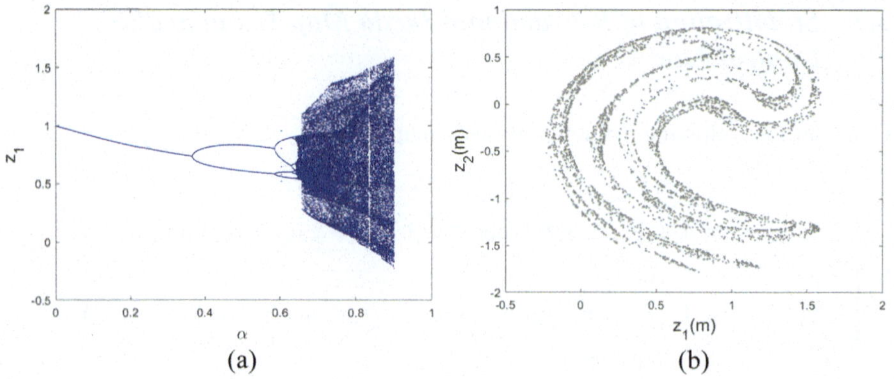

Fig. 1 **a** Bifurcation versus of the h-fractional Ikeda map (10). α **b** Chaotic attractor for $\alpha = 0.9$ and $\gamma = 0.9$

$$\begin{cases} {}_{h}^{C}\Delta_{\theta}^{\gamma} z_1(s) &= \Big(z_1(s+h\gamma)\cos f(s+h\gamma) - z_2(s+h\gamma)\sin f(s+h\gamma)\Big)\alpha - z_1(s+h\gamma), \\ {}_{h}^{C}\Delta_{\theta}^{\gamma} z_2(s) &= \Big(z_1(s+h\gamma)\sin f(s+h\gamma) + z_2(s+h\gamma)\cos f(s+h\gamma)\Big)\alpha - z_2(s+h\gamma), \end{cases} \tag{14}$$

Now we need to demonstrate that the zero fixed point of (14) is asymptotically stable. In that case, we can determine that the controlled system's states provided in (13) will necessarily converge to zero. However, to accomplish this goal, we can use the Lyapunov method, which was previously outlined in Theorem 1. To show this, consider the following Lyapunov function:

$$V(S) = \frac{1}{2}\left(z_1^2(s) + z_2^2(s)\right), \quad s \in (h\mathbb{N})_{\theta+(1-\gamma)h}. \tag{15}$$

So, using the Caputo h-difference operator on (15), we get the following statement:

$${}_{h}^{C}\Delta_{\theta}^{\gamma} V(s) = \frac{1}{2}\,{}_{h}^{C}\Delta_{\theta}^{\gamma} z_1^2(s) + \frac{1}{2}\,{}_{h}^{C}\Delta_{\theta}^{\gamma} z_2^2(s). \tag{16}$$

Using Lemma 1, we obtain

$$\begin{aligned} {}_{h}^{C}\Delta_{\theta}^{\gamma} V &\leq z_1(s+h\gamma)\,{}_{h}^{C}\Delta_{\theta}^{\gamma} z_1(s) + z_2(s+h\gamma)\,{}_{h}^{C}\Delta_{\theta}^{\gamma} z_2(s) \\ &= \Big(z_1^2(s+h\gamma)\cos f(s+h\gamma) - z_1(s+h\gamma)z_2(s+h\gamma)\sin f(s+h\gamma)\Big)\alpha - z_1^2(s+h\gamma) \\ &\quad + \Big(z_2(s+h\gamma)z_1(s+h\gamma)\sin f(s+h\gamma) + z_2^2(s+h\gamma)\cos f(s+h\gamma)\Big)\alpha - z_2^2(s+h\gamma) \\ &= \alpha z_1^2(s+h\gamma)\cos f(s+h\gamma) - z_1^2(s+h\gamma) + \alpha z_2^2(s+h\gamma)\cos f(s+h\gamma) - z_2^2(s+h\gamma) \\ &\leq (\alpha-1)z_1^2(s+h\gamma) + (\alpha-1)z_2^2(s+h\gamma) < 0, \quad \text{(That is because } \alpha = 0.9\text{).} \end{aligned}$$

Hence, we can conclude that the scalar control law stabilizes the controlled states of the fractional Ikeda map (10).

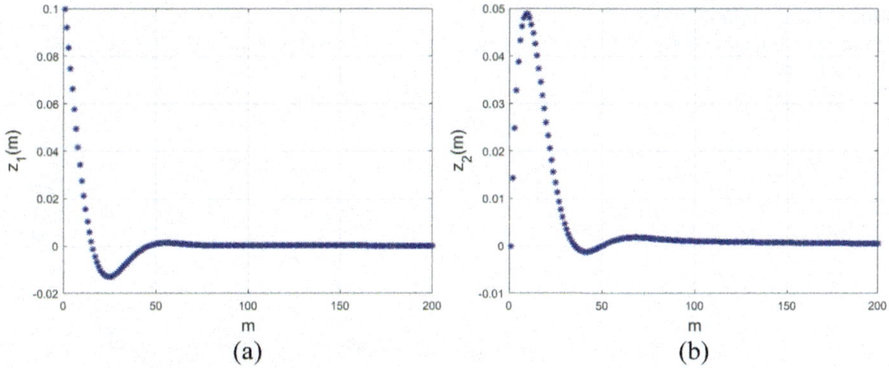

Fig. 2 Time evolution of the states of the controlled h-fractional Ikeda map (10)

In order to verify this result, The states of the controlled h-fractional Ikeda map are shown in Fig. 2. Clearly, the system is asymptotically stable and its states converge toward zero.

3.2 *Stabilization of h-Fractional Lorenz Maps Using One-Dimensional Linear Control Law*

Consider the following fractional discrete Lorenz system based on the Caputo h-difference operator [102]

$$\begin{cases} {}_{h}^{C}\Delta_{\theta}^{\gamma} z_1(s) &= \beta z_1(s+h\gamma) - \beta z_1(s+h\gamma) z_2(s+h\gamma), \\ {}_{h}^{C}\Delta_{\theta}^{\gamma} z_2(s) &= \beta \left(z_1^2(s+h\gamma) - z_2(s+h\gamma) \right). \end{cases} \tag{17}$$

Taking the system parameter $\beta = 0.95$, the fractional order $\gamma = 0.9$ and the initial conditions $(z_1(0), z_2(0)) = (0.2, 0.3)$. Figure 3 illustrates the phase attractors of the fractional system (17). It can be observed that the states of the map are chaotic.

Theorem 3 *The 2D fractional discrete system* (17) *is stabilized subject to the following control law*

$$C(s) = -(1+\beta) z_1(s), \quad s \in (h\mathbb{N})_{\theta+(1-\gamma)h}. \tag{18}$$

Proof The controlled fractional discrete system can be described as:

$$\begin{cases} {}_{h}^{C}\Delta_{\theta}^{\gamma} z_1(s) &= \beta z_1(s+h\gamma) - \beta z_1(s+h\gamma) z_2(s+h\gamma) + C(s+h\gamma), \\ {}_{h}^{C}\Delta_{\theta}^{\gamma} z_2(s) &= \beta \left(z_1^2(s-1+\gamma) - z_2(s+h\gamma) \right). \end{cases} \tag{19}$$

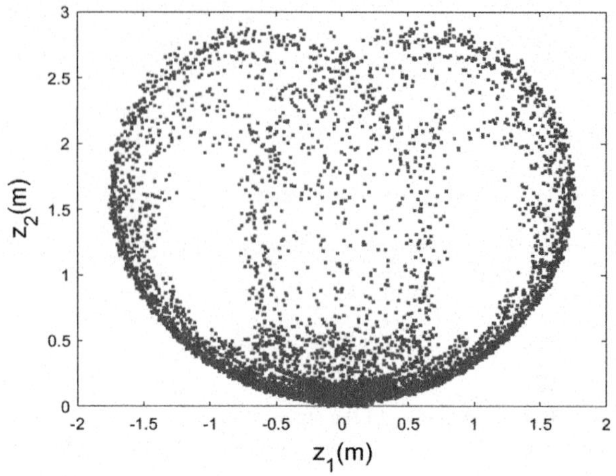

Fig. 3 Phase portrait of the 2D fractional map (17) for $\beta = 0.95$ and $\gamma = 0.9$

Substituting the control law (18) into (19), we obtain

$$\begin{cases} {}_{h}^{C}\Delta_{\theta}^{\gamma} z_1(s) & = -z_1(s+h\gamma) - \beta z_1(s+h\gamma) z_2(s+h\gamma) \\ {}_{h}^{C}\Delta_{\theta}^{\gamma} z_2(s) & = \beta\left(z_1^2(s-1+\gamma) - z_2(s+h\gamma)\right). \end{cases} \tag{20}$$

Now, Consider the following Lyapunov function:

$$V(S) = \frac{1}{2} z_1^2(s) + \frac{1}{2} z_2^2(s), \tag{21}$$

This implies:

$${}_{h}^{C}\Delta_{\theta}^{\gamma} V(s) = \frac{1}{2}\, {}_{h}^{C}\Delta_{\theta}^{\gamma} z_1^2(s) + \frac{1}{2}\, {}_{h}^{C}\Delta_{\theta}^{\gamma} z_2^2(s).$$

Using Lemma 1, we obtain

$$\begin{aligned} {}_{h}^{C}\Delta_{\theta}^{\gamma} V &\leq z_1(s+h\gamma)\, {}_{h}^{C}\Delta_{\theta}^{\gamma} z_1(s) + z_2(s+h\gamma)\, {}_{h}^{C}\Delta_{\theta}^{\gamma} z_2(s) \\ &= -z_1^2(s+h\gamma) - \beta z_1^2(s+h\gamma) z_2(s+h\gamma) + \beta z_1^2(s+h\gamma) z_2(s+h\gamma) - \beta z_2^2(s+h\gamma) \\ &= -z_1^2(s+h\gamma) - \beta z_2^2(s+h\gamma) < 0. \end{aligned}$$

Hence, according to Theorem 1, the system is asymptotically stable.

To illustrate the theoretical results in the previous theorem, numerical simulations are performed. We choose $h = 1$, $\gamma = 0.9$ and the initial values $(z_1(0), z_2(0)) = (0.3, 0.1)$. Figure 4 reports the time evolution of states of the controlled h-fractional system (19) clearly that the errors converge to zero, which confirms the stabilization results.

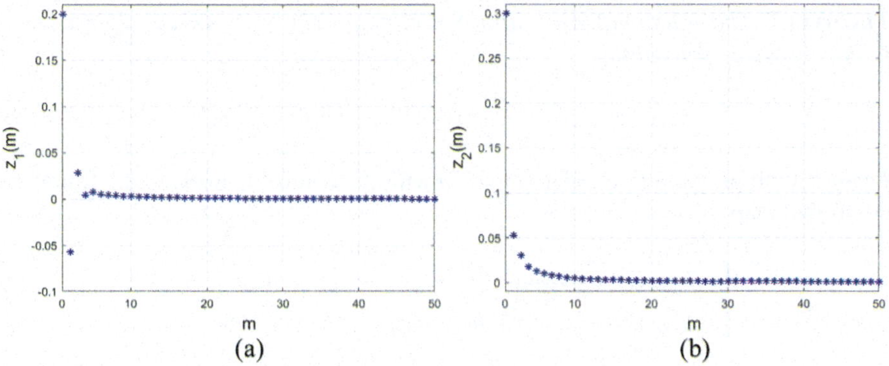

Fig. 4 Time evolution of the states of the controlled fractional map (20)

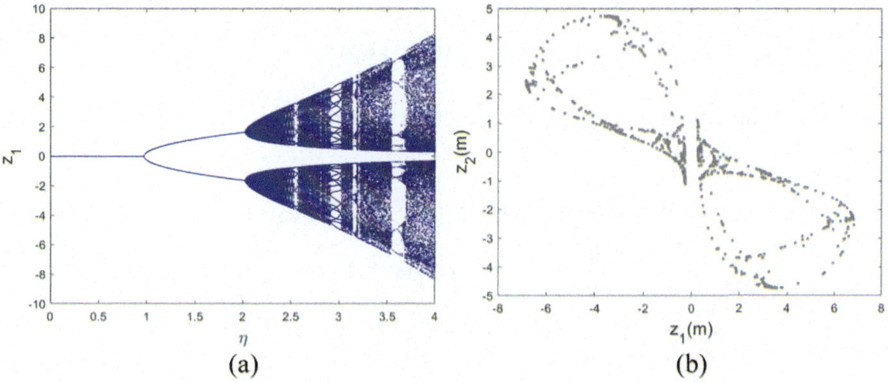

Fig. 5 **a** Bifurcation versus of the h-fractional Ikeda map (10). α **b** Chaotic attractor for $\alpha = 0.9$ and $\gamma = 0.98$

3.3 Stabilization of h-Fractional Zeraoulia–Sprott Rational Map Using One-Dimensional Linear Control Law

Consider the following fractional Zeraoulia–Sprott rational map [12]:

$$\begin{cases} {}_{h}^{C}\Delta_{\theta}^{\gamma} z_1(s) &= \frac{-\eta z_1(s+h\gamma)}{1+z_2^2(s+h\gamma)} - z_1(s+h\gamma), \\ {}_{h}^{C}\Delta_{\theta}^{\gamma} z_2(s) &= z_1(s+h\gamma) + (\alpha-1) z_2(s+h\gamma). \end{cases} \tag{22}$$

Considering parameters $\eta = 0.6$ and $\alpha = 3.8$, generate the phase portrait depicted in Fig. 5. As can be seen, the system does indeed have a chaotic attractor. Now, we intend to present a theorem that demonstrates a one-dimensional controller proposed for controlling the fractional Zeraoulia–Sprott map based on the h-difference operator given in (22).

Theorem 4 *The fractional Zeraoulia–Sprott map (22) is stabilized subject to the following linear control law*

$$C(s) = -z_2(s) - \eta z_1(s), \quad s \in (h\mathbb{N})_{\theta+(1-\gamma)h}. \tag{23}$$

Proof Combining the fractional map (22) with the control C leads to the following controlled map:

$$\begin{cases} {}^C_h\Delta^\gamma_\theta z_1(s) &= \frac{-\eta z_1(s+h\gamma)}{1+z_2^2(s+h\gamma)} - z_1(s+h\gamma) + C(s+h\gamma), \\ {}^C_h\Delta^\gamma_\theta z_2(s) &= z_1(s+h\gamma) + (\alpha-1)z_2(s+h\gamma). \end{cases} \tag{24}$$

By substituting the law C into the system (24), one can get the following simplified form:

$$\begin{cases} {}^C_h\Delta^\gamma_\theta z_1(s) &= \frac{-\eta z_1(s+h\gamma)}{1+z_2^2(s+h\gamma)} - (\eta+1)z_1(s+h\gamma) - z_2(s+h\gamma), \\ {}^C_h\Delta^\gamma_\theta z_2(s) &= z_1(s+h\gamma) + (\alpha-1)z_2(s+h\gamma). \end{cases} \tag{25}$$

By Taking the following Lyapunov function $V(S) = \frac{1}{2}z_1^2(s) + \frac{1}{2}z_2^2(s)$, and Using Lemma 1, we obtain

$$\begin{aligned} {}^C_h\Delta^\gamma_\theta V &\leq z_1(s+h\gamma)\,{}^C_h\Delta^\gamma_\theta z_1(s) + z_2(s+h\gamma)\,{}^C_h\Delta^\gamma_\theta z_2(s) \\ &= \frac{-\eta z_1^2(s+h\gamma)}{1+z_2^2(s+h\gamma)} - (\eta+1)z_1^2(s+h\gamma) - z_1(s+h\gamma)z_2(s+h\gamma) \\ &\quad + z_2(s+h\gamma)z_1(s+h\gamma) + (\alpha-1)z_2^2(s+h\gamma) \\ &\leq \frac{\eta z_1^2(s+h\gamma)}{1+z_2^2(s+h\gamma)} - (\eta+1)z_1^2(s+h\gamma) + (\alpha-1)z_2^2(s+h\gamma) \\ &\leq \eta z_1^2(s+h\gamma) - (\eta+1)z_1^2(s+h\gamma) + (\alpha-1)z_2^2(s+h\gamma) \\ &= -z_1^2(s+h\gamma) + (\alpha-1)z_2^2(s+h\gamma). \end{aligned}$$

Since $\alpha - 1 < 0$, then ${}^C_h\Delta^\gamma_\theta V < 0$. Hence, we can conclude that the fractional Zeraoulia–Sprott rational map (22) is asymptotically stable.

The plots of the phase space and the time evolution of the controlled map (23) are shown in Fig. 6 to illustrate the validity of the suggested linear control law. These graphs clearly illustrate that the one-dimensional linear control law presented in Theorem 4 can stabilize all chaotic states of the fractional Zeraoulia–Sprott map (22).

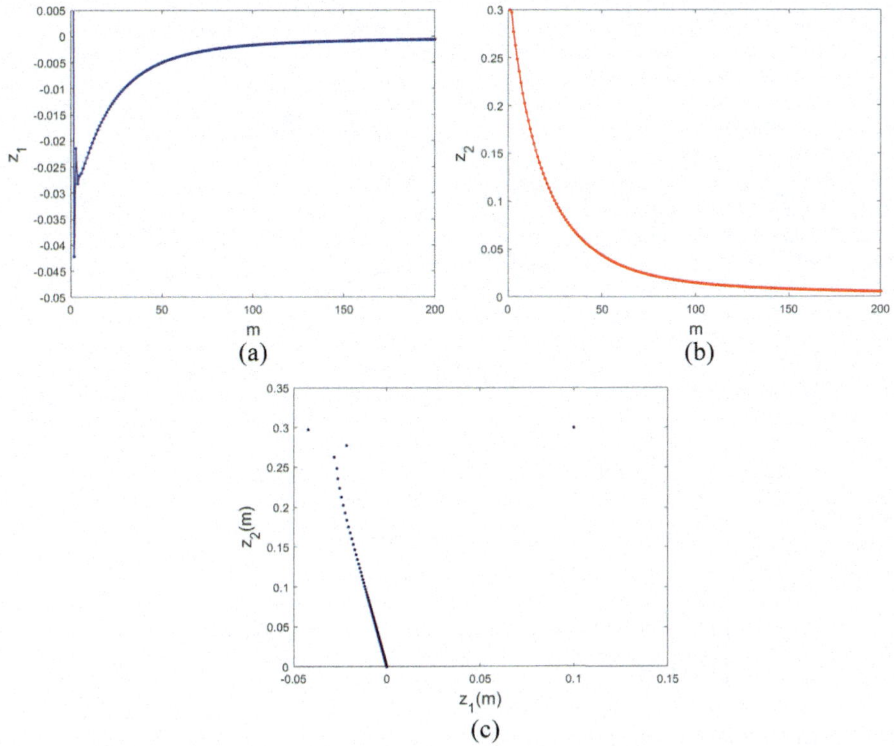

Fig. 6 Stabilization of the states of the controlled h-fractional map (23)

3.4 Stabilization of h-Fractional Generalized Hénon Map Using Two-Dimensional Linear Control Law

Consider the following 3D fractional generalized Hénon Map based on the Caputo h-difference operator [107]:

$$\begin{cases} {}_h^C\Delta_\theta^\gamma z_1(s) &= 1.4 + \beta z_3(s + h\gamma) + \alpha z_2(s + h\gamma) - z_1^2(s + h\gamma) - z_1(s + h\gamma), \\ {}_h^C\Delta_\theta^\gamma z_2(s) &= z_1(s - 1 + \gamma) - z_2(s + h\gamma). \\ {}_h^C\Delta_\theta^\gamma z_3(s) &= z_2(s - 1 + \gamma) - z_3(s + h\gamma). \end{cases} \tag{26}$$

Taking the system parameter $(\alpha, \beta) = (0.2, 0.1)$, the fractional order $\gamma = 0.98$ and the initial conditions $(z_1(0), z_2(0), z_3(0)) = (0.3, 0.3, 0.3)$. Figure 7 illustrates the phase attractors of the 3D fractional system (26). It can be observed that the states of the map are chaotic.

Theorem 5 *The 3D fractional discrete system* (26) *is stabilized subject to the following control law*

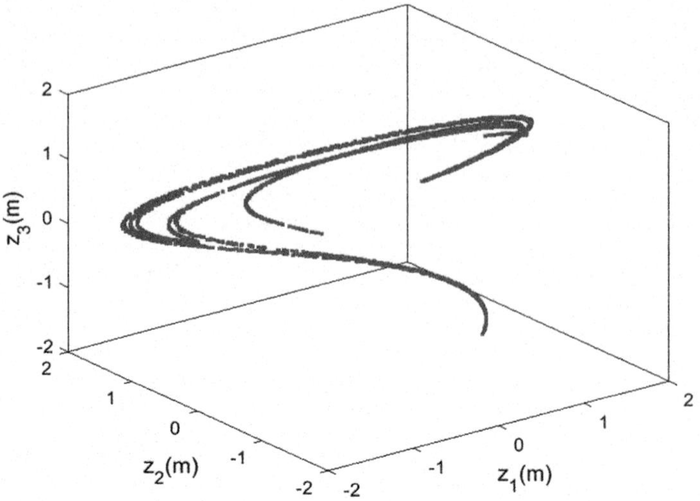

Fig. 7 Phase portraits of the 3D fractional discrete map (26) for $\beta = 0.95$ and $\gamma = 0.9$

$$\begin{cases} C_1(s) = -bz_1(s) - (\alpha + 1)z_2(s) - \beta z_3(s) - 1.4, \\ C_2(s) = -z_3(s), \end{cases} \tag{27}$$

where $|x(s)| \leq b$, $s \in (h\mathbb{N})_{\theta+(1-\gamma)h}$.

Proof The controlled fractional discrete system can be described as:

$$\begin{cases} {}_h^C\Delta_\theta^\gamma z_1(s) &= 1.4 + \beta z_3(s+h\gamma) + \alpha z_2(s+h\gamma) - z_1^2(s+h\gamma) - z_1(s+h\gamma) + C_1(s+h\gamma), \\ {}_h^C\Delta_\theta^\gamma z_2(s) &= z_1(s-1+\gamma) - z_2(s+h\gamma) + C_2(s+h\gamma). \\ {}_h^C\Delta_\theta^\gamma z_3(s) &= z_2(s-1+\gamma) - z_3(s+h\gamma). \end{cases} \tag{28}$$

Substituting the control law (27) into (28), we obtain

$$\begin{cases} {}_h^C\Delta_\theta^\gamma z_1(s) &= -z_2(s+h\gamma) - z_1^2(s+h\gamma) - (b+1)z_1(s+h\gamma), \\ {}_h^C\Delta_\theta^\gamma z_2(s) &= z_1(s-1+\gamma) - z_2(s+h\gamma) - z_3(s+h\gamma). \\ {}_h^C\Delta_\theta^\gamma z_3(s) &= z_2(s-1+\gamma) - z_3(s+h\gamma). \end{cases} \tag{29}$$

Now, Consider the following Lyapunov function:

$$V(s) = \frac{1}{2}z_1^2(s) + \frac{1}{2}z_2^2(s) + \frac{1}{2}z_3^2(s), \tag{30}$$

This implies:

$${}_h^C\Delta_\theta^\gamma V(s) = \frac{1}{2}\,{}_h^C\Delta_\theta^\gamma z_1^2(s) + \frac{1}{2}\,{}_h^C\Delta_\theta^\gamma z_2^2(s) + \frac{1}{2}\,{}_h^C\Delta_\theta^\gamma z_3^2(s).$$

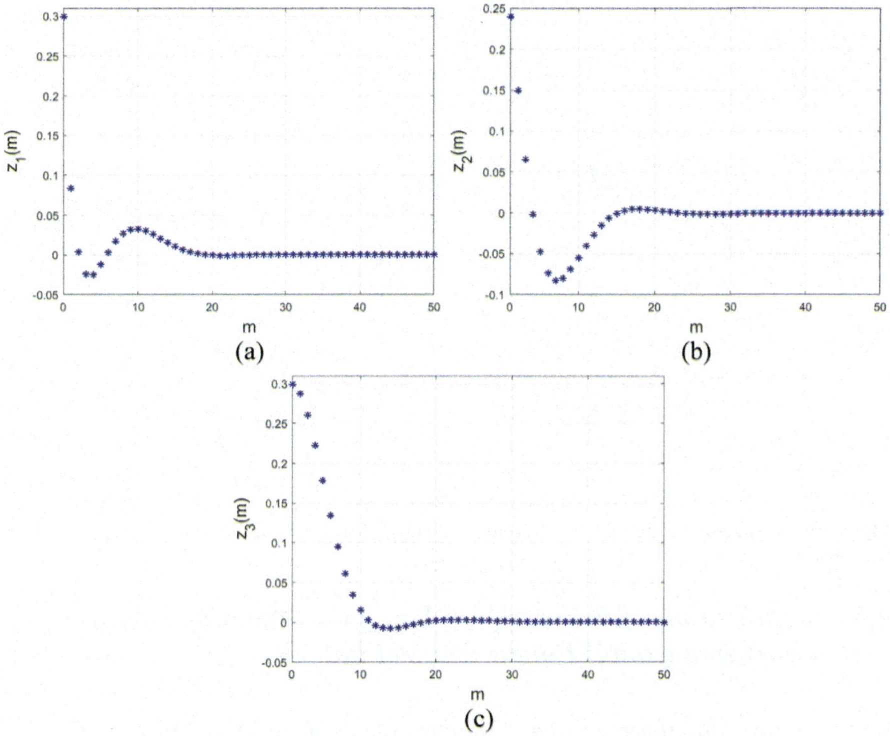

Fig. 8 Time evolution of the states of the controlled fractional map (29)

Using Lemma 1, we obtain

$$\begin{aligned}
{}_h^C\Delta_\theta^\gamma V &\leq z_1(s+h\gamma)_h^C\Delta_\theta^\gamma z_1(s) + z_2(s+h\gamma)_h^C\Delta_\theta^\gamma z_2^2(s) + z_3(s+h\gamma)_h^C\Delta_\theta^\gamma z_3^2(s) \\
&= -z_1(s+h\gamma)z_2(s+h\gamma) - z_1^3(s+h\gamma) - (b+1)z_1^2(s+h\gamma) + z_1(s+h\gamma)z_2(s+h\gamma) \\
&\quad - z_2^2(s+h\gamma) - z_2(s+h\gamma)z_3(s+h\gamma) + z_2(s+h\gamma)z_3(s+h\gamma) - z_3^2(s+h\gamma) \\
&= -(b+1)z_1^2(s+h\gamma) - z_1^3(s+h\gamma) - z_2^2(s+h\gamma) - z_3^2(s+h\gamma) \\
&\leq -(b+1)z_1^2(s+h\gamma) + |z_1(s+h\gamma)|z_1^2(s+h\gamma) - z_2^2(s+h\gamma) - z_3^2(s+h\gamma) \\
&\leq -(b+1)z_1^2(s+h\gamma) + bz_1^2(s+h\gamma) - z_2^2(s+h\gamma) - z_3^2(s+h\gamma) \\
&= -z_1^2(s+h\gamma) - z_2^2(s+h\gamma) - z_3^2(s+h\gamma) < 0.
\end{aligned}$$

Hence, according to Theorem 1, the system is asymptotically stable.

To illustrate the theoretical results in the previous theorem, numerical simulations are performed. Figure 8 reports the time evolution of states of the controlled h-fractional system (28). Clearly that the errors converge to zero indicating that the above results are effective.

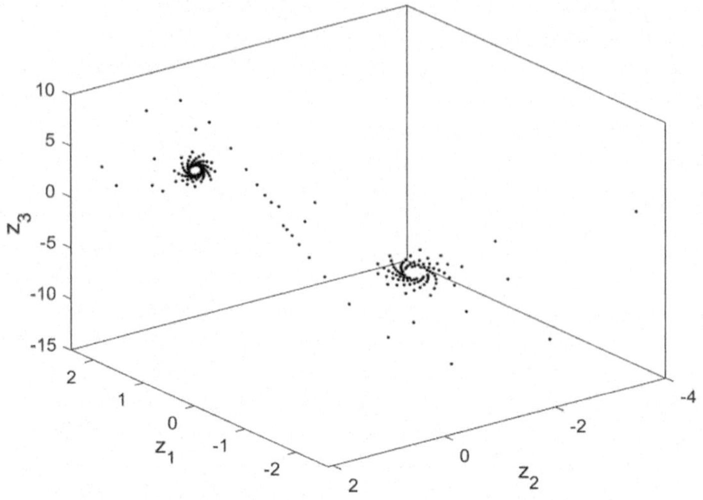

Fig. 9 Phase space portraits of the h-fractional Wang map for $\gamma = 0.97$

3.5 Stabilization of h-Fractional Wang Chaotic Map Using Two-Dimensional Linear Control Law

Consider the h-difference version of the fractional Wang map as follow [102]:

$$\begin{cases} {}_h^C\Delta_\theta^\gamma z_1(s) = a_1 z_1(s+h\gamma) + a_2 z_2(s+h\gamma), \\ {}_h^C\Delta_\theta^\gamma z_2(s) = b_1 z_1(s+h\gamma) + b_2 z_3(s+h\gamma), \\ {}_h^C\Delta_\theta^\gamma z_3(s) = c_1 z_3(s+h\gamma) + c_2 z_2(s+h\gamma) z_3(s+h\gamma) + c_3. \end{cases} \tag{31}$$

Considering the parameters system $(a_1, a_2, b_1, b_2, c_1, c_2, c_3) = (-2.3, 0.5 - 1.9, 0.2, -1.9, -0.6, 2)$ and the initial values $x(0) = 0.5$, $y(0) = 0.6$ and $z(0) = 0.02$, generate the phase portrait depicted in Fig. 9. As can be seen, the system does indeed have a chaotic attractor. Now, we intend to present a theorem that demonstrates a simple linear controller proposed for controlling the fractional Hopfield neural network system based on the h-difference operator given in (31).

Theorem 6 *The h-fractional Wang map* (31) *is stabilized subject to the following linear control law*

$$\begin{cases} C_1(s) = -(a_2 + b_1) z_2(s) - (a_1 + 1) z_1(s), \\ C_2(s) = -c_3 - (|c_2|\lambda) z_3(s) - b_2 z_2(s), \end{cases} \tag{32}$$

where $|y(s)| \leq \lambda$ *and* $s \in (h\mathbb{N})_{\theta+(1-\gamma)h}$.

Proof Combining the fractional map (31) with the control (32) leads to the following controlled system:

$$\begin{cases} {}_{h}^{C}\Delta_{\theta}^{\gamma} z_1(s) = a_1 z_1(s+h\gamma) + a_2 z_2(s+h\gamma) + C_1(s+h\gamma), \\ {}_{h}^{C}\Delta_{\theta}^{\gamma} z_2(s) = b_1 z_1(s+h\gamma) + b_2 z_3(s+h\gamma), \\ {}_{h}^{C}\Delta_{\theta}^{\gamma} z_3(s) = c_1 z_3(s+h\gamma) + c_2 z_2(s+h\gamma) z_3(s+h\gamma) + c_3 + C_3(s+h\gamma). \end{cases} \tag{33}$$

By substituting the laws $C_1,\ C_2,\ C_3$ into the system (33), one can get the following simplified form:

$$\begin{cases} {}_{h}^{C}\Delta_{\theta}^{\gamma} z_1(s) = -z_1(s+h\gamma) - b_1 z_2(s+h\gamma), \\ {}_{h}^{C}\Delta_{\theta}^{\gamma} z_2(s) = b_1 z_1(s+h\gamma) + b_2 z_3(s+h\gamma), \\ {}_{h}^{C}\Delta_{\theta}^{\gamma} z_3(s) = (c_1 - (|c_2|\lambda)) z_3(s+h\gamma) + c_2 z_2(s+h\gamma) z_3(s+h\gamma) - b_2 z_2(s+h\gamma). \end{cases} \tag{34}$$

By Taking the following Lyapunov function $V(S) = \frac{1}{2} z_1^2(s) + \frac{1}{2} z_2^2(s) + \frac{1}{2} z_3^2(s),$, and Using Lemma 1, we obtain

$$\begin{aligned} {}_{h}^{C}\Delta_{\theta}^{\gamma} V &\leq z_1(s+h\gamma) {}_{h}^{C}\Delta_{\theta}^{\gamma} z_1(s) + z_2(s+h\gamma) {}_{h}^{C}\Delta_{\theta}^{\gamma} z_2(s) + z_3(s+h\gamma) {}_{h}^{C}\Delta_{\theta}^{\gamma} z_3(s) \\ &= -z_1^2(s+h\gamma) - b_1 z_1(s+h\gamma) z_2(s+h\gamma) + b_1 z_2(s+h\gamma) z_1(s+h\gamma) \\ &\quad + b_2 z_2(s+h\gamma) z_3(s+h\gamma) + (c_1 - (|c_2|\lambda)) + c_2 z_3(s+h\gamma) z_2(s+h\gamma) z_3^2(s+h\gamma) \\ &\quad - b_2 z_3(s+h\gamma) z_2(s+h\gamma) \\ &= -z_1^2(s+h\gamma) + (c_1 - (|c_2|\lambda)) z_3^2(s+h\gamma) + c_2 z_2(s+h\gamma) z_3^2(s+h\gamma) \\ &\leq -z_1^2(s+h\gamma) + (c_1 - (|c_2|\lambda)) z_3^2(s+h\gamma) + |c_2||z_2(s+h\gamma)| z_3^2(s+h\gamma) \\ &\leq -z_1^2(s+h\gamma) + (c_1 - (|c_2|\lambda)) z_3^2(s+h\gamma) + |c_2|\lambda z_3^2(s+h\gamma) \\ &= -z_1^2(s+h\gamma) + c_1 z_3^2(s+h\gamma) < 0 \text{ (because } c_1 = -1.9). \end{aligned}$$

Hence, we can conclude that the controlled h-fractional Wang map (33) is asymptotically stable.

The plots of the phase space and the time evolution of the controlled system (33) are shown in Fig. 10 to illustrate the validity of the suggested linear control law. These graphs clearly illustrate that the two-dimensional linear control law presented in (32) can stabilize all chaotic states of the fractional Wanf map based on the h-difference operator given in (31).

4 Synchronization in h-Fractional Chaotic Maps with Linear Controllers

Synchronization properties in fractional discrete systems represent an important research subject. The goal of synchronization is to force the synchronization error between the states of the master chaotic system and the states of the slave chaotic systems to converge toward zero. In this section, by using some linear controllers,

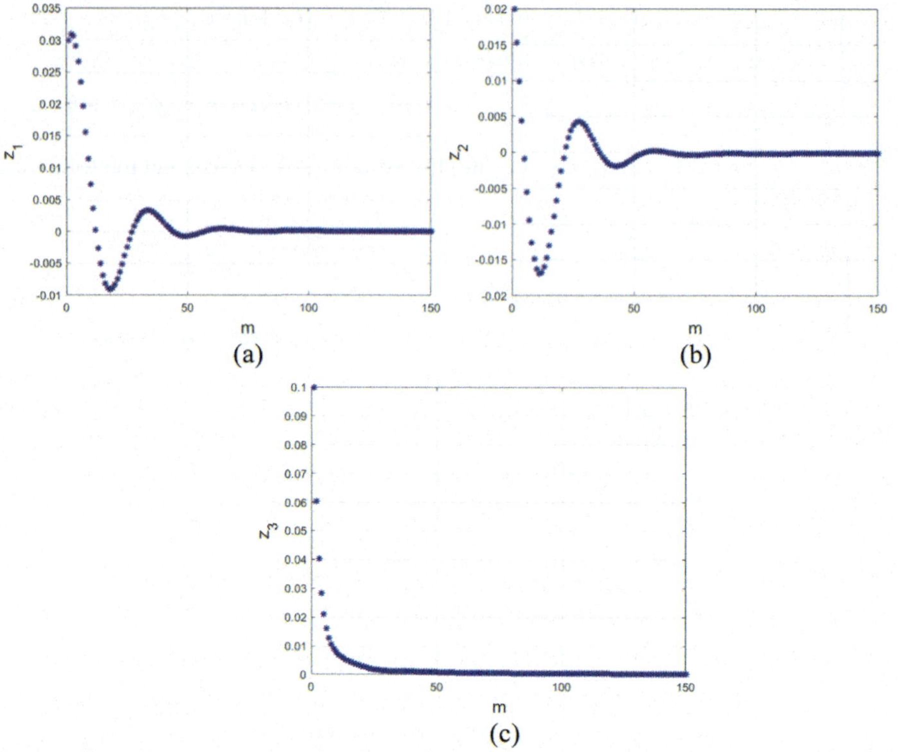

Fig. 10 Time evolution of the states of the controlled fractional map (33)

we present different synchronization types between fractional-order chaotic maps based on the Caputo h-difference operator.

In the following, the master and the slave systems are considered as follows

$$\begin{cases} {}_h^C\Delta_\theta^\gamma x(s) = F(x(s+h\gamma)), \quad s \in (h\mathbb{N})_{\theta+(1-\gamma)h}, \\ {}_h^C\Delta_\theta^\gamma z(s) = G(z(s+h\gamma)) + C, \end{cases} \tag{35}$$

where $x(s) = (x_1(s), x_2(s), \ldots, x_n(s)^T \in \mathbb{R}^n$ is the master state vector and $z(s) = (Z_1(s), z_2(s), \ldots, z_m(s)^T \in \mathbb{R}^m$ is the slave state vector. $F : \mathbb{R}^n \to \mathbb{R}^n$, $G : \mathbb{R}^m \to \mathbb{R}^m$ are nonlinear functions and C is the vector controller to be determined. In the following, we present different types of synchronization for fractional discrete system.

Definition 5 ([20]) If there exists a controller $C = (C_1, C_2, \ldots, C_m)$ and either constants $\nu \in \mathbb{R}^*$, a matrix $M = (m_{ij}) \in \mathbb{R}^{m\times n}$, a map $\Phi : \mathbb{R}^n \to \mathbb{R}^m$, a matrix $L = (L_{ij}) \in \mathbb{R}^{n\times m}$ or a map $\Psi : \mathbb{R}^m \to \mathbb{R}^n$ such that

$\lim_{s\to\infty} \|z(s) - \nu x(s)\| = 0$	$\Longrightarrow$	The pair (35) is projective synchronized (PS).
$\lim_{s\to\infty} \|z(s) - \Phi x(s)\| = 0$	$\Longrightarrow$	The pair (35) is generalized synchronized (GS).
$\lim_{s\to\infty} \|x(s) - \Psi z(s)\| = 0$	$\Longrightarrow$	The pair (35) is inverse generalized synchronized (IGS).
$\lim_{s\to\infty} \|z(s) - Mx(s)\| = 0$	$\Longrightarrow$	The pair (35) is full state hybrid projective synchronized (FSHPS).
$\lim_{s\to\infty} \|x(s) - Lz(s)\| = 0$	$\Longrightarrow$	The pair (35) is inverse full state hybrid projective synchronized (IFSHPS).

4.1 Synchronization Between h-Fractional Ikeda Maps

In this section, we will present an adaptive controller to synchronize the h-fractional Ikeda map. Let us consider the master system as

$$\begin{cases} {}_h^C\Delta_\theta^\gamma z_{1m}(s) &= 1 + \Big(z_{1m}(s+h\gamma)\cos f(s+h\gamma) - z_{2m}(s+h\gamma)\sin f(s+h\gamma)\Big)\alpha \\ &\quad - z_{1m}(s+h\gamma), \\ {}_h^C\Delta_\theta^\gamma z_{2m}(s) &= \Big(z_{1m}(s+h\gamma)\sin f(s+h\gamma) + z_{2m}(s+h\gamma)\cos f(s+h\gamma)\Big)\alpha \\ &\quad - z_{2m}(s+h\gamma). \end{cases} \tag{36}$$

Define the slave system as

$$\begin{cases} {}_h^C\Delta_\theta^\gamma z_{1s}(s) &= 1 + \Big(z_{1s}(s+h\gamma)\cos f(s+h\gamma) - z_{2s}(s+h\gamma)\sin f(s+h\gamma)\Big)\alpha \\ &\quad - z_{1s}(s+h\gamma) + U(s+h\gamma), \\ {}_h^C\Delta_\theta^\gamma z_{2s}(s) &= \Big(z_{1s}(s+h\gamma)\sin f(s+h\gamma) + z_{2s}(s+h\gamma)\cos f(s+h\gamma)\Big)\alpha \\ &\quad - z_{2s}(s+h\gamma), \end{cases} \tag{37}$$

where C is synchronization controllers. Note that the subscript m indicates the master system states and r signifies the slave system states.

The synchronization error is defined for $s \in \mathbb{N}_{\theta-1+\gamma}$ as

$$\begin{cases} e_1(s) = z_{1s}(s) - z_{1m}(s), \\ e_2(s) = z_{2s}(s) - z_{2m}(s), \end{cases} \tag{38}$$

then, the master system (36) and slave system (37) are said to be synchronized if

$$\lim_{t\to\infty} |e_i(s)| = 0, \quad for\ i = 1, 2. \tag{39}$$

The following theorem describes the suggested control law for achieving this synchronization scheme.

Theorem 7 *Subject to*

$$U(s) = -e_1(s), \tag{40}$$

then the master system (36) *and slave system* (37) *are synchronized.*

Proof On the error system (38), using the Caputo h-fractional differences, we get:

$$\begin{cases} {}_h^C\Delta_\theta^\gamma e_1(s) &= \Big(e_1(s+h\gamma)\cos f(s+h\gamma) - e_2(s+h\gamma)\sin f(s+h\gamma)\Big)\alpha + U(s+h\gamma) \\ & \quad -e_1(s+h\gamma), \\ {}_h^C\Delta_\theta^\gamma e_2(s) &= \Big(e_1(s+h\gamma)\sin f(s+h\gamma) + e_2(s+h\gamma)\cos f(s+h\gamma)\Big)\alpha - e_2(s+h\gamma). \end{cases} \tag{41}$$

Replacing U in system (41) leads to the following h-fractional discrete system:

$$\begin{cases} {}_h^C\Delta_\theta^\gamma e_1(s) &= \Big(e_1(s+h\gamma)\cos f(s+h\gamma) - e_2(s+h\gamma)\sin f(s+h\gamma)\Big)\alpha - 2e_1(s+h\gamma), \\ {}_h^C\Delta_\theta^\gamma e_2(s) &= \Big(e_1(s+h\gamma)\sin f(s+h\gamma) + e_2(s+h\gamma)\cos f(s+h\gamma)\Big)\alpha - e_2(s+h\gamma). \end{cases} \tag{42}$$

Consider now the following Lyapunov function

$$V(s) = \frac{1}{2}\left(e_1^2(s) + e_2^2(s)\right), \tag{43}$$

This implies, according to Lemma 1:

$$\begin{aligned} {}_h^C\Delta_\theta^\gamma V &\le e_1(s+h\gamma)\,{}_h^C\Delta_\theta^\gamma e_1(s) + e_2(s+h\gamma)\,{}_h^C\Delta_\theta^\gamma e_2^2(s) \\ &= \Big(e_1^2(s+h\gamma)\cos f(s+h\gamma) - e_1(s+h\gamma)e_2(s+h\gamma)\sin f(s+h\gamma)\Big)\alpha - 2e_1^2(s+h\gamma) \\ &\quad + \Big(e_2(s+h\gamma)e_1(s+h\gamma)\sin f(s+h\gamma) + e_2^2(s+h\gamma)\cos f(s+h\gamma)\Big)\alpha - e_2^2(s+h\gamma) \\ &= \alpha e_1^2(s+h\gamma)\cos f(s+h\gamma) - 2e_1^2(s+h\gamma) + \alpha e_2^2(s+h\gamma)\cos f(s+h\gamma) - e_2^2(s+h\gamma) \\ &\le \alpha e_1^2(s+h\gamma) - 2e_1^2(s+h\gamma) + \alpha e_2^2(s+h\gamma) - e_2^2(s+h\gamma) \\ &= (\alpha-2)e_1^2(s+h\gamma) + (\alpha-1)e_2^2(s+h\gamma). \end{aligned}$$

Since $\alpha = 0.9$, then ${}_h^C\Delta_\theta^\gamma V < 0$. So, according to Theorem 1, the zero solution of the h-fractional discrete error system (41) is asymptotic stable. Hence, the master system (36) and the slave system (37) are synchronized.

In order to verify this result, numerical simulations are performed. We choose $\gamma = 0.9$ and the initial values $(e_1(0), e_2(0)) = (0.3, 0.2)$. Figure 11 reports the time

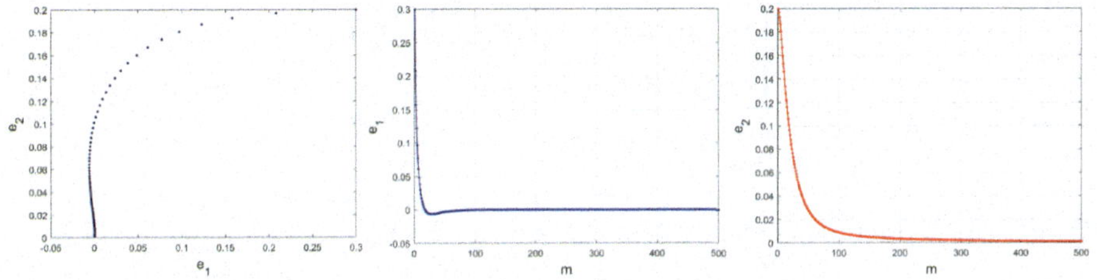

Fig. 11 Time evolution of the synchronization errors described by Eq. (41)

evolution of states of the error system (41). Clearly that the errors converge to zero, suggesting that the synchronization described above is effective.

4.2 Synchronization Between the h-Fractional Grassi–Miller Maps

In the following, linear controllers are used to synchronize a master-slave system using two identical chaotic h-fractional Grassi–Miller maps. The master system is expressed as follows [103]:

$$\begin{cases} {}_{h}^{C}\Delta_{\theta}^{\gamma}x_m(s) = a - x_m(s+h\gamma) - y_m^2(s+h\gamma) - bz_m(s+h\gamma), \\ {}_{h}^{C}\Delta_{\theta}^{\gamma}y_m(s) = x_m(s+h\gamma) - y_m(s+h\gamma), \\ {}_{h}^{C}\Delta_{\theta}^{\gamma}z_m(s) = y_m(s+h\gamma) - z_m(s+h\gamma). \end{cases} \tag{44}$$

The slave system is written as follows:

$$\begin{cases} {}_{h}^{C}\Delta_{\theta}^{\gamma}x_s(s) = a - x_s(s+h\gamma) - y_s^2(s+h\gamma) - bz_s(s+h\gamma) + U_1(s+h\gamma), \\ {}_{h}^{C}\Delta_{\theta}^{\gamma}y_s(s) = x_s(s+h\gamma) - y_s(s+h\gamma), \\ {}_{h}^{C}\Delta_{\theta}^{\gamma}z_s(s) = y_s(s+h\gamma) - z_s(s+h\gamma) + U_2(s+h\gamma), \end{cases} \tag{45}$$

where U_1 and U_2 are suitable linear controllers to be determined. To derive the error system, we take the difference between the master system (44) and the slave system (45), which is:

$$(e_1(s), e_2(s), e_3(s))^T = (x_s(s), y_s(s), z_s(s))^T - (x_m(s), y_m(s), z_m(s))^T. \tag{46}$$

We now establish a theorem based on the Lyapunov Method to achieve the desired master-slave systems synchronization using linear controllers U_1 and U_2.

Theorem 8 *The dynamics of the master system* (44) *and the slave system* (45) *are synchronized if the linear control laws* U_1 *and* U_2 *are chosen as:*

$$\begin{cases} U_1(s) = \left(1 - (\lambda + \frac{1}{2})^2\right) e_1(s), \\ U_2(s) = be_1(s) - e_2(s), \end{cases} \tag{47}$$

where $|y_m(s)| = |y_s(s)| \leq \lambda$, $s \in (h\mathbb{N})_{\theta+(1-\gamma)h}$

Proof The h-fractional error system cam be written as:

$$\begin{cases} {}_h^C\Delta_\theta^\gamma e_1(s) = -e_1(s+h\gamma) - be_3(s+h\gamma) + y_m^2(s+h\gamma) - y_s^2(s+h\gamma) + U_1(s+h\gamma), \\ {}_h^C\Delta_\theta^\gamma e_2(s) = e_1(s+h\gamma) - e_2(s+h\gamma), \\ {}_h^C\Delta_\theta^\gamma e_3(s) = e_2(s+h\gamma) - e_3(s+h\gamma) + U_2(s+h\gamma). \end{cases} \tag{48}$$

By substituting the control (47) into error system (48), we obtain

$$\begin{cases} {}_h^C\Delta_\theta^\gamma e_1(s) = -be_3(s+h\gamma) - (y_s(s+h\gamma) + y_m(s+h\gamma))e_2(s+h\gamma) - \left(\lambda + \frac{1}{2}\right)^2 e_1(s+h\gamma), \\ {}_h^C\Delta_\theta^\gamma e_2(s) = e_1(s+h\gamma) - e_2(s+h\gamma), \\ {}_h^C\Delta_\theta^\gamma e_3(s) = be_1(s+h\gamma) - e_3(s+h\gamma). \end{cases} \tag{49}$$

Now, by choosing a Lyapunov function $V(s) = \frac{1}{2}\left(e_1^2(s) + e_2^2(s) + e_3^2(s)\right)$ and applying Lemma 1, we get

$$\begin{aligned} {}_h^C\Delta_\theta^\gamma V &\leq e_1(s+h\gamma) {}_h^C\Delta_\theta^\gamma e_1(s) + e_2(s+h\gamma) {}_h^C\Delta_\theta^\gamma e_2^2(s) \\ &= -be_1(s+h\gamma)e_3(s+h\gamma) - (y_s(s+h\gamma) + y_m(s+h\gamma))e_1(s+h\gamma)e_2(s+h\gamma) \\ &\quad - \left(\lambda + \frac{1}{2}\right)^2 e_1^2(s+h\gamma) + e_2(s+h\gamma)e_1(s+h\gamma) - e_2^2(s+h\gamma) \\ &\quad + be_3(s+h\gamma)e_1(s+h\gamma) - e_3^2(s+h\gamma) \\ &\leq (1 + |y_s(s+h\gamma) + y_m(s+h\gamma)|)|e_1(s+h\gamma)||e_2(s+h\gamma)| - \left(\lambda + \frac{1}{2}\right)^2 e_1^2(s+h\gamma) \\ &\quad - e_2^2(s+h\gamma) - e_3^2(s+h\gamma) \\ &\leq (1+2\lambda)|e_1(s+h\gamma)||e_2(s+h\gamma)| - \left(\lambda + \frac{1}{2}\right)^2 e_1^2(s+h\gamma) - e_2^2(s+h\gamma) - e_3^2(s+h\gamma) \\ &= -\left(\left(\lambda + \frac{1}{2}\right)^2 e_1(s+h\gamma) - e_2(s+h\gamma)\right)^2 - e_3^2(s+h\gamma) < 0. \end{aligned}$$

As a result of Theorem 1, we see that the zero solution of the error system (48) is stabilized. Therefore, it is shown that the master system (44) and the slave system (45) establish synchronized dynamics by using linear control laws (47).

In order to demonstrate the efficacy of the suggested scheme, Fig. 12 depicts the evolution of the states of the master system (blue) and the slave system (red) when $a = 1$, $b = 0.5$, and $\gamma = 0.999$. These plots clearly demonstrate that two identical

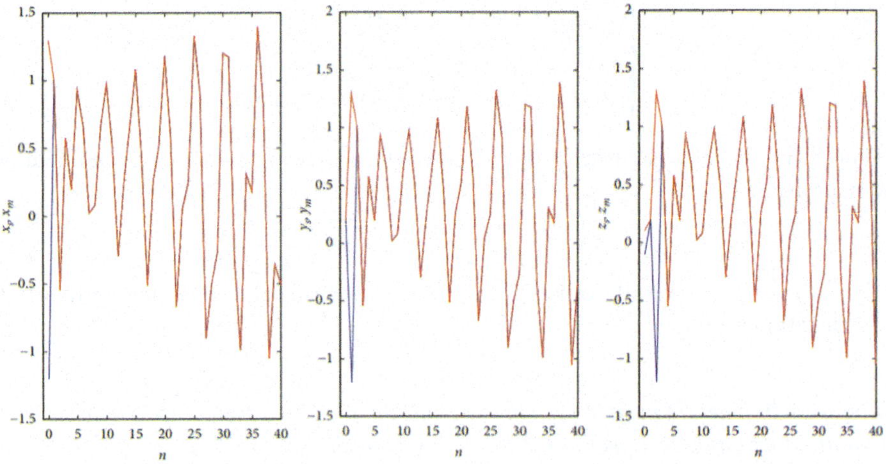

Fig. 12 The time evolution of states of the master system (blue) and the slave system (red)

Grassi–Miller maps can establish chaotic synchronization through linear controller laws.

4.3 Hybrid Synchronization Scheme Between h-Fractional Maps

In this part, we will use suitable synchronization scheme to synchronize the 2D h-fractional Lorenz map given in (17) and the 3D h-fractional Wang map given in (31), even though they have different dimensions. First, Let's consider the 2D h-fractional Lorenz map as the master system and refer to each of its states with the subscript m [102].

$$\begin{cases} {}_h^C\Delta_\theta^\gamma z_{1m}(s) &= \beta z_{1m}(s+h\gamma) - \beta z_{1m}(s+h\gamma)z_{2m}(s+h\gamma), \\ {}_h^C\Delta_\theta^\gamma z_{2m}(s) &= \beta z_{1m}^2(s+h\gamma) - \beta z_{2m}(s+h\gamma). \end{cases} \tag{50}$$

Now, we consider the 3D h-fractional Wang map as the slave system and refer to each of its states by another subscript s,

$$\begin{cases} {}_h^C\Delta_\theta^\gamma z_{1s}(s) = a_1 z_{1s}(s+h\gamma) + a_2 z_{2s}(s+h\gamma) + U_1(s+h\gamma), \\ {}_h^C\Delta_\theta^\gamma z_{2s}(s) = b_1 z_{1s}(s+h\gamma) + b_2 z_{3s}(s+h\gamma) + U_2(s+h\gamma), \\ {}_h^C\Delta_\theta^\gamma z_{3s}(s) = c_1 z_{3s}(s+h\gamma) + c_2 z_{2s}(s+h\gamma)z_{3s}(s+h\gamma) + c_3 + U_3(s+h\gamma), \end{cases} \tag{51}$$

where the functions U_1, U_2 and U_3 are synchronization controllers. The synchronization error is defined as

$$\begin{cases} e_1(s) = z_{1s}(S) - z_{1m}(s) \\ e_2(S) = z_{2s}(s) - z_{2m}(s) \\ e_3(s) = z_{3s}(s) - (z_{1m}(s) + z_{2m}(s)) \end{cases} \tag{52}$$

Then, The master system (50) and slave system (51) are said to be combination synchronized if

$$\lim_{s \to \infty} |e_i(s)| = 0, \quad for\ i = \overline{1, 3}. \tag{53}$$

The following theorem describes the suggested control law for achieving this synchronization scheme.

Theorem 9 *The 2D master system* (50) *and the 3D slave system* (51) *achieve synchronized dynamics Subject to the following control law:*

$$\begin{cases} U_1(s) = -\beta z_{2m}(s) z_{1m}(s) - a_2 z_{2s}(s) + (\beta - a_1) z_{1m}(s) \\ U_2(s) = -\beta z_{1m}^2(s) - b_2 z_{3s}(s) - b_1 z_{1s}(s) - \beta z_{2s}(s) \\ U_3(s) = -(c_1 + 1) z_{3s}(s) - c_2 z_{2s}(s) z_{3s}(s) - \beta z_{2m}(s) z_{1m}(s) - c_3 + \beta z_{1m}^2(s) \\ \qquad + (\beta + 1)(z_{2m}(s) + z_{1m}(s)). \end{cases} \tag{54}$$

$s \in (h\mathbb{N})_{\theta + (1-\gamma)h}$.

Proof On the error system (52), using the Caputo-type fractional h-difference operator, we get:

$$\begin{cases} {}_h^C\Delta_\theta^\gamma e_1(s) = a_1 z_{1s}(s + h\gamma) + a_2 z_{2s}(s + h\gamma) - \beta z_{1m}(s + h\gamma) \\ \qquad + \beta z_{1m}(s + h\gamma) z_{2m}(s + h\gamma) + U_1(s + h\gamma), \\ {}_h^C\Delta_\theta^\gamma e_2(s) = b_1 z_{1s}(s + h\gamma) + b_2 z_{3s}(s + h\gamma) - \beta z_{1m}^2(s + h\gamma) + \beta z_{2m}(s + h\gamma) \\ \qquad + U_2(s + h\gamma), \\ {}_h^C\Delta_\theta^\gamma e_3(s) = c_1 z_{3s}(s + h\gamma) + c_2 z_{2s}(s + h\gamma) z_{3s}(s + h\gamma) + c_3 - \beta z_{1m}(s + h\gamma) \\ \qquad + \beta z_{1m}(s + h\gamma) z_{2m}(s + h\gamma) - \beta z_{1m}^2(s + h\gamma) + \beta z_{2m}(s + h\gamma) + U_3(s + h\gamma). \end{cases} \tag{55}$$

Substituting the control law (54) into (55), we obtain

$$\begin{cases} {}_h^C\Delta_\theta^\gamma e_1(s) = -a_1 e_1(s + h\gamma), \\ {}_h^C\Delta_\theta^\gamma e_2(s) = -\beta e_2(s + h\gamma), \\ {}_h^C\Delta_a^\gamma e_3(t) = -e_3(s + h\gamma). \end{cases} \tag{56}$$

Now, consider the following Lyapunov function:

$$V(s) = \frac{1}{2} e_1^2(s) + \frac{1}{2} e_2^2(s) + \frac{1}{2} e_3^2(s) \tag{57}$$

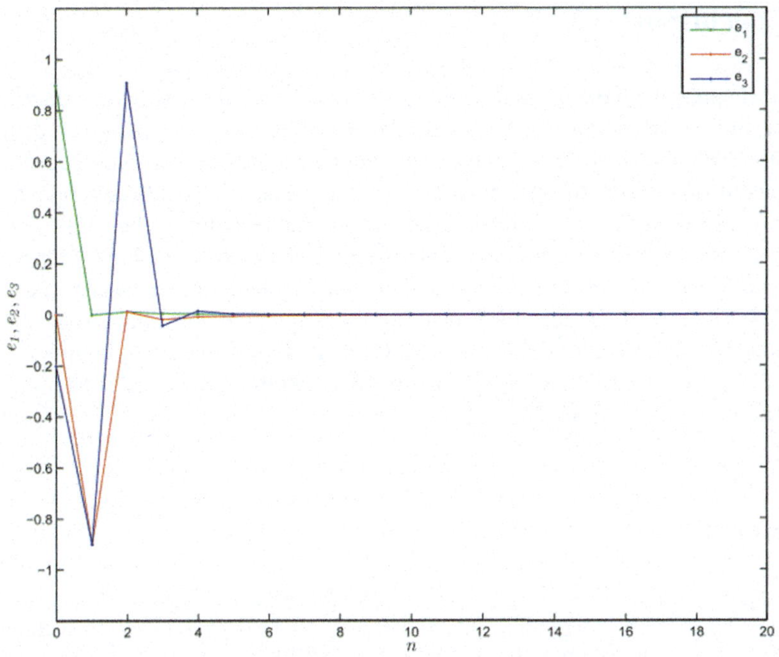

Fig. 13 Time evolution of the synchronization errors described by Eq. (55)

This implies:

$$
{}_{h}^{C}\Delta_{\theta}^{\gamma} V(t) = \frac{1}{2}\,{}_{h}^{C}\Delta_{\theta}^{\gamma} e_1^2(t) + \frac{1}{2}\,{}_{h}^{C}\Delta_{\theta}^{\gamma} e_2^2(t) + \frac{1}{2}\,{}_{h}^{C}\Delta_{\theta}^{\gamma} e_3^2(t)
$$

and using Lemma 1, we obtain

$$
\begin{aligned}
{}_{h}^{C}\Delta_{\theta}^{\gamma} V &\leq e_1(s+h\gamma)\,{}_{h}^{C}\Delta_{\theta}^{\gamma} e_1(t) + e_2(s+h\gamma)\,{}_{h}^{C}\Delta_{\theta}^{\gamma} e_2(t) + e_3(s+h\gamma)\,{}_{h}^{C}\Delta_{\theta}^{\gamma} e_3(t) \\
&= a_1 e_1^2(s+h\gamma) - \beta e_2^2(s+h\gamma) - e_3^2(s+h\gamma) < 0.
\end{aligned}
$$

So, we can conclude that the dynamics of the h-fractional error system (56) are stabilized at zero, according to Theorem 1. As a result, the linear control laws (54) are used to achieve synchronization between the 2D master system (50) and the 3D h-fractional slave system (51), and the proof of Theorem 9 is complete.

To illustrate the theoretical results in the previous theorem, numerical simulations are now performed using Matlab. We choose $h = 1$, $v = 0.8$ and the initial values $e_1(0) = 0.9$, $e_2(0) = 0.05$, $e_3(0) = -0.2$. Figure 13 depicts the numerical states of the h-fractional error system (55). Clearly that the errors converge to zero, and the h-fractional Lorenz map and h-fractional Wang map achieve hybrid synchronization.

5 Conclusion

In this chapter, we have proposed different linear control strategies to stabilize and synchronize the dynamics of the fractional maps based on the Caputo h-difference operator. New theorems have been developed in order to derive more effective methods to control and synchronize map dynamics. The stability and convergence of the states of each map were established by means of the stability theory of fractional discrete systems based on the Lyapunov approach. Besides, we have proposed linear control laws to stabilize the dynamics of the considered fractional maps at the origin. Furthermore, we have also proposed different schemes that enable synchronization to be achieved between master and slave systems based on identical chaotic maps. Numerical simulations were carried out throughout the chapter to support and confirm the theoretical results.

References

1. Abbes, A., Ouannas, A., Shawagfeh, N.: Synchronization in fractional discrete neural networks using linear control laws. In: 2021 International Conference on Recent Advances in Mathematics and Informatics (ICRAMI), pp. 1–5. IEEE (2021)
2. Abbes, A., Ouannas, A., Shawagfeh, N.: The incommensurate fractional discrete macroeconomic system: bifurcation, chaos and complexity. Chin. Phys. B (2022)
3. Abbes, A., Ouannas, A., Shawagfeh, N., Grassi, G.: The effect of the caputo fractional difference operator on a new discrete covid-19 model. Results Phys. **39**, 105797 (2022)
4. Abbes, A., Ouannas, A., Shawagfeh, N., Khennaoui, A.A.: Incommensurate fractional discrete neural network: chaos and complexity. Eur. Phys. J. Plus **137**(2), 235 (2022)
5. Abbes, A., Ouannas, A., Shawagfeh, N., Khennaoui, A.A.: Incommensurate fractional discrete neural network: chaos and complexity. Eur. Phys. J. Plus. **137**(2), 235 (2022)
6. Abbes, A., Ouannas, A., Shawagfeh, N., Jahanshahi, H.: The fractional-order discrete covid-19 pandemic model: stability and chaos. Nonlinear Dyn. **111**(1), 965–983 (2023)
7. Abdeljawad, T.: Different type kernel h- fractional differences and their fractional h- sums. Chaos Solitons & Fractals **116**, 146–156 (2018)
8. Ahlem, G., Ouannas, A.: A general control method for inverse hybrid function projective synchronization of a class of chaotic systems. Int. J. Math. Anal. **9**(9–12), 429–436 (2015)
9. Ahmad, I., Ouannas, A., Shafiq, M., Pham, V.-T., Baleanu, D.: Finite-time stabilization of a perturbed chaotic finance model. J. Adv. Res. **32**, 1–14 (2021)
10. Albadarneh, R.B., Batiha, I.M., Ouannas, A., Momani, S.: Modeling covid-19 pandemic outbreak using fractional-order systems. Int. J. Math. Comput. Sci **16**, 1405–1421 (2021)
11. Almatroud, A.O., Khennaoui, A.-A., Ouannas, A., Pham, V.-T.: Infinite line of equilibriums in a novel fractional map with coexisting infinitely many attractors and initial offset boosting. Int. J. Nonlinear Sci. Numer. Simul. (2021)
12. Almatroud, A.O., Ouannas, A., Grassi, G., Batiha, I.M., Gasri, A., Al-Sawalha, M.M.: Different linear control laws for fractional chaotic maps using lyapunov functional. Arch. Control Sci. **31**(4) (2021)
13. Almatroud, A.O., Khennaoui, A.-A., Ouannas, A., Grassi, G., Al-Sawalha, M.M., Gasri, A.: Dynamical analysis of a new chaotic fractional discrete-time system and its control. Entropy **22**(12), 1344 (2020)
14. Al-Shbeil, I., Djenina, N., Jaradat, A., Al-Husban, A., Ouannas, A., Grassi, G.: A new covid-19 pandemic model including the compartment of vaccinated individuals: global stability of the disease-free fixed point. Mathematics **11**(3), 576 (2023)

15. Azar, A.T., Vaidyanathan, S., Ouannas, A.: Fractional Order Control and Synchronization of Chaotic Systems, vol. 688. Springer, Berlin (2017)
16. Baleanu, D., Guo-Cheng, W., Bai, Y.-R., Chen, F.-L.: Stability analysis of caputo-like discrete fractional systems. Commun. Nonlinear Sci. Numer. Simul. **48**, 520–530 (2017)
17. Bastos, N.R.O., Ferreira, R.A.C., Torres, D.F.M.: Discrete-time fractional variational problems. Signal Process. **91**(3), 513–524 (2011)
18. Batiha, I.M., Ouannas, A., Emwas, J.A.: A stabilization approach for a novel chaotic fractional-order discrete neural network. J. Math. Comput. Sci. **11**(5), 5514–5524 (2021)
19. Bendoukha, S., Abdelmalek, S., Ouannas, A.: Secure communication systems based on the synchronization of chaotic systems. Math. Appl. Eng. Modell. Soc. Issues 281–311 (2019)
20. Bendoukha, S., Ouannas, A., Wang, X., Khennaoui, A.-A., Pham, V.-T., Grassi, G., Van Huynh, V.: The co-existence of different synchronization types in fractional-order discrete-time chaotic systems with non-identical dimensions and orders. Entropy **20**(9), 710 (2018)
21. Debbouche, N., Ouannas, A., Batiha, I.M., Grassi, G.: Chaotic dynamics in a novel covid-19 pandemic model described by commensurate and incommensurate fractional-order derivatives. Nonlinear Dyn. 1–13 (2021)
22. Debbouche, N., Ouannas, A., Momani, S., Cafagna, D., Pham, V.-T.: Fractional-order biological system: chaos, multistability and coexisting attractors. Eur. Phys. J. Spec. Top. 1–10 (2021)
23. Debbouche, N., Almatroud, A.O., Ouannas, A., Batiha, I.M.: Chaos and coexisting attractors in glucose-insulin regulatory system with incommensurate fractional-order derivatives. Chaos Solitons & Fractals **143**, 110575 (2021)
24. Djenina, N., Ouannas, A., Oussaeif, T.-E., Grassi, G., Batiha, I.M., Momani, S., Albadarneh, R.B.: On the stability of incommensurate h-nabla fractional-order difference systems. Fractal Fract. **6**(3), 158 (2022)
25. Ferreira, R.A.C., Torres, D.F.M.: Fractional h-difference equations arising from the calculus of variations. Appl. Anal. Discrete Math. **123**, 110–121 (2011)
26. Gasri, A., Ouannas, A., Khennaoui, A.A., Grassi, G., Oussaeif, T.-E., Pham, V.-T.: Chaotic fractional discrete neural networks based on the caputo h-difference operator: stabilization and linear control laws for synchronization. Eur. Phys. J. Spec. Top. **231**(10), 1815–1829 (2022)
27. Gasri, A., Khennaoui, A.-A., Ouannas, A., Grassi, G., Iatropoulos, A., Moysis, L., Volos, C.: A new fractional-order map with infinite number of equilibria and its encryption application. Complexity **1–18**, 2022 (2022)
28. Grassi, G., Ouannas, A., Pham, V.-T.: A general unified approach to chaos synchronization in continuous-time systems (with or without equilibrium points) as well as in discrete-time systems. Arch. Control Sci. **28**(1) (2018)
29. Hadjabi, F., Ouannas, A., Shawagfeh, N., Khennaoui, A.-A., Grassi, G.: On two-dimensional fractional chaotic maps with symmetries. Symmetry **12**(5), 756 (2020)
30. He, Z.-Y., Abbes, A., Jahanshahi, H., Alotaibi, N.D., Wang, Y.: Fractional-order discrete-time sir epidemic model with vaccination: chaos and complexity. Mathematics **10**(2), 165 (2022)
31. Hioual, A.: Novel results for the stability of h-discrete fractional neural networks with non-singular and nonlocal kernels. Innov. J. Math. (IJM) **1**(3), 1–13 (2022)
32. Hioual, A., Ouannas, A., Oussaeif, T.-E., Grassi, G., Batiha, I.M., Momani, S.: On variable-order fractional discrete neural networks: solvability and stability. Fractal Fract. **6**(2), 119 (2022)
33. Hioual, A., Oussaeif, T.-E., Ouannas, A., Grassi, G., Batiha, I.M., Momani, S.: New results for the stability of fractional-order discrete-time neural networks. Alex. Eng. J. **61**(12), 10359–10369 (2022)
34. Hioual, A., Ouannas, A., Grassi, G., Oussaeif, T.-E.: Nonlinear nabla variable-order fractional discrete systems: asymptotic stability and application to neural networks. J. Comput. Appl. Math. **423**, 114939 (2023)
35. Huynh, V.V., Ouannas, A., Wang, X., Pham, V.-T., Nguyen, X.Q., Alsaadi, F.E.: Chaotic map with no fixed points: entropy, implementation and control. Entropy **21**(3), 279 (2019)

36. Jouini, L., Ouannas, A., Khennaoui, A.-A., Wang, X., Grassi, G., Pham, V.-T.: The fractional form of a new three-dimensional generalized hénon map. Adv. Diff. Equ. **2019**(1), 1–12 (2019)
37. Khennaoui, A.-A., Ouannas, A., Bendoukha, S., Wang, X., Pham, V.-T.: On chaos in the fractional-order discrete-time unified system and its control synchronization. Entropy **20**(7), 530 (2018)
38. Khennaoui, A.-A., Ouannas, A., Bendoukha, S., Grassi, G., Lozi, R.P., Pham, V.-T.: On fractional-order discrete-time systems: chaos, stabilization and synchronization. Chaos Solitons & Fractals **119**, 150–162 (2019)
39. Khennaoui, A.-A., Ouannas, A., Odibat, Z., Pham, V.-T., Grassi, G.: On the three-dimensional fractional-order hénon map with lorenz-like attractors. Int. J. Bifur. Chaos **30**(11), 2050217 (2020)
40. Khennaoui, A.-A., Ouannas, A., Momani, S., Batiha, I.M., Dibi, Z., Grassi, G.: On dynamics of a fractional-order discrete system with only one nonlinear term and without fixed points. Electronics **9**(12), 2179 (2020)
41. Khennaoui, A.A., Almatroud, A.O., Ouannas, A., Al-sawalha, M.M., Grassi, G., Pham, V.-T., Batiha, I.M.: An unprecedented 2-dimensional discrete-time fractional-order system and its hidden chaotic attractors. Math. Probl. Eng. **1–10**, 2021 (2021)
42. Khennaoui, A.-A., Ouannas, A., Bekiros, S., Aly, A.A., Alotaibi, A., Jahanshahi, H., Alsubaie, H.: Hidden homogeneous extreme multistability of a fractional-order hyperchaotic discrete-time system: Chaos, initial offset boosting, amplitude control, control, and synchronization. Symmetry **15**(1), 139 (2023)
43. Mellah, M., Ouannas, A., Khennaoui, A.-A., Grassi, G.: Fractional discrete neural networks with different dimensions: coexistence of complete synchronization, antiphase synchronization and full state hybrid projective synchronization. Nonlinear Dyn. Syst. Theory **21**(4), 410–419 (2021)
44. Mesdoui, F., Ouannas, A., Shawagfeh, N., Grassi, G., Pham, V.-T.: Synchronization methods for the degn-harrison reaction-diffusion systems. IEEE Access **8**, 91829–91836 (2020)
45. Mesdoui, F., Shawagfeh, N., Ouannas, A.: Global synchronization of fractional-order and integer-order n component reaction diffusion systems: application to biochemical models. Math. Methods Appl. Sci. **44**(1), 1003–1012 (2021)
46. Mesdoui, F., Shawagfeh, N., Ouannas, A., et al.: Synchronization methods for chaotic systems involving fractional derivative with a non-singular kernel. J. Appl. Nonlinear Dyn. **11**(02), 375–386 (2022)
47. Ouannas, A., Abdelli, M., Odibat, Z., Wang, X., Pham, V.-T., Grassi, G., Alsaedi, A.: Synchronization control in reaction-diffusion systems: application to lengyel-epstein system. Complexity **2019** (2019)
48. Ouannas, A., Aljazaery, I.: A new method to generate a discrete chaotic dynamical systems using synchronization technique. Far East J. Dyn. Syst. **24**(1–2), 15–24 (2014)
49. Ouannas, A., Al-sawalha, M.M.: Synchronization between different dimensional chaotic systems using two scaling matrices. Optik **127**(2), 959–963 (2016)
50. Ouannas, A., Batiha, I.M., Khennaoui, A.-A., Zraiqat, A., Al-Nana, A.A.: Novel control law for the fractional-order chaotic duffing map. In: 2021 International Conference on Information Technology (ICIT), pp. 238–241. IEEE (2021)
51. Ouannas, A., Grassi, G., Azar, A.T., Khennaoui, A.A.: Synchronization control in fractional discrete-time systems with chaotic hidden attractors. In: Advanced Machine Learning Technologies and Applications: Proceedings of AMLTA 2020, pp. 661–669. Springer (2021)
52. Ouannas, A., Grassi, G., Azar, A.T.: A new generalized synchronization scheme to control fractional chaotic systems with non-identical dimensions and different orders. In: The International Conference on Advanced Machine Learning Technologies and Applications (AMLTA2019) 4, pp. 415–424. Springer (2020)
53. Ouannas, A., Grassi, G., Azar, A.T.: A new generalized synchronization scheme to control fractional chaotic systems with non-identical dimensions and different orders. In: The International Conference on Advanced Machine Learning Technologies and Applications (AMLTA2019), vol. 4, pp. 415–424. Springer (2020)

54. Ouannas, A., Khennaoui, A.-A., Momani, S., Grassi, G., Pham, V.-T., El-Khazali, R., Vo Hoang, D.: A quadratic fractional map without equilibria: Bifurcation, 0–1 test, complexity, entropy, and control. Electronics **9**(5), 748 (2020)
55. Ouannas, A., Khennaoui, A.-A., Momani, S., Pham, V.-T.: The discrete fractional duffing system: Chaos, 0–1 test, c 0 complexity, entropy, and control. Chaos Interdiscip. J. Nonlinear Sci. **30**(8), 083131 (2020)
56. Ouannas, A., Mahmoud, E.E.: Inverse matrix projective synchronization for discrete chaotic systems with different dimensions. J. Comput. Intell. Electr. Syst. **3**(3), 188–192 (2014)
57. Ouannas, A., Mesdoui, F., Momani, S., Batiha, I., Grassi, G.: Synchronization of fitzhugh-nagumo reaction-diffusion systems via one-dimensional linear control law. Arch Control Sci **31** (2021)
58. Ouannas, A., Wang, X., Pham, V.-T., Ziar, T.: Dynamic analysis of complex synchronization schemes between integer order and fractional order chaotic systems with different dimensions. Complexity **2017** (2017)
59. Ouannas, A., Zehrour, O., Laadjal, Z.: Nonlinear methods to control synchronization between fractional-order and integer-order chaotic systems. Nonlinear Studies **25**(1) (2018)
60. Ouannas, A.P.: Fractional Discrete Chaos: Theories. Methods and Applications. World Scientific Pub (2023)
61. Ouannas, A.: On full-state hybrid projective synchronization of general discrete chaotic systems. J. Nonlinear Dyn. **2014** (2014)
62. Ouannas, A.: Co-existence of complete synchronization and anti-synchronization in a class of discrete rational chaotic systems. Far East J. Dyn. Syst. **23**(1/2), 41 (2013)
63. Ouannas, A.: A new qs synchronization scheme for discrete chaotic systems. Far East J. Appl. Math. **84**(2), 89 (2013)
64. Ouannas, A.: Nonlinear control method of chaos synchronization for arbitrary 2d quadratic dynamical systems in discrete-time. Int. J. Math. Anal. **8**(53), 2611–2617 (2014)
65. Ouannas, A.: A synchronization criterion for a class of sinusoidal chaotic maps via linear controller. Int. J. Contemp. Math. Sci. **9**(14), 677–683 (2014)
66. Ouannas, A.: Synchronization of discrete chaotic systems via double scaling matrix in different dimensions. J. Appl. Math. Comput. **3**(7), 1–5 (2014)
67. Ouannas, A.: Synchronization and inverse synchronization of some different dimensional discrete-time chaotic dynamical systems via scaling matrices. Int. J. Chaos Control Model. Simul. **3**, 1–17 (2014)
68. Ouannas, A.: Chaos synchronization approach based on new criterion of stability. Nonlinear Dyn. Syst. Theory **14**(4), 395–401 (2014)
69. Ouannas, A.: A new synchronization scheme for general 3d quadratic chaotic systems in discrete-time. Nonlinear Dyn. Syst. Theory **15**(2), 163–170 (2015)
70. Ouannas, A., Abu-Saris, R.: A robust control method for qs synchronization between different dimensional integer-order and fractional-order chaotic systems. J. Control Sci. Eng. **55–55**, 2015 (2015)
71. Ouannas, A., Ahlem, G.: General synchronization approach for general chaotic systems in continuous-time. Far East J. Math. Sci. **96**(2), 231 (2015)
72. Ouannas, A., Al-sawalha, M.M.: On λ- ϕ generalized synchronization of chaotic dynamical systems in continuous-time. Eur. Phys. J. Special Top. **225**(1), 187–196 (2016)
73. Ouannas, A., Grassi, G.: A new approach to study the coexistence of some synchronization types between chaotic maps with different dimensions. Nonlinear Dyn. **86**(2), 1319–1328 (2016)
74. Ouannas, A., Al-Sawalha, M.M., Ziar, T.: Fractional chaos synchronization schemes for different dimensional systems with non-identical fractional-orders via two scaling matrices. Optik **127**(20), 8410–8418 (2016)
75. Ouannas, A., Azar, A.T., Vaidyanathan, S.: A robust method for new fractional hybrid chaos synchronization. Math. Methods Appl. Sci. **40**(5), 1804–1812 (2017)
76. Ouannas, A., Azar, A.T., Vaidyanathan, S.: A new fractional hybrid chaos synchronisation. Int. J. Modell. Ident. Control **27**(4), 314–322 (2017)

77. Ouannas, A., Grassi, G., Ziar, T., Odibat, Z.: On a function projective synchronization scheme for non-identical fractional-order chaotic (hyperchaotic) systems with different dimensions and orders. Optik **136**, 513–523 (2017)
78. Ouannas, A., Odibat, Z., Hayat, T.: Fractional analysis of co-existence of some types of chaos synchronization. Chaos Solitons & Fractals **105**, 215–223 (2017)
79. Ouannas, A., Wang, X., Khennaoui, A.-A., Bendoukha, S., Pham, V.-T., Alsaadi, F.E.: Fractional form of a chaotic map without fixed points: chaos, entropy and control. Entropy **20**(10), 720 (2018)
80. Ouannas, A., Khennaoui, A.-A., Bendoukha, S., Vo, T.P., Pham, V.-T., Huynh, V.V.: The fractional form of the tinkerbell map is chaotic. Appl. Sci. **8**(12), 2640 (2018)
81. Ouannas, A., Odibat, Z., Alsaedi, A., Hobiny, A., Hayat, T.: Investigation of qs synchronization in coupled chaotic incommensurate fractional order systems. Chin. J. Phys. **56**(5), 1940–1948 (2018)
82. Ouannas, A., Wang, X., Pham, V.-T., Grassi, G., Ziar, T.: Coexistence of identical synchronization, antiphase synchronization and inverse full state hybrid projective synchronization in different dimensional fractional-order chaotic systems. Adv. Diff. Equ. **2018**(1), 1–16 (2018)
83. Ouannas, A., Debbouche, N., Wang, X., Pham, V.-T., Zehrour, O.: Secure multiple-input multiple-output communications based on f-m synchronization of fractional-order chaotic systems with non-identical dimensions and orders. Appl. Sci. **8**(10), 1746 (2018)
84. Ouannas, A., Odibat, Z., Shawagfeh, N.: A new q-s synchronization results for discrete chaotic systems. Diff. Equ. Dynam. Syst. **27**, 413–422 (2019)
85. Ouannas, A., Khennaoui, A.-A., Grassi, G., Bendoukha, S.: On chaos in the fractional-order grassi-miller map and its control. J. Comput. Appl. Math. **358**, 293–305 (2019)
86. Ouannas, A., Khennaoui, A.-A., Bendoukha, S., Grassi, G.: On the dynamics and control of a fractional form of the discrete double scroll. Int. J. Bifur. Chaos **29**(06), 1950078 (2019)
87. Ouannas, A., Azar, A.T., Ziar, T.: Control of continuous-time chaotic (hyperchaotic) systems: F-m synchronisation. Int. J. Autom. Control **13**(2), 226–242 (2019)
88. Ouannas, A., Bendoukha, S., Khennaoui, A.-A., Grassi, G., Wang, X., Pham, V.-T.: Chaos synchronization of fractional-order discrete-time systems with different dimensions using two scaling matrices. Open Phys. **17**(1), 942–949 (2019)
89. Ouannas, A., Bendoukha, S., Karouma, A., Abdelmalek, S.: A general method to study the co-existence of different hybrid synchronizations in fractional-order chaotic systems. Int. J. Nonlinear Sci. Numer. Simul. **20**(3–4), 351–359 (2019)
90. Ouannas, A., Khennaoui, A.-A., Odibat, Z., Pham, V.-T., Grassi, G.: On the dynamics, control and synchronization of fractional-order ikeda map. Chaos Solitons & Fractals **123**, 108–115 (2019)
91. Ouannas, A., Wang, X., Pham, V.-T., Grassi, G., Van Huynh, V.: Synchronization results for a class of fractional-order spatiotemporal partial differential systems based on fractional lyapunov approach. Bound. Value Probl. **2019**(1), 1–12 (2019)
92. Ouannas, A., Khennaoui, A.-A., Wang, X., Pham, V.-T., Boulaaras, S., Momani, S.: Bifurcation and chaos in the fractional form of hénon-lozi type map. Eur. Phys. J. Spec. Top. **229**, 2261–2273 (2020)
93. Ouannas, A., Khennaoui, A.A., Momani, S., Pham, V.-T., El-Khazali, R.: Hidden attractors in a new fractional-order discrete system: chaos, complexity, entropy, and control. Chin. Phys. B **29**(5), 050504 (2020)
94. Ouannas, A., Khennaoui, A.-A., Bendoukha, S., Wang, Z., Pham, V.-T.: The dynamics and control of the fractional forms of some rational chaotic maps. J. Syst. Sci. Compl. **33**(3), 584–603 (2020)
95. Ouannas, A., Karouma, A., Grassi, G., Pham, V.-T., et al.: A novel secure communications scheme based on chaotic modulation, recursive encryption and chaotic masking. Alex. Eng. J. **60**(1), 1873–1884 (2021)
96. Ouannasz, M.M., Batiha, I.M., et al.: A general method for stabilizing the fractional-order discrete neural networks via linear control law. Nonlinear Stud. **29** (2022)

97. Pham, V.-T., Ouannas, A., Volos, C., Kapitaniak, T.: A simple fractional-order chaotic system without equilibrium and its synchronization. AEU-Int. J. Electr. Commun. **86**, 69–76 (2018)
98. Saadeh, R., Abbes, A., Al-Husban, A., Ouannas, A., Grassi, G.: The fractional discrete predator-prey model: Chaos, control and synchronization. Fractal Fract. **7**(2), 120 (2023)
99. Shatnawi, M.T., Abbes, A., Ouannas, A., Batiha, I.: Hidden multistability of fractional discrete non-equilibrium point memristor based map. Physica Scripta (2023)
100. Shatnawi, M.T., Abbes, A., Ouannas, A., Batiha, I.M.: A new two-dimensional fractional discrete rational map: chaos and complexity. Physica Scripta **98**(1), 015208 (2022)
101. Singh, S., Azar, A.T., Bhat, M.A., Vaidyanathan, S., Ouannas, A.: Active control for multi-switching combination synchronization of non-identical chaotic systems. In: Advances in System Dynamics and Control, pp. 129–162. IGI Global (2018)
102. Talbi, I., Ouannas, A., Khennaoui, A.-A., Berkane, A., Batiha, I.M., Grassi, G., Pham, V.-T.: Different dimensional fractional-order discrete chaotic systems based on the caputo h-difference discrete operator: dynamics, control, and synchronization. Adv. Diff. Equ. **1–15**, 2020 (2020)
103. Talbi, I., Ouannas, A., Grassi, G., Khennaoui, A.-A., Pham, V.-T., Baleanu, D.: Fractional grassi-miller map based on the caputo h-difference operator: Linear methods for chaos control and synchronization. Discret. Dyn. Nat. Soc. **1–10**, 2020 (2020)
104. Wang, B., Ouannas, A., Karaca, Y., Xia, W.-F., Jahanshahi, H., Alkhateeb, A.F., Nour, M., et al.: A hybrid approach for synchronizing between two reaction-diffusion systems of integer- and fractional-order applied on certain chemical models. FRACTALS (fractals) **30**(05), 1–11 (2022)
105. Wang, X., Ouannas, A., Pham, V.-T., Abdolmohammadi, H.R.: A fractional-order form of a system with stable equilibria and its synchronization. Adv. Diff. Equ. **2018**(1), 1–13 (2018)
106. Yousfi, H., Gasri, A., Ouannas, A.: Stabilization of chaotic h-difference systems with fractional order. Nonlinear Dyn. Syst. Theory 468 (2022)
107. Zarour, A., Ouannas, A., Latrous, C., Berkane, A.: Linear chaos control of fractional generalized hénon map. Nonlinear Dyn. Syst. Theory **21**(2), 216–224 (2021)

Artificial Neural Network Design for Non Linear Takagi–Sugeno Systems: Application to Tracking of Trajectory, State and Fault Estimation of MIABOT Robot

Dhouha Miri, Mohamed Guerfel, Atef Khedher, and Kamal Belkhiria

Abstract The problem of the trajectory tracking states and faults for nonlinear systems modeled using the Takagi Sugeno (T–S) approach is studied in this chapter. The the tracking require the state and fault estimation. In order to make this estimation, a ANN is conceived. An ANN with unknown inputs is used for the purpose of tracking the system state and faults affecting the system. The proposed method shows that it is possible to conceive the trajectory tracking of states and faults. We apply the proposed method to a robot modeled using the Takagi–Sugano framework. Indeed, in this work, the state and actuator errors affecting a nonlinear system modeled using the Takagi–Sugano structure, using 16 local models, are presented in the form of the Takagi–Sugano structure. We propose to estimate it by applying it to a robot model. The form a T–S structure of mobile robot design is formulated in term of linear matrix inequalities (LMI).

Keywords Artificial Neural Network · Takagi–Sugeno systems · Fault estimation · State estimation · Linear matrix inequalities · MIABOT robot

D. Miri (✉) · K. Belkhiria
LASEE, ENIM, Sousse, Tunisia
e-mail: dhouhamiri123@gmail.com

K. Belkhiria
e-mail: kamel.benothman@enim.rnu.tn

M. Guerfel
LARATSI, ENIM, Sousse, Tunisia

A. Khedher
LATIS, ENISO, Sousse, Tunisia

A. Ben Makhlouf et al. (eds.), *State Estimation and Stabilization of Nonlinear Systems*, Studies in Systems, Decision and Control 491,
https://doi.org/10.1007/978-3-031-37970-3_15

1 Introduction

Inspired by the functioning of the human brain, **A**rtificial **N**eural **N**etworks (**ANN**) today occupy a primordial place in several fields of Engineering Sciences. Advances in understanding how the human brain works have contributed significantly to the development of ANN. However scientists are still impressed by the architecture of the human neural system, his complexity, his efficiency and speed in handling certain complex problems (for example those related to perception) that the most powerful computer today cannot solve with the same efficiency.

Neural networks are used today for all kinds of applications in various fields [1, 4, 5, 7, 8, 10, 11, 13, 16, 18, 22, 31, 32].

For example, we have developed an autopilot for an airplane, or even a guidance system for automobiles, [13, 17, 28], systems have been designed for the automatic reading of bank checks and postal addresses, we produce signal processing systems for various military applications, a speech synthesis system, networks are also used to build computer vision systems, to make predictions on the money markets, to evaluate financial or insurance risk,for different manufacturing processes, for the medical diagnosis, for oil and gas exploration, in robotics, in telecommunications [17, 25, 28, 30].

The emergence of neural network shells has facilitated the proliferation of their applications (e.g., Brain-maker, Neuralyst, Neuroshell, and Professional II Plus) and complementary tools (for SAS, MATLAB, Python and Excel) that provide developers with the means specify the ANN architecture and train the neural network [19, 20, 32].

For example, let's quote: To efficiently recover the evolution of bifurcation phenomena in computational fluid dynamics problems using ANN, we use a multi-parameterized setup with different geometries to find the position of the bifurcation points. Analyzed the impact. In addition to discussing the model and its numerical approximations in the context of low-order modeling and building a base with his POD method for branch units, the latter eventually led to ANN and Joined. Francesco provides excellent approximation properties when considering such complex behavior [8].

Reference [31] describes a monitoring and evaluation method based on Intelligent Compaction (IC) and (**ANN**) to assess the compaction quality of highway subbases. Field compaction tests were performed with different roller speeds and vibration modes. Compactness was predicted using a particle swarm optimization backpropagation neural network (*PSO-BP-NN*) model. The model is integrated with the IC system to automatically assess compaction quality and efficiently control subsoil compaction. This model can be used for engineering practice and facilitates IC applications.

Approaches using neural networks can generalize to many results dealing with systems with noisy or partial inputs. They allow manipulation of continuous input data to solve diagnostic problems [11, 22]. Their adaptive learning ability allows them to be used efficiently for diagnostics and error detection [29].

In recent years, the novel coronavirus disease (**COVID-19**) outbreak has become one of the biggest global problems. Due to the paucity of global studies on spatio-temporal modeling of **COVID-19**, this study [22] is a potential explanatory variable associated with **COVID-19** ($n = 75$) aims to determine the relative importance of COVID-19 surveys. Prevalence and mortality using multi-layer perceptron ANN topologies.

Insights from this study may provide public health decision makers with common insights to monitor disease spread and aid decision making.

(**ANN**) are currently viewed as cheap predictive tools in automotive applications. In [11], work on **ANN** can mature and saturate with repeated application of the same approach using the same network type, training algorithm, and input/output parameters. This review article critically discusses the recent use of **ANN** in **ICE**. **ANN** has been successfully applied and is now becoming an indispensable tool for rapidly predicting engine performance, combustion and emissions characteristics.

At [7], research aimed at predicting the thermal efficiency and water volume of solar stills using a hybrid ANN model. The model is fine-tuned using a metaheuristic optimizer called the humpback optimizer. These results demonstrate the critical role of electric heaters in improving the thermal performance of solar stills and of the humpback optimizer in improving the predictive accuracy of traditional neural network models.

Approaches using neural networks can generalize to many results dealing with systems with noisy or partial inputs. They allow manipulation of continuous input data to solve diagnostic problems [1, 2, 19, 20]. Because of its adaptive learning ability, it can be used efficiently for diagnostics and error detection [6, 13, 29].

In this chapter, we propose strategies for ANN. The proposed ANN is designed based on the T–S concept and implemented as a state fault-tracking path. This method relies on the state estimation error and the error estimate. Under these conditions, state and error estimation is required, so ANN are designed to estimate the state and error of the system. The main purpose of this chapter is to propose ANN. The proposed ANN is intended for systems affected by actuator faults.

The chapter is organized as follows. Section 2 recalls elementary background on ANN. Section 3 presents the Neuron connection structures. Section 4 presents The learning that is the most interesting property of neural networks. Section 5 the principle of T–S models. Section 6 describes the Kinematic model of the Unicycle-type Robot. Studied the case of system the Mobile Robot an example of simulation showing the robustness of the proposed method for Neural Network design for State and fault estimation of the Mobile Robot.

2 Elementary Background on Artificial Neural Networks (ANN)

2.1 Historic

Today, neural networks thanks to their properties, in particular their learning capacity, are used in many fields [9, 24].

- In 1890 W. James, the famous American psychologist introduced the concept of memory and proposed what would become the law of neural network learning later known as Hebb's law.
- In1943 J. McCulloch and W. Pitts leave their names to a model of the biological neuron (a neuron with binary behavior). At least theoretically, they were the first to show that simple formal neural networks can perform complex symbolic arithmetic and logic functions.
- In1949 D. Hebb, American physiologist explains conditioning in animals through the properties of neurons themselves. The law of modification of the properties of connections between neurons that he proposes partly explains this type of experimental result.
- In1958 Thanks to the work of Rosenblatt, designer of the perceptron, the first ANN appeared. The latter is inspired by the visual system and has an input neuron layer, an output layer ("Decisional"). This network achieve to learn, to identify simple shapes and to calculate some logical functions.
- In1969 M. Minsky and S. Papert highlights the theoretical limitations of the perceptron, in particular concerning the impossibility of dealing with nonlinear problems with this model. They extend these limitations implicitly to any model of ANN, researchers are turning mainly to **AI** (Artificial Intelligence) and rule-based systems.
  It was not until the early 80*s* that interest in the field reappeared.
- In1982 Hopfield demonstrates the interest of using recurrent networks (known as "feedback") for process modeling. Recurrent networks then constitute the second major class of neural networks, with perceptron-type networks (known as "feed-forward").
  In parallel with Hopfield's work, Werbos designed his back-propagation algorithm, which was only popularized in 1986 by Rumelhart.

2.2 Biological Neuron

The neuron is a nerve cell. It is composed of a cellular body called "Soma" which contains the nucleus (where vital cellular activities take place) of extension called "Neuritis" fig. These are of two types, the Dendrites which serve as entrance channels and the axon, unique which is the output channel [24].

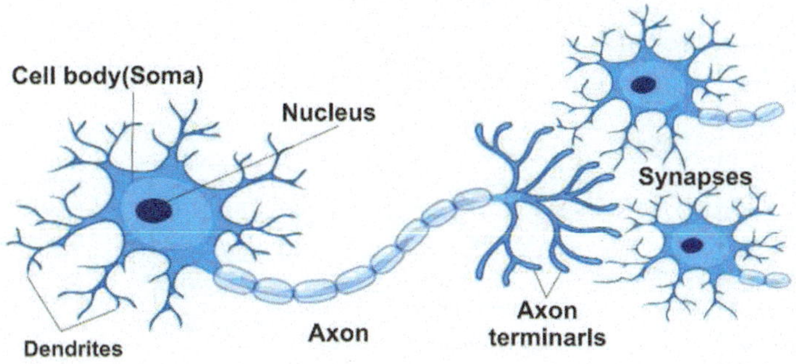

Fig. 1 Biological neuron

Functionally, we consider the neuron as a polarized entity, i.e. information is only transmitted in one direction, from the dendrites to the axon. The neuron will therefore receive information, coming from other neurons, thanks to its dendrites. There will then be a summation, at the level of the cell body, of all this information and via an action potential (electric signal). The result of the analysis will pass along the axon to the synaptic endings. At this location, when the signal arrives, synaptic vesicles will fuse with the cell membrane, which will allow the release of neurotransmitters (chemical mediators) in the synaptic cleft. The electric signal cannot pass the synapse (in the case of a chemical synapse), neurotransmitters therefore allow information to pass from one neuron to another.

Synapses have a "memory" output which allows them to adjust their functioning. According to their "history", i.e. their repeated activation or not between two neurons, they synaptic connections can therefore be modified. Thus, the synapse will or will not facilitate the passage of nerve impulses. This plasticity is the source of learning mechanisms [24] (Fig. 1).

2.3 *Formal Neuron*

Formal neurons are nonlinear, bounded algebraic functions whose values depend on parameters called coefficients or weights. The variables of this function are usually called the "inputs" of the neuron, and the values of the function are called its "outputs".

Neurons are therefore, above all, mathematical operators that can compute numbers in a few lines of software. I've become accustomed to graphing neurons as shown in the Fig. 2.

From observations of biological neurons, follows the formal neuron model proposed by ***W. M. Culloch*** and ***W. Pitts*** in 1943:
* The x_i represent the input vectors, they come either from the outputs of other neurons, or from sensory stimuli (visual, sound sensor, etc.);

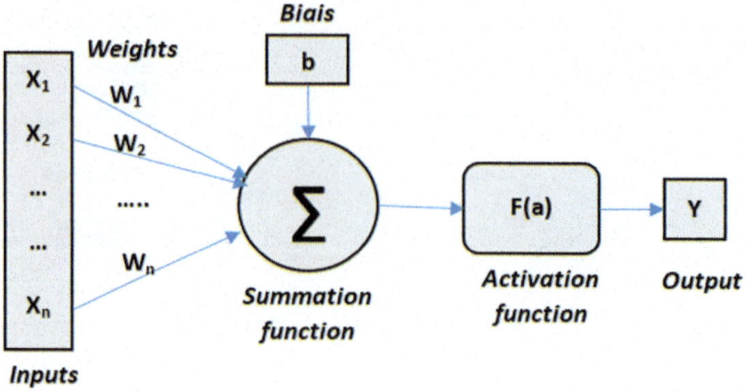

Fig. 2 Model of a formal neuron

* The W_{ij} are the synaptic weights of neuron j. They correspond to synaptic efficiency in biological neurons ($W_{ij} > 0$: excitatory synapse; $W_{ij} < 0$: inhibitory synapse). These weights weight the inputs and can be changed by learning;
* **Bias** b: input often takes the values -1 or $+1$ which allows to add flexibility to the network by allowing to vary the trigger threshold of the neuron by adjusting the weights and the bias during training;
* **Kernel**: Integrates all input and bias and calculates the output of the neuron according to an activation function that is often non-linear to give greater learning flexibility.

2.4 Activation Function

The activation function (or thresholding function, or transfer function) is used to introduce a non-linearity in the functioning of the neuron Threshold functions usually have three intervals:

- Below the threshold (frequently in this case, its output is 0 or -1) hence The neuron is not active.
- around the threshold, a phase of transition.
- above the threshold, (frequently in this case, its output is 1) hence the neuron is active.

In its first version, the formal neuron was implemented with a threshold function, but many versions exist. Thus, the **McCulloch** and **Pitts** neuron was generalized in different ways, by choosing other activation functions, like the functions classified in the table tab tab The three most commonly used functions are "*threshold*", "*linear*", "*sigmoid*" functions (Fig. 3).

Activation function	Equation	Example	1D Graph
Unit step (Heaviside)	$\phi(z) = \begin{cases} 0, & z < 0, \\ 0.5, & z = 0, \\ 1, & z > 0, \end{cases}$	Perceptron variant	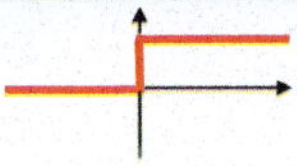
Sign (Signum)	$\phi(z) = \begin{cases} -1, & z < 0, \\ 0, & z = 0, \\ 1, & z > 0, \end{cases}$	Perceptron variant	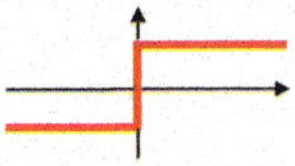
Linear	$\phi(z) = z$	Adaline, linear regression	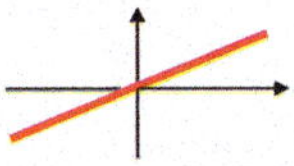
Piece-wise linear	$\phi(z) = \begin{cases} 1, & z \geq \frac{1}{2}, \\ z + \frac{1}{2}, & -\frac{1}{2} < z < \frac{1}{2}, \\ 0, & z \leq -\frac{1}{2}, \end{cases}$	Support vector machine	
Logistic (sigmoid)	$\phi(z) = \frac{1}{1 + e^{-z}}$	Logistic regression, Multi-layer NN	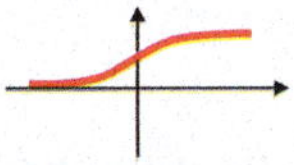
Hyperbolic tangent	$\phi(z) = \frac{e^{z} - e^{-z}}{e^{z} + e^{-z}}$	Multi-layer Neural Networks	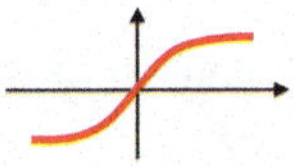
Rectifier, ReLU (Rectified Linear Unit)	$\phi(z) = max(0, z)$	Multi-layer Neural Networks	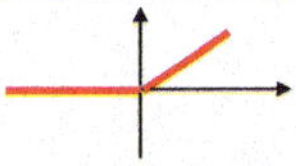
Rectifier, softplus Copyright © Sebastian Raschka 2016 (http://sebastianraschka.com)	$\phi(z) = \ln(1 + e^{z})$	Multi-layer Neural Networks	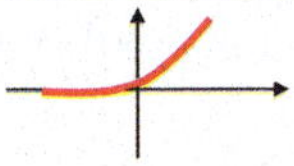

Fig. 3 Transfer function

3 Neuron Connection Structures

A neural network is designed to perform a task that the designer defines by a set of input values, and by a set of corresponding desired values for the activities of some neurons of the network called output neurons. There are two types of neural networks: non-looped networks and looped networks [14, 24, 26].

3.1 Non-looped Static Neural Networks

A network of non-looped neurons also called **feed-forward** is therefore graphically represented by a set of interconnected neurons, information flowing from inputs to outputs without going back [13].

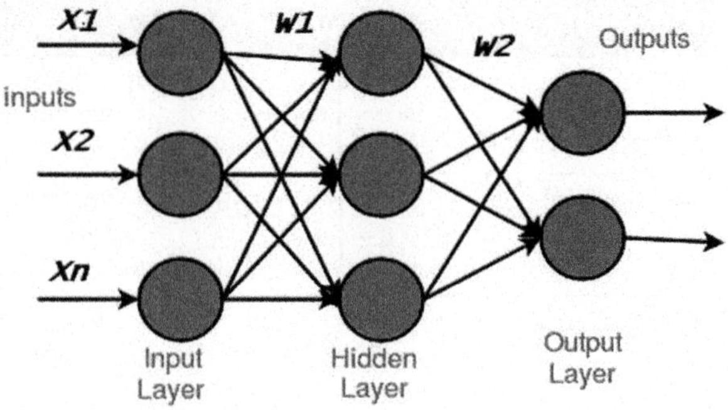

Fig. 4 A non-looped neural network

IF WE MOVE IN THE NETWORK, STARTING FROM ANY NEURON, FOLLOWING THE CONNECTIONS, WE CANNOT RETURN TO THE STARTING NEURON.

3.1.1 Non-looped Neural Network

The jth neuron of the hidden layer l at time k receives (Fig. 4):

$$a_j^l(k) = \sum_{i=1}^{N_{l-1}} w_{ji}^{l-1}(k) x_i^{l-1}(k) \tag{1}$$

The output is given by:

$$s_j^l(k) = g\left(a_j^l(k)\right) \; with \; 1 \le l \le L \tag{2}$$

g: The neuron's activation function.

The jth neuron in the output layer gets:

$$x_j^L(k) = \sum_{i=1}^{N_{L-1}} w_{ij}^{L-1}(k) s_i^{L-1}(k) \tag{3}$$

where
L: total number of layers, l: layer index.
N_i: Number of neurons present in layer l.
$a_j^l(k)$: Input to jth neuron in layer l at time k.
$s_j^l(k)$: Output to jth neuron in layer l at time k.
$w_{ij}^l(k)$: Weight of the connection going from the ith neuron of layer l to the jth neuron of layer $l+1$, at time k.

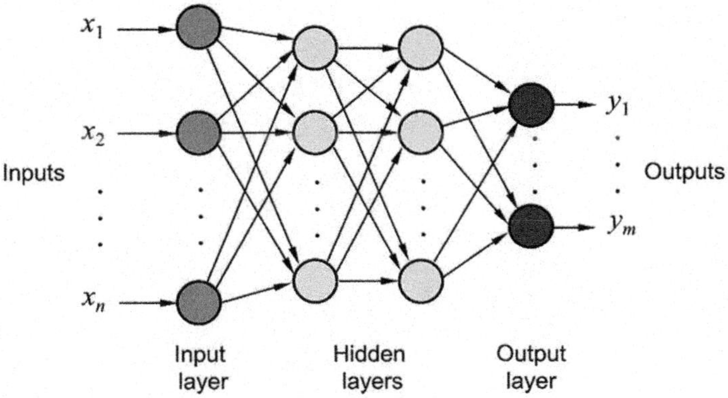

Fig. 5 Multi layer perceptron **MLP**

3.1.2 Multi-layer Neural Networks (Multi Layer Perceptron MLP)

This is an extension of the previous, with one or more hidden layers between input and output [13]. Each neuron in a layer is connected to all the neurons of the previous layer and of the following layer. There are no connections between the cells of the same layer.

It can solve non-linearly separable problems and more complicated logical problems, and especially the famous problems **XOR**. The activation functions used in this type of networks are mainly threshold or sigmoid functions (Fig. 5).

3.1.3 Radial Function Networks RBF

These networks are also called **RBF** (radial basic function). The architecture is the same as for the **MLP** However, the base functions used here are Gaussian functions. **RBF** will therefore be employed in the same types of problems as **MLP** namely, in particular, in classification and approximation of functions (Fig. 6).

3.2 Loop Dynamic Neural Network

Also known as recurring networks **feed-back** these are networks in which there is backtracking of information. The information in the **ANN** can be transmitted in the forward direction and in the reverse direction (backwards transmission), each neuron can receive its own values from the past or the past values from other neurons. Different dynamic neural network architectures are present in the literature. Totally recurrent **ANN**s, memory **ANN**s, networks of **Elman**... (Fig. 7).

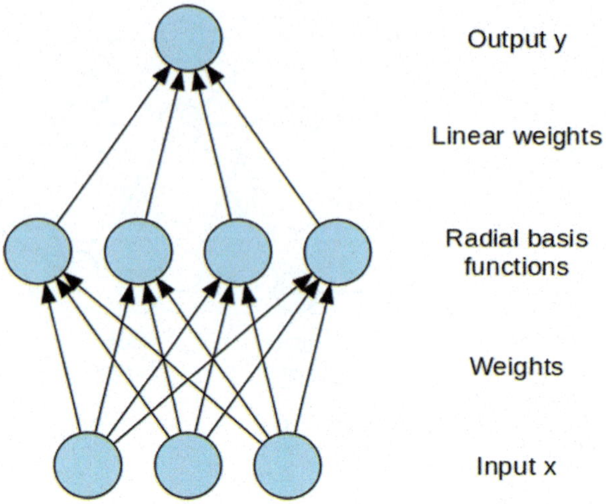

Fig. 6 Radial basic function **RBF**

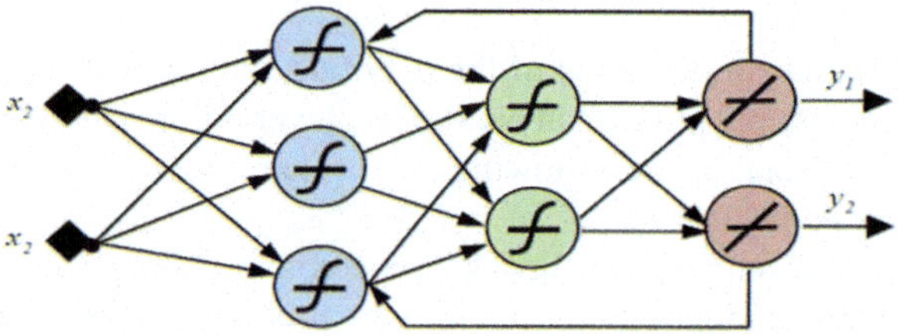

Fig. 7 Looped networks **feed-back**

4 The Learning

Learning is perhaps the most interesting property of neural networks. This is the stage of neural network development where the behavior of the network is modified until the desired behavior is achieved.

The learning is a dynamic and interactive process allowing the parameters of a network to be modified in reaction to the stimuli it receives from its environment. The Type of learning is determined by the way parameter changes occur [23, 26].

There are three types of learning [27]:

1. **Supervised learning:** It is based on a supervisor who provides values of outputs (desired values), in this case the learning consists in adjusting the network parameters to minimize the error between the desired output and the real output of the network.

   The Synaptic weights are those that optimize a performance criterion defined by a cost function see figure.
2. **Unsupervised learning:** In this case the network does not take into consideration the desired values, so there is no supervisor, contrary to, it modifies internal parameters by collecting information in its environment (classification), in other words, he is an autodidact.

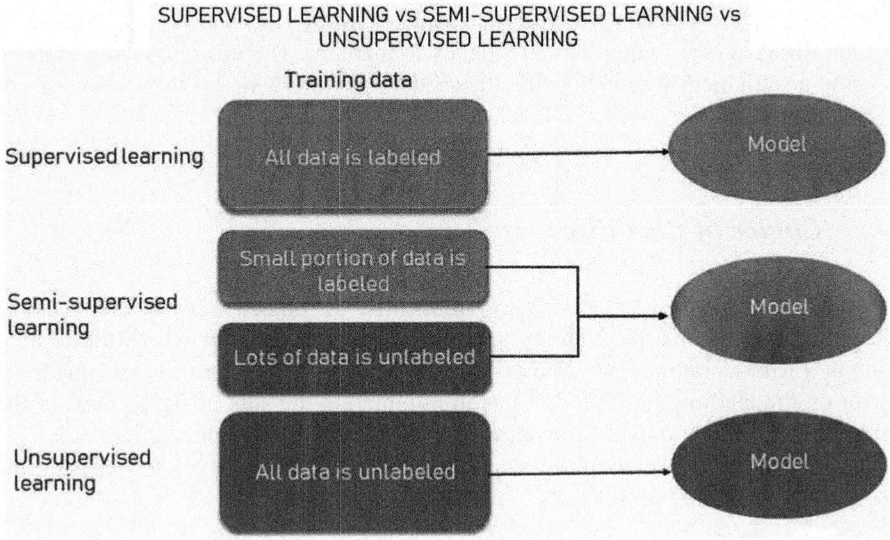

Fig. 8 The types of learning

3. **Semi-supervised learning:** It does not require a reference behavior (values desired), but only Gaussian information (faux, vrais, ...) about the behavior of the network (Fig. 8).

4.1 Concept of Generalization

After parameter learning is complete, the model output will still not reproduce the system output identically and there is still a modeling error. Two reasons are behind this error:

Firstly The model structure does not usually correspond to the exact structure of the system. This error is called bias error. We know in particular that a neural network is capable of approaching any nonlinear function with arbitrary precision. This precision depends on the number of parameters constituting the model. The more parameters introduced into the model, the lower the bias error will be, however, it is obvious that this error is unavoidable.

The second The origin of the modeling error is due to the fact that the training data set is noisy and of limited size. This implies that the parameters obtained during learning will be different from the optimal parameters. This error is called the variance error.

If the bias error decreases with increasing network size, this is not the case for the variance error which has a tendency to increase. Therefore, it is necessary to find a settlement when choosing the model structure between these two sources of error.

4.2 Choice of Cost Function

The choice of the cost function is conditioned by the objective to be achieved. For Regression Problems, the training set consists of examples for which the desired output Y_{di}^{p} is a continuous variable. The most used cost function is the quadratic error on the learning basis: it consists in minimizing the sum of the squares of the errors between the output of the network and the real value of the output.

By noting Y_{di}^{p} the value of the output i calculated by the network for the example p, the cost function becomes

$$E(w) = \sum_{p} E_{P} = \sum_{p} \sum_{i=1}^{N_{i+1}} \frac{1}{2} \left(Y_{di}^{p} - Y_{i}^{p}\right)^{T} \left(Y_{di}^{p} - Y_{i}^{p}\right) \tag{4}$$

This cost function comes from the principle of maximum likelihood with a Gaussian hypothesis on the distribution of outputs.

4.3 Learning Algorithm

The learning algorithm is the mathematical method that will modify the connection weights in order to converge towards a solution that will allow the network to perform the task of optimizing the values of the network weights. Several iterative algorithms can be implemented, including: **The Back-propagation algorithm**, the method of **Quasi-Newton**, Algorithm of **BFGS**, ...

4.4 Algorithms Principle

Let $E(w)$ be the cost function, (4). The algorithms used require that $E(w)$ be derivable compared to the weights. The principle of these methods is to place oneself at an initial point, to find a direction of descent of the cost in the space of the parameters w, then move one step in that direction. We are reaching a new point, and we iterate the procedure until a stopping criterion is satisfied.

Hence, at iteration k, we calculate:

$$\mathrm{w}_{k+1} = w_{k} + \Delta w_{k} = w_{k} + \eta_{k} p_{k} \tag{5}$$

η_k and p_k are the pitch and the direction of the descent. The different algorithms are distinguished by the choice of these two quantities.

4.5 Descent of the Gradient

The objective of the algorithm is to minimize the cost function (4). In other words the function $E(w)$ must decrease at each iteration:

$$\mathrm{E}(\mathrm{w}_{k+1}) < E(w_k) \tag{6}$$

Considering, **the first-order Taylor series expansion** of $E(w)$ in the neighborhood of w_k.

$$\begin{aligned} \mathrm{E}(\mathrm{w}_{k+1}) &= E(w_k + \Delta w_k) = E(w_k) + \nabla_k \Delta w_k \\ \mathrm{E}(\mathrm{w}_{k+1}) &- E(w_k) = \nabla_k \Delta w_k \end{aligned} \tag{7}$$

where, ∇_k is the evaluated gradient for w_k. In order to make sure that $\mathrm{E}(\mathrm{w}_{k+1})$ is less than Ew_k, the second term on the right side of (7) must be negative:

$$\nabla_k \Delta w_k = \eta_k \nabla_k p_k < 0 \tag{8}$$

And since the learning step is small but greater than zero, then:

$$\nabla_k p_k < 0 \tag{9}$$

Any vector p_k which satisfies this condition is called the direction of descent.

The function will decrease if we take a small enough step in this direction.

Which brings us back to another question: **In which direction the descent is faster?**

This happens when $\nabla_k p_k$ is the most negative, when the direction is opposite the gradient and of the same value.

$$p_k = -\nabla_k \tag{10}$$

Replaced in (5) we will have:

$$\begin{aligned} \mathrm{w}_{k+1} &= w_k - \eta_k \nabla_k \\ \Delta w_k &= -\eta_k \frac{\partial E(w)}{\partial w_k} \end{aligned} \tag{11}$$

The algorithm therefore consists in choosing as the direction of descent the opposite of the gradient of the cost function. This method is effective far from the minimum and only allows to approach it. For this reason, the determination of the pitch is not crucial: far from the minimum, you just have to check that the step is neither too small nor too big citeref.

4.6 *Design of a Neural Network*

Neural networks implement parameterized nonlinear functions. Their implementation requires:

- Determination of corresponding inputs and outputs.
- A collection of data required to train a neural network and evaluate its performance.
- Determination of hidden neurons needed to get an acceptable approximation.
- Learning outcomes.
- Analysis of neural network performance after training.

5 Takagi–Sugeno Models

The T–S model allows us to describe a large class of nonlinear systems using a suitable mathematical model. Its principle is to decompose the operating space of the system into a finite number of operating zones. A linear model describes the behavior of the system within each operating zone. The contribution of each submodel of the global model is quantified with a nonlinear weighting function that can have various structures [2, 32].

The T–S model is described by fuzzy IF-THEN rules that describe the locally linear input/output relationship of nonlinear systems. It has a rule base of M rules, each with p preconditions, and a ith rule expressed as:

$$\begin{aligned} R^i &: \text{IF } \xi_1 \text{ is } F_1^i \text{ and } \ldots \text{ and } \xi_p \text{ is } F_p^i \\ \text{THEN} &: \begin{cases} \dot{x}(t) = A_i x(t) + B_i u(t) \\ y = C_i x(t) \end{cases} \end{aligned} \tag{12}$$

where: $i \in \{1, \ldots, M\}$, $F_j^i (j \in \{1, \ldots, p\})$ are fuzzy sets and $\xi = [\xi_1, \xi_2, \ldots, \xi_p]$ is a known vector of premise variables [32] which may depend on the state, the input or the output. Variables ξ is called decision variable.

The global T–S fuzzy model is given by the aggregation of the sub-models using the weighting functions as follows:

$$\begin{cases} \dot{x}(t) = \sum_{i=1}^{M} \mu_i(\xi(t)) \left(A_i x(t) + B_i u(t)\right) \\ y(t) = \sum_{i=1}^{M} \mu_i(\xi(t)) C_i x(t) \end{cases} \tag{13}$$

where $x(t) \in \mathbb{R}^n$ is the state vector, $u(t) \in \mathbb{R}^r$ is the control vector, $y(t) \in \mathbb{R}^m$ is the vector of measures and A_i, B_i and C_i are known constant matrices with appropriate dimensions.

The weighting functions $\mu_i(\xi(t))$ assure a progressive passage between the local models and verifies the convex proprieties:

$\sum_{i=1}^{M} \mu_i(\xi(t)) = 1, \forall t$ and $0 \le \mu_i(\xi(t)) \le 1, \;\; \forall i = 1 \ldots M, \;\; \forall t$

If, in the equation of the output, it is supposed that $C_1 = C_2 = \cdots = C_M = C$, the output of the multiple model (13) is reduced to : $y(t) = Cx(t)$ and the multiple model state equation becomes:

$$\begin{cases} \dot{x}(t) = \sum_{i=1}^{M} \mu_i(\xi(t))(A_i x(t) + B_i u(t)) \\ y(t) = Cx(t) \end{cases} \tag{14}$$

6 The Kinematic Model of the Unicycle-Type Robot

The use of mobile robots has increased significantly in recent years. Today they are available for different applications and can be used in different types of environments such as: B. In education, home, industry, and medicine. Mobile autonomous robots must be able to navigate a priori little-known environments while accomplishing their designed tasks. Therefore, the robot must be able to create a model of the environment, use this model to estimate its current position and orientation in the environment, and navigate the environment to reach the target point [3, 21].

There are different control problems for the mobile robot:

- Path tracking where the objective is for a point linked to the robot to follow a predetermined curve by imposing a given speed on the robot.
- The stabilization of trajectories consisting of taking into account the temporal dimension: the reference trajectory depends on time and the speed of the robot is no longer fixed in advance.
- stabilization of fixed configurations: it is a question of asymptotically stabilizing the system in a given equilibrium position [12, 15].

In this work, we are only interested in the problem of trajectory tracking for a non-holonomic unicycle type mobile robot.

6.1 Modeling and Structural Properties

There are many models of mobile robots according to their structure, the type of wheels used,.... The two most frequently encountered classes of models: unicycle-type robots and car-type robots.

6.1.1 Hypotheses

Generally for the control of mobile robots, a speed control model is used rather than a control model. The main reasons for this choice are:

- The command calculation is easier for a kinematic model than for a dynamic model.
- There are no complicated geometric or inertial parameters to identify for a kinematic model.
- In the case of the miniature mobile robots used in our application (MIABOT robot), the inertia is low and the dynamics of the electric motorization is very fast.

For these reasons, we consider in the following only kinematic models taking into account the following simplifying assumptions:

- The mobile robot is considered as a rigid vehicle moving in a horizontal plane.
- Conventional wheels are supposed to be non-deformable, of radius noted thereafter r.
- Each wheel/ground contact is reduced to one point.
- The wheels roll without slipping on the ground.

We direct our research towards unicycle type mobile robots. The **MIABOT Mobile Robot** used in this application is a one-wheeled mobile robot. It consists of two independent drive wheels in the middle of the robot body. It consists of two independent drive wheels in the middle of the robot body. The mobile unicycle robot model corresponds to the model of wheels rolling on a plane, as shown in Fig. 9.

6.2 Kinematic Non Linear Model

The kinematic non linear model of this mobile robot can be presented in the case of non-slip bearing by the following formula [29]:

$$\begin{cases} \dot{x} = \nu cos(\theta) \\ \dot{y} = \nu sin(\theta) \\ \dot{\theta} = \omega \end{cases} \tag{15}$$

where ν and ω are the controls of the mobile robot, linear and angular speed respectively. x and y are the output variables and θ represents the orientation of the robot

Fig. 9 Robot mobile MIABOT

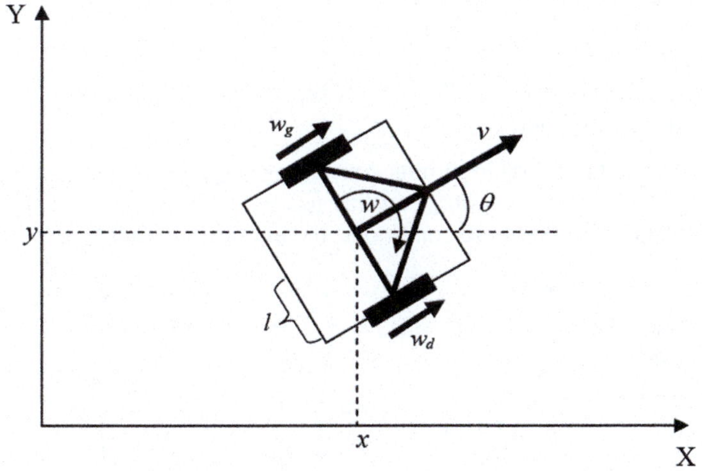

Fig. 10 Unicycle mobile robot

(angle between the X axis and the robot speed vector). The Fig. 10 gives the diagram of the unicycle mobile robot.

We denote by r the radius of the robot wheel and l the distance between the wheel and the center of the robot. Angular speeds of the left and right wheels, ω_g and ω_d, are given by the following formula:

$$\begin{bmatrix} \omega_d \\ \omega_g \end{bmatrix} = 1/r \begin{bmatrix} 1 & l \\ l & -l \end{bmatrix} \begin{bmatrix} \nu \\ \omega \end{bmatrix}$$

6.3 T–S Model of the Robot

The kinematic non linear model of the robot can be represented by a T–S model with 16 local models. This model is given by the following equation:

$$\begin{cases} \dot{x}(t) = \sum_{i=1}^{16} \mu_i(\xi(t))(A_i x(t) + B_i u(t)) \\ y(t) = Cx(t) \end{cases} \tag{16}$$

The state and control matrices are defined as follows for $i \in \{1, \ldots, 16\}$:

$$A_i = \begin{pmatrix} 0 & -\epsilon_i^1 \omega_{r,max} & 0 \\ \epsilon_i^1 \omega_{r,max} & 0 & v_i \\ 0 & 0 & 0 \end{pmatrix}, \quad B_i = \begin{pmatrix} -1 & \epsilon_i^2 e_{max} \\ 0 & \epsilon_i^3 e_{max} \\ 0 & -1 \end{pmatrix}. \tag{17}$$

with: $e_{max} = 0.1(m)$, $\epsilon_i^3 = (-1)^{i+1}$,

$\epsilon_i^1 = \begin{cases} +1 & for \;\; 1 \le i \le 8 \\ -1 & otherwise \end{cases}$,

$v_i = \begin{cases} \frac{2}{\pi} \nu_{r,min} & for \;\; 1 \le i \le 4 \;\; and \;\; 9 \le i \le 12 \\ \nu_{r,max} \end{cases}$

and $\epsilon_i^2 = \begin{cases} +1 & for \;\; i \in \{3, 4, 7, 8, 11, 12, 15, 16\} \\ -1 & otherwise \end{cases}$

We will also use the following limits concerning the reference trajectory: $v_{r\,\min} \le v_r \le v_{r\,\max}$, $-w_{r\,\min} \le w_r \le w_{r\,\max}$, With $v_{\min} = 0.755$ m.s, $v_{\max} = 0.02$ m.s and $w_{\max} = 2.55$ rad.s. Under these conditions, the matrices A_i and B_i are:

$$A_1 = A_2 = A_3 = A_4 = \begin{pmatrix} 0 & -2.55 & 0 \\ 2.55 & 0 & 1.185 \\ 0 & 0 & 0 \end{pmatrix},$$

$$A_5 = A_6 = A_7 = A_8 = \begin{pmatrix} 0 & -2.55 & 0 \\ 2.55 & 0 & 0.02 \\ 0 & 0 & 0 \end{pmatrix},$$

$$A_9 = A_{10} = A_{11} = A_{12} = \begin{pmatrix} 0 & 2.55 & 0 \\ -2.55 & 0 & 1.185 \\ 0 & 0 & 0 \end{pmatrix},$$

$$A_{13} = A_{14} = A_{15} = A_{16} = \begin{pmatrix} 0 & 2.55 & 0 \\ -2.55 & 0 & 0.02 \\ 0 & 0 & 0 \end{pmatrix},$$

$$B_1 = B_5 = B_9 = B_{13} = \begin{pmatrix} -1 & -0.1 \\ 0 & 0.1 \\ 0 & -1 \end{pmatrix},$$

$$B_2 = B_6 = B_{10} = B_{14} = \begin{pmatrix} -1 & -0.1 \\ 0 & -0.1 \\ 0 & -1 \end{pmatrix},$$

$$B_3 = B_7 = B_{11} = B_{15} = \begin{pmatrix} -1 & 0.1 \\ 0 & 0.1 \\ 0 & -1 \end{pmatrix},$$

$$B_4 = B_8 = B_{12} = B_{16} = \begin{pmatrix} -1 & 0.1 \\ 0 & -0.1 \\ 0 & -1 \end{pmatrix}.$$

7 Neural Network Design for State and Fault Estimation of the Mobile Robot

The main objective of this part is to use the neural network to apply the proposed techniques of state and fault estimation on the T–S model of the robot described in the previous section.

7.1 *State Estimation*

For the simulation, the considered system input $u(t)$ is: $u(t) = [u_1^T(t) \quad u_2^T(t)]^T$, with:

$$u_1(t) = 2 * sin(0.1\pi t) \quad \text{and}$$
$$u_2(t) = \begin{cases} 0.25 \quad pour \quad t \leq 30\,\text{s} \\ 0 \quad pour \quad 30\,\text{s} < t \leq 50\,\text{s} \\ 0.25 \quad pour \quad 50\,\text{s} < t \leq 70\,\text{s} \\ 0 \quad pour \quad t > 70\,\text{s} \end{cases}$$

We use the **MATLAB** application, we have as system inputs 5 inputs which are the 3 system states x_1, x_2 and x_3 and The system input $u(t) = [u_1(t), u_2(t)]$ applied to the system, and the desired neuron output are the 3 estimated states ($x_1$estimated, $x_2$estimated, $x_3$estimated).

When the "logsig" function is used, we generate a single hidden layer neural network containing 12 neurons, and the output layer of 3 neurons is chosen as the activation function for the first "logsig". The learning mechanism should be able to adjust the weights to reduce the error in the interval (0.0 to 1.0) and select the second "purelin" to optimize the error, as shown in the following Fig. 11.

The Figs. 12, 13 and 14 respectively illustrate the rate of following the path of the states and are estimated.

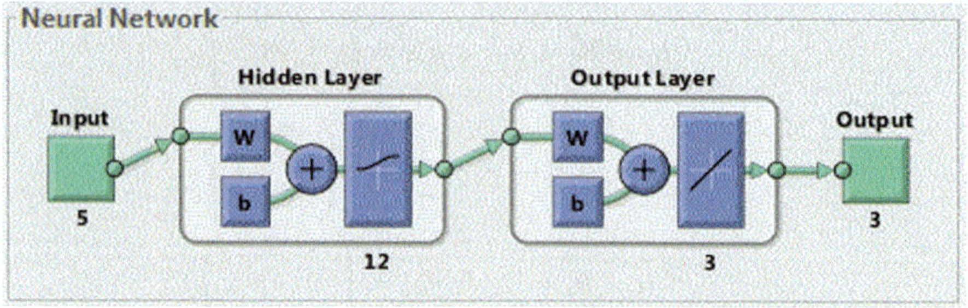

Fig. 11 The neural network

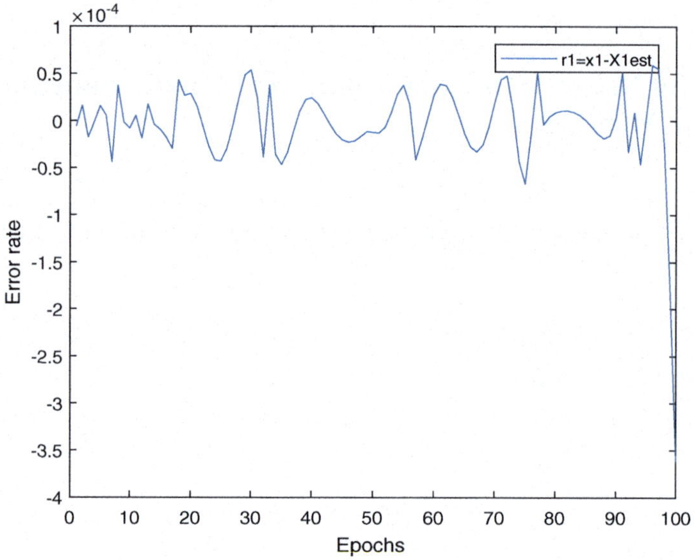

Fig. 12 Follow-up rate of $r1$ robot state

As shown by the curves the shape of the estimated state X_{iest} almost coincides with the shape of the state of the system X_i.

Then in can say that the network achieves the goal of following the path of the robot with an acceptable estimate.

7.2 Fault Estimation

It is assumed that the system (16) is affected by an actuator fault.

The goal of this section is to estimate an actuator fault affecting a nonlinear system (in our case the mobile robot) described by T–S model via an ANN. The considered system is submitted to an actuator fault $f(t)$. It is given by the following structure:

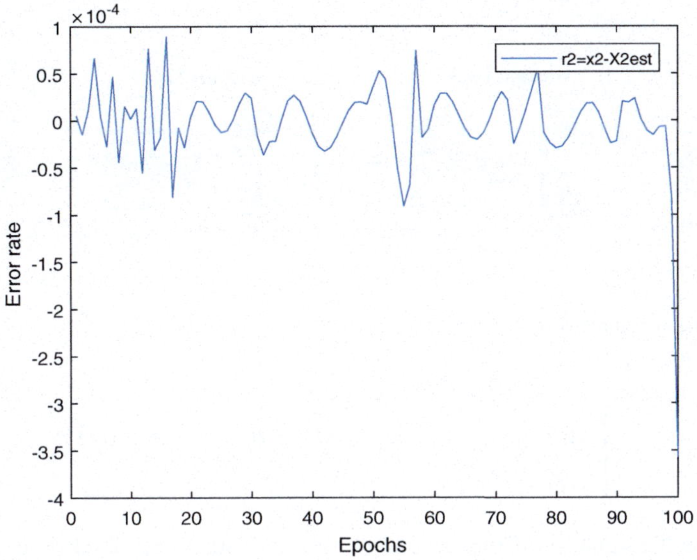

Fig. 13 Follow-up rate of $r2$ robot state

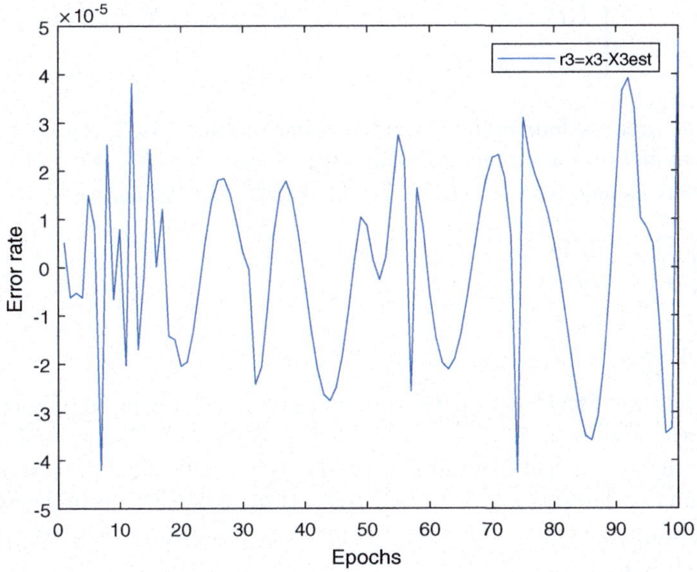

Fig. 14 Follow-up rate of $r3$ robot state

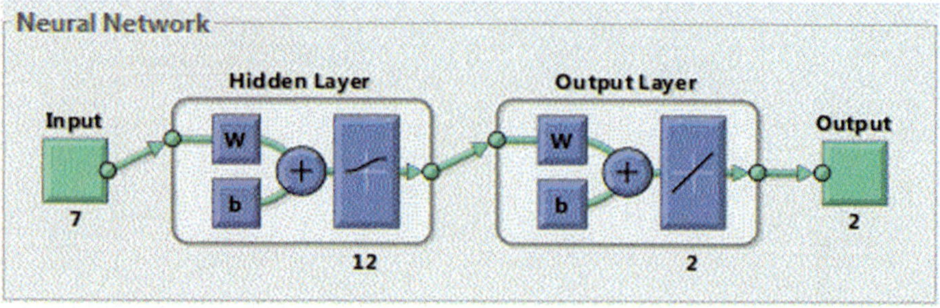

Fig. 15 The neural network

$$\begin{cases} \dot{x}(t) = \sum_{i=1}^{M} \mu_i(\xi(t))(A_i x(t) + B_i u(t) + E_i f(t)) \\ y(t) = Cx(t) \end{cases} \quad (18)$$

For the fault estimation, Consider the nonlinear system described by a T–S mode given by the following equations:

$$\begin{cases} \dot{x}(t) = \sum_{i=1}^{16} \mu_i(\xi(t))(A_i x(t) + B_i u(t) + E_i f(t)) \\ y(t) = Cx(t) \end{cases} \quad (19)$$

we keep the same system input $u(t)$ and the same matrices $A_1, \ldots A_{16}$, $B_1, \ldots B_{16}$ and C of T–S model for the mobile robot, and we choose $E_i = B_i, \ \forall i \in \{1, \ldots, 16\}$.

The faults signals $f(t)$ is: $f(t) = \begin{bmatrix} f_1(t) & f_2(t) \end{bmatrix}^T$ are defined as:

$$f_1 = \begin{cases} 0, & t \leq 0.4 \text{ s} \\ 0.1 * sin(\pi t), & t > 0.6 \text{ s} \end{cases}$$

$$f_2 = \begin{cases} 0, & t \leq 1.5 \text{ s} \\ 0.4, & t > 1.5 \text{ s} \end{cases}$$

We are using the MATLAB application, but this time as network inputs there are 7 inputs which are (the 3 system states x_1, x_2 and x_3 and The system input $u(t)$ applied to the system and 2 faults $f = (f_1(t), f_2(t))$ as then the desired output will be the estimated defects ($f_1(t)$ estimed, $f_2(t)$ estimed). Create a neural network with the same design (a single hidden layer contains twelve neurons and an output layer of two neurons).

For the activation function: the first chosen “logsig”, when using of the logsig function, the learning mechanism must be able to adjust the weights, to reduce the error in the intervals from (0, 0 to 1, 0) and the second “purelin” choose to optimize the error as shown in the following Fig. 15.

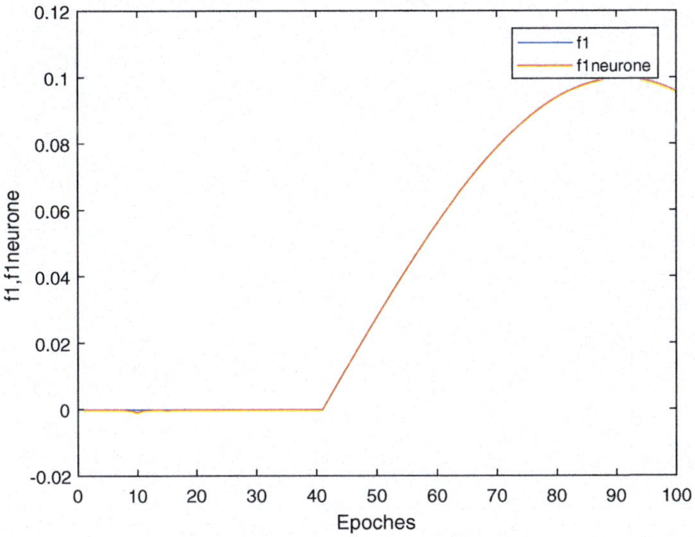

Fig. 16 The robot fault 1 and their estimation

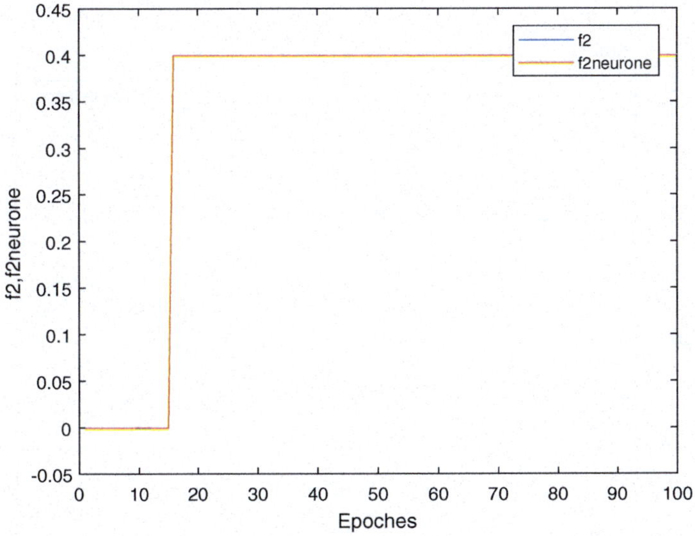

Fig. 17 The robot fault 1 and their estimation

As shown in the Figs. 16 and 17, the estimated error on the network follows the actually applied error. We can clearly see that the network responds quickly and can track the trajectory better from the first step of error (Figs. 18 and 19).

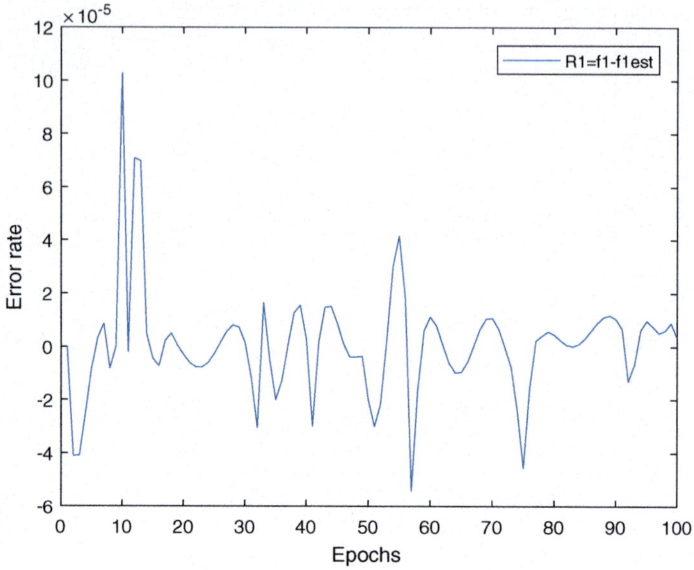

Fig. 18 Follow-up rate of f1 robot default

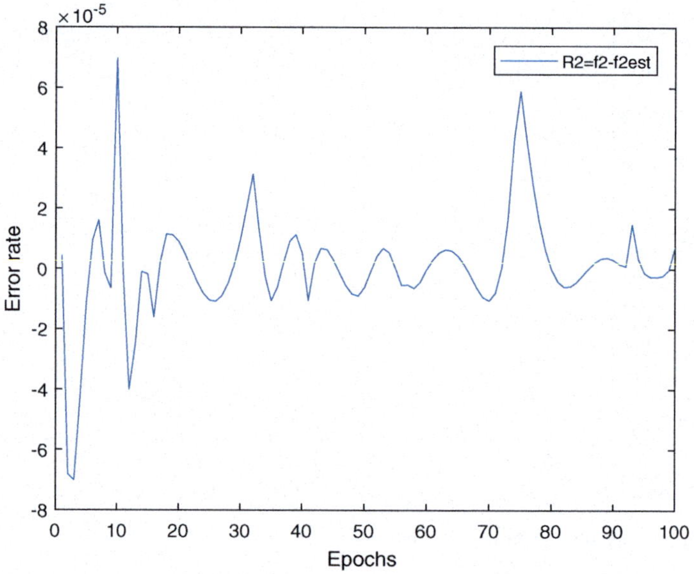

Fig. 19 Follow-up rate of f2 robot default

8 Conclusion

For nonlinear systems modeled according to the Takagi–Sugano approach, this chapter presents a procedure for estimating system states and actuator errors that can affect the system. The presented method uses an ANN that can track the trajectory of the system. You can also estimate system states and actuator errors.

The effectiveness of the proposed method is demonstrated by applying it to the Takagi–Sugano model of the mobile MIABOT robot. Simulation results show that the proposed method allows good estimation of actuator error and system state. In fact, the estimation error is less than 10^{-5}, which is less than 1% of error.

References

1. Aglodiya, A.: Application of artificial neural network (ANN) in chemical engineering: a review. Int. J. Adv. Res. Innov. Ideas Educ. **03**(02) (2017)
2. Akhenak, A., Chadli, M., Maquin, D., Ragot, J.: Estimation of state and unknown inputs of a non linear system represented by a multiple model. In: 11th IFAC Symposium on Automation in Mining, Mineral and Metal Processing (MMM'04) (2004). ISSN 2075-1702
3. Al-Sagban, M., Dhaouadi, R.: Neural based autonomous navigation of wheeled mobile robots. J. Autom. Mobile Robot. & Intell. Syst. **10**(2) (2016)
4. Dai, P., Yu, X.: An artificial neural network approach for solving space fractional differential equations. Symmetry **14**(03) (2022). https://doi.org/10.3390/sym14030535
5. El Jery, A., Khudhair, A.K., Abbas, S.Q., Abed, M.A., Khedher, K.M.: Numerical simulation and artificial neural network prediction of hydrodynamic and heat transfer in a geothermal heat exchanger to obtain the optimal diameter of tubes with the lowest entropy using water and Al2O3/water nanofluid. Geothermics **107**, Article 102605 (2023)
6. El-Hadi, G., Karim, B., SaŽo, B.: Tracking control for wheeled mobile robot based on delayed sensor measurements. Sensors **19**(23), Article 5177 (2019). https://doi.org/10.3390/s19235177
7. Essam, B.M., Ahmed, H.H., Ammar, H.E.: A new optimized artificial neural network model to predict thermal efficiency and water yield of tubular solar still. Case Stud. Therm. Eng. **30**, Article 101750 (2022)
8. Federico, P., Francesco, B., Gianluigi, R.: An artificial neural network approach to bifurcating phenomena in computational fluid dynamics. Comput. & Fluids **254**, Article **105813**, 3 (2023)
9. Frank, E., Zhen, Y., Han, F.,Shailesh, T., Matthias, D.: An introductory review of deep learning for prediction models with big data. Frontières de l'intelligence artificielle. **03** (2020). https://doi.org/10.3389/frai.2020.00004
10. Hatem Heriz, H., Mohammed Salah, H., Abu Abdu, S., El Sbihi, M., Abu-Naser, S.: English alphabet prediction using artificial neural networks. Int. J. Acad. Pedag. Res. (IJAPR) **02**(11), 08–14 (2018). ISSN: 2000-004X
11. Ibham, V., Asif, A., Mujtaba, M.A., Anh, T.H., Dhinesh, B., Manigandan, S., Fattah, I.M.R., Soudagar, M.E.M., Ahmed, I.ELS., Djamari, D.W., Hananto, A.L., Putra, N.R., Noreffendy, T.: Review of artificial neural networks for gasoline, diesel and homogeneous charge compression ignition engine. Alex. Eng. J. **61**(11), 8363–8391 (2022)
12. Jing-Jun, Z., Zhi-Li, F., Zhong-Qi, Z., Rui-Zhen, G., Shao-Bo, Z.: Trajectory tracking control of nonholonomic wheeled mobile robots using model predictive control subjected to lyapunov-based input constraints. Int. J. Control Autom. Syst. **20**, 1640–1651 (2022). https://doi.org/10.1007/s12555-019-0814-x

13. Kevin, L., Ellen, K.: A new family of constitutive artificial neural networks towards automated model discovery. Comput. Methods Appl. Mech. Eng. **403**, Part A, Article 115731, 1 (2023)
14. Khalil, K., Eldash, O., Kumar, A., Bayoumi, M.: An efficient approach for neural network architecture. In: 25th IEEE International Conference on Electronics, Circuits and Systems (ICECS), pp. 745–748 (2018). https://doi.org/10.1109/ICECS.2018.8617887
15. Li, L., Weiqun, C., Hongjiu, Y., Qing, G.: Trajectory tracking control for a wheel mobile robot on rough and uneven ground. Mechatronics **83**, Article 102741 (2022)
16. Luis, E., Jesus, H.-D., Pablo, G.-de-S.: Toward autonomous mobile robot navigation in early-stage crop growth. In: 19th International Conference on Informatics in Control, Automation and Robotics (ICINCO 2022), 14–16 July, 2022. Science and Technology Publication, Lisbon–Portugal. ISSN: 2184-2809
17. Malu, K., Majumdar, J.: Kinematics, localization and control of differential drive mobile robot. Global J. Res. Eng.: H Robot. & Nano-Tech **14**(01) (2014)
18. Martin, V., Aguilar, G., Pèrez, S., Pèrez, H., Jimènez, E.: Artificial neural networks: a practical review of applications involving fractional calculus. Eur. Phys. J. Special Topics 2059–2095 (2022). https://doi.org/10.1140/epjs/s11734-022-00455-3
19. Miri, D., Khedher, A., Belkhiria, K.: State and faults estimation via artificial neural networks. In: 20th International Conference on Sciences and Techniques of Automatic Control and Computer Engineering, pp. 57–62. https://doi.org/10.1109/STA50679.2020.9329322. Accessed 20–22 Dec 2020
20. Miri, D., Khedher, A., Belkhiria, K.: Tracking of trajectory, state and fault estimation of MIABOT robot using an artificial neural network. In: 18th International Multiconference on Systems Signals & Devices SSD, pp. 1296–1301. https://doi.org/10.1109/SSD52085.2021.9429397. Accessed 22–25 March 2021
21. Nguyen, D.D., Roan, V.H., Pham, V.M., Pham, V.H., Le Van, A., Tran Duc, C.: Research, design and control mobile robots for intelligent navigation based on ROS programming. J. Mech. Eng. Res. Devel. **45**(1), 34–45 (2022). ISSN: 1024-1752
22. Nima, K., Mohammad, S.M., Abolfazl, M., Mehrdad, K.: Spatio-temporal modeling of COVID-19 prevalence and mortality using artificial neural network algorithms. Spatial Spatio-Temp. Epidemiol. **40**, Article 100471 (2022)
23. Nunes da Silva, I., Hernane Spatti, D., Andrade Flauzino, R., Helena Bartocci Liboni, L., Franco dos Reis Alves, S.: Artificial Neural Networks. Springer International Publishing Switzerland, eBook (2017). https://doi.org/10.1007/978-3-319-43162-8
24. Pam, M., John, Th., Design of neural networks. In: The 14th International Conference on Axiomatic Design (ICAD 2021) (2021). https://doi.org/10.1088/1757-899X/1174/1/012023
25. Pappalardo, C.M., Guida, D.: Forward and inverse dynamics of a unicycle-like mobile robot. In: The 18th International Conference Research and Development in Mechanical Industry (RaDMI-2018) (2018). ISSN 2075-1702
26. Roza, D., Mohsen, S.: Artificial neural network systems. Int. J. Imaging Robot. **21**(2) (2021)
27. Sathya, R., Annamma, A.: Comparison of supervised and unsupervised learning algorithms for pattern classification. Int. J. Adv. Res. Art. Intell. (IJARAI) **02**(02) (2013)
28. Shivam, Ch., Vicenç, P.: Autonomous vehicle state estimation and mapping using Takagi–Sugeno modeling approach. Sensors **22**(9), Article 3399 (2022). https://doi.org/10.3390/s22093399
29. Wang, F., Qin, Y., Guo, F., Ren, B., Yeow, J.T.W.: Adaptive visually servoed tracking control for wheeled mobile robot with uncertain model parameters in complex environment. Hindawi **2020**, ID 8836468, 13 (2020). https://doi.org/10.1155/2020/8836468
30. Xu, T., Zhang, S., Jiang, Z., Liu, Z., Cheng, H.: Collision avoidance of high-speed obstacles for mobile robots via maximum-speed aware velocity obstacle method. IEEE Access **08**(04), 138493–138507 (2020). https://doi.org/10.1109/ACCESS.2020.3012513

31. Xuefei, W., Chi, C., Jiale, L., Jianmin, Z., Guowei, M., Jinzhao, J.: Automated monitoring and evaluation of highway subgrade compaction quality using artificial neural networks. Autom. Constr. **145**, Article 104663 (2023)
32. Yadaiah, N., Sowmya, G.: Neural network based state estimation of dynamical systems. In: International Joint Conference on Neural Networks, July 16–21 (2006)

Sliding Mode Fault Tolerant Control Against Actuator Failures for UAVs

Chaymaa Targhi, Abdellah Benaddy, Moussa Labbadi, and Mostafa Bouzi

Abstract In this study, the difficulty of developing a sliding mode control method based on fault tolerant against actuator failures for a quadrotor UAV is presented. The system of quadrotors is divided into two subsystems: the position subsystem and the attitude subsystem. First, the actuator fault model is determined, a sliding mode nonlinear controller is analyzed with the objective of designing a fault-tolerant law, the concept is to introduce faults to the system during operation and to control the stability of the quadrotor. Then, the system's stability is ensured using the Lyapunov theory. The simulation results show the efficacy of the sliding mode control approach proposed.

Keywords Quadrotor · Fault-tolerant control · Lyapunov function · Nonlinear control · Actuator fault

1 Introduction

Small unmanned aerial vehicles (UAVs), particularly quadrotors, have become very important in various applications [1], thanks to their ability to hover, fly at high and low speed, move in all directions and take off and land vertically [2]. However, their reliability must be ensured before they can be used [3]. In recent years the great challenge for researchers is to develop algorithms capable of ensuring a flight path tracking of high precision in the presence of perturbations [4]. The robustness of

C. Targhi (✉) · A. Benaddy · M. Bouzi
MIET Laboratory, Faculty of Science and Technology, Hassan First University of Settat, Settat, Morocco
e-mail: c.targhi@uhp.ac.ma

A. Benaddy
e-mail: a.beanddy@uhp.ac.ma

M. Labbadi
GIPSA-lab, CNRS, Grenoble INP, Université Grenoble Alpes, 38000 Grenoble, France
e-mail: moussa.labbadi@gipsa-lab.grenoble-inp.fr

A. Ben Makhlouf et al. (eds.), *State Estimation and Stabilization of Nonlinear Systems*, Studies in Systems, Decision and Control 491,
https://doi.org/10.1007/978-3-031-37970-3_16

the quadrotor system against external disturbances can be ensured by applying a law of integral terminal sliding mode control [5]. A controller based on the second order sliding mode approach is a stable control law for tracking the trajectory of a disturbed quadrotor [6, 7]. The authors in [8] propose gain optimization procedures for PID controllers. However, many control methods have been realized to maintain quadrotor stability in the existence of failures or external disturbances. Such as, the modified supertwisting fast nonlinear sliding mode controller (MSTFNSMC) [9], the adaptive control law ANFTSM [10]. Second order sliding mode controllers [11]. A sliding mode control technique-based nonlinear control approach is employed to control the position and attitude of the UAV in [12]. A sliding mode controller with a high degree of robustness is used to stabilize a linked nonlinear system of second order, after that it is generalized to command a group of multiple input and output nonlinear systems with model uncertainties. To stabilize the quadrotor's attitude and complete the trajectory tracking task, a nonlinear controller that use integral backstepping in conjunction with sliding mode control (integral BS-SMC) is proposed in [13, 14]. A design of a fractional operator-based global sliding mode control scheme for tracking the trajectory of a quadrotor is studied by the authors in [15]. A fast terminal non-singular adaptive fractional order sliding mode controller (AFONFTSMC) is proposed in [16].

In this paper, a sliding mode setup for compensating for actuator failure applied to a quadrotor is proposed. To manage operational failures, a sliding mode surface is created and a control strategy is constructed based on the ideal control rule, under proposed actuator failure scenarios. Lyapunov theory is employed to analyze the system stability. Our proposed control approach successfully handles the actuator faults while improving the performance and reliability of the control system.

This test is structured as follows. The quarotor is described in Sect. 2 of the paper. The proposed fault-tolerant SMC controller's design is detailed in Sect. 3. Section 4 presents and discusses the simulation results. Finally, Sect. 5 presents the paper's major conclusion as well as suggestions for future work.

2 Prboblem Statement

2.1 The Quadrator Dynamics

Figure 1 shows the configuration of the UAV four rotors. The reference frames {E} represented by (o_e, x_e, y_e, z_e), and the frames attached to the body (o_b, x_b, y_b, z_b), We can define the quadrotor's absolute position as follows: $\varrho = [x, y, z]^T$. The Euler angles roll, pitch, and yaw can be described as follows: $\Gamma = [\phi, \theta, \psi]^T$. Let $\Omega = [p, q, r]^T$ the quadrotor's angular velocity, however, the vector $\vartheta = [\vartheta_1, \vartheta_2, \vartheta_3]^T$ is the linear velocity. The transformation matrix of conversion of the coordinates set to the earth in inertial frame is as seen below:

Fig. 1 Four-rotor drone configuration

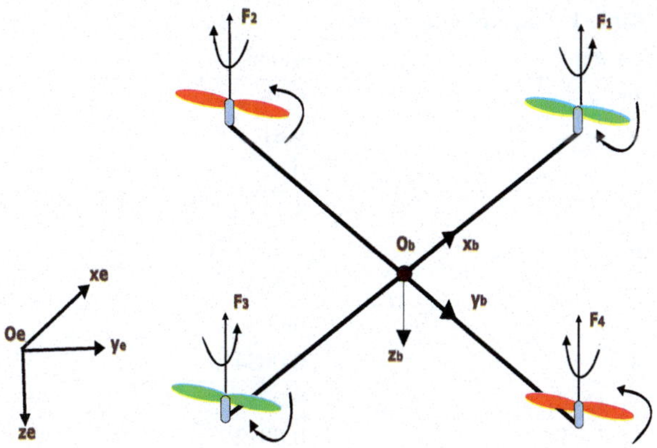

$$R = \begin{bmatrix} c\psi c\theta & s\phi s\theta c\psi - c\phi s\psi & c\phi s\theta c\psi + s\phi s\psi \\ s\psi c\theta & s\phi s\theta s\psi + c\phi c\psi & c\phi s\theta s\psi - s\phi c\psi \\ -s\theta & s\phi c\theta & c\phi s\theta \end{bmatrix} \quad (1)$$

Where c and s stand for the cos and sin functions, respectively. The rotation matrix $R.T$ is defined as follows:

$$R.T = \begin{bmatrix} 1 & 0 & -\sin(\theta) \\ 0 & \cos(\phi) & \sin(\phi)\cos(\theta) \\ 0 & -\sin(\phi) & \cos(\phi)\cos(\theta) \end{bmatrix} \quad (2)$$

Defining the state space of the plant by the vector: $x = [x\ \dot{x}\ y\ \dot{y}\ z\ \dot{z}\ \phi\ \dot{\phi}\ \theta\ \dot{\theta}\ \psi\dot{\psi}]^T = [x_1\ x_2\ x_3\ x_4\ x_5\ x_6\ x_7\ x_8\ x_9\ x_{10}\ x_{11}\ x_{12}]^T$.

$Y = [x_1\ x_3\ x_5\ x_7\ x_9\ x_{11}]^T$ the output representation of the system is as follows:

$$\begin{cases} \dot{x}_1 = x_2 \\ \dot{x}_2 = l_1 x_2 + w_1 \\ \dot{x}_3 = x_4 \\ \dot{x}_4 = l_2 x_4 + w_2 \\ \dot{x}_5 = x_6 \\ \dot{x}_6 = l_3 x_6 + w_3 \\ \dot{x}_7 = x_8 \\ \dot{x}_8 = f_1 x_{10} x_{12} + f_4 x_7^2 + f_7 \omega_r x_{10} + r_1 \Lambda_2 \\ \dot{x}_9 = x_{10} \\ \dot{x}_{10} = f_2 x_8 x_{12} + f_5 x_{10}^2 + f_8 \omega_r x_8 + r_2 \Lambda_3 \\ \dot{x}_{11} = x_{12} \\ \dot{x}_{12} = f_3 x_8 x_{10} + f_6 x_{12}^2 + r_3 \Lambda_4 \end{cases} \quad (3)$$

Table 1 Quadrotor parameters

Symbol	Value	Unit
m	0.74	kg
g	9,81	m.s^{-2}
l	0,5	m
b	4.147.10^{-5}	N/rad/s
d	2.6024.10^{-6}	N.M/rad/s
I_x	0.004	kg.m^2
I_y	0.004	kg.m^2
I_z	0.0084	kg.m^2
k_1, k_2, k_3	0.001	m^{-1}.s.N
k_4, k_5, k_6	0.0012	rad^{-1}.s.N

with

$$l_1 = \frac{-k_1}{m},\ l_2 = \frac{-k_2}{m},\ l_3 = \frac{-k_3}{m},\ f_1 = \frac{I_y-I_z}{I_x},\ f_2 = \frac{I_z-I_x}{I_y},\ f_3 = \frac{I_x-I_y}{I_z},\ f_4 = \frac{-k_4}{I_x}$$
$$f_5 = \frac{-k_5}{I_y},\ f_6 = \frac{-k_6}{I_z},\ f_7 = \frac{-J_r}{I_x},\ f_8 = \frac{-J_r}{I_y},\quad r_1 = \frac{l}{I_x},\ r_2 = \frac{l}{I_y},\ r_3 = \frac{l}{I_z}.$$

The virtual control signals are defined as follows [7]

$$\begin{cases} w_1 = (c\phi s\theta c\psi + s\phi s\psi).\frac{\Lambda_1}{m} \\ w_2 = (c\phi s\theta s\psi - s\phi c\psi).\frac{\Lambda_1}{m} \\ w_3 = \quad -g + (c\phi.c\theta).\frac{\Lambda_1}{m} \end{cases} \tag{4}$$

The required angles (ϕ_d, θ_d) are not entirely reliant on yaw, the stresses in roll and pitch have increased according to the non-linear equation, and the overall thrust Λ_1 can be represented as follows:

$$\begin{cases} \Lambda_1 = m.\sqrt{w_1^2 + w_2^2 + (w_3+g)^2} \\ \phi_d = \arctan(\cos\theta_d(\frac{w_1 \sin\psi_d - w_2 \cos\psi_d}{w_3+g})) \\ \theta_d = \arctan((\frac{w_1 \cos\psi_d + w_2 \sin\psi_d}{w_3+g})) \end{cases} \tag{5}$$

Table 1 presents the physical characteristics of the quadrotor.

2.2 *Actuator Faults Modeling*

The model of actuator failure described can be utilized to mimic a large number of failure cases [17]. When there are actuator problems, $\Lambda(t)$ can be written as:

$$\Lambda = \rho\nu(t) + \sigma(\Lambda_i(t) - \rho\nu(t)) = \rho(I - \sigma)\nu(t) + \sigma\overline{\Lambda}_i(t) \tag{6}$$

where Λ is the system input, ν is the controller output, $\overline{\Lambda}(t)$ represents the additive fault signal in flight control systems such as propellers, $\sigma = \text{diag}\ \{\sigma_1, \sigma_2, \sigma_3\}$, with $\sigma_i = 1$ if the ith actuator has failed and $\sigma_i = 0$ otherwise. $\rho = \text{diag}\ \{\rho_1, \rho_2, \rho_3\}$. For $\rho = 0$; for diminished effectiveness, $0 < \rho_j < 1$, $j = 1{:}\ 6$ and $\overline{\Lambda}(t) = 0$.

By taking: $\overline{\Lambda}(t) = \rho(I - \sigma)$ and $\bar{\text{g}}(\text{x}, \Lambda) = \sigma\overline{\Lambda}(t)$ we can write:

$$\Lambda(t) = b\nu + \bar{\text{g}}(\text{x}, \Lambda). \tag{7}$$

Assumption 1 We will use (7) to describe the control system that is subjected to actuator problems for the rest of this study.

3 SMC Controller Design

In this section a nonlinear controller based on the sliding mode is proposed to ensure the trajectory tracking, in the presence of actuator faults, the general scheme of the proposed controller is presented in Fig. 2.

Consider the following system:

$$\begin{cases} \dot{\text{x}}_1 = x_2 \\ \dot{\text{x}}_2 = f(x) + \Lambda \end{cases} \tag{8}$$

with the existence of actuator failures as defined by (7), Eq. (8) can be rephrased as:

$$\begin{cases} \dot{\text{x}}_1 = x_2 \\ \dot{\text{x}}_2 = f(x) + \bar{\text{b}}\nu + \bar{\text{g}}(\text{x}, \Lambda) \end{cases} \tag{9}$$

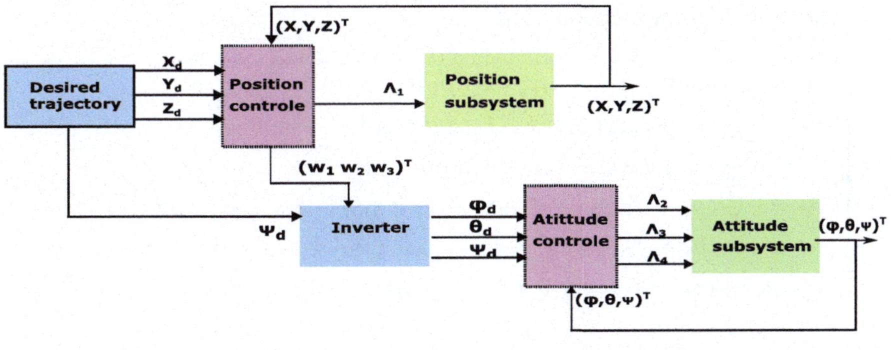

Fig. 2 The control structure

with $\Lambda(t) = \bar{b}\nu(t) + \bar{g}(x, \Lambda)$ and $\bar{b} = \rho(1-\sigma)b$; $\bar{g}(x, \Lambda) = \bar{b}g(x)$; $\bar{b} \geq 0$.

Consider the output ψ, the tracking error is defined as $e_\psi = \psi - \psi_r$. Let us define the following sliding surface and its time derivative:

$$s_\psi = \dot{e}_\psi + \lambda_6 e_\psi \quad \Rightarrow \quad \dot{s}_\psi = \ddot{\psi} - \ddot{\psi}_r + \lambda_6 \dot{e}_\psi \tag{10}$$

Denoting: $\ddot{\psi} = f_3 x_8 x_{10} + f_6 x_{12}^2 + \bar{b}\nu_{eq}$

$$\dot{s}_\psi = f_3 x_8 x_{10} + f_6 x_{12}^2 + \bar{b}\nu_{eq} + \bar{g}(x, \Lambda) + p_6 \tag{11}$$

with $p_6 = \lambda_6\, \dot{e}_\psi - \ddot{\psi}_r$.

Supposing that g = 0, then the equivalent control law is:

$$\nu_{eq} = -\frac{1}{\bar{b}_6}(p_6 + f_3 x_8 x_{10} + f_6 x_{12}^2) \tag{12}$$

Switching control:

$$\nu_{sw} = -k_{11}.s - (k_{12} + \alpha)sign(s) \tag{13}$$

where $k_{11} > 0$, $k_{12} > |g| > 0$, and α is a small parameter to be designed.

Control: $\nu_\psi = \nu_{\psi eq} + \nu_{\psi sw}$

$$\nu_\psi = -\frac{1}{\bar{b}_6}(p_6 + f_3 x_8 x_{10} + f_6 x_{12}^2) - k_{11}s - k_{12}sign(s) \tag{14}$$

Assumption 2 The control law presented in (14) is implemented if the following condition is satisfied $g < |g| < k_{12}$.

The same way for the other controls. To avoid the chattering phenomen as a result of hight frequency switching of the sliding mode control, sign function is replaced the hyperbolic tangent function $tanh(s/\epsilon)$.

$$\begin{cases} \nu_x = -\frac{1}{\bar{b}_1}(p_1 + l_1 x_2) - k_1 s - k_2 tanh(s/\epsilon) \\ \nu_y = -\frac{1}{\bar{b}_2}(p_2 + l_2 x_4) - k_3 s - k_4 tanh(s/\epsilon) \\ \nu_z = -\frac{1}{\bar{b}_3}(p_3 + l_3 x_6) - k_5 s - k_6 tanh(s/\epsilon) \\ \nu_\phi = -\frac{1}{\bar{b}_4}(p_4 + f_1 x_{10} x_{12} + f_4 x_7^2 + f_7 \omega_r x_{10}) - k_7 s - k_8 tanh(s/\epsilon) \\ \nu_\theta = -\frac{1}{\bar{b}_5}(p_5 + f_2 x_8 x_{12} + f_5 x_{10}^2 + f_8 \omega_r x_8) - k_9 s - k_{10} tanh(s/\epsilon) \\ \nu_\psi = -\frac{1}{\bar{b}_6}(p_6 + f_3 x_8 x_{10} + f_6 x_{12}^2) - k_{11} s - k_{12} tanh(s/\epsilon) \end{cases} . \tag{15}$$

4 Main Result

The major goal of this work is to test the proposed controller's stability properties in the face of actuator defects in a quadrotor system. The following theorem expresses these properties.

Assumption 3 Consider a quadrotor system (9). Given the planned sliding surface, and the yaw subsystem is asymptotically stable.

Let us consider the Lyapunov function:

$$V = \frac{1}{2} s_\psi^2 \tag{16}$$

It's time derivative is given as follows

$$\dot{V} = s_\psi \dot{s}_\psi = s_\psi.(f_3 x_8 x_{10} + f_6 x_{12}^2 + \bar{b}_6 \nu_{eq} + \bar{g}(x, \Lambda) + \lambda_6 \dot{e_\psi} - \ddot{\psi}_r) \tag{17}$$

$$\begin{aligned} \dot{V} &= s_\psi.(f_3 x_8 x_{10} + f_6 x_{12}^2 + \bar{b}_6 \left(-\frac{1}{\bar{b}_6}(\lambda_6 \dot{e}_\psi - \ddot{\psi}_r + f_3 x_8 x_{10} + f_6 x_{12}^2 + \bar{g}(x, \Lambda)) \right) \\ &+ \bar{g}(x, \Lambda) + \lambda_6 \dot{e}_\psi - \ddot{\psi}_r - k_{11}.s_\psi - (k_{12} + \alpha) sign(s_\psi)) \end{aligned} \tag{18}$$

$$V = -k_{11} s_\psi^2 + \bar{g}(x, \Lambda) - (k_{12} + \alpha) s_\psi sign(s_\psi) \tag{19}$$

$$V \leq -k_{11} s_\psi^2 - \alpha |s_\psi| < 0 \tag{20}$$

$$\dot{V} = s_\psi.(-k_{11}.s_\psi - k_{10}.sign(s_\psi)) = -k_{11}.s_\psi^2 - k_{12}.|s_\psi| \leq 0 \tag{21}$$

The condition of stability is guaranteed. Yaw subsystem errors will converge to zero in limited time, according to the Lyapunov equation $\dot{V} \leq 0$.

Assumption 4 The controllers in (15) applied to the investigated system (3) ensure the stability of the quadrotor system.

The Lyapunov function candidate for the quadrotor system as follow:

$$V = \frac{1}{2} s_x^2 + \frac{1}{2} s_y^2 + \frac{1}{2} s_z^2 + \frac{1}{2} s_\phi^2 + \frac{1}{2} s_\theta^2 + \frac{1}{2} s_\phi^2 \tag{22}$$

Its time derivatives:

$$\begin{aligned} \dot{V} &= s_x.(-k_1.s_x - k_2.sign(s_x)) + s_y.(-k_3.s_y - k_4.sign(s_y)) \\ &+ s_z.(-k_5.s_z - k_6.sign(s_z)) + s_\phi.(-k_7.s_\phi - k_8.sign(s_\phi)) \\ &+ s_\theta.(-k_9.s_\theta - k_{10}.sign(s_\theta)) + s_\psi.(-k_{11}.s_\psi - k_{12}.sign(s_\psi)) \end{aligned} \tag{23}$$

$$\dot{V} = -k_1.s_x^2 - k_2.|s_x| - k_3.s_y^2 - k_4.|s_y| - k_5.s_z^2 - k_6.|s_z| \\ - k_7.s_\phi^2 - k_8.|s_\phi| - k_9.s_\theta^2 - k_{10}.|s_\theta| - k_{11}.s_\psi^2 - k_{12}.|s_\psi| \leq 0. \quad (24)$$

The stability of the quadrotor is guaranteed.

5 Simulations and Discussion

We consider three different types of faults to evaluate the efficiency of the suggested controller. The Table 2 shows the controller parameters utilized in the simulation.

In simulations the initial conditions of position and attitude and their desired values are chosen by $[x_0 \quad y_0 \quad z_0] = [0 \quad 0 \quad 0]$m, $[x_d \quad y_d \quad z_d] = [0.8 \quad 0.4 \quad 1]$, $[\psi_0 \quad \psi_d] = [0 \quad 0.6]$rad.

Three sets of comparative simulations are run in order to produce a persuasive result.

The constant perturbations are added in the position subsystem, Fig. 3 represents the results of the three simulations performed on x, y and z, for three different types of faults, and the simulation is performed for 80s.

Case 1: The first example considered is the case without defect with $[X_\rho \quad Y_\rho \quad Z_\rho] = [0.9 \quad 0.9 \quad 0.9]$, which leads the quadrotor to follow the desired flight trajectory faster and more accurately.

Case 2: $[X_\rho \quad Y_\rho \quad Z_\rho] = [0.5 \quad 0.5 \quad 0.5]$, for positions x and y at $5 < t < 10$ the signal starts to oscillate uncontrollably, and recovers at time $t = 10s$, however $Z_\rho = 0.5$ gives a slower response than $Z_\rho = 0.9$.

Case 3: $[X_\rho \quad Y_\rho \quad Z_\rho] = [0.3 \quad 0.3 \quad 0.3]$ in the last case one can notice the errors of follow-up of the trajectory of the position on the figure.

On the same figure the attitudes ϕ, θ and ψ are represented, we can notice the problem of the chattering, the curves of the angles of follow-up show that the precision of the follow-up is affected by various forms of defect $\rho = 0.5; 0.3$. The response curves are unstable, as can be seen in the comparison of the curves presented in Fig. 4. In order to demonstrate the efficacy of the proposed control strategy, the quadrotor trajectory's performance in 3D space is exposed in Fig. 4.

Table 2 The control parameters

Parameter	Value	Parameter	Value
$\lambda_1, \lambda_2, \lambda_3$	1	$\lambda_4, \lambda_5, \lambda_6$	0.9
σ_z	0.001	σ_x, σ_y	0.0012
k_5	10	$\sigma_\phi, \sigma_\theta, \sigma_\psi$	0.0001
k_2, k_4, k_6	0.99	k_8, k_{10}, k_{12}	100
k_1, k_3	0.02	k_7, K_9, k_{11}	0.05

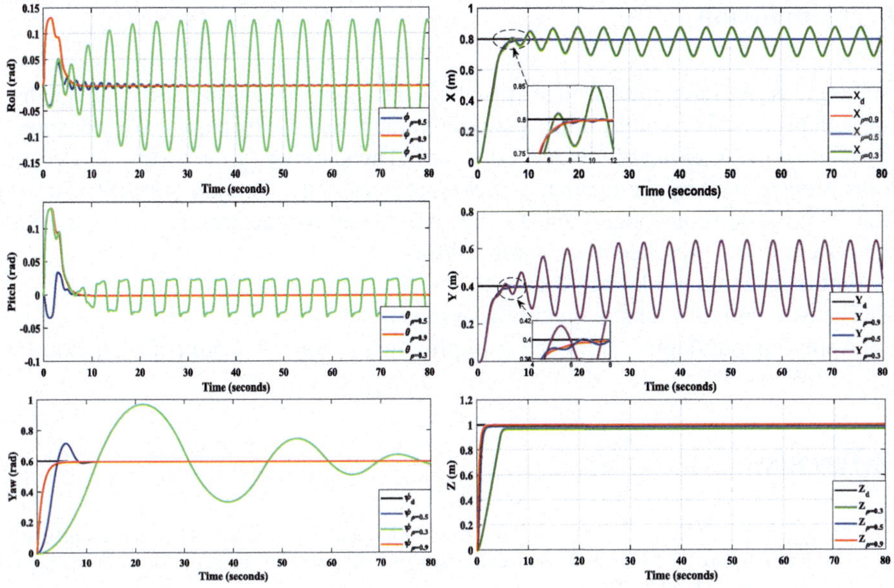

Fig. 3 Response of position and attitude

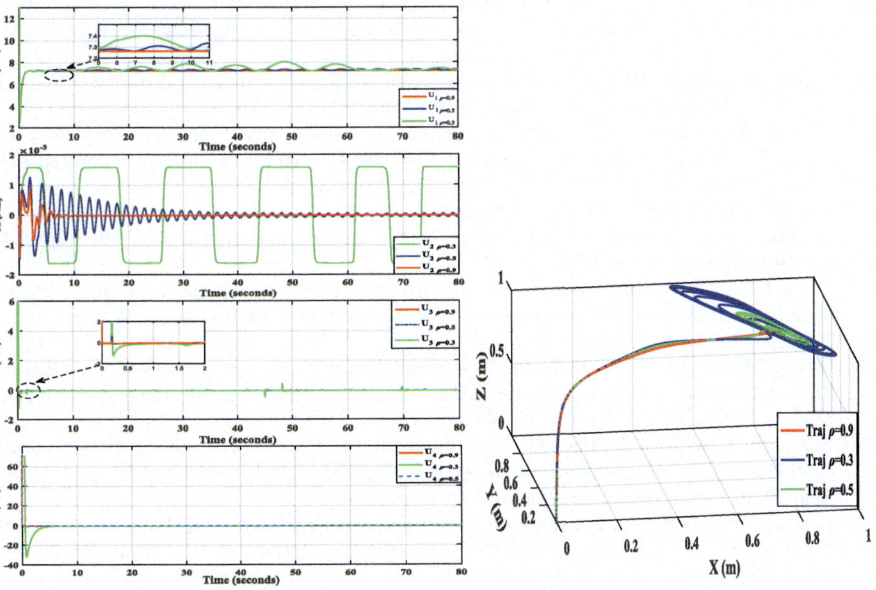

Fig. 4 Control inputs and position 3D trajectory

6 Conclusion

In this study, we have addressed the problem of tracking the flight path of the quadrotor with proposed actuator faults. A sliding mode based controller has been designed to stabilize the tracking of the desired values. The stability of the quadrotor subsystems is performed by the Lyapunov theory. However, the simulation results showed that the proposed controller is able to improve the control performance of the quadrotor system in the presence of actuator failures.

Future research works will focus on:

(i) The detection of faults in quadrotor system.

(ii) Applying the adaptive approach in improving the control system of the quadrotor.

References

1. Zhao, L., Dai, L., Xia, Y., Li, P.: Attitude control for quadrotors subjected to wind disturbances via active disturbance rejection control and integral sliding mode control. Mech. Syst. Signal Process. **129**, 531–545 (2019)
2. Nascimento, T.P., Saska, M.: Position and attitude control of multi-rotor aerial vehicles: a survey. Annu. Rev. Control. **48**, 129–146 (2019)
3. Nascimento, T.P., Saska, M.: Position and attitude control of multi-rotor aerial vehicles: a survey. Annu. Rev. Control. **48**, 129–146 (2019). https://doi.org/10.1016/j.arcontrol.2019.08.004
4. Chang, K., Xia, Y., Huang, K., Ma, D.: Obstacle avoidance and active disturbance rejection control for a quadrotor. Neurocomputing **190**, 60–69 (2016). May
5. Labbadi, M., Cherkaoui, M.: Robust integral terminal sliding mode control for quadrotor UAV with external disturbances. Int. J. Aerosp. Eng. **2019**, 1–10 (2019). https://doi.org/10.1155/2019/2016416
6. Labbadi, M., Cherkaoui, M.: Novel robust super twisting integral sliding mode controller for a quadrotor under external disturbances. Int. J. Dyn. Control **8**(3), 805–815 (2019). https://doi.org/10.1007/s40435-019-00599-6
7. Benaddy, A., Labbadi, M., Bouzi, M.: Robust flight control for a quadrotor under external disturbances based on generic second order sliding mode control. IFAC-PapersOnLine **55**(12), 270–275 (2022)
8. Freire, F.P., Martins, N.A., Splendor, F.: A simple optimization method for tuning the gains of PID controllers for the autopilot of Cessna 182 aircraft using model-in-the-loop platform. J. Control Autom. Electr. Syst. **29**(4), 441–450 (2018)
9. El Houm, Y., Abbou, A., Labbadi, M., Cherkaoui, M.: Optimal new sliding mode controller combined with modified supertwisting algorithm for a perturbed quadrotor UAV. Int. J. Aerosp. Eng. **2020**, 1–10 (2020). https://doi.org/10.1155/2020/9753870
10. Labbadi, M., Cherkaoui, M.: Robust adaptive nonsingular fast terminal sliding-mode tracking control for an uncertain quadrotor UAV subjected to disturbances. ISA Trans. **99**, 290–304 (2020). https://doi.org/10.1016/j.isatra.2019.10.012
11. Muñoz, F., González-Hernández, I., Salazar, S., Espinoza, E.S., Lozano, R.: Second order sliding mode controllers for altitude control of a quadrotor UAS: real-time implementation in outdoor environments. Neurocomputing **233**, 61–71 (2017). https://doi.org/10.1016/j.neucom.2016.08.111
12. Labbadi, M., Cherkaoui, M., Houm, Y.E., Guisser, M.: Modeling and robust integral sliding mode control for a quadrotor unmanned aerial vehicle. In: 2018 6th International Renewable and Sustainable Energy Conference (IRSEC) (2018)

13. Jia, Z., Yu, J., Mei, Y., Chen, Y., Shen, Y., Ai, X.: Integral backstepping sliding mode control for quadrotor helicopter under external uncertain disturbances. Aerosp. Sci. Technol. **68**, 299–307 (2017)
14. Guo, Y., Ma, B.-L.: Global sliding mode with fractional operators and application to control robot manipulators. Int. J. Control **92**(7), 1497–1510 (2017)
15. Labbadi, M., Boukal, Y., Cherkaoui, M., Djemai, M.: Fractional-order global sliding mode controller for an uncertain quadrotor UAVs subjected to external disturbances. J. Franklin Inst. **358**(9), 4822–4847 (2021)
16. Labbadi, M., Cherkaoui, M.: Adaptive fractional-order nonsingular fast terminal sliding mode based robust tracking control of quadrotor UAV with gaussian random disturbances and uncertainties. IEEE Trans. Aerosp. Electron. Syst. **57**(4), 2265–2277 (2021)
17. Benchaita, H., Ladaci, S.: Fractional adaptive SMC fault tolerant control against actuator failures for wing rock supervision. Aerosp. Sci. Technol. **114**, 106745 (2021). https://doi.org/10.1016/j.ast.2021.106745

Frequency Stabilization in Microgrid Using Super Twisting Sliding Mode

Ark Dev, Urvashi Chauhan, Kunalkumar Bhatt, and Mrinal Kanti Sarkar

Abstract In this chapter, continuous second order sliding mode technique commonly known as super twisting sliding mode is used for frequency stabilization in microgrids (MGs) under load disturbance. The MGs are well known for reliable, efficient, low cost and clean energy sources. They are often looked as a solution for future energy crisis, zero-emission energy sources, and a mean to optimally utilize the energy present in the nature. However, the stability of MGs has always been a matter of concern due to uncertain behavior of renewable energy sources like wind and solar. Any mismatch in the generation and load demand lead to frequency deviation from its scheduled value. Offset frequency lead to system instability and at worse may lead to large scale blackouts. In this work, a PI type sliding surface is used to obtain better transient behavior. Frequency stabilization is also observed with the presence of typical energy system nonlinearities like generation rate constraint (GRC) and governor deadband (GDB), and with actuator time delay. The present work is compared to commonly used PI controller tuned using Ziegler Nichols method and ANN-GA algorithm. It is found that the frequency deviation obtained using continuous second order sliding mode has a better transient behavior compared to PI controller. The stability of the controller is verified theoretically using Lyapunov candidate function and validated using Simulink in MATLAB.

A. Dev (✉) · K. Bhatt
Department of Electrical Engineering, Parul Institute of Engineering and Technology, Parul University, Waghodia, Gujarat, India
e-mail: ark.dev24287@paruluniversity.ac.in

U. Chauhan
Department of Electronics and Communication Engineeirng, Parul Institute of Engineering and Technology, Parul University, Waghodia, Gujarat, India

M. K. Sarkar
Department of Electrical Engineering, National Institute of Technology, Jamshedpur, Jharkhand, India

A. Ben Makhlouf et al. (eds.), *State Estimation and Stabilization of Nonlinear Systems*, Studies in Systems, Decision and Control 491,
https://doi.org/10.1007/978-3-031-37970-3_17

1 Introduction

Energy is an important parameter in the growth of any economy. But the rising demand of energy has led to the depletion of fossil fuels and consequently increasing pollution level day by day. The energy sector worldwide is witnessing a paradigm shift because of increasing carbon emission, depleting fossil fuels, increasing global temperature and frequent blackouts. Renewable energies based MGs have been found as an important part of this dramatic transition. A MG is a standalone set of loads and small power sources (distributed generation) in the form of a single control system that provide generation capacity for a local area (10 MW) [1, 2]. MGs are modern, reliable, and often viewed as small scale electrical power distribution systems. Renewable energies based MGs include wind turbine generators (WTGs), diesel engine generators (DEGs), photovoltaics (PVs), fuel cells (FCs), battery energy systems (BESs), and flywheel energy system (FES). These sources are clean and abundantly present. However, due to uncertain nature of wind speed and solar irradiation, the frequency stability of MGs has always been a matter of interest for researchers and control engineers. Any mismatch between the generation and demand lead to frequency deviation from its schedule value. Off normal frequency causes system instability and at worse can cause blackouts. Therefore, a robust control technique is required to achieve frequency stability in presence of uncertainties and disturbance. Several control strategies have been proposed and can be referred from [3, 4]. Classical control techniques such as proportional –integral (PI), proportional-integral-derivative (PID) controllers are often used [5, 6]. The main limitation with these controllers is that they are tuned for one operating point and may fail under variable operating points. Several soft computing techniques have also been reported to tune PI and PID controllers for optimum results [7–10]. Model predictive control (MPC) with its fast transient response and ability to accommodate different constraints in individual and interconnected MGs is reported in [11–13]. Robust control techniques such as H_∞ and μ-synthesis controllers, LMI based H_∞ controller in secondary control loop is reported in [1, 14, 15]. Sliding mode control (SMC) for frequency stabilization in the presence of disturbance and uncertainties is proposed in [16–18].

In this work, super twisting sliding mode control (ST-SMC) approach is used for frequency stability in microgrid. Sliding mode control (SMC) approach is well known for its robustness, faster convergence, insensitivity to disturbances and easier design [19, 20]. However, all this is at a cost of oscillation in the control input often known as chattering. The problem of chattering leads to reduced control action, wear and tear of moving mechanical parts and high heat losses in the system. The second order sliding mode strategy commonly known as super twisting sliding mode (also known as continuous sliding mode) has been found as a solution to this problem without affecting the system stability [21, 22]. The robustness of the controller is validated under random change in load disturbance, typical energy system nonlinearities like generation rate constraints (GRC), governor dead band (GDB) and actuator delay in form of time delay.

The outline of the chapter is as follow: In Sect. 2, modelling of MG is studied with its state dynamics and control objectives. In Sect. 3, design of super twisting SMC is studied and its stability is theoretically proved using Lyapunov theorem. In Sect. 4, the stability of MG is validated under different scenarios using MATLAB simulation. Finally, in Sect. 5, few conclusions are drawn and scope for future work is discussed.

2 System Modelling

Figure 1 depicts the frequency control loop of an islanded MG. The MG in Fig. 1 contains the following power generators: WTG, solar PV, diesel engine generator, fuel cell system, flywheel and battery energy storage systems. The primary control action is the droop control action and the secondary control action is provided using the ST-SMC. Since the power generated by renewable energy sources are highly uncertain, thus, are not considered in the secondary control loop. The fuel cell system and diesel generator is connected in the secondary control loop to compensate the frequency deviation. The battery storage systems are connected in feedback with the load so as to compensate the frequency deviation in case of any load disturbance or excess generation. The system dynamics are as follows:

$$\Delta \dot{f} = -\frac{D}{2H}\Delta f + \frac{\Delta P_{PV}}{2H} + \frac{\Delta P_{WTG}}{2H} + \frac{\Delta P_{FC}}{2H} + \frac{\Delta P_{G}}{2H} \pm \frac{\Delta P_{BESS}}{2H} \pm \frac{\Delta P_{FESS}}{2H} - \frac{\Delta P_{LOAD}}{2H} \tag{1}$$

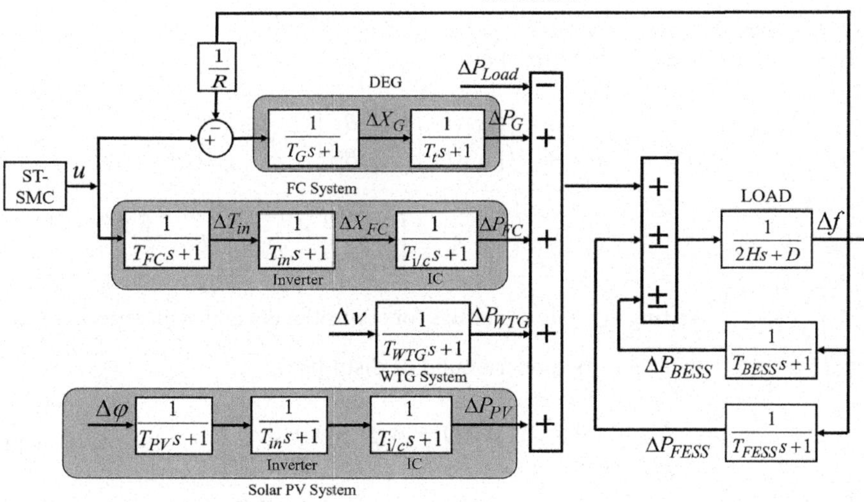

Fig. 1 Load frequency model of microgrid [23]

$$\Delta \dot{P}_{FESS} = -\frac{\Delta P_{FESS}}{T_{FESS}} + \frac{\Delta f}{T_{FESS}} \tag{2}$$

$$\Delta \dot{P}_{BESS} = -\frac{\Delta P_{BESS}}{T_{BESS}} + \frac{\Delta f}{T_{BESS}} \tag{3}$$

$$\Delta \dot{P}_{G} = -\frac{\Delta P_{G}}{T_{t}} + \frac{\Delta X_{G}}{T_{t}} \tag{4}$$

$$\Delta \dot{X}_{G} = -\frac{\Delta f}{RT_{G}} - \frac{\Delta X_{G}}{T_{G}} + \frac{u}{T_{G}} \tag{5}$$

$$\Delta \dot{P}_{FC} = -\frac{\Delta P_{FC}}{T_{1/c}} - \frac{\Delta X_{FC}}{T_{1/c}} \tag{6}$$

$$\Delta \dot{X}_{FC} = -\frac{\Delta X_{FC}}{T_{in}} + \frac{\Delta T_{in}}{T_{in}} \tag{7}$$

$$\Delta \dot{T}_{in} = -\frac{\Delta T_{in}}{T_{FC}} + \frac{u}{T_{FC}} \tag{8}$$

The state Eqs. (1)–(8) can be rewritten in matrix form as follows,

$$\begin{aligned} \dot{X}(t) &= AX(t) + Bu(t) + Dd(t) \\ y(t) &= \begin{bmatrix} 1\,0\,0\,0\,0\,0\,0\,0 \end{bmatrix} X(t) \end{aligned} \tag{9}$$

where $A^{(8\times8)}$, $B^{(8\times1)}$, and $D^{(8\times1)}$ represent system, input, and disturbance matrices respectively. The state vector $X(t)$ is of form, $\begin{bmatrix} \Delta f \; \Delta P_{FESS} \; \Delta P_{BESS} \; \Delta P_{G} \; \Delta X_{G} \; \Delta P_{FC} \; \Delta X_{FC} \; \Delta T_{in} \end{bmatrix}^{T}$.

Rewriting (9) in augmented form as,

$$\begin{bmatrix} \dot{x}_1(t) \\ \dot{x}_2(t) \end{bmatrix} = \begin{bmatrix} A_{11} & A_{12} \\ A_{21} & A_{22} \end{bmatrix} \begin{bmatrix} x_1(t) \\ x_2(t) \end{bmatrix} + \begin{bmatrix} B_1 \\ B_2 \end{bmatrix} u(t) + \begin{bmatrix} D_1 \\ D_2 \end{bmatrix} d(t) \tag{10}$$

where $X(t) = \begin{bmatrix} x_1^{(1\times1)}(t) \; x_2^{(7\times1)}(t) \end{bmatrix}^{T}$, $A = \begin{bmatrix} A_{11}^{(1\times1)} & A_{12}^{(1\times7)} \\ A_{21}^{(7\times1)} & A_{22}^{(7\times7)} \end{bmatrix}$, $B = \begin{bmatrix} B_1^{(1\times1)} \\ B_2^{(7\times1)} \end{bmatrix}$, and $D = \begin{bmatrix} D_1^{(1\times1)} \\ D_2^{(7\times1)} \end{bmatrix}$. Writing the state dynamics for controller designs with an assumption that $|D_1 d| \le \xi$, where ξ is the upper bound of disturbance.

$$\left.\begin{aligned} \dot{x}_1(t) &= A_{11}x_1(t) + A_{12}x_2(t) + \xi \\ \dot{x}_2(t) &= A_{21}x_1(t) + A_{22}x_2(t) + B_2u(t) \end{aligned}\right\} \tag{11}$$

The matrices in (11) are given as below:

$$A_{11} = -\frac{D}{2H};\ A_{12} = \left[-\tfrac{1}{2H}\ -\tfrac{1}{2H}\ \tfrac{1}{2H}\ 0\ \tfrac{1}{2H}\ 0\ 0\right];$$

$$A_{21} = \left[\tfrac{1}{T_{FESS}}\ \tfrac{1}{T_{BESS}}\ 0\ -\tfrac{1}{RT_G}\ 0\ 0\ 0\right]^T;$$

$$A_{22} = \begin{bmatrix} -\frac{1}{T_{FESS}} & 0 & 0 & 0 & 0 & 0 & 0 \\ 0 & -\frac{1}{T_{BESS}} & 0 & 0 & 0 & 0 & 0 \\ 0 & 0 & -\frac{1}{T_t} & \frac{1}{T_t} & 0 & 0 & 0 \\ 0 & 0 & 0 & -\frac{1}{T_G} & 0 & 0 & 0 \\ 0 & 0 & 0 & 0 & -\frac{1}{T_{1/c}} & \frac{1}{T_{1/c}} & 0 \\ 0 & 0 & 0 & 0 & 0 & -\frac{1}{T_{in}} & \frac{1}{T_{in}} \\ 0 & 0 & 0 & 0 & 0 & 0 & \frac{1}{T_{FC}} \end{bmatrix};$$

$$B = \left[0\ 0\ 0\ 0\ \tfrac{1}{T_G}\ 0\ 0\ \tfrac{1}{T_{FC}}\right]^T;\ D = \left[-\tfrac{1}{2H}\ 0\ 0\ 0\ 0\ 0\ 0\ 0\right]^T.$$

The work in this chapter has following main control objective:

- Frequency stabilization $\left(\underset{t\to\infty}{\Delta f(t) = 0}\right)$ under load disturbance.

The system parameters are described in Table 1.

Table 1 System variable description

Variable	Description
Δf	Frequency deviation
ΔP_{PV}	Output power change of photovoltaic generator
ΔP_{WTG}	Output power change of wind turbine generator
ΔP_{BESS}	Output power change of battery energy storage system
ΔP_{FESS}	Output power change of flywheel energy storage system
ΔP_G	Output power change of diesel generator output
ΔX_G	Governor valve position
ΔP_{FC}	Output power change of fuel cell system
ΔP_{LOAD}	Load disturbance
$u(t)$	Control signal
$X(t)$	State vector

3 Frequency Control Strategy: Design of ST-SMC

The uncertain nature of renewable energy sources like wind and solar make MG frequency stability very questionable. Thus, a frequency control technique is required to maintain frequency stability of MGs. There are basically two control loops: primary and secondary. Primary control loop is mainly a fixed gain control loop or commonly termed as droop control. Droop control methods have been reported many times in literature, however, the main limitation with them is the offset error and poor transient behavior. Thus, a secondary control loop is required which will ensure frequency stabilization and better transient behavior. Since, PVs and WTGs are weather dependent, they are not included in the secondary control loop. The secondary control action is performed using micro-turbine (MT) and fuel cell (FC).

In this work, the secondary control action in provided by a powerful second order continuous super twisting sliding mode control. The main limitation with conventional or first order SMC is the problem of chattering. Chattering is high, fixed amplitude oscillations in the control signal. The problem of chattering causes wear and tear of moving mechanical parts, heat losses and often reduces the control accuracy. Many methods have been reported in literature to reduce the problem of chattering. Higher order SMC is one of the methods. However, the limitation with higher order SMC is that it acts on higher order time derivatives of sliding surface to mitigate chattering. This limitation of higher order SMC is overcome by the use of continuous second order SMC commonly known as super twisting SMC. The super twisting SMC reduces chattering by acting only on the first time derivative of sliding surface without any loss in control accuracy.

Designing of SMC involves two fundamental steps: firstly, a user defined sliding surface is selected and secondly, a control action is designed to drive the system states from any operating point to the sliding surface in finite time. The sliding surface $s(t)$ is selected as below [23],

$$s(t) = \lambda_1 x_1(t) + G x_2 + \lambda_2 \int x_1(t) \tag{12}$$

where λ_1 and λ_2 are design constants and $G^{(1\times 7)}$ is the gain matrix to be designed such that GB_2 is non-singular. Differentiating (12), one can obtain,

$$\dot{s}(t) = \lambda_1 \dot{x}_1(t) + G\dot{x}_2(t) + \lambda_2 x_1(t) \tag{13}$$

Substituting (11) in (13),

$$\dot{s}(t) = \lambda_1\{A_{11}x_1(t) + A_{12}x_2(t)\} + G\{A_{21}x_1(t) + A_{22}x_2(t) + B_2 u(t)\} + \lambda_2 x_1(t) \tag{14}$$

Here the control input $u(t)$ consist of three parts namely, equivalent control $\left(u_{eq}(t)\right)$ obtained by substituting $\dot{s}(t) = 0$, the continuous state function and the discontinuous state function with an integrator [21, 24].

Substituting $\dot{s}(t) = 0$ to obtain equivalent control $\left(u_{eq}(t)\right)$,

$$u_{eq}(t) = \frac{-1}{GB_2}\{(\lambda_1 A_{11} + GA_{21})x_1(t) + (\lambda_1 A_{12} + GA_{22})x_2(t) + \lambda_2 x_1(t)\} \quad (15)$$

The final control input that drives the system state to the desired trajectory is given by,

$$u(t) = u_{eq}(t) - \alpha_1 |s(t)|^{0.5} sign(s(t)) - \alpha_2 \int_0^t sign(s(t))dt \quad (16)$$

where α_1, α_2 are positive constants and $\text{sgn}(s(t)) = \left.\begin{matrix} 1, & s(t) > 0 \\ 0, & s(t) = 0 \\ -1, & s(t) < 0 \end{matrix}\right\}$. Recalling, $|D_1 d| \leq \xi$, the constant are selected as follows: $\alpha_1 = 1.5\sqrt{\xi}$ and $\alpha_2 = 1.1\xi$.

In order to examine the existence of stability of sliding surface, consider a Lyapunov candidate function as,

$$V(t) = \frac{1}{2}s^2(t) \quad (17)$$

Taking the derivative of (17),

$$\dot{V}(t) = s(t)\dot{s}(t) \quad (18)$$

Substituting (13) in (18),

$$\dot{V}(t) = s(t)\{\lambda_1 \dot{x}_1(t) + G\dot{x}_2(t) + \lambda_2 x_1(t)\} \quad (19)$$

Substituting (11) in (19),

$$\dot{V}(t) = s(t)\{\lambda_1(A_{11}x_1(t) + A_{12}x_2(t) + \xi) + \lambda_2 x_1(t) + G(A_{21}x_1(t) + A_{22}x_2(t) + B_2 u(t))\} \quad (20)$$

Using the control law (16) in (20),

$$\dot{V}(t) = s(t)\left\{-\alpha_1 |s(t)|^{0.5} sign(s(t)) - \alpha_2 \int_0^t sign(s(t))dt + \xi\right\} \quad (21)$$

Equation (21) will always be negative i.e. $\dot{V}(t) < 0$ irrespective of the $s(t)$, provided that the controller tuning gain is selected satisfying $\alpha_1 = 1.5\sqrt{\xi}$ and $\alpha_2 = 1.1\xi$. This ensures asymptotic stability and thus system states tend to zero.

4 Simulation Results

Systems parameter description and values are given in Table 2. It should be noted that lower order dynamic models are considered for study because only small changes in loads are expected to occur during normal operation

Case I—In this case, the designed controller is validated with random load disturbance as shown in Fig. 2. The wind speed and solar irradiance is shown in Figs. 3 and 4. The insensitivity to disturbance and uncertainty property of sliding mode is able to compensate this uncertainty and ensure frequency stabilization in finite time (see Fig. 5). It can be seen from figure that frequency deviation is well within acceptable limits under disturbance. It can also be seen that the control input with ST-SMC is less affected by the problem of chattering compared to that of SMC (see Fig. 6).

Case II—In this case, the MG is modelled with generation rate constraint (GRC) and governor deadband (GDB). These are two typical nonlinearities and cannot be ignored in system modelling. The GRC and GDB limits the immediate response of the system to reject disturbance. The GRC is a physical constraint on the rate of change in the generating unit whereas the GDB is the band where no action occurs for the governor [25]. In the work, the value of GRC is 10% p.u./min and GDB is 0.06%. Figure 7 depicts the modelling of MG with GRC and GDB. It is observed from

Table 2 Syestem parameter description and values

Symbol	System parameters	
	Description	Value
D	Load damping constant	0.015 p.u./Hz
H	Intertia constant	0.0833 p.u.-sec
T_{BESS}	Time constant (Battery energy storage system)	0.1 s
T_{FESS}	Time constant (Fly wheel energy storage system)	0.01 s
T_{WTG}	Time constant (Fuel cell system)	1.5 s
T_{FC}	Time constant (Fuel cell system)	0.26 s
T_G	Time constant (Governor)	0.08 s
$T_{1/c}$	Time constant	0.004 s
T_{in}	Time constant	0.04 s
T_t	Time constant (Turbine)	0.4 s
R	Speed regulation coefficient	3 Hz/p.u
τ	Actuator time delay	75 ms

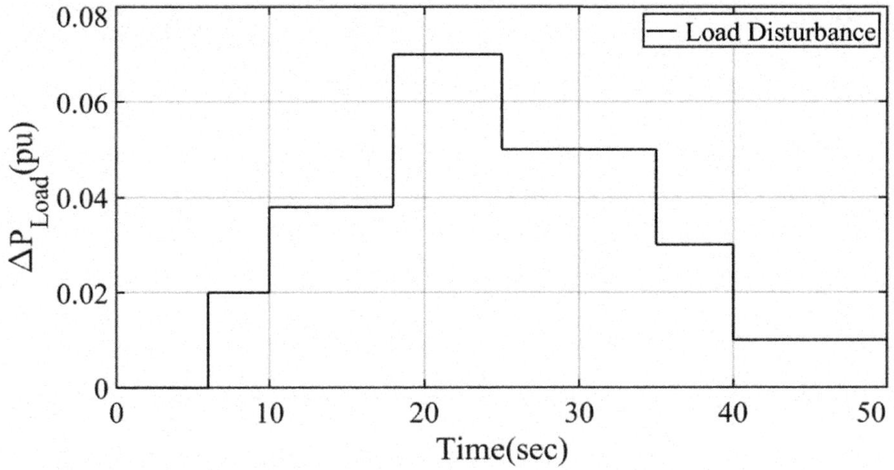

Fig. 2 Random load disturbance

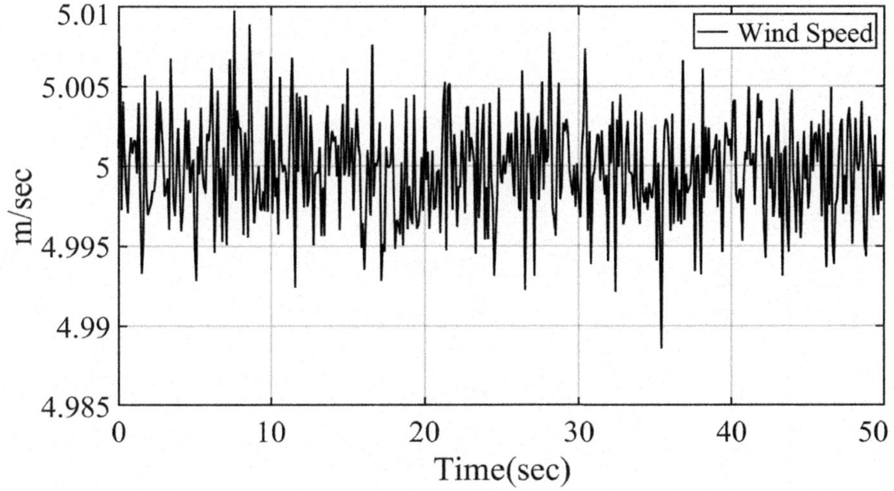

Fig. 3 Wind speed

Fig. (8) that the frequency deviation with GRC and GDB is affected by overshoot, however, the controller is able to stabilize frequency under permissible limits. The frequency deviation is also observed with actuator time delay (τ) in the control loop and it is found that the frequency deviation is within acceptable limits although it is affected by oscillations (see Fig. 9). This confirms robustness of the designed controller.

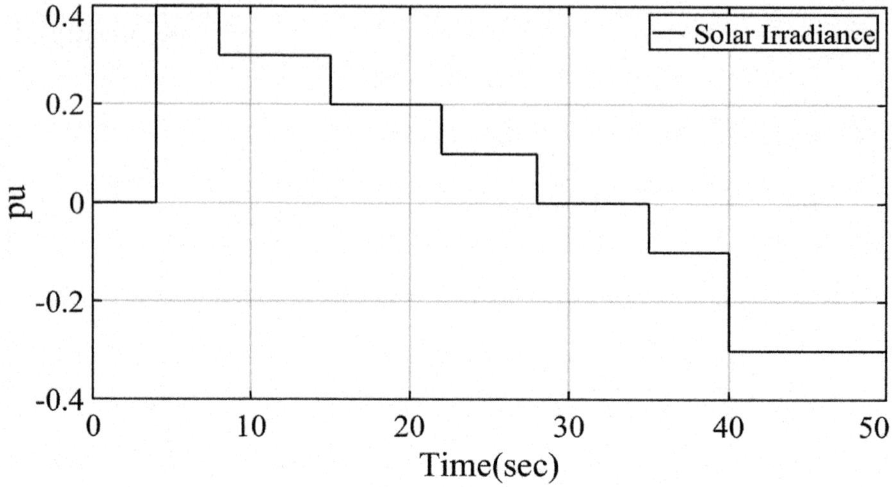

Fig. 4 Solar irradiance

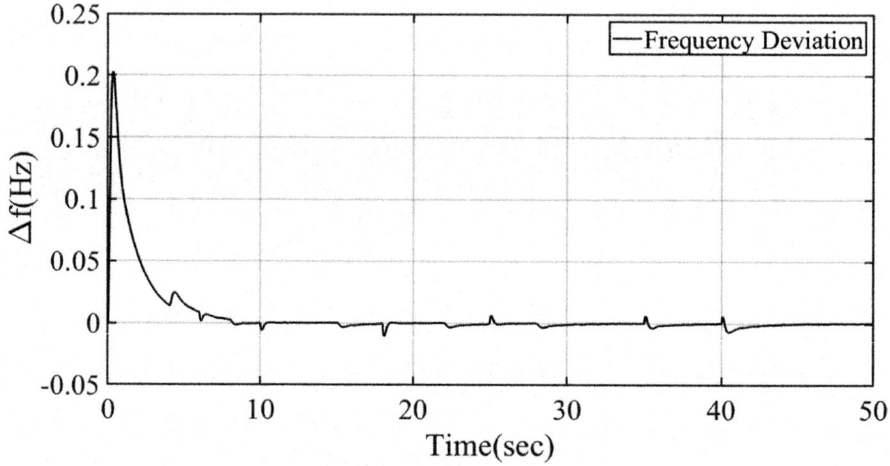

Fig. 5 Frequency deviation under case 1

Case III—In this case, frequency deviation is compared with commonly used PI controller tuned using Ziegler Nichols method and ANN-GA algorithm [26]. It can be seen from the Fig. 10 that the frequency stabilization achieved by using ST-SMC has better transient behavior compared to that of obtained using PI controller.

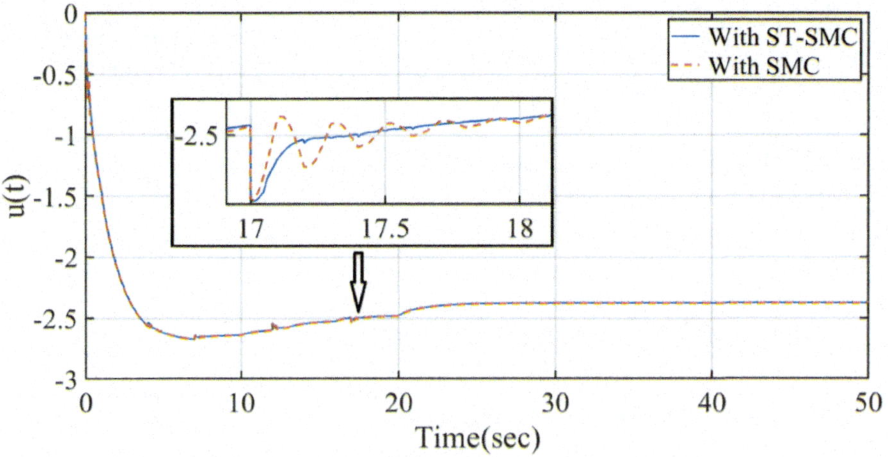

Fig. 6 Control input comparrsion

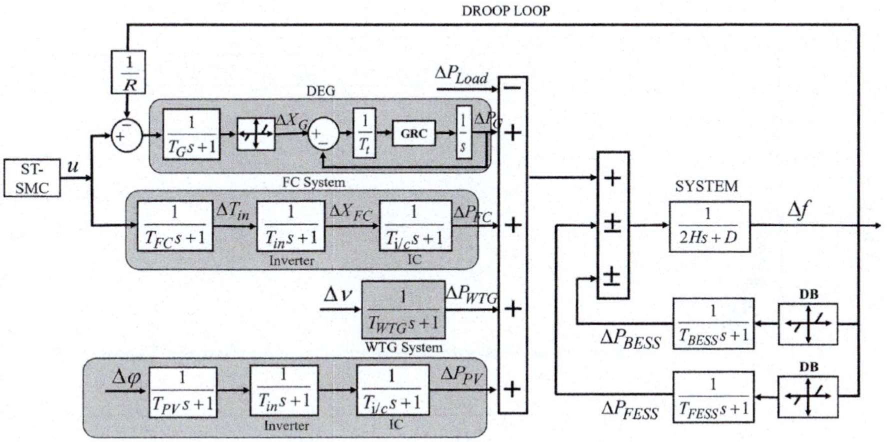

Fig. 7 Load frequency model of microgrid with GRC and GDB

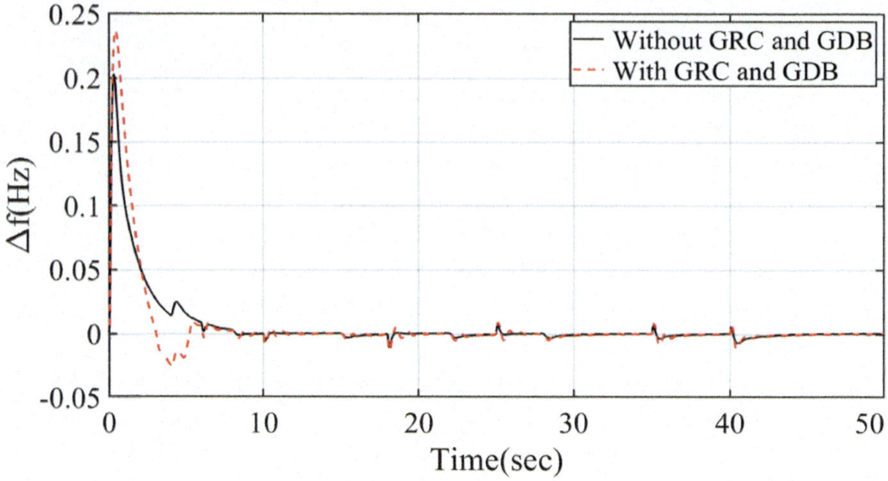

Fig. 8 Frequency deviation with GRC and GDB

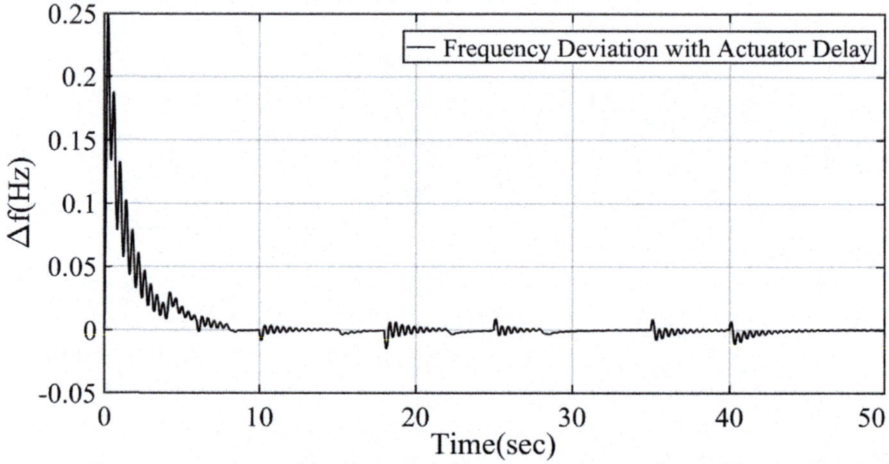

Fig. 9 Frequency deviation with actuator time delay

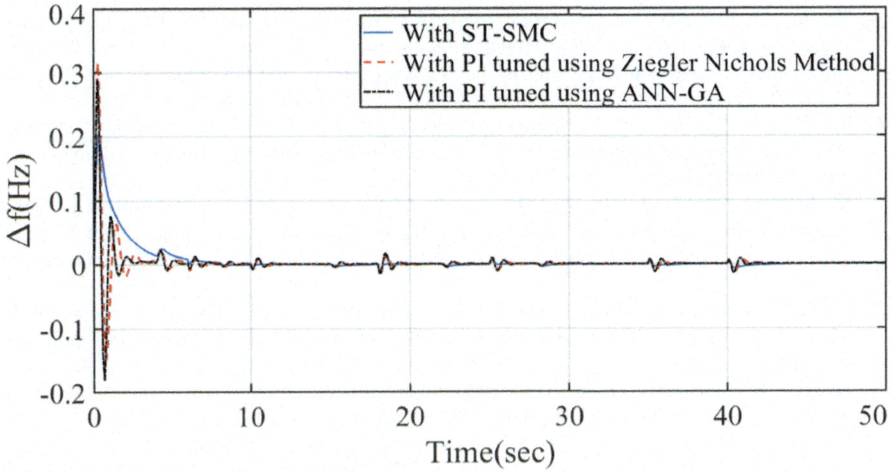

Fig. 10 Frequency deviation comparison

5 Conclusion

In this chapter, frequency stability of MG is studied using sliding mode. It should be noted that frequency is an important aspect of power quality. Thus, its stabilization over wide operating range is necessary. The ST-SMC is used to stabilize frequency under random load disturbance and in the presence of uncertain wind and solar energy penetration. It is observed that the frequency deviation obtained using ST-SMC is within permissible limits in the presence of uncertainties, nonlinearities and actuator time delay. It is also observed that the frequency deviation obtained using ST-SMC has better transient response compared to commonly used PI controller. In future, the stability of MG can be studied using sliding mode under cyberattacks in form of time varying delays.

References

1. Bevrani, H., Feizi, R., Ataee, S.: Robust frequency control in an islanded microgrid: H_∞ and μ-synthesis approaches. IEEE Trans. Smart Grid **7**(3), 706–717 (2015)
2. Rajesh, K.S., Dash, S.S., Rajagopal, R.: Load frequency control of microgrid: a technical review. In: Drück, H., Mathur, J., Panthalookaran, V., Sreekumar, V. (eds.), Green Buildings and Sustainable Engineering. Springer Transactions in Civil and Environmental Engineering. Springer, Singapore (2020)
3. Heidary, J., Gheisarnejad, M., Rastegar, H., Khooban, M.H.: Survey on microgrids frequency regulation: modeling and control systems. In: Electric Power Systems Research, vol. 213. Springer (2022)

4. Veronica, A.J., Kumar, N.S.: Control strategies for frequency regulation in microgrids: a review. Wind Eng. **43**(1), 107–122 (2019). Baudoin, S., Vechiu, I., Camblong, H.: A review of voltage and frequency control strategies for islanded microgrid. In: 16th International Conference on System Theory, Control and Computing (ICSTCC), Sinaia, Romania, pp. 1–5 (2012)
5. Maneesh.: Frequency control of a microgrid by using PI controller. In: International Conference on Energy, Power and Environment: Towards Sustainable Growth (ICEPE), Shillong, India, pp. 1–5 (2015)
6. Michael, C., Tola, O.J., Nwohu, M.N., Ambafi, J.G.: Load frequency control of a micro-grid using fractional order pid controller. In: IEEE Nigeria 4th International Conference on Disruptive Technologies for Sustainable Development (NIGERCON), Lagos, Nigeria, pp. 1–6 (2022)
7. Dashtdar, M., Flah, A., Hosseinimoghadam, S.M.S., Fergany, A.E.: Frequency control of the islanded microgrid including energy storage using soft computing. In Scientific Report, vol. 12 (2022)
8. Akula, S.K., Salehfar, H.: Frequency control in microgrid communities using neural networks. In: North American Power Symposium (NAPS), Wichita, KS, USA, pp. 1–6 (2019)
9. Rehimi, S., Mirzaei, R., Bevrani, H.: ANN-based frequency and tie-line power control in interconnected microgrids. In: 6th International Conference on Control, Instrumentation and Automation (ICCIA), Sanandaj, Iran, pp. 1–6 (2019)
10. Kayalvizhi, S., Vinod Kumar, D.M.: Load frequency control of an isolated micro grid using fuzzy adaptive model predictive control. IEEE Access **5**, 16241–16251 (2017)
11. Ma, L., Zhang, X., Rao, Y., Fang, Z., Rao, Y., Yang, J.: A robust model predictive control for multi-microgrids frequency control. In: IEEE 4th Conference on Energy Internet and Energy System Integration (EI2), Wuhan, China, pp. 1637–1641 (2020)
12. Liu, T., Chen, A., Gao, F., Liu, X., Li, X., Hu, S.: Double-loop control strategy with cascaded model predictive control to improve frequency regulation for islanded microgrids. IEEE Trans. Smart Grid **13**(5), 3954–3967 (2022)
13. Shahzad, S., Abbasi, M.A., Chaudhry, M.A., Hussain, M.M.: Model predictive control strategies in microgrids: a concise revisit. IEEE Access **10**, 122211–122225 (2022)
14. Armin, M. et al.: Robust extended H_∞ control strategy using linear matrix inequality approach for Islanded microgrid. IEEE Access, 135883–135896 (2020)
15. Azizi, S.M., Khajehoddin, S.A.: Robust decentralized voltage and frequency control of generators in islanded microgrids using μ-synthesis. In: IEEE Energy Conversion Congress and Exposition (ECCE), Milwaukee, WI, USA, pp. 1–8 (2016)
16. Setyawan, L., Peng, W., Jianfang, X.: Implementation of sliding mode control in DC microgrids. In: 9th IEEE Conference on Industrial Electronics and Applications, Hangzhou, China, pp. 578–583 (2014)
17. Somarin, H.M., Parvari, R.: Micro-grid stabilizer design using sliding mode controller. Int. J. Electrical Power Energy Syst. **116** (2020)
18. Cucuzzella, M., Incremona, G.P., Ferrara, A.: Decentralized sliding mode control of islanded ac microgrids with arbitrary topology. IEEE Trans. Indust. Electron. **64**(8), 6706–6713 (2017)
19. Vrdoljak, K., Peric, N., Petrovic, I.: Sliding mode-based load-frequency control in power systems. Electric Power Syst. Res. **80**, 514–527 (2009)
20. Bandyopadhyay, B., Deepak, F., Kim, K.S.: Sliding Mode Control Using Novel Sliding Surfaces. Springer, Berlin (2009)
21. Heng, C.T., Jamaludin, Z., Hashim, A.Y.B., Abdullah, L., Rafan, N.A.: Design of super twisting algorithm for chattering suppression in machine tools. Int. J. Control Autom. Syst. **15**(3), 1259–1266 (2017)
22. Levant, A.: Higher order sliding modes, differentiation and output feedback control. Int. J. Control **76**(9–10), 924–941 (2003)
23. Ibraheema, M.I., Edrisi, M., Gholipour, M., Alhelou, H.H.: A novel frequency regulation in islanded microgrid using sliding mode control with disturbance observers considering storages and EVs. Comput. Electrical Eng. **105** (2023)

24. Dev, A., Léchappé, V., Sarkar, M.K.: Prediction-based super twisting sliding mode load frequency control for multi area interconnected power systems with state and input time delays using disturbance observer. Int. J. Control, Taylor and Francis Online **94**(7), 1751–1764 (2019)
25. Tan, W., Chang, S., Zhou, R.: Load frequency control of power systems with non linearities. IET Generat., Trans. Distrib. **11**(17), 4307–4313 (2017)
26. Dashtdar, M., Flah, A., Hosseinimoghadam, S.M.S., El-Fergany, A.: Frequency control of the islanded microgrid including energy storage using soft computing. Sci. Rep. **12** (2022)

Determination of the Dynamic Parameters of the Planar Robot with 2 Degrees of Freedom by the Method of Least Squares and Instrumental Variables

Saada Fadwa, Chabir Karim, and Abdelkrim Mohamed Naceur

Abstract The identification of robot dynamic parameters is usually based on the use of the reverse model which is linear in relation to the dynamic parameters to be identified. To get an over determined system, this model is sampled while the robot is trained with fascinating movements. The optimum solution is obtained with linear least squares (LCL). The efficacy of this approach has been demonstrated through the experimental identification of numerous industrial prototypes and robots. It has been expanded to other systems including compactors, cars, engines, and haptic interfaces. Though, this method requires accurate measurement at a high rate of torque and position image. It also requires estimating speeds and accelerations by band-pass filtering positions. This results in a noisy observation matrix. Furthermore, the identification is done in a closed loop due to the unstable nature of the robots' dual integrator. The observation matrix being noisy, and the identification being carried out in a closed loop, we have adapted the method of the instrumental variable (IV) to remedy the problem of the noisy observation matrix. This article highlights this technique and proposes an extension of this method which is applied to a flat robot of 2 degrees of freedom (DOF).

Keywords Dynamic model · Least squares instrumental variable · Identification

S. Fadwa (✉) · C. Karim · A. M. Naceur
Research Laboratory Modeling, Analysis and Control of Systems (MACS) LR16ES22, National Engineering School of Gabes, University of Gabes, Gabes, Tunisia
e-mail: saada.fadwa@gmail.com

A. M. Naceur
e-mail: naceur.abdelkrim@enig.rnu.tn

A. Ben Makhlouf et al. (eds.), *State Estimation and Stabilization of Nonlinear Systems*, Studies in Systems, Decision and Control 491,
https://doi.org/10.1007/978-3-031-37970-3_18

1 Introduction

A haptic device is a robotic system designed to establish a link between a human operator and a virtual environment. It provides users with the ability to interact with objects within the virtual environment using both tactile feedback (sense of touch) and kinesthetic feedback (sense of force). In general, and for any type of control architecture used, the precise reproduction of the movements made by the operator and the telepresence are the ultimate objectives of the teleoperation systems. The ultimate goal of kinesthetic feedback is to enable users to experience a sense of direct interaction with the handling environment. For this, the human operator must be provided with sufficient feedback that allows him to have the feeling of presence in the site remotely. The system can combine visual, audible and/or haptic feedback depending on the intended application.

The links utilized for the Internet of Things (IoT), referred to as "All-IP" connections, lack determinacy as the transmission of information is not guaranteed to be received within a controlled timeframe. As a result, these connections may cause delays in the exchange of real-time information. The goal is to maintain the accuracy and dependability of digital data throughout its lifespan.

The EDUFORCE device is an ergonomic haptic interface that utilizes two active degrees of freedom and includes a return of effort feature. Its structure comprises articulated mechanics, position sensors, and motors, all of which culminate in an ergonomic handle. The control electronics incorporates a calculation card that ensures real-time control of the motors according to the instructions received by a fast digital link [1].

The identification approach, which generally employs the reverse model and linear least squares, has proven to be successful on a wide range of prototypes, as well as in the context of industrial robots, automobiles, compactors, electrical machines, and haptic interfaces [2, 14]. The results achieved were interesting and coherent. However, in any case, to calculate joint speeds and accelerations, a differential bandwidth filter is needed. In addition, the identification procedure is performed when the robot is controlled in position and speed as the robots have an unstable dual integrator behavior in an open loop. So, we find ourselves in a closed-circuit identification situation. Under such circumstances, the residual matrix and the observational matrix may become statistically dependent, resulting in a violation of their independence. Consequently, the MCL estimator becomes incoherent. The instrument variable method addresses with the problem of the noisy observation matrix and can be statistically optimum [1, 15–19] for example.

The straightforwardness of this approach makes it highly appealing. Specifically, it is designed for linear systems that rely on identified parameters and does not require identification of the noise model. Moreover, this method remains reliable even when dealing with colored noise [15]. However, this approach is usually applied unchanged to linear discrete systems, which is not the case with robotics. Furthermore, Method IV is exclusively intended for open-loop identification. As a result, algorithms that are well-suited to our environment have emerged over time, particularly in the context of closed-loop continuous-time systems.

Specifically, the SRIVC algorithm suggested in [16] for open-loop identification, which employs auxiliary models as instrumental variables, appears to be well-suited for our problem. Recently, this algorithm has also been modified to identify closed-loop systems [18, 19], and this approach is implemented within the CONTSID toolbox developed by the CRAN team [20]. However, it should be noted that this algorithm is limited to linear systems in their current state.

In [21], the authors triumphantly adapted the IV method for determining the 1-DDL cable interface. However, the robustness of method IV for data filtering and for optimal solution initialization and computation has not been discussed. This paper discusses these problems, and the IV method is applied to a two-plane DDL robot developed by the IRCCyN robotics team.

The structure of this document is organized in the following manner: the second section outlines the dynamic model and its identification process, while the third section covers the presentation of the linear least squares method. In the fourth section, the instrumental variables method is discussed, and the document concludes with a final section that includes a comparison between the two methods mentioned earlier and a discussion of the results obtained.

2 Dynamic Model and Identification

A. Direct Dynamic Model

The direct dynamic model expresses the articular accelerations according to the positions, speeds and torque. It is therefore represented by the following relationship:

$$\ddot{\theta} = \mathbf{M}^{-1}(\tau - \mathbf{c}(\theta, \theta)) \tag{1}$$

So, we can write (1) in a form of state. We then pose:

$$\dot{\mathbf{x}} = \begin{pmatrix} \dot{\theta} \\ \ddot{\theta} \end{pmatrix} = \begin{pmatrix} \dot{\theta} \\ \mathbf{M}^{-1}(\theta)(\tau - \mathbf{c}(\theta, \dot{\theta})) \end{pmatrix} = \begin{pmatrix} \dot{\theta} \\ -\mathbf{M}^{-1}(\theta)\mathbf{c}(\theta, \dot{\theta}) \end{pmatrix} + \begin{pmatrix} 0 \\ \mathbf{M}^{-1}(\theta) \end{pmatrix}\tau \tag{2}$$

where, $\boldsymbol{\theta}$ is the joint position vector, $\dot{\theta}$ is the joint velocity vector, $\ddot{\theta}$ is the vector of joint accelerations, $\mathbf{M(\theta)}$ is the kinetic energy matrix (nxn), also called the robot inertia matrix, which is symmetrical and positive defined, $\mathbf{C}(\boldsymbol{\theta}, \dot{''})$ is the dimension matrix (nxn) representing Coriolis torque/forces and centrifugal forces and $\boldsymbol{\tau}$ is the vector of the torques/forces of the actuators. After the calculations: We find the following results:

$$\begin{bmatrix}\tau_1\\ \tau_2\end{bmatrix}=\begin{bmatrix}m_1 l_{G1}^2+\mathbf{I}_{13}+\mathbf{m}_2\left(\mathbf{l}_1^2+l_{G2}^2+2l_1\mathbf{l}_{G2}\cos(\theta_2)\right)+\mathbf{I}_{23}m_2\left(l_{G2}^2+l_1\mathbf{l}_{G2}\cos(\theta_2)\right)+\mathbf{I}_{23}\\ m_2\left(l_{G2}^2+l_1\mathbf{l}_{G2}\cos(\theta_2)\right)+\mathbf{I}_{23}m_2\mathbf{l}_{G2}+\mathbf{I}_{23}\end{bmatrix}$$

$$\cdot\begin{bmatrix}\ddot{\theta}_1\\ \ddot{\theta}_2\end{bmatrix}+\begin{bmatrix}-2m_2\mathbf{l}_1\mathbf{l}_{G2}\sin(\theta_2)\theta_1\theta_2-m_2\mathbf{l}_1\mathbf{l}_{G2}\sin(\theta_2))+\theta_2^2\\ m_2\mathbf{l}_1\mathbf{l}_{G2}\sin(\theta_2)\dot{\theta}_1^2\end{bmatrix}$$

$$+\begin{bmatrix}m_1\mathbf{g}_{G1}\cos(\theta_1)+m_2 g(\mathbf{l}_1\cos(\theta_1)+\mathbf{l}_{G2}\cos(\theta_1+\theta_2)\\ m_2\mathbf{g}\mathbf{l}_{G2}\cos(\theta_1+\theta_2)\end{bmatrix} \tag{3}$$

In our case we suppose the gravitation does not intervene in the movement because our robot is a 2R plane. What makes our MDD (3) equation will be

$$\begin{bmatrix}\tau_1\\ \tau_2\end{bmatrix}=\begin{bmatrix}m_1 1_{G2}^1+I_{13}+m_2\left(1_1^2+1_{G2}^2+21_1 1_{G2}\cos(\theta_2)\right)+I_{23} & m_2\left(1_{C2}^2+1_1 1_{G2}\cos(\theta_2)\right)+I_{23}\\ m_2\left(1_{G2}^2+1_1 1_{G2}\cos(\theta_2)\right)+I_{23} & m_2 1_{G2}^2+I_{23}\end{bmatrix}$$

$$\cdot\begin{bmatrix}\overline{\theta}_1\\ \theta_2\end{bmatrix}+\begin{bmatrix}-2m_2 1_1 1_{G2}\sin(\theta_2)\theta_1\theta_2-m_2 1_1 1_{G2}\sin(\theta_2)\theta_2^2\\ m_2 1_1 1_{G2}\sin(\theta_2)\theta_1^2\end{bmatrix} \tag{4}$$

With: m_1: the mass of arm 1, m_2: the mass of arm 2, $l_{g2}=l_2/2$ (center of gravity of arm 2), I_{13}: moment of inertia of arm 1, I_{23}: moment of inertia arm 2.

B. Inverse dynamic model

The inverse dynamic model (IDM) describes the joint position, velocity, and acceleration as a function of the torques. And this is what we will model in the form of a state representation, by taking out the angles according to the couples of our previous equation:

$$\dot{X}=AX+Bu \tag{5}$$

$$Y=CX+Du$$

In our case, we assumed the following notation:

X: state vector $(\theta_1, \theta_2, d\theta_1, d\theta_2)$ of size$(n*1)\rightarrow(4*1)$.
u: command vector (τ_1, τ_2) of size $(m*1)\rightarrow(2*1)$.
Y: measure vector, in our case$Y=X=(\theta_1, \theta_2, d\theta_1, d\theta_2)$.
A: state matrix $(n*n)\rightarrow(4*4)$.
B: command matrix $(n*m)\rightarrow(4*2)$.
C: dimension observation matrix $(p*n)=(n*n)$ because $Y=X$ in our case.
D: dimension feedforward matrix $(p*m)=(n*m)$, in our case $=0$.

The inverse dynamic model of a robot with n degrees of freedom is written in the following form [2]:

$$\tau=\mathbf{M}(\theta)\,\ddot{\theta}+\mathbf{C}(\theta,\dot{\theta})\,\dot{\theta}+\mathbf{G}(\theta) \tag{6}$$

where, $\boldsymbol{\theta}$ is the vector (n × 1) of joint positions, $\dot{\theta}$ and $\ddot{\theta}$ its time derivatives, $\boldsymbol{\tau}$ the vector (n × 1) of the joint torques, $\mathbf{M}(\boldsymbol{\theta})$ the symmetrical inertia matrix (n × n), **C(θ,** $\dot{''}$) the dimension matrix (n × n) representing Coriolis torque/forces and centrifugal forces, $\mathbf{G}(\boldsymbol{\theta})$ the vector of torques/forces of gravity.

C. Identification

The inverse dynamic version of a robotic with n levels of freedom is written withinside the following form [22]:

$$\tau = \mathbf{M}(\theta)\,\ddot{\theta} + \mathbf{N}(\theta, \ddot{\theta}) + \text{offset} \tag{7}$$

where θ is the vector (n × 1) of joint positions, $\dot{\theta}$ and $\dot{\theta}$ its time derivatives, τ the vector (n × 1) of joint torques, $M(\theta)$ the symmetric (nxn) inertia matrix, $N(\theta, \dot{\theta})$ the vector (n × 1) grouping the centrifugal forces, Coriolis, gravity, and friction, offset being the vector (n × 1) of the torque offsets.

Using Denavit and Modified Hartenberg (DHM) geometric descriptions, we obtain a linear inverse dynamics model according to standard dynamic parameters [20]:

$$\tau = D_{STD}(\theta, \dot{\theta}, \ddot{\theta})\chi_{STD} \tag{8}$$

D_{STD} $(\theta, \dot{\theta}, \ddot{\theta})$ is the standard linear directory of nxc size, χ_{STD} is the cx1 column vector of the standard dynamic parameters.

For each object j, these parameters are the coefficients XX_j, XY_j, XZ_j, YY_j, YZ_j, ZZ_j (Kgm^2) of the inertia tensor, denoted ${}^jJ_j$, the mass is m_j, and the first moment vector about the origin of the marker j is denoted ${}^jM_j = [MX_j\ MY_j\ MZ_j]$ (kgm), the viscous and dry friction parameters are denoted as Fv_j (Nms/rad) and Fs_j (Nm), respectively, the inertial Ia_j and torque offset of the actuator are denoted as $Offset_j$.

To facilitate the identification process, it is crucial to identify the fundamental parameters. These standard parameters are combined to form the inverse dynamic model, and only their grouping or combination can be identified. This search is reduced to calculating the rank of D_{STD} and detecting the linear combinations between its columns. There are mainly two ways to calculate the minimum inertial parameters: a literal method based on calculating energies and a numeric method based on the QR extension [23, 23].

The minimum system is based on the following general relation:

$$\tau = D(\theta, \dot{\theta}, \ddot{\theta})\chi \tag{9}$$

D $(\theta, \dot{\theta}, \ddot{\theta})$ is the minimum linear regressor of dimension n × b, χ the column vector of the base parameters of dimension b × 1. The inverse dynamic model is sampled while the robot is turned on to achieve an overdetermined system such as:

$$Y(\tau) = W(\theta, \dot{\theta}, \ddot{\theta})\chi + \rho \tag{10}$$

where $Y(\tau)$ is the vector of measurements of dimension $r \times 1$, W is the observation matrix of dimension $r \times b$ where $r = n_e \times n$ where n_e is the number of samples retrieved. Estimation theory offers a wide range of methods. Solving overdetermined systems using classical methods gives good results if the elements of W are handled appropriately.

$$\hat{\chi} = \min_{\hat{\chi}} \|\rho\|_2 \tag{11}$$

Since we consider both fundamental parameters and exciting trajectories [23], W has the largest rank. Then we have an explicit and unique $\widehat{\chi}$ solution:

$$\hat{\chi} = \left(\left(w^T W\right)^{-1} W^T\right) Y = W + Y \tag{12}$$

Practically, we compute an estimate of the standard deviation of the identified values by considering that W is deterministic and ρ is a central random vector of independent components, the standard deviation σ_ρ, and the covariance matrix C_ρ such that:

$$C_\rho = E\left(\rho \rho^T\right) = \sigma_\rho^2 I_r \tag{13}$$

I_r is the identity matrix of dimension r. Assuming the error vectors are centered, have independent components, and are equally distributed, the standard deviation σ_ρ can be computed by the following unbiased estimator:

$$\sigma_\rho^2 = \frac{\|Y - W_{\hat{\chi}}\|^2}{(r - b)} \tag{14}$$

The expression of the estimated error covariance matrix is:

$$C_{\hat{\chi}} = \sigma_\rho^2 \left(w^T w\right)^{-1} \tag{15}$$

We deduce the standard deviation:

$$\sigma_{\hat{\chi}_j} = \sqrt{C_{\hat{\chi}}(j, j)} \tag{16}$$

This interpretation is often used but must be considered with caution because, in our case, the assumptions are not satisfied, W not being deterministic [6]. Indeed, the measurements are noisy, and the proposed model is not perfect, which requires prior processing. The relative standard deviation is estimated by:

$$\sigma_{\hat{\chi}_{j\%}} = 100\frac{\sigma_{\hat{\chi}_j}}{|\hat{\chi}_j|} \tag{17}$$

This criterion is often accustomed assess the standard of the identification. However, the experimental information area unit clattery and W isn't settled. This drawback is solved by the filtering of Y and by the filtering of the columns of W as delineated in [9, 24] for instance.

The use of MCL is especially fascinating because of it's not necessary to integrate a differential system, and there's no downside of initial conditions. However, the calculation of velocities and accelerations via bandpass filtering of the position is needed.

3 Linear Least Squares

As the inverse dynamic model is linear in relation to the parameter, the approach generally adopted in robotics is to use this inverse model associated with linear least squares. This approach has been widely used for industrial robots [9, 10]. Figure 1 illustrates the principle of the method [11].

The identification process aims to confirm the accuracy of the Dynamic model that has been theoretically calculated. By comparing the results obtained from the theoretical and experimental approaches, the identification process seeks to achieve an optimal dynamic model. For this we will use the least squares method; the principle of the latter consists in sampling the inverse dynamic model of the robot along exciting movements. For a moving robot subjected to n joint torques Γ, with Ne observations per joint, we obtain an overdetermined linear system of $(n \times N_e)$ equations with unknown N_f, corresponding to the number of parameters to be identified.

Fig. 1 Identification method by inverse model and linear least squares

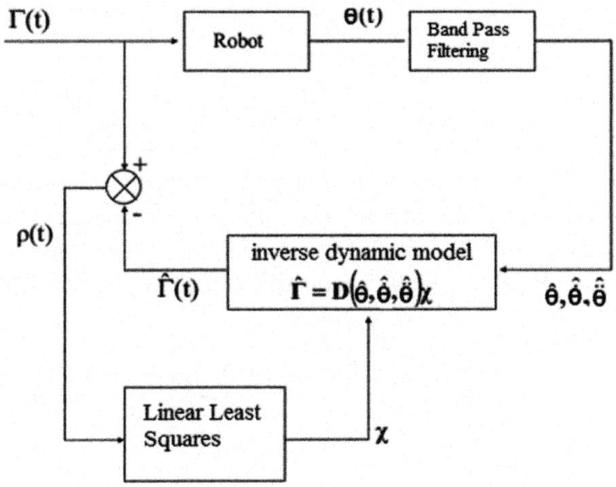

The least squares algorithm use to make the identification. This algorithm comes from the work of [9]. Consider the following discrete system:

$$y(k+1) = -a \cdot y(k) + bu(k) = \theta^T \varphi(k) \quad (18)$$

where $\boldsymbol{\theta^T}$ is the parameter vector and $\boldsymbol{\phi(k)} = \mathbf{[y(k),\ u(k)]}^{\mathbf{T}}$ is the vector of measurements with u(k) is the command and y(k) are the output.

The a priori prediction model will be:

$$\hat{y}(k+1) = -\hat{a} \cdot y(k) + \hat{b} \cdot u(k) = \hat{\theta}(k)^T \cdot \varphi(k) \quad (19)$$

where $\hat{y}(k+1)$ represents the a priori prediction depending on the values of the estimated parameters at time k and $\hat{\theta}^T = \left[\hat{a}(k), \hat{b}(k)\right]$ is the vector of the estimated parameters.

The objective is to find a parameter estimation algorithm that minimizes the error between the output of the process and the output of the prediction model in the sense of the "least squares".

$$\min_{\hat{\theta}(k)} J(k) = \sum_{i=1}^{k} \left(y(i) - \theta(k)^T \varphi(i-1)^2\right) \quad (20)$$

With

$$\hat{\theta}(k)^T \phi(i-1) = -\hat{\mathbf{a}}(\mathbf{k}).(\mathbf{y}-1) + \hat{\mathbf{b}}(\mathbf{k}).\mathbf{u}(\mathbf{i}-1) = \hat{\mathbf{y}}\left(\mathbf{i}/\hat{\theta}(\mathbf{k})\right)$$

It is therefore the prediction of the output at time i ($i \le k$) based on the estimation of the parameters at time k obtained using k measurements [29]. The quantity within the parenthesis represents the prediction error between the process and the model at time i, based on the estimation of the parameters at time k:

$$\in \left(\frac{\mathbf{i}}{\hat{\theta}(\mathbf{k})}\right) = \mathbf{y}(\mathbf{i}) - \hat{\theta}(\mathbf{k})^{\mathbf{T}} \phi(\mathbf{i}-1) \quad (21)$$

The first step is to estimate a parameter θ at time k so that it minimizes the sum of the squares of errors between the process and the prediction model over a range of k measures. The value of $\hat{\theta}(k)$ which minimizes the criterion is obtained by searching for the value which cancels the derivative δJ(k)/δθ(k).

$$\frac{\delta \mathbf{J}(\mathbf{k})}{\delta\theta} | \theta = \hat{\theta}(\mathbf{k}) = -2\sum_{\mathbf{i}=1}^{\mathbf{k}} \left(\mathbf{y}(\mathbf{i}) - \hat{\theta}(\mathbf{k})^{\mathbf{T}} \phi(\mathbf{i}-1)\right) \phi(\mathbf{i}-1) = 0 \quad (22)$$

We can say that:

$$\left(\hat{\theta}(\mathbf{k})^{\mathbf{T}}\phi(\mathbf{i}-1)\right)\phi(\mathbf{i}-1)=\phi(\mathbf{i}-1)\phi(\mathbf{i}-1)^{\mathbf{T}}\hat{\theta}(\mathbf{k}) \tag{23}$$

So, the equation becomes:

$$\sum_{\mathbf{i}=1}^{\mathbf{k}}(\phi(\mathbf{i}-1)\phi(\mathbf{i}-1)^{\mathbf{T}})\hat{\theta}(\mathbf{k})=\sum_{\mathbf{i}=1}^{\mathbf{k}}\mathbf{y}(\mathbf{i})\phi(\mathbf{i}-1)$$

Thus, we find:

$$\hat{\theta}(k)=\mathrm{inv}\sum_{i=1}^{k}\left(\varphi(i-1)\varphi(i-1)^{T}\right)\sum_{i=1}^{k}y(i)\varphi(i-1) \tag{24}$$

D. Dynamic Model by Least Square

Using the least squared method, we will write our dynamic model, to identify and validate the parameters of the 2 robots.

According to the dynamic model (4) and to apply the least squares algorithm we must have a linear system with a relation of: MX = V such that X does not depend on time (constant throughout the movement), and it is indeed the vector of the parameters that one seeks to estimate, and it is the same vector (θ in the least squares algorithm). M and V known and come from the data retrieved. V is the matrix of couples calculated from the data retrieved:

$$\begin{bmatrix}\tau_1\\ \tau_2\end{bmatrix}=J^{-1}.\begin{bmatrix}F_x\\ F_y\end{bmatrix} \tag{25}$$

And the matrix M is the matrix that contains everything that depends on time (everything that is related to the angles because the angles change over time and that we have recovered) in our equation, we find it as follows:

$$\begin{bmatrix}\tau_1\\ \tau_2\end{bmatrix}=\begin{bmatrix}A+B\cos(\theta_2)C+D\cos(\theta_2)\\ E+F\cos(\theta_2)G\end{bmatrix}\cdot\begin{bmatrix}\ddot{\theta}_1\\ \ddot{\theta}_2\end{bmatrix}+\begin{bmatrix}H\sin(\theta_2)-\left(2\dot{\theta}_1\dot{\theta}_2-\dot{\theta}_2^2\right)\\ H\sin(\theta_2)\cdot\dot{\theta}_1^2\end{bmatrix} \tag{26}$$

After simplified the equation:

$$\begin{bmatrix}\tau_1\\ \tau_2\end{bmatrix}=\begin{bmatrix}\ddot{\theta}_1 & \ddot{\theta}_1\cos(\theta_2) & \ddot{\theta}_2 & \ddot{\theta}_1\cos(\theta_2) & 0 & 0 & 0 & \left(-2\dot{\theta}_1\dot{\theta}_2-\dot{\theta}_2^2\right)\sin(\theta_2)\\ 0 & 0 & 0 & 0 & \ddot{\theta}_1 & \ddot{\theta}_1\cos(\theta_2) & \ddot{\theta}_2 & \dot{\theta}_1{}^2\sin(\theta_2)\end{bmatrix}\cdot\begin{bmatrix}A\\ B\\ C\\ D\\ E\\ F\\ G\\ H\end{bmatrix} \tag{27}$$

4 Instrumental Variable Method

A. Theoretical Approach

In practical applications, the mathematical assumptions made to ensure the MCL computer's cost-effectiveness from a theoretical standpoint may not hold. Indeed, in Eq. (12), the observation matrix W is constructed with the joint positions θ and of $\dot{\theta}$, $\ddot{\theta}$ which are numerically calculated from θ. Thus, W is noisy. Furthermore, the identification process is performed within a control system. The violations of these assumptions can lead to inconsistencies in the MCL estimator. Indeed, from (10) it comes that:

$$W^T Y = W^T W \chi + W^T \rho = W^T W \hat{\chi} \tag{28}$$

Using Sultsky's theorem, we get:

$$\lim_{n\to\infty}\hat{\chi} = \chi + \lim_{n\to\infty}\left(\frac{1}{n}W^T W\right)^{-1}\lim_{n\to\infty}\left(\frac{1}{n}W^T\rho\right) \tag{29}$$

Under the classical assumptions:
$\lim_{n\to\infty}\left(\frac{1}{n}W^T W\right) = E\left(W^T W\right)$ is positive definite, $\lim_{n\to\infty}\left(\frac{1}{n}W^T\rho\right)$ exist.
As ρ depends on the observation matrix W, $E(W^T\rho) \neq 0$. Thus, the estimator is not consistent.

The instrumental variable method deals with this downside and may be statistically optimum [1]. This method proposes an estimator consisting of constructing an instrumental matrix V such that (28) becomes:

$$V^T Y = V^T W \chi + V^T \rho = V^T W \hat{\chi}_v \tag{30}$$

The solution in the sense of the instrumental variable is:

$$\hat{\chi}_V = (V^T W)^{-1} V^T Y \tag{31}$$

Subsequently, V is calculated as a function of $\hat{\chi}_v$. This defines an iterative procedure such as:

$$\hat{\chi}_{Vk} = (V_k^T W)^{-1} V_k^T Y \tag{32}$$

where:

$$V_k = V(\hat{\chi}_{Vk}) \tag{33}$$

For this to be possible, it is essential that the matrix $V_k^T W$ is invertible.

Assuming:

$$\lim_{n\to\infty}\left(\frac{1}{n}V_k^T W\right) \text{ is nonsingular} \tag{34}$$

$$\lim_{n\to\infty}\left(\frac{1}{n}V_k^T \rho\right) = E(V_k^T \rho) = 0 \tag{35}$$

For each k, $\hat{\chi}_{Vk}$ converges to χ with n. Indeed, with Slutsky's theorem, we get:

$$\lim_{n\to\infty} \hat{\chi}_{Vn} = \chi + \lim_{n\to\infty}\left(\frac{1}{n}V_k^T W\right)^{-1} \lim_{n\to\infty}\left(\frac{1}{n}V_k^T \rho\right) \tag{36}$$

Which finally gives us:

$$\lim_{n\to\infty} \hat{\chi}_{Vn} = \chi \tag{37}$$

Therefore, the IV estimator is highly reliable.

B. Calculation of an Instrumental Matrix

The primary challenge is to identify an instrumental matrix V. One possible solution is to generate an observation matrix using simulated data instead of actual measurements. The simulated data, also known as instruments, are the outcomes of an auxiliary system [16] that approximates the noise-free model of the system that needs to be identified. In robotics, the MDD of the robot given by (27) serves as the auxiliary model. The algorithm for the IV method designed for the dynamic identification of robots is summarized in the following steps, as shown in Fig. 1. The algorithmic program is initialized (see C).

At every iteration θ_s, $\dot{\theta}_s$ and $\ddot{\theta}_s$ are calculated by simulation and integration of the robot's MDD with the parameters identified at the previous iteration. W_S is obtained as a V sampling and that we select the instrumental matrix as follows:

$$V(\hat{\chi}_{Vk}) = W_s(\theta_s, \dot{\theta}_s, \ddot{\theta}_s, \hat{\chi}_{Vk}) \tag{38}$$

Y (τ) is calculated by sampling and filtering of measurements τ and W $(\theta_s, \dot{\theta}_s, \ddot{\theta}_s)$ is calculated with the MDI using θ sampling.

$\hat{\chi}_{Vk+1}$ is given by (31). The algorithm terminates when the relative errors become insignificant.

As stipulated in [19], the instrumental matrix should be on the point of the W_{nf} matrix outlined by:

$$W_{nf} = W(\theta_{nf}, \dot{\theta}_{nf}, \ddot{\theta}_{nf}) \tag{39}$$

where $\theta_{nf}, \dot{\theta}_{nf}, \ddot{\theta}_{nf}$ are the values without noise of $\theta, \dot{\theta}, \ddot{\theta}$.

This presupposes a well-adjusted management loop that keeps $\theta_{nf}, \dot{\theta}_{nf}, \ddot{\theta}_{nf}$ about to the states of reference $\theta_r, \dot{\theta}_r, \ddot{\theta}_r$ From (39), it comes that the simulated states that $\theta_s, \dot{\theta}_s, \ddot{\theta}_s$ should stay about to $\theta_r, \dot{\theta}_r, \ddot{\theta}_r$ at every iteration of the rule. This can be achieved by simulating a management law structure that remains constant, like the one applied to the robot with fixed data measurements. As the vector $\hat{\chi}_{Vk}$ varies with each iteration, the gains in the simulator must be adapted according to $\hat{\chi}_{Vk}$.

For example, to regulate a PD command in position for every articulation j, the MDD is simplified by a double integrator $1/ZZ_{RJ}S^2$ as illustrated Fig. 2.

The θ_j/θ_{jr} transfer function is:

$$H_j(s) = \frac{1}{\frac{ZZ_{Rj}}{K_{pj}K_{vj}}s^2 + \frac{1}{K_{pj}}s + 1} = \frac{1}{\frac{1}{\omega_{nj}^2}s^2 + \frac{2\eta_j}{\omega_{nj}}s + 1} \tag{40}$$

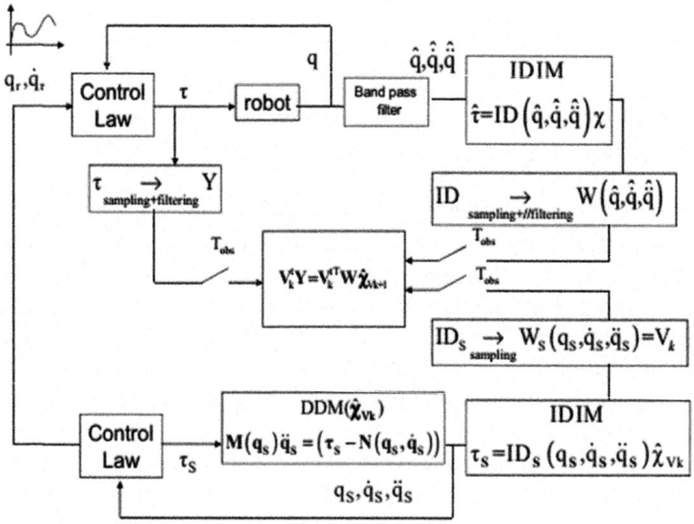

Fig. 2 Instrumental variable procedure

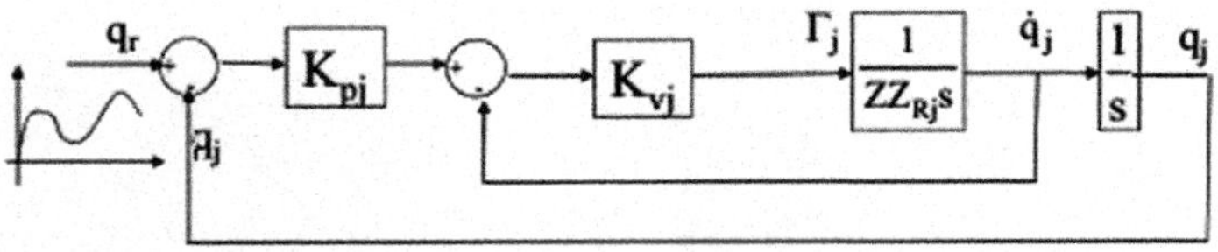

Fig. 3 Order diagram for the J axis

Avec: $K_{pj} = \omega_{nj}/2\eta_j$ et $K_{vj} = 2\eta_j\ \omega_{nj}\ ZZ_{Rj}$.

It is recalled that ω_{nj} is the natural pulsation and η_j is the depreciation constant of the transfer operate of the second order of the control system. To keep the bandwidth of the closed loop, ω_{nj} and η_j must be constant. So K_{vj} should be updated with (Fig. 3). For example, to regulate a PD command in position for every articulation j, the MDD is simplified by a double integrator $1/\ ZZ_{RJ}S^2$ as illustrated (Fig. 3):

$$K_{vj} = 2\eta_j\omega_{nj}ZZ_{Rj/k} \tag{41}$$

where $ZZ_{RJ/K}$ is the IV estimate of ZZ_{RJ} at the iteration k.

Thanks to this, V_k tends towards an independent deterministic matrix of ρ. Supposing that ρ is an independent noise with zero average value, we get:

$$E\left(V_k^T\rho\right) = E\left(V_k^T\right)E(\rho) = 0 \tag{42}$$

The final equation guarantees the fulfillment of hypothesis (35), while hypothesis (34) remains satisfied. Therefore, the IV estimator proposed is consistent. In IV methods, the columns of V are filtered using an optimal filter denoted as F(s) [20]. The application of this filter is justified by the fact that the frequency range in which the model we intend to identify is accurate can be explicitly calculated for linear systems. In our case, and within the broader scope of robotics, there is no need to use the F(s) filter due to the following reasons. First, the V columns are deterministic. Second, the MDD has double integrator behavior. Therefore, high frequency noises are filtered. Finally, the frequency range that interests us is determined by the bandwidth of the closed loop. Therefore, the incorporation of MDD functions as an adaptable filter. This critical aspect will be investigated during the experimental identification process.

C. Prototype to Identify

Utilizing the instrumental variable method for the dynamic identification of a two-degree-of-freedom planar robot. For the geometric description, we use the modified notation of Dennavit and Hartenberg [26]. The robot is devoid of a gear reduction mechanism and instead operates through direct current machines (Fig. 4).

The inverse dynamic model depends on 8 minimal parameters: $\chi = [ZZ_{1R}\ fv_1\ fs_1\ ZZ_2\ MX_2\ MY_2\ fv_2\ fs_2]^T$.

Closed loop is the PD instruction described in paragraph 3.B. Its bandwidth is 2 Hz. The acquisition frequency is 200 Hz. The torque is obtained with the current

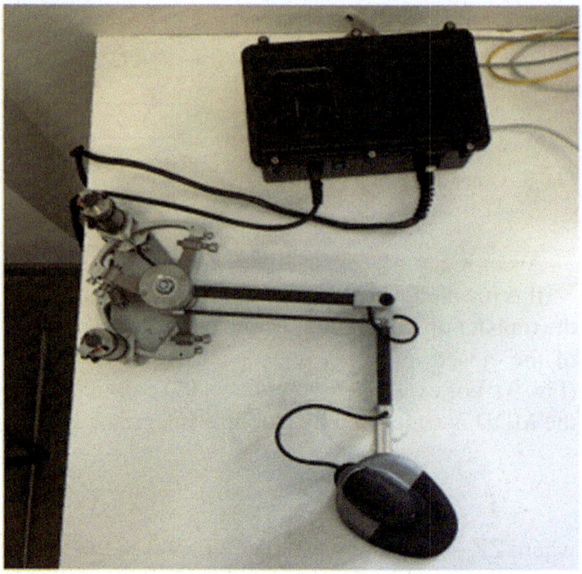

Fig. 4 Prototype 2DDL plan to identify

reference v_{ir} assuming a wide bandwidth of the current loop (1 kHz):

$$\tau_j = gt_j v_{irj} \tag{43}$$

gt_j being the j-axis actuation gain. Robot simulations are performed using the same exciting trajectories and control structures as real robots.

5 Experimental Results

To conduct the initial test, the Butterworth low-pass filter and the decimation filter were set to have a cutoff frequency near 5 Hz. This frequency was determined by performing spectral analysis on the W column. The maximum bandwidth allowed is 0.5 Hz, but to preserve all relevant data, the cutoff frequency was increased to 5 Hz by multiplying the maximum bandwidth by 10. The velocity close to zero is disregarded due to the friction model being employed.

By programming the least squares algorithme on Matlab, we find the vector X which estimates the parameters. We replace the parameters found in the equation of the dynamic model to estimate the torques. The estimated torques are compared with the calculated torques using the recovered forces. The measured torques are almost equal to 0 and the majority of the estimated torques are negative.

The values identified with the least squares method are listed in Table 1. They are close to the nominal values. Direct comparisons were made. The results are close to those shown in Fig. 5. We can tell that the estimated torque and measured torque are

Table 1 Values identified with least squares

Parameters	Nominal values	MCL values	$\sigma_{\chi j}\%$
ZZ_{1R}	3.45	3.45	0.50
Fv_1	X	0.002	52.0
Fs_1	X	0.82	6.0
ZZ_2	0.061	0.061	0.51
MX_2	0.247	0.247	0.56
MY_2	0.014	0.014	5.0
Fv_2	X	0.001	23.0
Fs_2	X	0.135	0.24

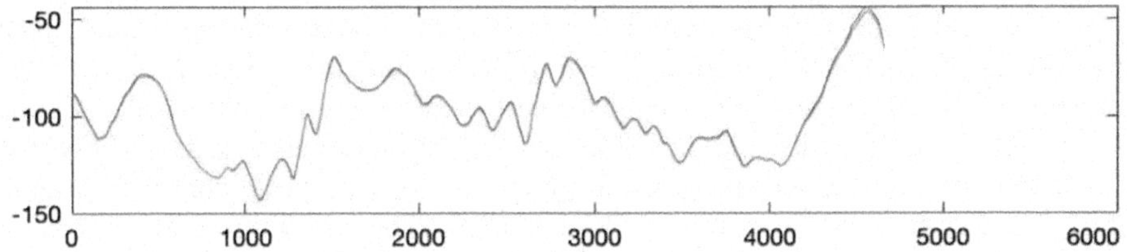

Fig. 5 Direct comparison made with the IV method

very close by looking at the data. We notice that the viscous terms contribute very little to the dynamics of the model. Indeed, they have a high relative standard deviation. Moreover, when they are eliminated from the dynamic model, the identified values of the other parameters do not vary, similarly norm of the residual.

A. IV Method Initialized with MC Values and V Column Filtering

We will now utilize the IV method, outlined in part 3, for data analysis. Like the W column, the V column is filtered with a decimation filter set at 5 Hz. The algorithm successfully converges after 4 iterations, and the results are presented in Table 2, summarizing the values obtained.

Table 2 Values identified after 4 iterations with method IV

Parameters	Nominal values	IV values	$\sigma_{\chi j}\%$
ZZ_{1R}	3.45	3.45	0.52
Fv_1	0.002	0.002	40.0
Fs_1	0.82	0.82	3.0
ZZ_2	0.061	0.061	0.49
MX_2	0.247	0.248	0.52
MY_2	0.014	0.014	3.5
Fv_2	0.001	0.001	30.0
Fs_2	0.135	0.133	3.0

Table 3 Values identified with method IV after 5 iterations and without V filtering

Parameters	Nominal values	IV values	$\sigma_{\chi j}\%$
ZZ_{1R}	1.0	3.45	0.2
Fv_1	0.0	0.002	33.0
Fs_1	0.0	0.82	1.0
ZZ_2	1.0	0.061	0.3
MX_2	0.0	0.248	0.2
MY_2	0.0	0.014	1.5
Fv_2	0.0	0.001	20.0
Fs_2	0.0	0.133	0.7

These values are near the initial values. Direct comparisons were made (see Fig. 5). The torque values estimated using the IV method are very close to the measured values. These results suggest that the IV method only provides significant improvements when the data filtering is accurately performed. Notably, the IV method produces values that are highly consistent with those identified through the MCLs, with relative variations of less than 2%. These findings align with those previously presented in [21].

B. Method IV with Bad Initialization and no Column Filtering of V

Next, we will evaluate the robustness of our algorithm in the absence of filtering on the V column and with a poorly chosen initialization. To compensate for these factors, gains are adjusted in the simulator to ensure that the simulated trajectories are comparable to the reference trajectories. In terms of initialization, ZZ_{1R} and ZZ_2 are initialized to 1, while the others are initialized to 0.

After only 5 iterations, one obtains the identified values exposed Table 3. it's noted that these values are equal to those exposed Table 1 (except the viscous terms because of their weak inputs).

Direct comparisons were made. The results are illustrated in Fig. 6. Note that the estimated torque is close to that measured and that the instrumental variable method acts as an adaptive filter.

The variable system's adaptability as a filter is mainly attributable to two factors. Firstly, the direct dynamic model possesses a dual integrating nature, effectively

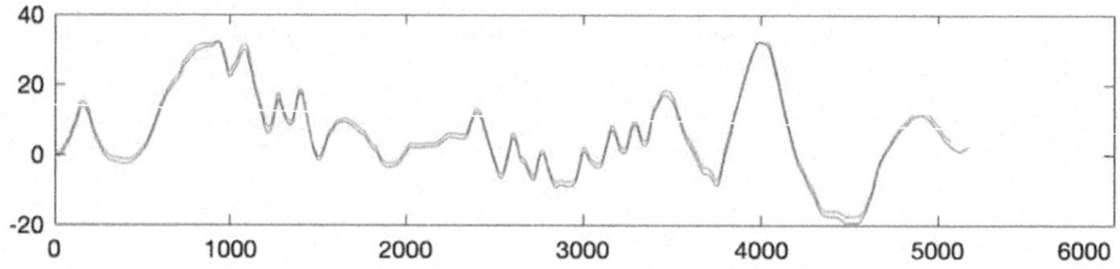

Fig. 6 Direct comparison performed with method IV without filtering of V

filtering out high-frequency noise during simulation. Secondly, by adjusting the control gains during each replication, the closed-loop bandwidth is maintained when simulating the MDD, further enhancing the filter's adaptability.

Consequently, the frequency band of interest is automatically defined by the closed-loop bandwidth, eliminating the need for filtering of the V columns.

Regarding the initial values, we tested different values for ZZ_j, such as $ZZ_{1R} = 300Kgm^2$ and $ZZ_2 = 6Kgm^2$. In all cases, the algorithm converged after 5 iterations, with the resulting values shown in Table 3.

It is crucial to emphasize that our algorithm is not sensitive to the initial conditions since the control gains are adapted in the simulator. This process guarantees that the simulated state closely matches the reference trajectory. In the event that the gains are not appropriately matched, the control loop bandwidth of the simulated system may differ from the real system, depending on the initial values. Consequently, V is no longer a reliable instrument, as the property $E(VT\rho) = 0$ can no longer be ensured.

6 Conclusion

The inverse dynamic model is noteworthy for its linearity with respect to the dynamic parameters, which enables the application of linear least squares. Despite the validation of the system modeling, the identification results obtained through the least squares method were unsatisfactory. This can be attributed to imperfect data caused by issues such as sampling errors, noise, and the use of excitation trajectories that differ from those commonly found in literature (which are usually given in position, speed, and acceleration, while the excitation command used here is force-based).

One of the primary benefits of our algorithm is that it eliminates the need to calculate joint accelerations, which is a major challenge in MCL. In the general case, this issue is resolved by appropriately filtering the data. However, selecting the appropriate cutoff frequency is crucial, as a frequency that is too low may result in a poorly conditioned system. Conversely, if the frequency is too high, the observation matrix becomes exceedingly noisy, resulting in biased MCL estimates. Consequently, techniques have been developed to overcome these issues, such as integrating the dynamic model [27] or using the power model [28]. However, the MDI method's drawback is the complex expressions required for this model. One of the disadvantages of the power model is the loss of information due to the mixing of torques. This is because the wrist's dynamics are typically insignificant compared to the wearer's dynamics. However, our algorithm is not susceptible to these issues as the equations of the MDI and MDD are maintained in their original form. Furthermore, the instrumental variable technique exhibits significantly lower sensitivity to the selection of the cutoff frequency. Our approach can be viewed as an implicit adaptive filtering method, whereby the gains are adjusted during each iteration to guarantee the bandwidth of the simulator's closed loop. Thus, the frequency range of interest is determined by this closed loop's bandwidth.

Therefore, this paper presented an expansion and implementation of the instrumental variable method, which was effectively utilized on a 2 DDL prototype. Results from these experiments demonstrate that our recursive algorithm is resilient to filtering and initialization. In addition, this algorithm allows for the simultaneous validation of both dynamic inverse and direct models, resulting in significant time savings when compared to alternative identification methods.

References

1. Janot, A.: Contribution à la modélisation et à l'identification des interfaces haptiques. Thèse de l'Université de Nantes (in French) (2007)
2. Atkeson, C.G., An, C.H., Hollerbach, J.M.: Estimation of inertial parameters of manipulator loads and links. Int. J. Robot. Res. **5**(3), 101–119 (1986)
3. Swevers, J., Ganseman, C., Tückel, D.B., de Schutter, J.D., Van Brussel, H.: Optimal robot excitation and identification. IEEE Trans. Robot. Autom. **13**(5), 730–740 (1997)
4. Khosla, P.K., Kanade, T.: Parameter identification of robot dynamics. In: Proceedings of the 24th IEEE Conference on Decision Control, Fort Lauderdale (1985)
5. Kozlowski, K.: Modelling and Identification in Robotics. Springer Verlag London Limited, Great Britain (1998)
6. Raucent, B., Bastin, G., Campion, G., Samin, J.C., Willems, P.Y.: Identification of barycentric parameters of robotic manipulators from external measurements. Automatica **28**(5), 1011–1016 (1992)
7. Gautier, M.: Identification of robot dynamics. In: Proceedings of the IFAC Symposium on Theory of Robots. Vienne, Autriche, pp. 351–356 (1986)
8. Gautier, M., Khalil, W., Restrepo, P.: Identification of the dynamic parameters of a closed loop robot. In: Proceedings of the IEEE on Int. Conference on Robotics and Automation. Nagoya, pp. 3045–3050 (1995)
9. Gautier, M.: Dynamic identification of robots with power model. In: Proceedings of the IEEE International Conference on Robotics and Automation. Albuquerque, pp. 1922–1927 (1997)
10. Khalil, W., Gautier, M., Lemoine, P.: Identification of the payload inertial parameters of industrial manipulators. In: IEEE International Conference on Robotics and Automation. Roma Italia, pp. 4943–4948 (2007)
11. Gautier, M., Poignet, P.: Identification en boucle fermée par modèle inverse des paramètres physiques de systèmes mécatroniques. JESA, J. Européen Des Systèmes Automatisés **36**(3), 465–480 (2002)
12. Gautier, M., Poignet, P.H.: Extended Kalman filtering and weighted least squares dynamic identification of robot. Control Eng. Pract. **9**, 1361–1372 (2001)
13. Venture, G., Ripert, P.J., Khalil, W., Gautier, M., Bodson, P.: Modeling and identification of passenger car dynamics using robotics formalism. IEEE Trans. Intell. Transp. Syst. **7**(3), 349–355 (2006)
14. Lemaire, C.-E., Vandanjon, P.-O., Gautier, M.: Dynamic identification of a vibratory asphalt compactor for contact efforts estimation. In: Proceedings of the 14th IFAC Symposium on System Identification, SYSID. Newcastle, Australia (2006)
15. Söderström, T., Stoica, P.: System Identification: Prentice Hall International Series in Systems and Control Engineering. Great Britain (1989)
16. Young, P.C., Jakeman, A.J.: Refined instrumental variable methods of time-series analysis: Part 1, SISO systems. Int. J. Control **29**, 1–30 (1979)
17. Young, P.C.: An instrumental variable approach to ARMA model identification and estimation. In: 14th IFAC Symposium on System Identification, SYSID 2006. Newcastle Australia, pp. 410–415 (2006)

18. Gilson, M., Garnier, H., Young, P.C., Van den Hof, P.: A refined IV method for closed loop system identification. In: 14th IFAC Symposium on System Identification, SYSID 2006. Newcastle Australia, pp. 903–908 (2006)
19. Young, P.C., Garnier, H., Gilson, M.: Simple refined IV methods of closed-loop system identification. In: Presented at 15th IFAC Symposium on System Identification. Saint-Malo, France, pp. 1151–1156, 6–8 July 2009
20. Garnier, H., Gilson, M., Cervellin, O.: Latest developments for the MATLAB CONTSID toolbox. In: Presented at 14th IFAC Symposium on System Identification, SYSID 2006. Newcastle Australia (2006)
21. Vandanjon, P.-O., Janot, A., Gautier, M., Khatounian, F.: Comparison of two identification techniques: theory and application. In: ICINCO 2007, 4th International Conference on Informatics in Control, Automation and Robotics; SS on Fractional Order Systems. Angers, pp. 341–347 (2007)
22. Khalil, W., Dombre, E.: Modeling, Identification and Control of Robots. In: Hermes Penton, London (2002)
23. Gautier, M., Khalil, W.: Exciting trajectories for the identification of the inertial parameters of robots. Int. J. Robot. Res. **11**(4), 362–375 (1992)
24. Pham, M.T., Gautier, M., Poignet, P.: Identification of joint stiffness with bandpass filtering. In: IEEE International Conference on Robotics and Automation, ICRA 2001. Seoul, Korea, pp. 2867–2872, 21–26 May 2001
25. Gautier, M., Khalil, W.: On the identification of the inertial parameters of robots. In: Proceedings of the 27th IEEE Conference on Decision and Control. Austin, pp. 2264–2269 (1988)
26. Khalil, W., Kleinfinger, J.F.: A new geometric notation for open and closed loop robots. In: IEEE International Conference on Robotics and Automation. San Francisco USA, pp. 1147–1180 (1986)
27. Middleton, H., Goodwin, G.C.: Adaptive computed torque control for rigid link manipulators. Syst. Control Lett. (10), 9–16 (1988)
28. Slotine, J.J.E., Li, W.: Adaptive manipulator control: a study. In: Proc. IEEE International Conference on Robotics and Automation. Raleigh, pp. 1312–1400 (1987)
29. Gautier, M., Janot, A., Vandanjon, P.-O.: A new identification method for mechatronic systems in closed-loop from only control data. In: Proceedings of the IFAC World Congress 2008. Seoul (2008)
30. Landau, I.D.: Identification des systèmes. Hermès, Paris (1998). ISBN: 2-86601-683-1

Design and Analysis of Nonsingular Terminal Super Twisting Sliding Mode Controller for Lower Limb Rehabilitation Exoskeleton Contacting with Ground

Mohammad A. Faraj, Boutheina Maalej, and Nabil Derbel

Abstract This chapter presents a non-singular terminal super-twisting control (NTSTC) for the problem of trajectory tracking a lower limb rehabilitation exoskeleton with uncertainties in parameters. The modeling of the lower limb rehabilitation exoskeleton has been implemented in the case of contact with the ground. The augmented Lagrange equations with Lagrange multiplier and Jacobian vectors for lower limb exoskeleton in case of contacting with ground have been developed. The non-singular terminal super-twisting surface has been employed to achieve finite-time convergence, non-singularity and reducing chattering phenomenon in control input. The selected nonlinear switching surface has been designed based on an exponent that avoid the singularity problem which occurs when only the position error converges to zero. The stability analysis of the closed-loop has been verified using Lyapunov theory. A comparison study has been conducted with the classical super-twisting algorithm (STA) for explaining the effectivity (NTSTC). The results of simulations for nominal and uncertain cases demonstrate the superiority of NTSTC controller over classical super twisting sliding mode controller (STA) in terms of tracking error and robustness towards the uncertainties in the parameters.

Keywords Rehabilitation robots · Lower limb exoskeletons · Super twisting sliding mode control · Nonsingular terminal sliding mode control · Holonomic constraints

M. A. Faraj (✉)
College of Engineering, University of Anbar, Ramadi, Iraq
e-mail: mohammed.faraj@uoanbar.edu.iq

M. A. Faraj · B. Maalej · N. Derbel
Control & Energy Management Laboratory (CEM Lab), National Engineering School of Sfax (ENIS), University of Sfax, Sfax, Tunisia
e-mail: maalej.boutheina@gmail.com

N. Derbel
e-mail: nabil.derbel@enis.tn

A. Ben Makhlouf et al. (eds.), *State Estimation and Stabilization of Nonlinear Systems*, Studies in Systems, Decision and Control 491,
https://doi.org/10.1007/978-3-031-37970-3_19

1 Introduction

In the past, the physical manual therapy methods which completely depends on the experience of physiotherapists has been considered as traditional method for restoring the walking ability and resume a normal life of lower limb extremity of patient after physical disability. However, the traditional manual therapy normally make the therapists stressed while helping patients to recover their normal motion [24, 38]. In last two decades, lower limb rehabilitation exoskeletons (LLE) have been adopted as an adequate and effective rehabilitation tool for recovering the gait rehabilitation of patients that suffer from motor dysfunction or permanent disability after the stroke or spinal cord injuries [18]. Hence, a variety of types of LLE have been created and developed by academic and industrial institutes for instance: CUHK-EXO [9], XOR [17], HAL [29], ALEX [36], LOPES [34], LOKOMAT [5].

In literature, the dynamic equations of lower limb robotic exoskeletons have been achieved using popular Lagrange equations [31]. In fact, the Lagrange multiplier and associated Jacobian vectors must be added to original Lagrange dynamic equation in case of contacting of robotic systems with vertical or horizontal surface i.e. when will be in constraint motion [7, 26]. In this chapter, the modelling of the LLE in case of contact with the ground has been accomplished.

The performance of LLE mainly depends on the nature of the control algorithm. Different control schemes have been constructed for tracking a trajectory for nonlinear lower limb rehabilitation exoskeletons systems, such as: Impedance control [25]. L1 adaptive control [22], Computed Torque Control [12], time delay based control [1, 15] and sliding mode control (SMC) [11, 13, 39]. Because of having an excellent attributes represented by insensitivity towards the uncertainties in parameters and the external disturbances, sliding mode control (SMC) has been proved as powerful tool for controlling uncertain and time varying systems [6, 33].

The main drawback of sliding mode controller is a having popular phenomenon which is known as chattering. To solve the problem of chattering, various types of sliding mode based controllers were formulated by scholars [10, 21, 35]. Second order sliding mode controllers (HOSM) were formulated to settle the chattering effect [19, 28]. In fact, the SOSM control law needs the measurement of the first time derivative of the designed sliding surface, which is usually not available. So, this issue makes the implementation of SOSM is very complicated. Super-Twisting sliding-mode algorithm (STA) is one of new types of recently SOSM controllers [14, 16]. The STA has the ability to make the states of the system reaching the equilibrium point infinite time and it does not need the derivatives of sliding surface in their structures [35].

However, classical STA control algorithms include traditional linear sliding surface that guarantees asymptotic convergence of the states of the system to the equilibrium point, but not in finite-time [14, 3]. Hence, terminal sliding-mode (TSM) which considered nonlinear sliding surfaces has been developed for achieve the finite-time convergence and enhance the convergence properties of dynamical systems. In fact, terminal sliding-mode sliding surface suffer from the singularity problem because of existing negative fractional powers. Consequently, nonsingular terminal sliding-mode (NTSM) has been formulated to eliminate the singularity problems. Compared to other sliding-mode manifolds, NTSM has finite-time convergence, singularity avoidance and chattering reduction [2, 6].

Inspired by the above-mentioned discussions, the excellent attributes of using terminal sliding surface with avoiding the singularity problem and second-order super twisting sliding mode controller (STA), the dynamic system of LLE in constraint motion has been developed and controlled in this study. The modeling of lower limb rehabilitation exoskeleton will be achieved in case of contacting with. This chapter is constructed as follows: The derivation of dynamic equations of the LLE in constrained motions has been achieved in Sect. 2. The nonsingular terminal super twisting sliding mode control law is developed in Sect. 3. The results of the chapter have been demonstrated and explained in Sect. 4. The conclusion of this chapter is achieved in Sect. 5.

2 Modeling of Lower Limb Rehabilitation Exoskeleton

2.1 Modelling of Lower Limb Exoskeleton in Free Case

In free case, traditional Lagrange equations are usually used to formulate the dynamic equations of LLE where study as indicated in Fig. 1. O is the coordinate origin and h refer to distance from the coordinate origin O at hip joint to the point of contacting of LLE with the ground and $\varnothing(q)$ refer to an algebraic constraint equation that

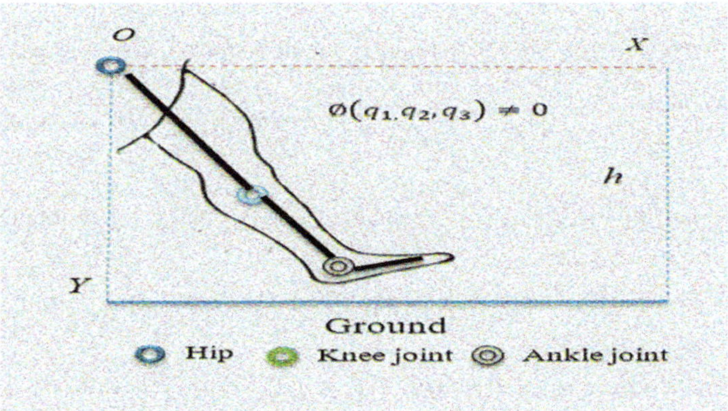

Fig. 1 Three-link exoskeleton motion in free motion case [13]

represent the constraint motion in joint space [8, 37]. The Euler–Lagrange formulas are normally used to derive the dynamic model equations of lower limb exoskeleton which can be expressed [31]

$$M(q)\ddot{q} + C(q, \dot{q})\dot{q} + G(q) + \tau_d = \tau \tag{1}$$

where: $q, \dot{q}, \ddot{q} \in R^3$ are angular joint position, velocity and acceleration vectors respectively, $M(q) \in R^{3*3}$ stands for positive definite inertia matrix, $C(q, \dot{q})\dot{q} \in R^3$ represnts Carioles, centrifugal forces and torques, $G(q) \in R^3$ is the vector of torques of the gravity, $\tau \in R^3$ is the vector of joint torques and $\tau_d \in R^3$ is the bounded unknown external disturbances. The structure of $M(q)$, $C(q, \dot{q})\dot{q}$ and $G(q)$ employed in Eq. (1) are:

$$M_{11} = a_1 + a_2 + a_4 + 2a_3\cos(q_2) + 2a_5\cos(q_2 + q_3) + 2a_6\cos(q_3)$$

$$M_{12} = M_{21} = a_2 + a_4 + a_3\cos(q_2) + a_5\cos(q_2 + q_3) + 2a_6\cos(q_3)$$

$$M_{13} = M_{31} = a_4 + a_5\cos(q_2 + q_3) + 2a_6\cos(q_3), M_{22} = a_2 + a_4 + 2a_6\cos(q_3)$$

$$M_{23} = M_{23}a_4 + 2a_6\cos(q_3), M_{33} = a_4 \tag{2}$$

$$C_{11} = -a_5(\dot{q}_2 + \dot{q}_3)\sin(q_2 + q_3) - a_3\dot{q}_2\sin(q_2) - a_6\dot{q}_3\sin(q_3)$$

$$C_{12} = -a_5(\dot{q}_1 + \dot{q}_2 + \dot{q}_3)\sin(q_2 + q_3) - a_3(\dot{q}_1 + \dot{q}_2)\sin(q_2) - a_6\dot{q}_3\sin(q_3)$$

$$C_{13} = -a_5(\dot{q}_1 + \dot{q}_2 + \dot{q}_3)\sin(q_2 + q_3) - a_3\sin(q_2) - a_6(\dot{q}_1 + \dot{q}_2 + \dot{q}_3)\sin(q_3)$$

$$C_{21} = -a_5\dot{q}_1 sin(q_2 + q_3) + a_3\dot{q}_1\sin(q_2) - a_6\dot{q}_3\sin(q_3), C_{22} = -a_6\dot{q}_3\sin(q_3)$$

$$C_{23} = -a_6(\dot{q}_1 + \dot{q}_2 + \dot{q}_3)\sin(q_3)$$

$$C_{31} = a_5\dot{q}_1\sin\left(q_2 + q_3\right) + a_6(\dot{q}_1 + \dot{q}_2)\sin(q_3), C_{32} = a_6(\dot{q}_1 + \dot{q}_2)\sin(q_3), C_{33} = 0$$

$$G_1 = -[b_1\sin(q_1) + b_2\sin(q_1 + q_2) + b_3\sin(q_1 + q_2 + q_3)]$$

$$G_2 = -[b_2\sin(q_1 + q_2) + b_3\sin(q_1 + q_2 + q_3)], G_3 = -[b_3\sin(q_1 + q_2 + q_3)]$$

$$b_1 = (m_1 d_{1+} m_2 l_{1+} m_3 l_1)g, b_2 = (m_2 d_{2+} m_3 l_2)g, b_3 = m_3 d_3 g$$

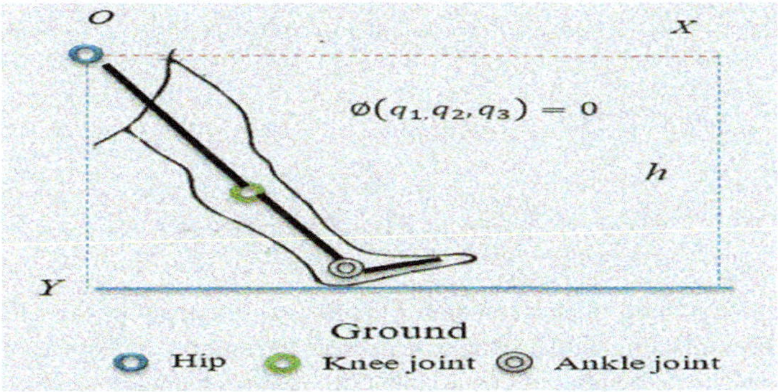

Fig. 2 Three-link exoskeleton motion for contacting with ground case [13]

$$a_1 = I_1 + m_1 d_1^2 + (m_2 + m_3) l_1^2, a_2 = I_2 + m_2 d_2^2 + m_3 l_2^2, a_3 = (m_2 d_2 + m_3 l_2) l_1$$

$$a_4 = I_3 + m_3 d_3^2, a_5 = m_3 d_3 l_1, a_6 = m_3 d_3 l_2$$

where: $m_1, m_2, m_3, l_1, l_2, l_3,$ are masses and lengths of thigh, shank, and foot of LLE and the human lower limb respectively, d_1, d_2, d_3 refer to the position of the canter of mass of thigh, shank and foot of LLE and human lower limb respectively, I_1, I_2, I_3 are the moments of inertia of thigh, shank and foot of the exoskeleton and human lower limb respectively, g is the gravity acceleration.

2.2 Modeling of Lower Limb Exoskeleton in Constraint Motion Case

When LLE contact with ground, the LLE is in constrained motion as depicted in Fig. 2. In this case, a holonomic constraint explained by an algebraic equation in joint space is used to describe the contacting of LLE with ground [7, 20]:

$$\varnothing(q) = l_1 \cos(q_1) + l_2 \cos(q_1 + q_2) + l_3 \cos(q_1 + q_2 + q_3) - h \tag{3}$$

These constraint case makes the term $\tau = J(q)^T \lambda$ existed in (1). So, the constrained dynamic equation of LLE can be written as [26]:

$$M(q)\ddot{q} + C(q, \dot{q})\dot{q} + G(q) + \tau_d = \tau + J(q)^T \lambda \tag{4}$$

where: $J(q) = \frac{\partial \varnothing(q)}{\partial q}$ is the Jacobian of the equation of constraint $\varnothing(q)$ with:

$$J_1 = \frac{\partial \varnothing(q)}{\partial q_1} = -l_1 \sin(q_1) - l_2 \sin(q_1 + q_2) - l_3 \sin(q_1 + q_2 + q_3),$$

$$J_2 = \frac{\partial \varnothing(q)}{\partial q_2} = -l_2 \sin(q_1 + q_2) - l_3 \sin(q_1 + q_2 + q_3) \quad (5)$$

$$J_3 = \frac{\partial \varnothing(q)}{\partial q_3} = -l_3 \sin(q_1 + q_2 + q_3)$$

The Lagrange multiplier λ stands for the forces of contact of the LLE with the ground. The contacting with ground case decrees the number of degrees of freedoms of LLE as the same number of constraints [26]. In this chapter, Eq. 3 has been used as constraint equation that describe the contacting of LLE with ground. Thus, joint coordinates of LLE has been decoupled into two category: the first class includes the independent joints coordinates $q_z = [q_1, q_2]^T, q_z \in R^2$. The second category includes dependent joint coordinates. $q_N \in R^1, q_N = q_3$. Consequently, a reduced order model has been obtained which depends on independent joints only (Hip and Knee joints). In fact, the motion of ankle joint will depend on the independents joint motion. A reduced order model can be obtained by differentiating the algebraic constraint in (3) with respect to time as:

$$\dot{\varnothing}(q) = \frac{d\varnothing}{dt} = \frac{\partial \varnothing(q)}{\partial q}\frac{dq}{dt} = J\dot{q} = [J_1\dot{q}_1 + J_2\dot{q}_2 + J_3\dot{q}_3] = 0 \quad (6)$$

The dependent velocity $\dot{q}_3$ can be expressed in terms of independent velocities $\dot{q}_1, \dot{q}_2$ as:

$$\dot{q}_3 = -J_3(q)^{-1}J_1(q) - J_3(q)^{-1}J_2(q) \quad (7)$$

Hence, velocity coordinates $(\dot{q}_1, \dot{q}_2, \dot{q}_3)$ can be written as:

$$\begin{bmatrix} \dot{q}_1 \\ \dot{q}_2 \\ \dot{q}_3 \end{bmatrix} = \begin{bmatrix} 1 & 0 \\ 0 & 1 \\ -J_3(q)^{-1}J_1(q) & -J_3(q)^{-1}J_2(q) \end{bmatrix} \begin{bmatrix} \dot{q}_1 \\ \dot{q}_2 \end{bmatrix} \quad (8)$$

$$\text{Thus, we can obtain}: \quad \dot{q} = H(q)\,q_z \quad (9)$$

where: $\dot{q} = \left[\dot{q}_1\ \dot{q}_2\ \dot{q}_3\right]^T, \dot{q}_z = \left[\dot{q}_1\ \dot{q}_2\right]^T, H = \begin{bmatrix} 1 & 0 \\ 0 & 1 \\ J_{z1} & J_{z2} \end{bmatrix}, J_{z1} = \frac{J_{1(q)}}{J_{3(q)}}, J_{z2} = \frac{J_{2(q)}}{J_{3(q)}}$.

The generalized acceleration can be expressed as:

$$\ddot{q} = H(q)\ddot{q}_z + \dot{H}(q)\dot{q}_z \quad (10)$$

If we sub (10) in (4), we get;

$$M(q)H(q)\ddot{q}_z + C_1(q,\dot{q})\dot{q}_z + G(q) + \tau_d = \tau + J(q)^T\lambda \tag{11}$$

$$\text{where}: C_1(q,\dot{q}) = C(q,\dot{q})H(q) + M(q)\dot{H}(q) \tag{12}$$

And:

$$\dot{H} = \begin{bmatrix} 1 & 0 \\ 0 & 1 \\ \dot{j}_{z1} & \dot{j}_{z2} \end{bmatrix}, \dot{j}_{z1} = -\frac{J_{3(q)}\dot{J}_{1(q)} - J_{1(q)}\dot{J}_{3(q)}}{(J_{3(q)})2} \dot{j}_{z2} = -\frac{J_{3(q)}\dot{J}_{2(q)} - J_{2(q)}\dot{J}_{3(q)}}{(J_{3(q)})2}$$

$$\dot{J}_{1(q)} = \psi_1\dot{q}_1 + \psi_2\dot{q}_2 + \psi_3\dot{q}_3, \dot{J}_{2(q)} = \psi_2\dot{q}_1 + \psi_2\dot{q}_2 + \psi_3\dot{q}_3, \dot{J}_{3(q)} = \psi_3\dot{q}_1 + \psi_3\dot{q}_2 + \psi_3\dot{q}_3$$

$$\psi_1 = -[l_1\cos(q_1) + l_2cos(q_1+q_2) + l_3\cos(q_1+q_2+q_3)] \tag{13}$$

$$\psi_2 = -[l_2cos(q_1+q_2) + l_3\cos(q_1+q_2+q_3)]$$

$$\psi_3 = -[l_3\cos(q_1+q_2+q_3)]$$

Equation (11) is the reduced order dynamic model for LLE when it contacting with ground. This equation includes $J(q)^T\lambda$ term in its structure. The deleting of this term is very essential for developing an efficient controller. The $J(q)^T\lambda$ term from (11) can be eliminated, by using the expression $J(q)^T H^T(q) = 0$ which is popular property in literature which is used for different fields [7]. Borrowing this property and pre multiply Eq. (10) by H^T, we can get [20]:

$$\overline{M}(q)\ddot{q}_z + \overline{C}(q,\dot{q})\dot{q}_z + \overline{G}(q) + \overline{\tau}_d = \overline{\tau} \tag{14}$$

With: $\overline{M}(q) = H^T M(q)H$, $\overline{C}(q,\dot{q}) = H^T C_2(q,\dot{q})$, $\overline{G}(q) = H^T G(q)$,$\overline{\tau} = H^T\tau$, $\overline{\tau}_d = H^T\tau_d$.

If we undertake the parameter uncertainties and external disturbances of the system, the model of the LLE system in (14) can be modified as

$$\Delta\overline{M}(q) + \overline{M}(q)\ddot{q}_z + \Delta\overline{C} + \overline{C}(q,\dot{q})\dot{q}_z + \Delta\overline{G} + \overline{G}(q) + \overline{\tau}_d = \overline{\tau} \tag{15}$$

where: $\Delta\overline{M}$, $\Delta\overline{C}$ and $\Delta\overline{G}$ denote the uncertain parts of $\overline{M}$, $\overline{C}$ and $\overline{G}$.

Hence, Eq. (15) can be expressed as:

$$\overline{M}(q)\ddot{q}_z + \overline{C}(q,\dot{q})\dot{q}_z + \overline{G}(q) + \Psi(q,\dot{q},\dot{q_z},\ddot{q}_z) = \overline{\tau} \tag{16}$$

$$\Psi(q,\dot{q},\dot{q_z},\ddot{q}_z) = \Delta\overline{M}(q)\ddot{q}_z + \Delta\overline{C}(q,\dot{q})\dot{q}_z + \Delta\overline{G}(q) + \overline{\tau}_d \tag{17}$$

$\Psi(q, \dot{q}, \dot{q_z}, \ddot{q}_z)$ is the uncertainty of the lumped parameter and disturbance of the system,the derivative of uncertainty and external disturbance satisfy $|\dot{\Psi}| \leq \rho$, where ρ is the upper bound of the derivative of the uncertainty and disturbance, which is a positive constant [3]

3 Development of NTSTC Controller

To ensure a fast finite-time convergence of the error and its derivative to zero. A Super-twisting control algorithm based the non-singular terminal sliding surface for uncertain LLE systems is designed in this Section. The first step in is the selection of the new non-singular terminal sliding surface. Firstly, we present the following sliding manifold [3] :

$$S_z = \dot{e_z} + \Lambda|e_z|^{\beta}\mathrm{sgn}(e_z) \tag{18}$$

where: $e_z = q_z - q_z^d$ represents the error between position of independent joints q_z and the desired position q_z^d. $\dot{e_z} = \dot{q}_z - \dot{q}_z^d$ stands for the error between velocity of independent joints $\dot{q}_z$ and the desired velocity of $\dot{q}_z^d$.$S_z = [S_{z1}, S_{z2}]^T$isthevectorofslidingsurfaceofindependentjoints and$\Lambda \in R^{2*2}$,are positive diagonal matrices. The subscript z stands for independent joint coordinates $\mathrm{sgn}(e_z)$ is the signum function which is defined by:

$$\mathrm{sgn}(e_z) = \begin{cases} 1 \ if e_z > 0 \\ 0 \ if e_z = 0 \\ -1 \ if e_z < 0 \end{cases} \tag{19}$$

And β is defined by [3]:

$$\begin{cases} 1 \ if \ |e_z| < \epsilon \\ \beta_0 \ if \ |e_z| \geq \epsilon \end{cases} \tag{20}$$

With $0 < \beta_0 < 1$ and ϵ is a small positive constant which is used to provide limitation when $|e_z|$ is very small. By taking the derivative of (18), one obtains:

$$\dot{S_z} = \ddot{e_z} + \Lambda\beta|e_z|^{\beta-1}\dot{e_z} \tag{21}$$

Hence, we obtain:

$$\dot{S_z} = \ddot{q_z} - \ddot{q}_z^d + \Lambda\beta|e_z|^{\beta-1}\dot{e_z} \tag{22}$$

By re-writing (14) and dropping dependency, we can get:

$$\ddot{q}_z = \overline{M}^{-1}[\overline{\tau} - \overline{C}\dot{q}_z - \overline{G} \tag{23}$$

Sub (23) in (22), we can obtain:

$$\dot{S}_z = \overline{M}^{-1}\left[\overline{\tau} - \overline{C}\dot{q}_z - \overline{G}\right] - \ddot{q}_z^d + \Lambda\beta|e_z|^{\beta-1}\dot{e}_z \tag{24}$$

For super twisting controller algorithm, the control signal consists of equivalent and switching control laws as follows:

$$\overline{\tau}_{\mathbf{NTSTC}} = \overline{\tau}_e + \overline{\tau}_{stw} \tag{25}$$

The equivalent part $\overline{\tau}_e$ can be obtained by setting $\dot{S}_z = 0$. By using the nominal model of LLE and without taking in to consideration the uncertainties and external disturbance and multiplying Eq. (24) by $\overline{M}$, we get:

$$\left[\overline{\tau} - \overline{C}\dot{q}_z - \overline{G}\right] - \overline{M}\ddot{q}_z^d + \overline{M}\Lambda\beta|e_z|^{\beta-1}\left(\dot{q}_z - \dot{q}_z^d\right) = 0 \tag{26}$$

Therefore, the equivalent component can be determined;

$$\overline{\tau}_e = \overline{C}\dot{q}_z + \overline{G} + \overline{M}\ddot{q}_z^d - \overline{M}\Lambda\beta|e_z|^{\beta-1}\left(\dot{q}_z - \dot{q}_z^d\right) \tag{27}$$

In addition, to drive the trajectories to the selected nonsingular terminal sliding surface (18) in finite-time and to ensure robustness against uncertainties and outside disturbances, the Super-Twisting control term $\overline{\tau}_{stw}$ is added which is given as:

$$\overline{\tau}_{stw} = -\overline{M}k_1|S_z|^{0.5}sgn(S_z) - \overline{M}k_2\int sgn(S_z)dt \tag{28}$$

where: $k_1, k_2 \in R^{2*2}$ are diagonal positive matrices. Consequently, we get an NTSTC controller which can be expressed as:

$$\overline{\tau}_{NTSTC} = \overline{C}\dot{q}_z + \overline{G} + \overline{M}\ddot{q}_z^d - \overline{M}\Lambda\beta|e_z|^{\beta-1}\left(\dot{q}_z - \dot{q}_z^d\right) - \overline{M}k_1|S_z|^{0.5}sgn(S_z) - \overline{M}k_2\int sgn(S_z)dt \tag{29}$$

Figure 3 depict the block diagram of NTSTC.

Theorem 1.1 *Consider the nonlinear LLE system in Eq. (14), if the super-twisting gains k_1 and k_2 are chosen as:*

$$k_1 > 2\xi, \xi > 0, k_2 = \delta k_1, \delta > \frac{\left(\xi^2 - \xi k_1\right)^2 + 4\xi^4 + \rho^2 - 2\xi^3 k_1}{2\xi k_1{}^2 - 4\xi^2 k_1} \tag{30}$$

Then, the proposed controller in Eq. (29) ensures the convergence of the sliding surface to zero in a maximal finite-time

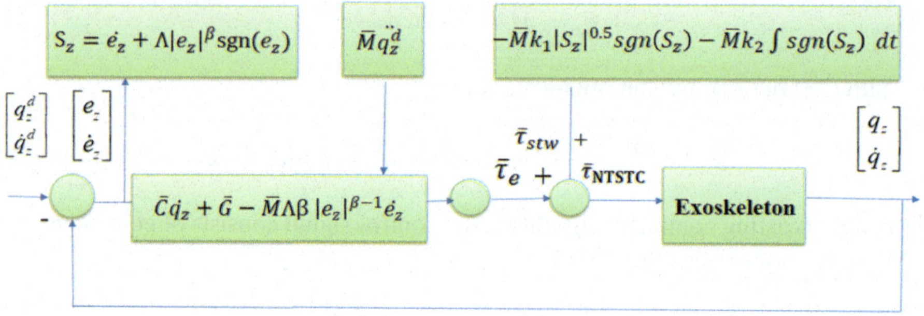

Fig. 3 Block diagram of Nonsingular Terminal Super twisting sliding mode controller

$$T_r \leq 2\frac{\sqrt{\lambda_{max}\{Q\}}}{\lambda_{min}\{W\}}\sqrt{V(0)} \tag{31}$$

Proof If we substitute Eq. (29) in to Eq. (21), the following Closed-loop error dynamics is obtained:

$$\dot{S}_z = k_1|S_z|^{0.5}sgn(S_z) + \eta$$

$$\dot{\eta} = -k_2 sgn(S_z) + \dot{\Psi}(q, \dot{q}, \dot{q_z}, \ddot{q}_z) \tag{32}$$

The stability of the closed-loop system can be proved by selection of the following positive definite, Lyapunov function [4]:

$$V = \sigma^T Q\sigma \tag{33}$$

where $\sigma = \left[|S_z|^{0.5}sgn(S_z), \eta\right]^T$ and Q is a symmetric positive definite matrix than can selected as:

$$Q = \begin{bmatrix} 2\xi^3 + 2\delta\xi k_1 & -2\xi^2 \\ -2\xi^2 & \xi \end{bmatrix} \tag{34}$$

The above matrix is symmetric positive definite if $\delta > 0$ and$\xi > 0$. If we take the first derivative of σ, we get:

$$\dot{\sigma} = \frac{1}{|\sigma_1|}A\sigma + \frac{1}{|\sigma_1|}B\dot{\Psi}|\sigma_1| \tag{35}$$

where $|\sigma_1| = |S_z|^{0.5}sgn(S_z)$ and the matraix A and B are defined as:

$$A = \begin{bmatrix} -0.5k_1 & 0.5 \\ -\delta k_1 & 0 \end{bmatrix}, B = \begin{bmatrix} 0 \\ 1 \end{bmatrix} \tag{36}$$

$$\text{Hence, } \dot{V} \text{canbeobtainedas} : \dot{V} = \dot{\sigma}^T Q\sigma + \sigma^T Q\dot{\sigma} \tag{37}$$

$$= \frac{1}{|\sigma_1|}\sigma^T(QA + A^T Q)\sigma + 2\dot{\Psi}B^T Q\sigma$$

$$\leq \frac{1}{|\sigma_1|}\sigma^T(QA + A^T Q + QBB^T Q + R\rho^2)\sigma$$

$$\leq -\frac{1}{|\sigma_1|}\sigma^T W\sigma$$

The matrices R and W can be defined as:

$$R = \begin{bmatrix} 1\,0 \\ 0\,0 \end{bmatrix}, W = \xi\begin{bmatrix} W_{11}\ W_{12} \\ W_{21}\ W_{22} \end{bmatrix} \tag{38}$$

With: $W_{11} = k_1(2\delta k_1 - 4\xi\delta + 2\xi^2) - 4\xi^3 - \rho^2, W_{12} = W_{21} = \xi(\xi - k_1), W_{22} = \xi$. We can get W as symmetric positive definite matrix if the diagonal elements W_{11} and W_{22} are positive definite and if the matrix determinant is definite positive. The verifying of these conditions can be obtained if the gains k_1 and k_2 are chosen as:$k_1 > 2\xi, \xi > 0, k_2 = \delta k_1$. Hence, if the conditions in Eq. (30) are verified, then $\dot{V}$ is negative definite, then the closed-loop stability of the system is proved. In order to prove the finite-time convergence, at first we recall that the Lyapunov function V is bounded

$$\lambda_{min}\{Q\}\|\sigma\|^2 \leq V \leq \lambda_{max}\{Q\}\|\sigma\|^2 \tag{39}$$

With $\lambda_{min}\{Q\}$, $\lambda_{max}\{Q\}$ represent the minimum and maximum eigenvalues of Q then, Eq. (39) can be rewritten as:

$$\frac{\sqrt{V}}{\sqrt{\lambda_{max}\{Q\}}} \leq \|\sigma\| \leq \frac{\sqrt{V}}{\sqrt{\lambda_{min}\{Q\}}} \tag{40}$$

In addition, Eq. (37) gives

$$\dot{V} \leq \frac{-1}{|\sigma_1|}\sigma^T W\sigma \leq \frac{-1}{|\sigma_1|}\lambda_{min}\{W\}\|\sigma\|^2 \tag{41}$$

$\lambda_{min}\{W\}$ is the minimum eigenvalue of W. It is clear that$|\sigma_1| \leq \|\sigma\|$, then:

$$\dot{V} \leq -\frac{\lambda_{min}\{W\}}{\sqrt{\lambda_{max}\{Q\}}}\sqrt{V} \tag{42}$$

Hence, according to Eq. (42), the system trajectories will converge to the sliding surface in finite-time and this completes the proof.

4 Results and Discussion

The performance analysis of developed nonsingular terminal super twisting sliding mode controller NTSTC in Eq. (29) has been investigated in this section. For comparison purpose, a traditional super twisting algorithm (STA) has been chosen which has the following formula [27]:

$$\overline{\tau}_{STSM} = \overline{C}\dot{q}_z + \overline{G} + \overline{M}\ddot{q}_z^d - \overline{M}\Lambda\left(\dot{q}_z - \dot{q}_z^d\right) - \overline{M}k_1|S_z|^{0.5}sgn(S_z) - \overline{M}k_2\int sgn(S_z)dt \quad (43)$$

The desired trajectories have been adopted from [23]. Whereas, the values of human and exoskeleton parameters have been undertaken adopted from [30, 32]. In addition, the integral absolute value error (IAE) has been used for measuring the tracking performances of NTSTC and STA controllers:

$$\text{IAE} = \int |e|dt \quad (44)$$

In order to obtain fair and accurate comparison, the simulations have been performed with the same initial conditions for both controllers. The values of parameters of STA controller are selected as $\Lambda=$ diag(5, 5), k_1 = diag(10, 10), k_2 = diag(10, 10). whereas the parameters values for NTSTC have been assigned as:$\Lambda=$ diag(5, 5), k_1 = diag(10, 10), k_2 = diag(10, 10), β_0 = diag(0.7, 0.7),ϵ = 0.01. In fact, the parameters of each controller are selected according to the conditions of stability. Two tests have been performed in this chapter. In the first test, the nominal values only have been considered. The uncertainty in masses and lengths of person wearing LLE have been undertaken in the second test.

Case 1: Nominal Test

The trajectory tracking performance, speed, tracking error for two controllers in case of nominal values are depicted in Figs. 4, 5, and 6. It can be seen that both strategies offer good performance in position trajectory tracking and speed. However, based on the error signals for in Fig. 7, it is clear that the smallest values of errors are obtained using NTSTC. This is a good indicator of the effectivity of nonsingular terminal sliding surface adopted in this chapter. Moreover, it is observed an excellent produced torque signals have been obtained for two controllers as shown in Fig. 7 due to the presence the switching term for NTSTC and STA controllers which reflects the important role of switching term to eliminate the chattering effect adequately. The results of IAE index for nominal values for NTSTC and STA controllers are shown in Fig. 8 and Table 1 It is evident from results that NTSTC controller has the lowest values of IAE index as compared STA controller.

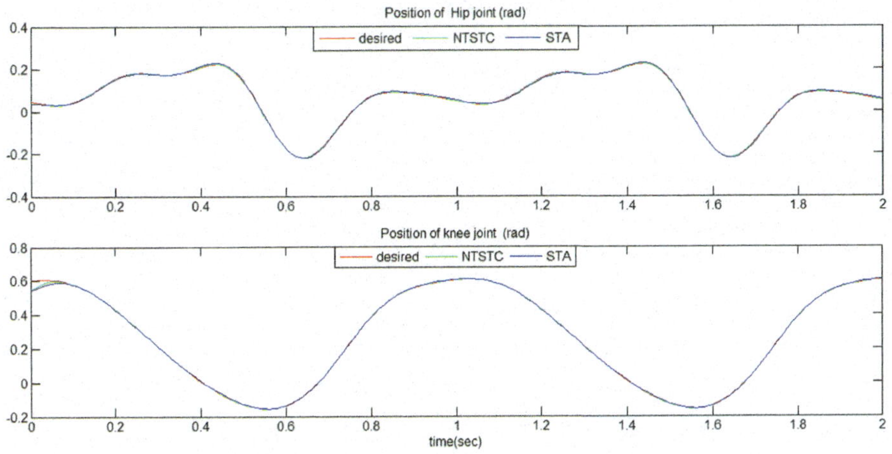

Fig. 4 Position tracking of both joints for two controllers with nominal values

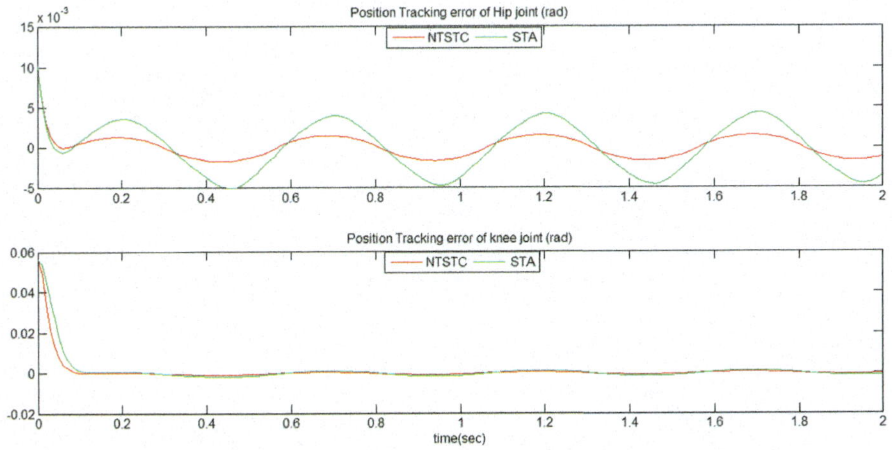

Fig. 5 Tracking error of both joints for two controllers with nominal values

Case 2: Uncertainty Test

Owing to changing the morphologies of patients under training are usually varied from person to person, the original values of masses and lengths of thigh, shank, foot of the human and exoskeleton system will be changed. Hence, it's important to investigate the performance of the NTSTC and STA controllers in case of parameter variations. It is assumed that the masses and lengths of the patients which wear the exoskeleton have been modified by +20% compared to their nominal quantities. The simulation results of this case have been depicted in Figs. 9, 10, 11, and 12. It's observed from these figures that the trajectory tracking performance, trajectory tracking errors, speeds and the input control signals of all controllers have been

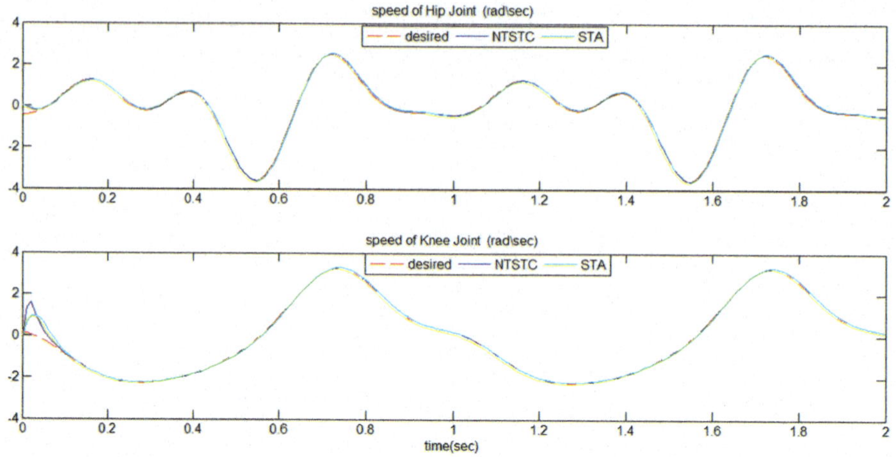

Fig. 6 Speed tracking of both joints for two controllers with nominal values

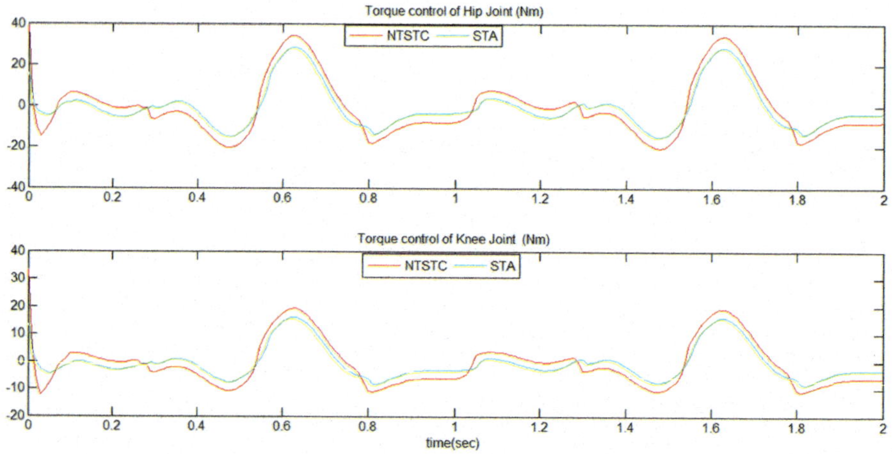

Fig. 7 Produced torques of both joints for two controllers with nominal values

remained excellent and robust against system uncertainty in spite of the uncertainties in parameters. The superiority of NTSTC controller over STA in handling the parametric uncertainties in terms of tracking performance index IAE has been indicated in Fig. 13 and Table 2. It is observed from this figure and table that the STA controller has been witnessed remarkable changes in IAE index values as compared with NTSTC controller.

In general speaking, we can observe from examining the results of nominal and uncertain tests that the performance NTSTC controller has the better results in tracking the trajectory of position and speeds, minimization the tracking error, robustness towards the uncertainties over STA controller.

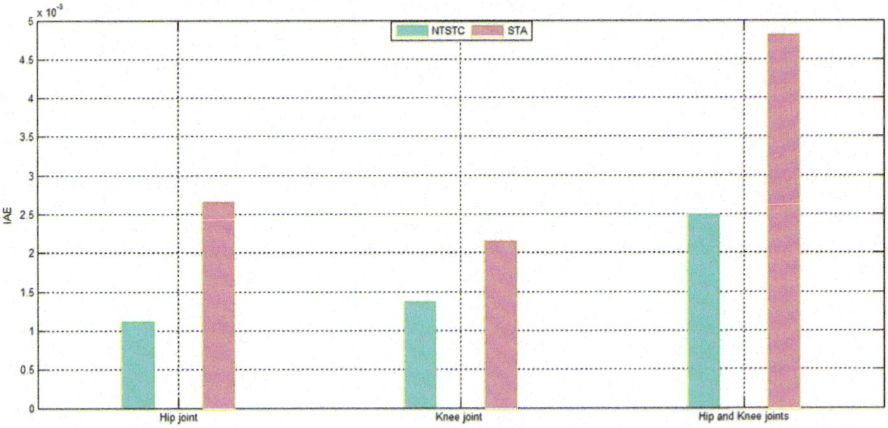

Fig. 8 Tracking performance (IAE) Index for two controllers with nominal values

Table 1 IAE performances index of NTSTC, STA controllers under nominal test

Joint	NTSTC	STA
Hip joint	0.0011	0.0027
Knee joint	0.0014	0.0022
Both joints	0.0025	0.0048

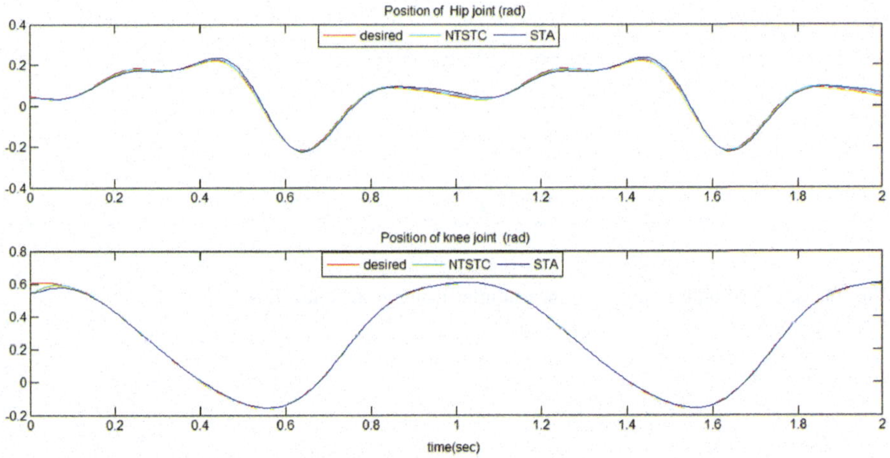

Fig. 9 Position tracking of both joints with uncertainties using both controllers

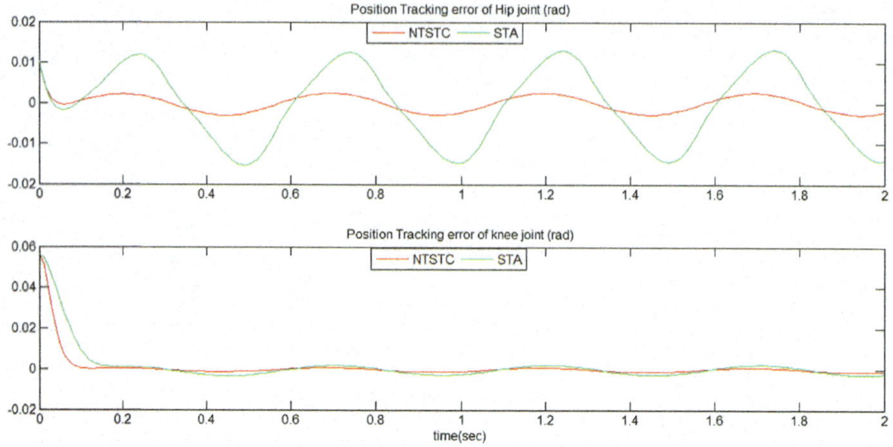

Fig. 10 Tracking error of both joints with uncertainties using both controllers

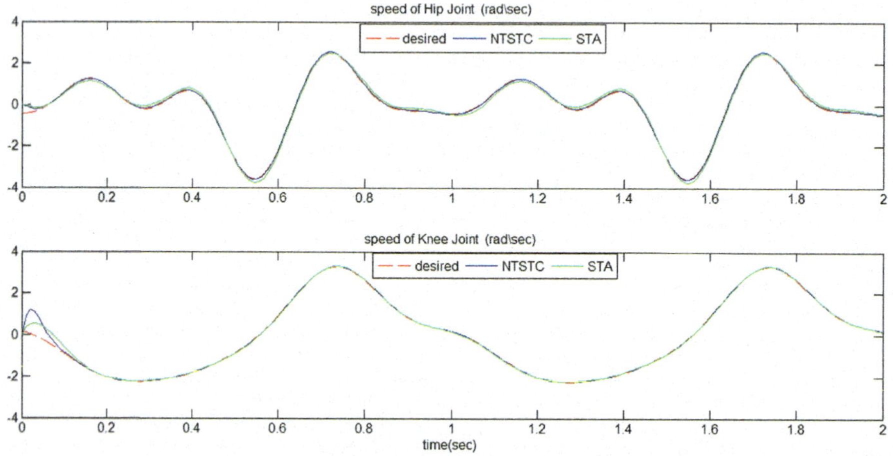

Fig. 11 Speed of both joints with uncertainties using both controllers

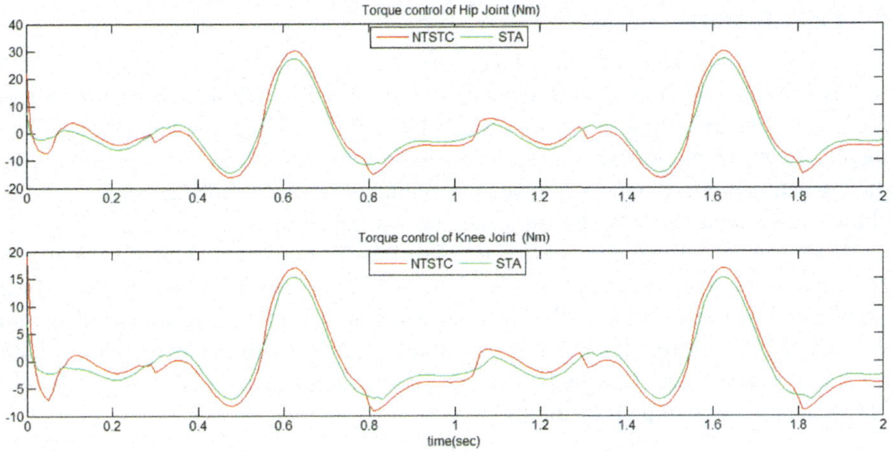

Fig. 12 Produced torques for both joints with uncertainties using both controllers

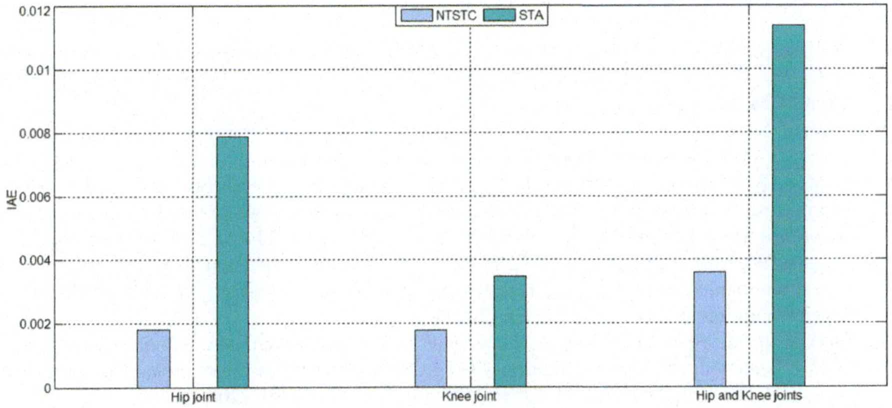

Fig. 13 Tracking performance (IAE) Index for two controllers with uncertainties

Table 2 IAE performances index of NTSTC, STA controllers under uncertainty test

Joint	NTSTC	STA
Hip joint	0.0018	0.0079
Knee joint	0.0018	0.0035
Both joints	0.0036	0.0114

5 Conclusion

In this work, a mathematical model and effective control laws have been formulated for lower limb rehabilitation exoskeleton. The augmented Lagrange equations which include Lagrange multiplier and Jacobian vectors for lower limb exoskeleton in case of contacting with ground have been formulated. A second-order super-twisting sliding mode controller based on a nonsingular terminal sliding surface has been developed and implemented on lower limb rehabilitation exoskeleton. Then, the chosen control algorithm has been derived and a new stability conditions have been established to reduce the chattering phenomenon. Simulation results explain that the nonsingular terminal super twisting sliding mode controller gives an excellent tracking performance, fast response and small tracking errors as compared with classical super twisting algorithm STA.

References

1. Ahmed, S., Wang, H., Tian, Y.: Model-free control using time delay estimation and fractional-order non-singular fast terminal sliding mode for uncertain lower-limb exoskeleton. J. Vib. Control **24**(22), 5273–5290 (2018)
2. Ahmed, S., Wang, H., Tian, Y.: Robust adaptive fractional-order terminal sliding mode control for lower-limb exoskeleton. Asian J. Control **21**(1), 473–482 (2019)
3. Alqaisi, W., Kali, Y., Ghommam, J., Saad, M., Nerguizian, V.: Position and attitude tracking of uncertain quadrotor unmanned aerialvehicles based on non-singular terminal super-twisting algorithm. Proc. Inst. Mech. Eng. Part I: J. Syst. Control Eng. **234**(3), 396–408 (2020)
4. Alqaisi, W., Kali, Y., El-Bayeh, C.Z.: Modified fast terminal super-twisting control for uncertain robot manipulators. In: 2021 18th International Multi-conference on Systems, Signals & Devices (SSD), pp. 1142–1147. IEEE (2021)
5. Bernhardt, M., Frey, M., Colombo, G., Riener, R.: Hybrid force position control yields cooperative behaviour of the rehabilitation robot LOKOMAT. In: 9th International Conference on Rehabilitation Robotics, 2005. ICORR 2005, pp. 536–539. IEEE (2005)
6. Boukattaya, M., Mezghani, N., Damak, T.: Adaptive non-singular fast terminal sliding-mode control for the tracking problem of uncertain dynamical systems. ISA Trans. **77**, 1–19 (2018)
7. Burghardt, A., Skwarek, W.: Modeling the dynamics of two cooperating robots. Appl. Sci. **11**(13), 6019 (2021)
8. Chemori, A., Loría, A.: Control of a planar under actuated biped on a complete walking cycle. IEEE Trans. Autom. Control **49**(5), 838–843 (2004)
9. Chen, B., Zhong, C.-H., Zhao, X., Ma, H., Qin, L., Liao, W.-H.: Reference joint trajectories generation of CUHK-EXO exoskeleton for system balance in walking assistance. IEEE Access **7**, 33809–33821 (2019)
10. Cheng, N.B., Guan, L.W., Wang, L.P., Han, J.: Chattering reduction of sliding mode control by adopting nonlinear saturation function. In: Advanced Materials Research, vol. 143, pp. 53–61. Trans Tech Publications (2011)
11. Ezhilarasi, D., Nair, A.S.: Modelling and evaluation of adaptive super twisting sliding mode control in lower extremity exoskeleton. Int. J. Precis. Eng. Manuf.-Green Technol. **8**(3), 901–915 (2021)
12. Faraj, M.A., Maalej, B., Derbel, N., Deriche, M.: Modelling and robust computed torque control for lower limb exoskeleton contacting with ground. In: 2022 19th International Multi-Conference on Systems, Signals& Devices (SSD), pp. 571–579. IEEE (2022)

13. Faraj, M.A., Maalej, B., Derbel, N.: Optimal sliding mode controller for lower limb rehabilitation exoskeleton in constrained environments. Indonesian J. Electr. Eng. Comput. Sci. **30**(3), 1458–1469 (2023)
14. Haghighi, D.A., Mobayen, S.: Design of an adaptive super-twisting decoupled terminal sliding mode control scheme for a class of fourth-order systems. ISA Trans. **75**, 216–225 (2018)
15. Han, S., Wang, H., Tian, Y.: Model-free based adaptive non-singular fast terminal sliding mode control with time-delay estimation for a 12 dof multi-functional lower limb exoskeleton. Adv. Eng. Softw. **119**, 38–47 (2018)
16. Humaidi, A.J., Hasan, A.F.: Particle swarm optimization–based adaptive super-twisting sliding mode control design for 2-degree-of-freedom helicopter. Meas. Control **52**(9–10), 1403–1419 (2019)
17. Hyon, S.-H., Morimoto, J., Matsubara, T., Noda, T., Kawato, M.: XOR: Hybrid drive exoskeleton robot that can balance. In: 2011 IEEE/RSJ International Conference on Intelligent Robots and Systems, pp. 3975–3981. IEEE (2011)
18. Iandolo, R., Marini, F., Semprini, M., Laffranchi, M., Mugnosso, M., Cherif, A., De Michieli, L., Chiappalone, M., Zenzeri, J.: Perspectives and challenges in robotic neurorehabilitation. Appl. Sci. **9**(15), 3183 (2019)
19. Kali, Y., Saad, M., Benjelloun, K., Khairallah, C.: Super-twisting algorithm with time delay estimation for uncertain robot manipulators. Nonlinear Dyn. **93**(2), 557–569 (2018)
20. Kumar, N., Rani, M.: Neural network-based hybrid force/position control of constrained reconfigurable manipulators. Neurocomputing **420**, 1–14 (2021)
21. Lee, H., Utkin, V.I.: Chattering suppression methods in sliding mode control systems. Annu. Rev. Control. **31**(2), 179–188 (2007)
22. Maalej, B., Khlif, R.J., Mhiri, C., Elleuch, M.H., Derbel, N.: Adaptive fractional control optimized by genetic algorithms with application to poly articulated robotic systems. Math. Probl. Eng. (2021)
23. Mefoued, S.: Commande robuste référencée intention d'une orthèse active pour l'assistance fonctionnelle aux mouvements du genou. Ph.D. thesis, Paris Est (2012)
24. Meng, W., Liu, Q., Zhou, Z., Ai, Q., Sheng, B., Xie, S.S.: Recent development of mechanisms and control strategies for robot-assisted lower limb rehabilitation. Mechatronics **31**, 132–145 (2015)
25. Mokhtari, M., Taghizadeh, M., Mazare, M.: Impedance control based on optimal adaptive high order super twisting sliding mode for a 7-DOF lower limb exoskeleton. Meccanica **56**(3), 535–548 (2021)
26. Murray, R.M., Li, Z., Sastry, S.S.: A Mathematical Introduction to Robotic Manipulation. CRC Press (2017)
27. Nair, A.S., Ezhilarasi, D.: Performance analysis of super twisting sliding mode controller by adams–matlab co-simulation in lower extremity exoskeleton. Int. J. Precis. Eng. Manuf.-Green Technol. **7**(3), 743–754 (2020)
28. Rivera, J., Garcia, L., Mora, C., Raygoza, J.J., Ortega, S.: Super twisting sliding mode in motion control systems. Sliding Mode Control **1**, 237–254 (2011)
29. Sankai, Y.: HAL: Hybrid assistive limb based on cybermobs. In: Robotics Research: The 13th International Symposium ISRR, pp. 25–34. Springer (2011)
30. Soleimani Amiri, M., Ramli, R., Ibrahim, M.F., Abd Wahab, D., Aliman, N.: Adaptive particle swarm optimization of PID gain tuning for lower-limb human exoskeleton in virtual environment. Mathematics **8**(11), 2040 (2020)
31. Spong, M.W., Hutchinson, S., Vidyasagar, M.: Robot Modeling and Control. Wiley (2020)
32. Tözeren, A.: Human Body Dynamics: Classical Mechanics and Human Movement. Springer Science & Business Media (1999)
33. Utkin, V., Guldner, J., Shi, J.: Sliding Mode Control in Electromechanical Systems. CRC Press (2017)
34. Veneman, J.F., Kruidhof, R., Hekman, E.E., Ekkelenkamp, R., Van Asseldonk, E.H., Van Der Kooij, H.: Design and evaluation of the LOPES exoskeleton robot for interactive gait rehabilitation. IEEE Trans. Neural Syst. Rehabil. Eng. **15**(3), 379–386 (2007)

35. Wang, Y., Chen, J., Yan, F., Zhu, K., Chen, B.: Adaptive super-twisting fractional-order non-singular terminal sliding mode control of cable-driven manipulators. ISA Trans. **86**, 163–180 (2019)
36. Zanotto, D., Stegall, P., Agrawal, S.K.: Alex III: a novel robotic platform with 12 DOFs for human gait training. In: 2013 IEEE International Conference on Robotics and Automation, pp. 3914–3919. IEEE (2013)
37. Zhang, J.-F., Dong, Y.-M., Yang, C.-J., Geng, Y., Chen, Y., Yang, Y.: 5-link model based gait trajectory adaption control strategies of the gait rehabilitation exoskeleton for post-stroke patients. Mechatronics **20**(3), 368–376 (2010)
38. Zhang, X., Yue, Z., Wang, J.: Robotics in lower-limb rehabilitation after stroke. Behav. Neurol. (2017)
39. Zhang, Y., Cao, G., Li, W., Chen, J., Li, L., Diao, D.: A self-adaptive-coefficient-double-power sliding mode control method for lower limb rehabilitation exoskeleton robot. Appl. Sci. **11**(21), 1032 (2021)

Generalized Predictive Control Design of Benchmark Distillation Columns: A Case Study for Multi-input Multi-output System

Anuj Abraham and Pranjal Vyas

Abstract This article describes the study of Multi-input Multi-output (MIMO) distillation column processes applied to several benchmark problems (2-by-2) namely, Wood and Berry (WB), Vinante and Luyben (VL), Wardle and Wood (WW), and Ogunnaike and Ray (OR) using Model Predictive Control (MPC) strategy. The main objective of the chapter deals with the chain of actions involving teaching and practicing MPC controller design to researchers. The control of the top and bottom compositions in the benchmark distillation column using reflux and steam flow rates has shown to be a particularly difficult problem, due to the significant time delays, non-minimum phases, and nonlinear interactions within the process. In this study, the predictive control algorithm uses the Controlled Auto-Regressive Integrated Moving Average (CARIMA) model to address this problem. The performance of the predictive control scheme gives better setpoint tracking when compared with the conventional controllers available in state-of-the-art, and the simulation result shows the effectiveness of the predictive algorithm to the chosen benchmarks.

1 Introduction

Noticeably, multivariable and multiloop Multi-input Multi-output (MIMO) processes are normally encountered in process control applications like refineries, concrete mixing industries, fertilizers industries, and wastewater treatment industries. Multivariable processes have more than one input or output required for proper control [1]. These complex processes find exceptionally tight control for the best product quality. Moreover, distillation columns are widely used in chemical processes, such as crude oil refineries and hydrocarbon process industries [2]. The distillation column has three

A. Abraham (✉)
Technology Innovation Institute, 9639 Masdar City, Abu Dhabi, United Arab Emirates
e-mail: anuj1986aei@gmail.com

P. Vyas
Advanced Remanufacturing and Technology Centre (ARTC), A*STAR, Singapore 138632, Singapore

A. Ben Makhlouf et al. (eds.), *State Estimation and Stabilization of Nonlinear Systems*, Studies in Systems, Decision and Control 491,
https://doi.org/10.1007/978-3-031-37970-3_20

main features: material balance, product quality, and maintenance of constraints. The material balance control is important on any column and must be consistent with the adjacent process equipment. Whereas product quality control must function satisfactorily for possible variations in the feed flow rate, feed composition, steam supply pressure, supply temperature, ambient temperature, thermal conditions, etc. Maintenance of constraints is essential for the smooth operation of the column. The most common automated multivariable control technology is Model-Predictive Control (MPC), which uses dynamic process models to predict future process responses [3].

MPC is an advanced popular advanced strategy for optimal control of multivariable systems with decades of a successful application to process industries. While satisfying a set of constraints, the predictive control algorithms are suitable for non-minimum phase or unstable processes, systems with delays, with different numbers of controlled and manipulated variables even nonlinear processes [3]. This facilitates the prediction of control variables in future time horizons and the calculation of control actions to minimize a cost function for dynamic models.

The control strategies based on MPC are preferred over classical Proportional-Integral-Derivative (PID) controllers due to the advantage of optimal control actions while considering the constraints on manipulated variables and controlled outputs [4], inherent loop interactions, and decoupler designs, the future response of a process with a finite horizon at each sampling instants, other performance metrics, etc. Over the decades, a popular control design method mainly, Dynamic Matrix Control (DMC) was developed for the control of the distillation columns to improve disturbance rejection capabilities. But there is a trade-off in using the DMC approach, it is observed that the regulatory response or load-rejection capability becomes worse as product purity increases [5].

The Biggest Log Modulus Tuning (BLT) method first proposed by Luyben is an approach to decentralize or multi-loop PI/PID controllers, applicable for both stable and unstable processes. It is regarded as the benchmark method for decentralized control. In the BLT method, Ziegler–Nichols (ZN) tuning rule is modified with detuning factors, which determines the trade-off between the stability and performance of the system [4, 6].

Other control strategies are robust MPC, adaptive MPC, nonlinear MPC based on mixed-integer optimization, intelligent control strategies, etc. The popular loop-sharing methods are Relative Gain Array (RGA), Niederlinski Index (NI), Relative Interaction (RI), dynamic Relative Interaction (dRI), Generalized Relative Dynamic Gain (GRDG), Decomposed Relative Gain Array (DRGA), Effective Relative Gain Array (ERGA), IMC Interaction measure, Rijnsdrop's Interaction measure, Nyquist Array (NA), Generalized Interaction (GI), Gramian based Interaction measure, etc. One of the main requirements of the classical decentralized control system is that a closed loop system should remain stable as sub-system controllers are dealt-with and even when input saturates [7].

In this chapter, we will study the benchmark problems of distillation columns namely, Wood and Berry (WB), Vinante and Luyben (VL), Wardle and Wood (WW), and Ogunnaike and Ray (OR) to improve the quality and performance of their control systems in a better way with predictive algorithms. Here, the MPC optimization is formulated as a convex Quadratic Program (QP). This encompasses the use of an advanced control strategy with Generalized Predictive Control (GPC), where the future set-points or disturbances can be handled as well if available [8].

The paper is structured as the following: Sect. 2 introduces model predictive control. In Sect. 3, we formulate a generalized predictive control design for transfer function and state space models and in Sect. 4 provide case studies of 2-by-2 benchmark distillation columns. In Sect. 5, results and discussions were summarized for GPC as a class of MPC. Finally, the work is concluded in Sect. 6 by discussing some ongoing research aspects and extensions to wider applications.

2 Introduction to MPC

For any general predictive control design, the future control errors, and control action changes are penalized in the cost function. In our proposed formulation, the cost function can be seen as a function of future control actions utilizing the process model in a form of the predictor and anticipating deviations from the reference.

To start with, the finite horizon quadratic criterion is considered in this section with Single-input Single-output (SISO) system cost function formulation. Here, the task is to minimize a quadratic cost function of control error and the manipulated variable increments during the corresponding prediction horizons [9, 10] as described by,

$$\mathrm{J} = r\sum_{j=N_1}^{N_2}\left[\hat{\mathrm{y}}(k+j) - \mathrm{w}(k+j)\right]^2 + q\sum_{j=1}^{N_3}[\Delta \mathrm{u}(k+j-1)]^2 \tag{1}$$

where $\hat{\mathrm{y}}(k+j)$ is an optimum j-step ahead prediction of the system output, N_1 is the starting point, N_2 is the prediction horizon and $\mathrm{w}(k+j)$ is a future set-point or reference for the controlled variable. The parameters, r and q are the weighting coefficient for control error and control increments respectively.

The cost function consists of two parts, mainly costs due to control error during the control error horizon N_2, which is also called the optimization or prediction horizon [11], and another is costs to penalize the control signal increments during the manipulated variable horizon N_3, which is also called the control horizon. For simplicity in the following text, we use $N_1 = 1$, $N_2 = N_3 = N$.

The basic structure of MPC algorithm [1] is shown in Fig. 1. At each time instant k,

i. The future outputs on the prediction horizon N are predicted using the process model. Predicted outputs $\mathrm{y}(k+i)$ for $i = 1\ldots N$ depend on the past inputs and outputs and on the future control signals $\mathrm{u}(k+i)$ for $\mathrm{i} = 0\ldots N-1$, which are to be calculated.

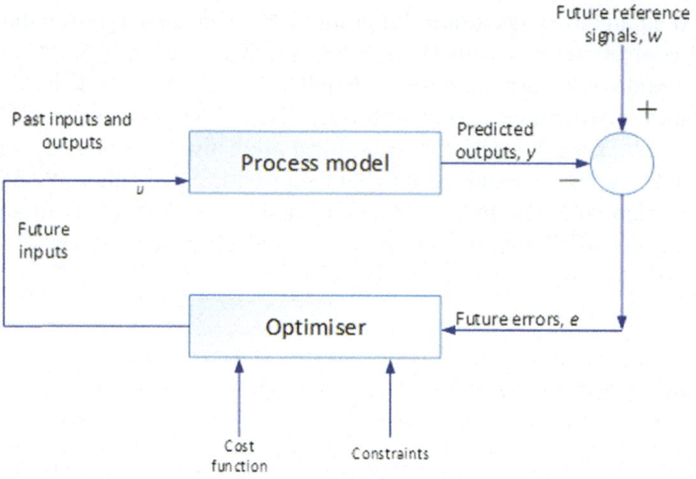

Fig. 1 Structure of MPC [1]

ii. The set of future control signals is calculated by criterion optimizing to keep the process as close as possible to the reference trajectory. If the model is linear and there are no constraints an explicit solution can be obtained, otherwise an iterative optimization method has to be used.
iii. The control signal u(k) is given to the process while the next calculated control signals are rejected. At the next sampling instant y($k + 1$) is known and step 1 is repeated with this new value using the receding horizon concept.

In MPC, the receding strategy concept means that at every sample time instant, only first the control action from the optimal vector is used and the horizon is shifted towards the future and the procedure is repeated for updated system state [12, 13]. The various predictive algorithms differ amongst themselves in the model used to represent the process, the cost function to be minimized and optimization method [14].

3 Generalized Predictive Control (GPC)

The most common transfer function model with MPC is the CARIMA model. When a plant operates around a particular working point and after linearization, it can be described by a CARIMA model. The chosen model plays importance when the uncertainty is included in a way that is a good representation of slow varying disturbances under non-zero steady-state conditions [15].

This model is appropriate for applications in which disturbances are non-stationary and the resulting controller has an integrating behaviour [16]. The reason a CARIMA model is used is that it incorporates a disturbance estimate and gives unbiased prediction in steady-state, irrespective of some parameter uncertainty i.e., it determines feasible combinations of all manipulated variables such that all control variables steady-state targets are at or within their operating limits [17].

The control objective is to compute a future control sequence that the future plant output is driven as like the reference trajectory. This is accomplished by minimizing the quadratic cost function defined over the prediction horizon [1, 18]. GPC with transfer function and state space model is formulated below along with their cost function and constraint representation for a SISO system.

3.1 GPC with Transfer Function Model

Following derivation seems to be easiest for the researchers understanding. We are considering process and disturbance model as mentioned below,

$$\mathrm{A}(z^{-1})\,\mathrm{y}(k) = \mathrm{B}(z^{-1})\,\mathrm{u}(k-1) + \frac{\mathrm{C}(z^{-1})}{\Delta}\mathrm{e}(k)$$

where, $\mathrm{e}(k) = \mathrm{y}(k) - \hat{\mathrm{y}}(k)$ is prediction error, and A (z^{-1}), B (z^{-1}) and C (z^{-1}) are polynomials defined as,

$$\begin{aligned}
\mathrm{A}\left(z^{-1}\right) &= 1 + a_1 z^{-1} + a_2 z^{-2} + \cdots + a_n z^{-n} \\
\mathrm{B}\left(z^{-1}\right) &= b_1 + b_2 z^{-1} + b_3 z^{-2} + \cdots + b_n z^{-n+1} \\
\mathrm{C}\left(z^{-1}\right) &= 1 + c_1 z^{-1} + c_2 z^{-2} + \cdots + c_{n_c} z^{-n_c}
\end{aligned}$$

The operator Δ is defined as $\Delta = 1 - z^{-1}$. The variables y(k), u(k), e(k) are output, input, and noise. The noise is supposed to be white noise random signal with zero mean.

Polynomials A (z^{-1}) and B (z^{-1}) are denominator and numerator of transfer function respectively. Polynomial C (z^{-1}) is disturbance model polynomial or can be seen as filtering polynomial. Rearranging the terms from (4),

$$\begin{aligned}
\Delta \mathrm{A}(z^{-1})\mathrm{y}(k) &= \mathrm{B}(z^{-1})\Delta \mathrm{u}(k-1) + \mathrm{C}(z^{-1})\mathrm{e}(k) \\
\left(1 - z^{-1}\right)\mathrm{A}(z^{-1})\,\mathrm{y}(k) &= \mathrm{B}(z^{-1})\Delta \mathrm{u}(k-1) + \mathrm{C}(z^{-1})\mathrm{e}(k)
\end{aligned} \tag{2}$$

$$\begin{aligned}
&\left(1 - z^{-1}\right)\left(1 + a_1 z^{-1} + a_2 z^{-2} + \cdots + a_n z^{-n}\right)\mathrm{y}(k) \\
&= \left(b_1 + b_2 z^{-1} + b_3 z^{-2} + \cdots + b_n z^{-n+1}\right)\Delta u(k-1) \\
&\quad + \left(1 + c_1 z^{-1} + c_2 z^{-2} + \cdots + c_{n_c} z^{-n_c}\right)\mathrm{e}(k)
\end{aligned} \tag{3}$$

The difference equation for step $k + 1$ is obtained as,

$$\hat{\mathrm{y}}(k+1) + (a_1 - 1)\mathrm{y}(k) + (a_2 - a_1)\mathrm{y}(k-1) + \cdots + (-a_n)\mathrm{y}(k-n)$$

$$= b_1\Delta \mathrm{u}(k) + b_2\Delta \mathrm{u}(k-1) + \cdots + b_n\Delta \mathrm{u}(k-n+1) + \mathrm{e}(k+1) + c_1\mathrm{e}(k) + \cdots + c_{n_c}\mathrm{e}(k-n_c+1) \tag{4}$$

The difference equation for step $k+2$ is obtained as,

$$\hat{\mathrm{y}}(k+2) + (a_1-1)\hat{\mathrm{y}}(k+1) + (a_2-a_1)\mathrm{y}(k) + \cdots + (-a_n)\mathrm{y}(k-n+1) = b_1\Delta \mathrm{u}(k+1) + b_2\Delta \mathrm{u}(k) + \cdots + b_n\Delta \mathrm{u}(k-n+2) + \mathrm{e}(k+2) + c_1\mathrm{e}(k+1) + \cdots + c_{n_c}\mathrm{e}(k-n_c+2) \tag{5}$$

Similarly, the difference equation for step $k+N$ is obtained as,

$$\hat{\mathrm{y}}(k+N) + (a_1-1)\hat{\mathrm{y}}(k+N-1) + (a_2-a_1)\hat{\mathrm{y}}(k+N-2) + \cdots + (-a_n)\hat{\mathrm{y}}(k+N-n-1) = b_1\Delta \mathrm{u}(k+N-1) + b_2\Delta \mathrm{u}(k+N-2) + \cdots + b_n\Delta \mathrm{u}(k+N-n) + \mathrm{e}(k+N) + c_1\mathrm{e}(k+N-1) + c_2\mathrm{e}(k+N-2) + \cdots + c_{n_c}\mathrm{e}(k+N-n_c) \tag{6}$$

Assuming, $\mathrm{e}(k+1) = \mathrm{e}(k+2) = \ldots.. = \mathrm{e}(k+N) = 0$.

By separating variables into future and past (together with present) and rewriting (4), (5), (6) in matrix form we get,

$$\mathrm{A}\underbrace{\begin{bmatrix} \hat{\mathrm{y}}(k+1) \\ \hat{\mathrm{y}}(k+2) \\ \vdots \\ \hat{\mathrm{y}}(k+N) \end{bmatrix}}_{\mathrm{Y}} = \mathrm{B}\underbrace{\begin{bmatrix} \Delta u(k) \\ \Delta u(k+1) \\ \vdots \\ \Delta u(k+N-1) \end{bmatrix}}_{\mathrm{U}} + \mathrm{A_p}\underbrace{\begin{bmatrix} \mathrm{y}(k) \\ \mathrm{y}(k-1) \\ \vdots \\ \mathrm{y}(k-n) \end{bmatrix}}_{\mathrm{Y_p}} + \mathrm{B_p}\underbrace{\begin{bmatrix} \Delta u(k-1) \\ \Delta u(k-2) \\ \vdots \\ \Delta u(k-n+1) \end{bmatrix}}_{\mathrm{U_p}} + \mathrm{C_p}\underbrace{\begin{bmatrix} \mathrm{e}(k) \\ \mathrm{e}(k-1) \\ \vdots \\ \mathrm{e}(k-n_c+1) \end{bmatrix}}_{\mathrm{E_p}} \tag{7}$$

By multiplying $\mathbf{A}^{-1}$ on both sides in (7), we get the predicted output defined by $\mathbf{Y}$,

$$\mathbf{Y} = \underbrace{\underbrace{\mathbf{A}^{-1}\mathbf{B}}_{\mathbf{G}}\mathbf{U}}_{\textbf{Forced response}} \underbrace{+ \underbrace{\mathbf{A}^{-1}\mathrm{A_p}}_{\mathbf{F_y}}\mathbf{Y_p} + \underbrace{\mathbf{A}^{-1}\mathrm{B_p}}_{\mathbf{F_u}}\mathbf{U_p} + \underbrace{\mathbf{A}^{-1}\mathbf{C_p}}_{\mathbf{F_e}}\mathbf{E_p}}_{\textbf{Free response}}$$

The prediction of the output is separated into two parts - free response and forced response. The simplified representation of the combined predicted output is as shown,

$$\mathbf{Y} = \mathbf{GU} + \underbrace{\left[\mathbf{F}_\mathrm{y}\ \mathbf{F}_\mathrm{u}\ \mathbf{F}_\mathrm{e}\right]}_{\mathbf{F}} \underbrace{\begin{bmatrix} \mathbf{Y}_\mathrm{p} \\ \mathbf{U}_\mathrm{p} \\ \mathbf{E}_\mathrm{p} \end{bmatrix}}_{\mathbf{X}}$$
$$\mathbf{Y} = \mathbf{GU} + \underbrace{\mathbf{FX}}_{\mathbf{f}} \tag{8}$$

3.2 GPC with State Space Model

State space models also can be used to formulate the predictive control problem [18]. A general discrete-time state-space model is given as,

$$\begin{aligned} \mathbf{x}(k+1) &= \mathbf{A}\,\mathbf{x}(k) + \mathbf{B}\,\mathrm{u}(k) \\ \mathrm{y}(k) &= \mathbf{C}\,\mathbf{x}(k) \end{aligned} \tag{9}$$

An incremental state space model [18, 19] can also be used if the model input is the control increment $\Delta\mathrm{u}(k)$ instead of $\mathrm{u}(k)$, where $\Delta\mathrm{u}(k) = \mathrm{u}(k)\text{-}\mathrm{u}(k\text{-}1)$

$$\begin{aligned} \underbrace{\begin{bmatrix} \mathbf{x}(k+1) \\ \mathrm{u}(k) \end{bmatrix}}_{\overline{\mathbf{x}}(k+1)} &= \underbrace{\begin{bmatrix} \mathbf{A} & \mathbf{B} \\ 0 & 1 \end{bmatrix}}_{\mathbf{M}} \underbrace{\begin{bmatrix} \mathbf{x}(k) \\ \mathrm{u}(k-1) \end{bmatrix}}_{\overline{\mathbf{x}}(k)} + \underbrace{\begin{bmatrix} \mathbf{B} \\ 1 \end{bmatrix}}_{\mathbf{N}} + \Delta\mathrm{u}(k) \\ \mathrm{y}(k) &= \underbrace{\left[\mathbf{C}\ 0\right]}_{\mathbf{Q}} \underbrace{\begin{bmatrix} \mathbf{x}(k) \\ \mathrm{u}(k-1) \end{bmatrix}}_{\overline{\mathbf{x}}(k)} \end{aligned} \tag{10}$$

The predicted output representation of state space model in matrix form is described as,

$$\underbrace{\begin{bmatrix} \hat{\mathrm{y}}(k+1) \\ \hat{\mathrm{y}}(k+2) \\ \vdots \\ \hat{\mathrm{y}}(k+N) \end{bmatrix}}_{\mathbf{Y}} = \underbrace{\begin{bmatrix} \mathbf{QN} & 0 & \cdots & 0 \\ \mathbf{QMN} & \mathbf{QN} & \cdots & 0 \\ \vdots & \vdots & \ddots & \vdots \\ \mathbf{QM}^{N-1}\mathbf{N} & \mathbf{QM}^{N}\mathbf{N} & \cdots & \mathbf{QN} \end{bmatrix}}_{\mathbf{G}} \underbrace{\begin{bmatrix} \Delta\mathrm{u}(k) \\ \Delta\mathrm{u}(k+1) \\ \vdots \\ \Delta\mathrm{u}(k+N-1) \end{bmatrix}}_{\mathbf{U}} + \underbrace{\begin{bmatrix} \mathbf{QM} \\ \mathbf{QM}^2 \\ \vdots \\ \mathbf{QM}^N \end{bmatrix}}_{\overline{\mathbf{F}}} \overline{\mathbf{x}}$$
$$\mathbf{Y} = \mathbf{GU} + \underbrace{\overline{\mathbf{F}}\overline{\mathbf{x}}}_{\mathbf{f}} \tag{11}$$

3.3 Cost Funtion

The cost function in (1) can be represented in matrix format as,

$$\mathbf{J} = (\mathbf{Y} - \mathbf{W})^{\mathrm{T}}\mathbf{R}(\mathbf{Y} - \mathbf{W}) + \mathbf{U}^{\mathrm{T}}\mathbf{Q}\mathbf{U} \tag{12}$$

where, **R** and **Q** are diagonal matrix with diagonal elements r and q respectively.

$$\mathbf{W} = \begin{bmatrix} \mathrm{w}(k+1) \\ \mathrm{w}(k+2) \\ \vdots \\ \mathrm{w}(k+N) \end{bmatrix}$$

Substituting **Y**, from (8) or (11),

$$\mathbf{J} = (\mathbf{GU} + \mathbf{f} - \mathbf{W})^{\mathrm{T}}\mathbf{R}(\mathbf{GU} + \mathbf{f} - \mathbf{W}) + \mathbf{U}^{\mathrm{T}}\mathbf{Q}\mathbf{U}$$

$$\mathbf{J} = \mathbf{U}^{\mathrm{T}}\mathbf{M}\mathbf{U} + 2\mathbf{m}^{\mathrm{T}}\mathbf{U} + \mathbf{k} \tag{13}$$

where,

$$\mathbf{M} = \mathbf{G}^{\mathrm{T}}\mathbf{R}\mathbf{G} + \mathbf{Q},$$

$$\mathbf{m}^{\mathrm{T}} = (\mathbf{f} - \mathbf{W})^{\mathrm{T}}\mathbf{R}^{\mathrm{T}}\mathbf{G}, \text{ and}$$

$$\mathbf{k} = \mathbf{f}^{\mathrm{T}}\mathbf{R}\mathbf{f} - 2\mathbf{f}^{\mathrm{T}}\mathbf{R}\mathbf{W} + \mathbf{W}^{\mathrm{T}}\mathbf{R}\mathbf{W}$$

The control law for unconstrained MPC cost function minimization is,

$$\mathbf{U} = -\mathbf{M}^{-1}\mathbf{m} = \underbrace{\left(\mathbf{G}^{\mathrm{T}}\mathbf{R}\mathbf{G} + \mathbf{Q}\right)^{-1}\mathbf{G}^{\mathrm{T}}\mathbf{R}}_{\mathbf{L}}(\mathbf{W} - \mathbf{f}) \tag{14}$$

First row in **L** corresponds to control action at time k,

$$\Delta\mathrm{u}(k) = \mathbf{K}(\mathbf{W} - \mathbf{f})$$

$$\begin{bmatrix} \Delta\mathrm{u}(k) \\ \Delta\mathrm{u}(k+1) \\ \vdots \\ \Delta\mathrm{u}(k+N-1) \end{bmatrix} = \underbrace{\begin{bmatrix} [\ \mathbf{K}\ ] \\ \\ \\ \end{bmatrix}}_{\mathbf{L}}(\mathbf{W} - \mathbf{f}) \tag{15}$$

3.4 Constraints

In a long-range predictive control, the controller has to anticipate constraint violation and correct them in an appropriate way. The output constraints are mainly due to safety reasons and must be controlled in advance because output variables are affected by process dynamics. Input/manipulated variables can always be kept in bound, by limiting the control action to an appropriate value satisfying the constraints. The input and output constraints are given by,

$$\mathrm{u}_{\min} \leq \mathrm{u}(k+i-1) \leq \mathrm{u}_{\max}, \text{ and}$$

$$\mathrm{y}_{\min} \leq \mathrm{y}(k+i) \leq \mathrm{y}_{\max}, \text{ where, } i = \{1, N\}$$

The implementation of MPC with constraints involves the minimization of a quadratic cost function subject to linear inequalities: Quadratic Programming (QP)– in MATLAB *quadprog* function from optimization toolbox [20]. The Hessian matrix **M** and gradient vector **m** are obtained from (13). The inequality and equality constraints are given by,

$$\mathbf{AU} \leq \mathbf{b},$$

$$\mathbf{A}_{eq}\mathbf{U} = \mathbf{b}_{eq}, \text{ and}$$

$$\mathbf{lb} \leq \mathbf{U} \leq \mathbf{ub}$$

Step-by-step implementation procedure for transfer function model and state space model are described below:

i. Define system continuous time process model and initialize sampling time, constraints, polynomial filter coefficients, weighting coefficient for control error and control increment.
ii. Discretize the system transfer function model or state space model.
iii. Compute prediction matrices (**G**, **F**).
iv. Compute **H**, **g** and other matrices for cost function by quadratic programming.
v. Initialize Kalman filter parameters for state space model.
vi. MPC iteration, at each time instant k recursively,

- Compute $\mathrm{y}(k)$ (simulated output or measured output).
- Find prediction error, $\mathrm{e}(k) = \mathrm{y}(k) - \hat{\mathrm{y}}(k)$.
- Find Δu from cost function by considering constraints using quadratic programming.
- Determine predicted output, $\hat{\mathrm{y}}(k)$.

Sections 3.1–3.4 are mathematical formulations described for SISO systems. Similarly, we can extend the same formulation to GPC design for the MIMO process (for example, 2-by-2 or Two-input-Two-output (TITO)), as described in [1] and readers can have a look.

4 Case Studies of 2-by-2 Distillation Columns

The predictive controller design with predictor based on transfer function model and another one based on state space process model are designed in simulations. State is estimated by Kalman filter. Constraints were considered and assumed that the future course of the set-point is known [21]. Active set algorithm was used in quadratic programming to minimize the cost function. All the benchmark models mentioned below are considered as first order SISO system.

4.1 Wood-Berry (WB) Distillation Column Model

Wood and Berry (1973) Distillation Column (DC) consist of an eight-tray column plus reboiler separating methanol and water. The schematic diagram of WB distillation column model is shown in Fig. 2, where F is the feed flow rate [11]. The output products are taken out and new feed is added to the column, but the column process is disturbed due to variation in ambient temperature and change in feed and this has resulted in difficulties in control of product quality. The composition of the top and bottom products expressed in weight % of methanol are the controlled variables [22]. The reflux and the reboiler steam flow rates are the manipulated inputs expressed in lb/min.

The open loop process transfer function for WB DC model is expressed as,

$$\begin{bmatrix} y_1(s) \\ y_2(s) \end{bmatrix} = \begin{bmatrix} \frac{12.8e^{-s}}{16.7s+1} & \frac{-18.9e^{-3s}}{21s+1} \\ \frac{6.6e^{-7s}}{10.9s+1} & \frac{-19.4e^{-3s}}{14.4s+1} \end{bmatrix} \begin{bmatrix} u_1(s) \\ u_2(s) \end{bmatrix}$$

where,

$y_1(s) = x_D(s)$, is the top composition, mol % methanol.
$y_2(s) = x_B(s)$, is the bottom composition, mol % methanol.
$u_1(s) = R(s)$ is the reflux flow rate to the reboiler, lb/min.
$u_2(s) = S(s)$ is the steam flow rate to the reboiler, lb/min.

From the literature, the Effective Transfer Function (ETF) method gives the best output results.

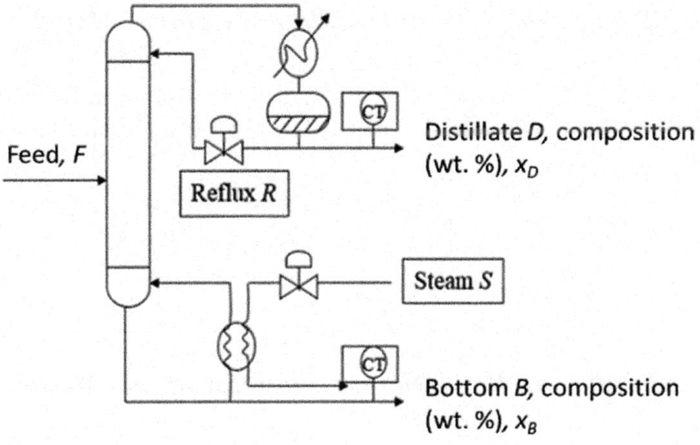

Fig. 2 Schematic diagram of WB distillation column model

4.2 *Vinante and Luyben (VL) Distillation Column Model*

Vinante and Luyben (developed in 1972) distillation column describes the existent dynamics between reflux and vapor flow and the temperatures. VL column is model relating temperature of plates 4 and 17 of a 24-tray tower separating a mixture of methanol and water, examined by Luyben has the following transfer function matrix [23, 24],

$$\begin{bmatrix} T_{17}(s) \\ T_4(s) \end{bmatrix} = \begin{bmatrix} \frac{-2.2e^{-s}}{7s+1} & \frac{1.3e^{-0.3s}}{7s+1} \\ \frac{-2.8e^{-1.8s}}{9s+1} & \frac{4.3e^{-0.35s}}{9.2s+1} \end{bmatrix} \begin{bmatrix} R(s) \\ S(s) \end{bmatrix}$$

where,

$T_{17}(s) =$ Temperature (^{0}C) of tray no. 17
$T_4(s) =$ Temperature (^{0}C) of tray no. 4
$R(s) =$ Reflux flow rate to the reboiler (kg/h)
$S(s) =$ Stream flow rate to the reboiler (kg/h).

From the transfer function matrix, it is easy to understand that the VL distillation does not exhibit open loop column diagonal dominance. If the off-diagonal elements in the transfer function model are large enough, the performance with classical decentralized control may be poor. The main reason is because of no attempt was made to counteract the complex loop interaction and failure of multi-loop and diagonal control designs. Classical decentralized control may not work in strongly coupled systems, and an optimal solution is always challenging. The interaction measure and loop paring may also cause stability problems.

4.3 Wardle and Wood (WW) Distillation Column Model

Wardle and Wood (WW) (presented in 1969) distillation are used to separate water and ethanol and the transfer function matrix form [24] is represented as below:

$$\begin{bmatrix} y_1(s) \\ y_2(s) \end{bmatrix} = \begin{bmatrix} \frac{0.126e^{-6s}}{60s+1} & \frac{-0.101e^{-12s}}{2160s^2+93s+1} \\ \frac{0.094e^{-8s}}{38s+1} & \frac{-0.12e^{-8s}}{35s+1} \end{bmatrix} \begin{bmatrix} u_1(s) \\ u_2(s) \end{bmatrix}$$

4.4 Ogunnaike and Ray (OR) Distillation Column Model

Ogunnaike and Ray (OR) column, a multi-product plant distillation column for separation of a binary ethanol–water mixture, was modeled experimentally in Ogunnaike et al. [25]. The transfer function matrix of the OR column is given by,

$$\begin{bmatrix} y_1(s) \\ y_2(s) \end{bmatrix} = \begin{bmatrix} \frac{22.89e^{-0.2s}}{4.572s+1} & \frac{-11.64e^{-0.4s}}{1.807s+1} \\ \frac{4.689e^{-0.2s}}{2.174s+1} & \frac{5.8e^{-0.4s}}{1.801s+1} \end{bmatrix} \begin{bmatrix} u_1(s) \\ u_2(s) \end{bmatrix}$$

5 Simulation Results and Discussion

GPC as a class of MPC was studied in this section. GPC design provides many benefits in the area of MPCs such as the ability of the controller to control a plant with various parameters, dead time, and higher model order, with a simple structure and robust dynamics. These types of designs are typically incorporated into plants that have non-minimum phases, unstable open loops, or models that are over-parameterized or under-parameterized with the estimation methods. In this work, the GPC algorithm is based on the CARIMA model and can be described after the linearization method is applied to the process and then formulating either in transfer function or state space models.

Considering operation around a particular set point of a TITO system, the response is obtained for multiple-step changes in the reference in a GPC with a prediction horizon of 50, a control horizon of 50, and weighting factors applied to the system for control error and control increments, respectively. The cost function is formulated as a function of future control actions utilizing the process model in a form of the predictor.

The controller parameters for benchmark problems on the binary distillation column are summarized in Table 1. Also, the linearization points applied to the benchmark distillation column represented for the 2-by-2 MIMO system are up_1 and

up_2, yp_1 and yp_2 respectively, and are tabulated in Table 1. The control experiment starts from the calculated steady-state input and output of a model.

Multiple positive and negative set-point steps were applied for the bottom product (output1) at time t = 50 h, 75 h, and t = 125 h, respectively. Similarly, multiple positive and negative set-point steps were also applied for the top product (output2) at time t = 25 h, 75 h, and t = 175 h, respectively. The same setpoints were given to all the benchmark distillation columns for controlling the bottom product and top product.

Figure 3 shows the behavior of the multivariable process of the WB distillation column in the presence of changes in reflux (*R*) and steam flow rates (*S*) with solid blue and redlines, which represent the linear model predictive controller action, respectively using the transfer function model. It is observed that the output follows the reference employing the contribution of the two manipulated variables R and S. It is also seen that any change affects all the variables with a slight deviation from its reference during the transient stage.

Table 1 Generalized Predictive Controller parameters for binary distillation column

Controller parameters	WB model	VL model	WW model	OR model
Sampling time	1 s	1 s	1 s	1 s
Linearization points	$up_1 = 0.55638$ and $up_2 = 0.53407$, $yp_1 = 0.0313$ and $yp_2 = 0.9645$	$up_1 = 0.55638$ and $up_2 = 0.53407$, $yp_1 = 0.0313$ and $yp_2 = 0.9645$	$up_1 = 55.638$ and $up_2 = 63.407$, $yp_1 = 0.0313$ and $yp_2 = 0.9645$	$up_1 = 0.007$ and $up_2 = 0.003$, $yp_1 = 0.6$ and $yp_2 = 0.95$
Start of costing horizon	1	1	1	1
End of costing horizon	50	50	50	50
End of control horizon	50	50	50	50
Weighting coefficient for control error	1	1	100	0.001
Weighting coefficient for control increments	0.1	0.1	0.001	1
Input constraints	$u_{max} = [1-up_1\ 1-up_2]$ and $u_{min} = [-up_1\ -up_2]$	$u_{max} = [1-up_1\ 1-up_2]$ and $u_{min} = [-up_1\ -up_2]$	$u_{max} = [100-up_1\ 100-up_2]$ and $u_{min} = [-up_1\ -up_2]$	$u_{max} = [1-up_1\ 1-up_2]$ and $u_{min} = [-5-up_1\ -5-up_2]$
Output constraints	$y_{max} = [1-yp_1\ 1-yp_2]$ and $y_{min} = [-yp_1\ -yp_2]$	$y_{max} = [1-yp_1\ 1-yp_2]$ and $y_{min} = [-yp_1\ -yp_2]$	$y_{max} = [1-yp_1\ 1-yp_2]$ and $y_{min} = [-yp_1\ -yp_2]$	$y_{max} = [1-yp_1\ 1-yp_2]$ and $y_{min} = [-yp_1\ -yp_2]$

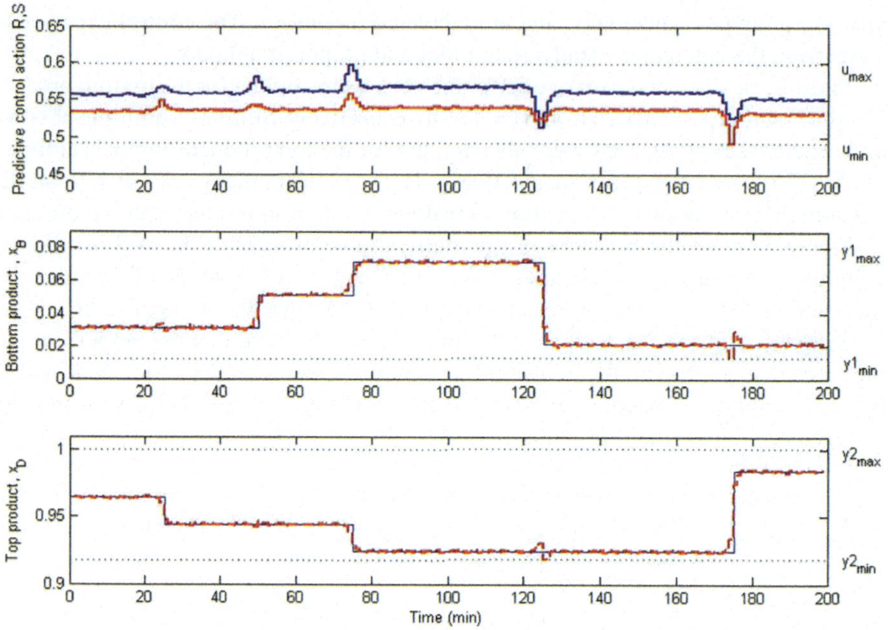

Fig. 3 Constrained predictive control response of WB distillation column model

Figure 4 illustrates the simulation responses of the VL distillation column in the presence of changes in T_4 and T_{17} with solid blue and redlines, which represents the linear model predictive controller action, respectively. It is observed that the output follows the reference utilizing the contribution of the two manipulated variables V and R. It is also seen that any change affects all the variables with a slight deviation from its reference during the transient stage. Satisfactory closed-loop responses are obtained for the multivariable process with the controller designed using the linear model.

Figures 5 and 6 demonstrates the behavior of the multivariable process of the WW and OR distillation column in the presence of changes in L and V with solid blue and redlines, which represent the linear model predictive controller action, respectively. It is observed that the output follows the reference through the contribution of the two manipulated variables $x_B(s)$ and $x_D(s)$. It is also seen that any change affects all the variables with a slight deviation from its reference during the transient stage.

From the literature, it is noticed that there are several comparative analyses performed on the classical decentralized methods of PI/PID controller based on the direct synthesis method, Integral Mode Controller (IMC) method, ETF method, etc. The proposed predictive controller using CARIMA models outperforms the classical controllers in terms of its performance metrics. The classical controllers need to have their parameters tuned effectively and adjusted for optimal performance when the dynamics or time delays of the system are varied or change during the plant operation, especially for nonlinear or time-variant systems.

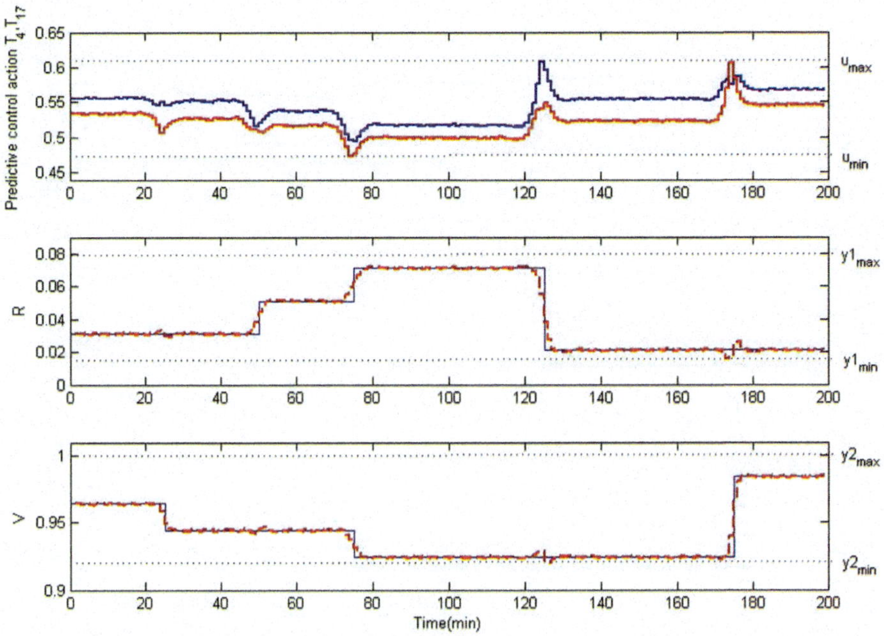

Fig. 4 Constrained predictive control response of VL distillation column model

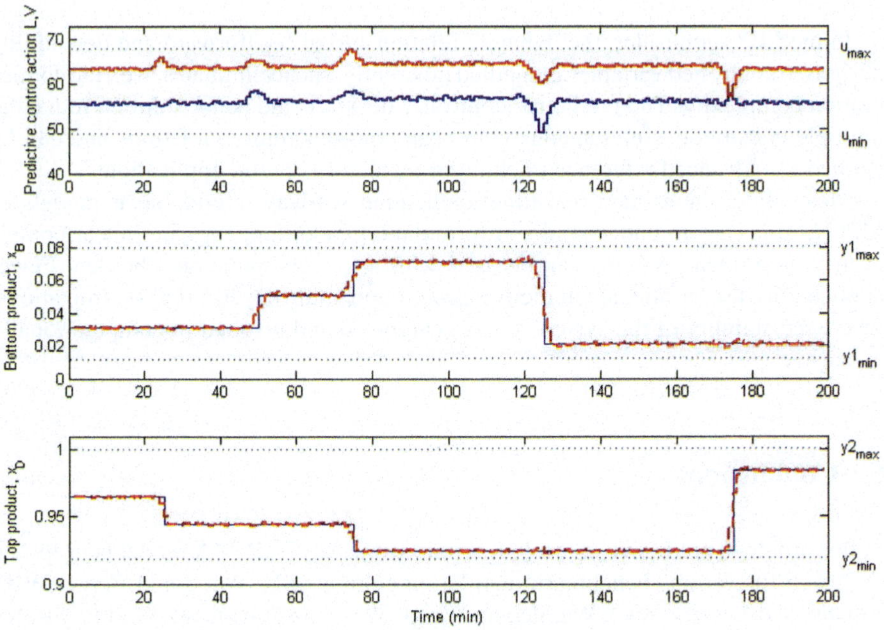

Fig. 5 Constrained predictive control response of WW distillation column model

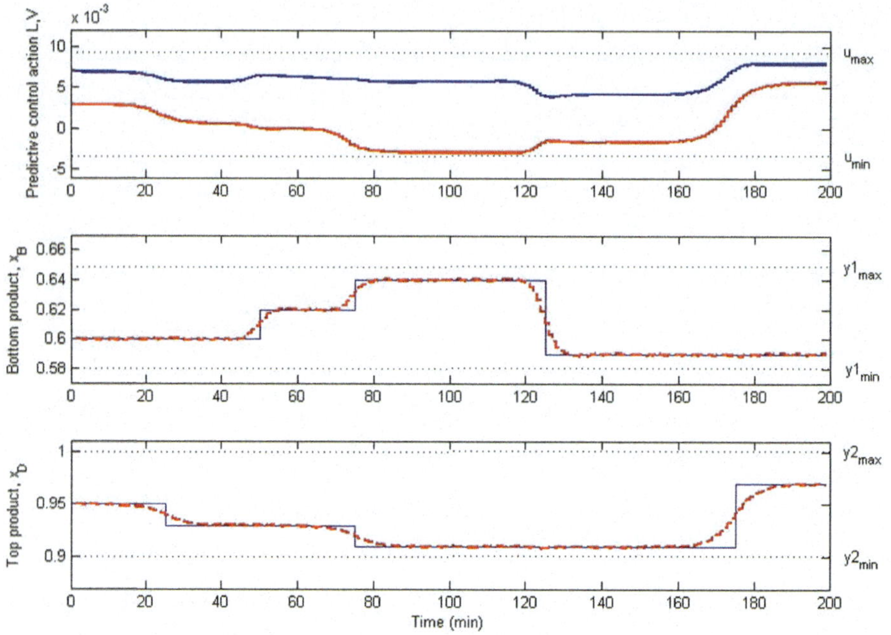

Fig. 6 Constrained predictive control response of OR distillation column model

In the MPC controller, the dynamic relationship of feedforward and decoupling designs to controlled variables is handled inherently, and it eliminates override configuration complexity. The predictive controller design is the better solution to handle multiple constraints simultaneously and can consider future violations and can be applied for advanced reference tracking or regulatory control applications.

Meanwhile, theory and real-time application software prefer linear models of MPC as it is easy to understand and formulate. In addition, applying linear control even to non-linear systems can be used with a low computational burden. Fewer applications use nonlinear predictive design and multi-layer MPC optimizations, where the stability of the system is not guaranteed and resulted in computationally expensive.

6 Conclusions

In this article, a predictive control algorithm using the CARIMA model is proposed to design for various benchmark distillation columns like Wood and Berry (WB), Vinante and Luyben (VL), Wardle and Wood (WW), and Ogunnaike and Ray (OR). The generalized predictive control design with transfer function and state space model is formulated along with their cost function and constraint representation such that any researcher can easily follow the steps and apply them to their domain application. It is observed that the predictive control response of the transfer function and state space model are similar in characteristics. Simulation and predictive control experiments were carried out and paves a good foundation for researchers in acquiring practical skills for understanding the basic concepts applied to process control applications. The generalized predictive controller design gives tight control with the best product quality for MIMO systems by considering the dynamic interactions and loop pairing within the multivariable process. The numerical results indicate that the method is simple, efficient, and can serve as a benchmark system for simulation and control.

Acknowledgements The authors gratefully acknowledge the contribution of Prof. Daniel Honc and Dr. Rahul Sharma, Faculty of Electrical Engineering and Informatics Department of Process Control, University of Pardubice, Czech Republic, and Prof. N. Pappa, Department of Instrumentation Engineering, Madras Institute of Technology (MIT) campus, Anna University, India for their timely guidance.

References

1. Abraham, A., Pappa, N., Honc, D., et al.: Reduced order modelling and predictive control of multivariable nonlinear process. Sādhanā **43**, 41 (2018). https://doi.org/10.1007/s12046-018-0798-x
2. Mishra, R.K., Khalkho, R., Rajesh Kumar, B.: Effect of tuning parameters of a model predictive binary distillation column. In: Proceedings of the IEEE International Conference on Emerging Trends in Computing, pp. 660–665 (2013)
3. Camacho, E.F., Bordons, C.: Model Predictive Control, Advanced Textbooks in Control and Signal Processing. Springer (2007)
4. Yang, J.S.: Optimization-based PI/PID control for a binary distillation column. In: Proceedings of the American Control Conference, pp. 3650–3655 (2005)
5. Georgiou, A., Georgakis, C.T., Luyben, W.L.: Nonlinear dynamic matrix control for high-purity distillation columns. AIChE J. **34**, 1287–1298 (1988)
6. Chidambaram, M.: Tuning of multivariable PI controllers by BLT method for TITO systems. Chem. Eng. Commun. **203**(4), 527–538 (2016). https://doi.org/10.1080/00986445.2015.1039121
7. Manzano, J.M., Limon, D., de la Peña, D.M., Calliess, J.-P.: Robust learning based MPC for nonlinear constrained systems. Automatica **117**, 108948 (2020). https://doi.org/10.1016/j.automatica.2020.108948
8. Cheng, Y., Chen, Z., Sun, M., Sun, Q.: Active disturbance rejection generalized predictive control for a high purity distillation column process with time delay. Can. J. Chem. Eng. **97**(11), 2941–2951 (2019)
9. Mahfouf, M., Linkens, D.A.: Constrained multivariable generalized predictive control (GPC) for anaesthesia: the quadratic-programming approach (QP). Int. J. Contr. **67**(4), 507–528 (1997)
10. Abraham, A., Pappa, N.: An optimal centralized predictive control scheme applied to a Non-Square MIMO benchmark problem. Proc. IFAC-PapersOnLine **49**(13), 59–64 (2016). https://doi.org/10.1016/j.ifacol.2016.07.927
11. Minh, V.T., Abdul Rani, M.: Modelling and control of distillation column in a petroleum process. Math. Probl. Eng. **2009**, 404702 (2009)

12. Abraham, A., Pappa, N.: Experimental validation of an enhanced optimal adaptive control scheme using dominant poles for a variable area process. J. Control Eng. Appl. Inf. Romania **18**(2), 28–36 (2016)
13. Canale, M., Casale-Brunet, S.: "A multidisciplinary approach for model predictive control education: a Lego Mindstorms NXT-based framework". Regular Papers Control Applications. Int. J. Control Autom. Syst. **12**(5), 1030–1039 (2014)
14. Clarke, D.W., Scattolini, R.: Constrained receding horizon predictive control. Proc. Inst. Electr. Eng. **138**, 347–354 (1991)
15. Jeffy, F.J., Abraham, A., Pappa, N.: Decoupler design using dynamic state feedback approach for a binary distillation column. In: Proceedings of the Trends in Industrial Measurement and Automation (TIMA), Chennai, India, pp. 1–7 (2017). https://doi.org/10.1109/TIMA.2017.8064823
16. Muralidharan, G., Abraham, A., Pappa, N.: Estimation of variation in Vapour Liquid Equilibrium constants for modeling a multicomponent crude distillation column. In: Proc. IFAC-PapersOnLine **49**(13), 199–204 (2016). https://doi.org/10.1016/j.ifacol.2016.07.951
17. Drgoňa, J., Klaučo, M., Janeček, F., Kvasnica, M.: Optimal control of a laboratory binary distillation column via regionless explicit MPC. Comput. Chem. Eng. **96**, 139–148 (2017)
18. Honc, D., Sharma, R., Abraham, A., Dusek, F., Pappa, N.: Teaching and practicing model predictive control. In: Proceedings of the 11th IFAC Symposium on Advances in Control Education (ACE), Bratislava-Slovakia, IFACPapersOnline, vol. 49, no. 6, pp. 34–39 (2016)
19. Tsang, T.T.C., Clarke, D.W.: Generalized predictive control with input constraints. Proc. Inst. Electr. Eng. **35**(6), 451–460 (1998)
20. Honc, D., Dušek, F., Sharma, R.: "GUNT RT 010 experimental unit modelling and predictive control application", Nostradamus 2014: Prediction, Modeling and Analysis of Complex Systems. Springer International Publishing, pp. 175–184 (2014)
21. Clarke, D.W., Mohtadi, C., Tuffs, P.S.: Generalized predictive control, part I: the basic algorithm. Automatica **23**, 137–148 (1987)
22. Abraham, A., Pappa, N., Shanmugha Priya, M., Hexy, M.: Predictive control design for a MIMO multivariable process using order reduction techniques. Int. J. Model. Simul. **37**(4), 199–207 (2017)
23. Yadav, E.S., Indiran, T., Priya, S.S.: Optimal energy consumption of the distillation process and its product purity analysis using ultraviolet spectroscopy. ACS Omega 31 **6**(2), 1697–1708 (2020)
24. Chen, D., Seborg, D.E.: Design of decentralized PI control systems based on Nyquist stability analysis. J. Process Control **13**(1), 27–39 (2003). https://doi.org/10.1016/S0959-1524(02)00021-5
25. Ogunnaike, B.A., Ray, W.H.: Process Dynamics Modeling and Control. Oxford University Press, New York (1994)

Robust EV's Speed Tracking Using Fractional Order Controller

Amina Mseddi, Omar Naifar, and Ahmed Abid

Abstract The purpose of this study is to construct a robust Fractional Order PI (FOPI) controller to regulate the speed of an electric vehicle (EV), which includes a single electric switched reluctance motor (SRM). The robustness of such an application appears to be critical. In fact, the nonlinearity of the SRM characteristics prevents typical controllers from ensuring stable performance under different operating situations over the medium and long term. Furthermore, in order to overcome the key issue with this type of vehicle, which is autonomy, the control law must be relevant, robust against the crucial load torque and the vehicle's parameter change, and capable of guaranteeing the optimum performances for the contemplated vehicle.

Keywords SRM · Electric vehicle · FOPI regulator · Speed Tracking

1 Introduction

Electric vehicles (EVs) are becoming increasingly popular due to their potential to reduce greenhouse gas emissions and improve air quality [1, 2]. There are three types of electric vehicles: battery electric vehicles (BEVs), plug-in hybrid electric vehicles (PHEVs), and hybrid electric vehicles (HEVs). EVs have several advantages over traditional gasoline-powered vehicles, including lower operating costs, reduced maintenance, and zero tailpipe emissions [3]. In recent years, there have been significant developments in EV technology, including improvements in battery technology, charging infrastructure, and range. For example, Tesla's new Model S

A. Mseddi
Engineering Preparatory Department, North American Private University, Sfax, Tunisia

O. Naifar (✉)
Control and Energy Management Laboratory, National Engineering School of Sfax, University of Sfax, Sfax, Tunisia
e-mail: email.naifar@gmail.com

A. Abid
ISE Company, Sfax, Tunisia

A. Ben Makhlouf et al. (eds.), *State Estimation and Stabilization of Nonlinear Systems*, Studies in Systems, Decision and Control 491,
https://doi.org/10.1007/978-3-031-37970-3_21

Plaid has a range of up to 390 miles, and charging times have been reduced to as little as 20 min. Many governments are providing incentives for the purchase of EVs, such as tax credits and rebates. Additionally, many countries are setting targets for the phase-out of gasoline-powered vehicles, with some aiming for all new vehicles to be electric by 2035. Despite their many benefits, there are still challenges facing the adoption of electric vehicles, such as range anxiety, charging infrastructure, and higher upfront costs and robust speed tracking.

There are several control strategies used for speed tracking of electric vehicles (EVs). Here is an overview of some of the most common control strategies along: Proportional Integral Derivative (PID) Control: PID control is one of the most commonly used control strategies for speed tracking of EVs. PID control is a feedback control system that adjusts the control input based on the difference between the desired speed and the actual speed. Recent research has focused on improving the performance of PID control by using advanced tuning methods such as fuzzy logic and neural networks [6]. The Model Predictive Control (MPC) is also used for speed tracking in EVs: MPC is a control strategy that uses a model of the system to predict future behavior and optimize control inputs. MPC has been applied to speed tracking of EVs to improve performance and reduce energy consumption. Recent research has focused on using MPC for speed tracking of EVs in complex driving scenarios such as urban traffic [7–9]. Moreover, the adaptive Control [10] which is a control strategy that adjusts the control input based on changes in the system dynamics. Adaptive control has been applied to speed tracking of EVs to improve performance in variable driving conditions such as changing road conditions or varying loads. Recent research has focused on developing adaptive control algorithms that are robust to uncertainties and disturbances in the system. Finally, Siding Mode Control (SMC) [11, 12] is a control strategy that uses a sliding surface to regulate the system output. SMC has been applied to speed tracking of EVs to improve performance in the presence of disturbances and uncertainties. Recent research has focused on using SMC for speed tracking of EVs with improved energy efficiency.

In this chapter a fractional order controller is used to the speed tracking of an EV. The first stage in building a controller is to create a realistic mathematical model that can represent the dynamic behavior of an EV. This model ranged from one degree of freedom to a complicated multibody model. To synthesize a regulator, the model should be simple enough to allow for easy implementation of the controller but still capable of defining the basic EV behavior. Furthermore, the concept of such a model is founded on generic simplifications and assumptions. As a result, the system must be viewed as a combination of an optimal model and another component that encompasses all structured and unstructured uncertainties.

EV is a system subject to the electric parameter fluctuations of its motor as well as external variables such as climatic conditions that make the load torque very big. As a result, it is necessary to appeal a sophisticated synthesizing method capable of establishing the proper control law. This latter cannot be considered optimal and robust unless it is founded on the previously stated logic.

Fractional order PI controllers are an extension of the traditional integer-order PI controllers, where the derivative and integral components have non-integer orders.

The use of fractional-order controllers has shown promising results in improving the control performance of EVs.

This work focuses on the design of a FOPI controller to improve the dynamic and energetic elements of the investigated EV and control its speed tracking. Finally, the effectiveness, robustness, and performance degrees of the two proposed controllers will be tested and evaluated.

2 Traction Chain

A monomotor structure electric vehicle is a type of electric vehicle that uses a single electric motor to power the vehicle. This motor is typically located on the rear axle of the vehicle and is used to drive the rear wheels. The battery provides the electric power to the powertrain, which is responsible for converting the electrical energy into mechanical energy to drive the motor. The motor and drive unit work together to turn the transmission, which in turn drives the wheels and tires to propel the vehicle forward.

The monomotor structure means that there is only one motor and drive unit powering the vehicle, unlike a multi-motor electric vehicle which may have multiple motors and drive units.

One recent example of a monomotor structure electric vehicle is the Volkswagen ID.3 [13]. The ID.3 uses a single electric motor with a maximum output of 150 kW (204 PS) to power the rear wheels of the vehicle. The vehicle also features a relatively large battery pack with a capacity of up to 77 kWh, which provides a range of up to 550 km (WLTP) on a single charge.

Another example is the Tesla Model 3 [14], which uses a single electric motor to power the rear wheels (or both front and rear wheels in the dual-motor version). The Model 3's motor has a maximum output of up to 258 kW (351 PS) and provides a range of up to 614 km (WLTP) on a single charge with the long-range battery pack.

The monomotor structure is a popular choice for electric vehicles due to its simplicity and cost-effectiveness compared to other structures such as dual-motor or quad-motor structures. However, it may have some limitations in terms of traction and handling, especially in certain driving conditions such as off-road or winter driving.

Overall, the monomotor structure is a viable option for electric vehicles, and it is expected that more electric vehicles with this structure will be introduced in the future.

The battery provides the electric power to the powertrain, which is responsible for converting the electrical energy into mechanical energy to drive the motor. The motor and drive unit work together to turn the transmission, which in turn drives the wheels and tires to propel the vehicle forward. A switched reluctance motor is used for driving the vehical. the switched reluctance motor (SRM) has been used in electric vehicles (EVs). SRMs are known for their high efficiency, low cost, and simple structure compared to other types of electric motors. These characteristics make SRMs a promising candidate for use in EVs.

Several studies have investigated the use of SRMs in EVs. For example, a study by [15] proposed a SRM-based traction motor for EVs. The proposed motor was able to achieve a high power density and efficiency, making it suitable for use in EVs. Another study [16] investigated the performance of a SRM-based EV compared to a conventional permanent magnet synchronous motor (PMSM)-based EV. The study found that the SRM-based EV was able to achieve similar performance to the PMSM-based EV while being more cost-effective.

Despite these promising results, there are still some challenges in using SRMs in EVs. One of the main challenges is controlling the torque and speed of the motor, as SRMs have a highly nonlinear torque-speed characteristic. Additionally, SRMs tend to produce more acoustic noise and vibration compared to other types of electric motors.

Overall, while there are some challenges to overcome, the use of SRMs in EVs has shown potential and is an area of active research. For these reasons, we choosed to deal with this motor in the present reaserch. Figure 1 shows the implemented Matlab Simulink model.

In this paragraph, we will describe the mathematical equations governing our Matlab model.

The electrical equation at the terminals of each of each stator phase of the SRM is:

$$u_x = \mathrm{R}i_x + \frac{\mathrm{d}\Psi_x}{\mathrm{dt}} \tag{1}$$

where:

R	represents the resistance of the winding.
u_x and i_x	are the voltage and current of phase x (x is a, b or c).
Ψ_x	is the total flow through the winding of phase x, defined by:

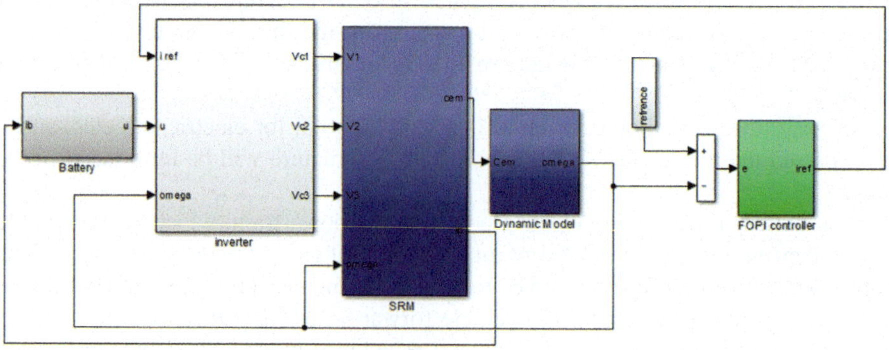

Fig. 1 The implemented Matlab simulink model for the monomotor EV

$$\psi_x(\theta, I) = L_x(\theta) \cdot i_x \tag{2}$$

In the material linearity assumption, the inductance is a function only of the angular position θ. In this case, and assuming negligible magnetic coupling between the phases, Eq. (1) becomes.

The total electromagnetic torque calculated from the virtual work method is written:

$$C_{em} = \frac{1}{2} \cdot \sum_{x=a}^{c} \frac{dL_x(\theta)}{d\theta} i_x^2 \tag{3}$$

The useful torque C_w transmitted to the wheels is expressed by (4):

$$C_w = r_m C_{em} \tag{4}$$

r_m is the reduction ratio.

The application of the dynamics fundamental law on the EV wheels gives the following motion equation

$$C_w = C_a + C_r \tag{5}$$

$C_a = \sigma M_v R_{wheel}^2 \frac{d\Omega}{dt}$: is the resistant torque corresponding to the acceleration with R_{wheel} is the wheel radius and M_v is the vehicle mass.

C_r: is the total load torque applied to the EV.

Assuming that K_t is the friction coefficient of the wheels, C_w may be expressed by:

$$C_w = M_v R_{wheel} \frac{d\Omega}{dt} + K_t \Omega + C_r \tag{6}$$

Since the SRM model is strongly nonlinear; an experimental identification of Ω(I) is performed as in the next figure for different operation points and an average model (7) is condacted (Fig. 2).

$$\frac{\Omega(s)}{I(s)} = \frac{19}{3.6 \times s + 1} \tag{7}$$

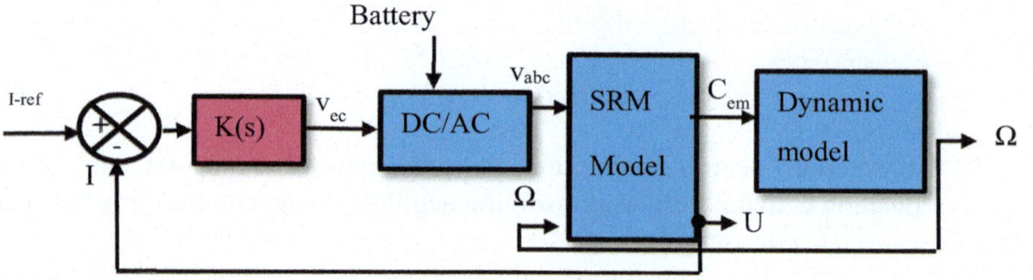

Fig. 2 Open loop experimental identification of the monomotor EV

3 Fractional order control of the EV

Fractional order controllers (FOCs) are a type of controller that can provide more flexible and accurate control compared to traditional integer-order controllers. However, designing and tuning FOCs can be more challenging due to their increased complexity. Here are some common methodologies for tuning FOCs:

- Analytical Tuning: This approach involves deriving an analytical expression for the controller parameters using mathematical equations. For example, the Ziegler-Nichols method can be used to determine the controller parameters based on the system's time response characteristics. Analytical tuning can be simple and fast, but it may not always provide optimal tuning.
- Empirical Tuning: This approach involves experimenting with different controller parameters and observing the system's response. Trial and error or manual tuning can be used to find the optimal controller parameters. However, empirical tuning can be time-consuming and may not always guarantee optimal tuning.
- Optimization-based Tuning: This approach involves using optimization techniques to find the optimal controller parameters that minimize a predefined objective function, such as steady-state error, rise time, overshoot, or energy consumption. Optimization-based tuning can provide a more systematic and efficient approach to tuning FOCs, but it can be computationally expensive.
- Intelligent Tuning: This approach involves using artificial intelligence techniques, such as neural networks, fuzzy logic, or evolutionary algorithms, to automatically tune the controller parameters. Intelligent tuning can provide a more adaptive and robust approach to tuning FOCs, but it requires significant computational resources and expertise in artificial intelligence.

In this work, empirical tuning is performed. In fact, empirical tuning is a trial-and-error approach for tuning controllers, where the controller parameters are adjusted based on observations of the system's response to different inputs. Empirical tuning involves changing the controller parameters and observing the system's behavior to determine the effect of the changes on the system's performance. The process is repeated until an acceptable performance is achieved.

Here are the steps for empirical tuning:

1 Set the controller gains to initial values.
2 Apply a step input to the system.
3 Observe the system's response, such as steady-state error, rise time, overshoot, and settling time.
4 Adjust the controller gains based on the observed response.
5 Repeat steps 2–4 until the desired performance is achieved.

The adjustments to the controller gains can be made in a variety of ways, such as by increasing or decreasing the gains, using proportional or derivative actions, or changing the integral time constant. For example, if the system exhibits overshoot, the proportional gain can be decreased, while if the system exhibits slow response, the integral gain can be increased.

The adjustment of the parameters of the controller will be done so that it insures for the open-loop:

- A high gain in low frequencies in order to refine the frequency tracking and to guaranty the disturbance rejection, so that the EV behavior won't be widely affected by an important load torque or any perturbation. W_1 should thereby present an integral action
- A low gain in high frequencies in order to avoid the effect of the neglected dynamics,
- The bandwidth should not be very high; else, the phase and delay margins won't be sufficient enough to guaranty the system stability

The Matlab Fomcon toolbox is used to tunes the final FOPI controller. The PI fractional order controlle can be modelled as follows (Fig. 3):

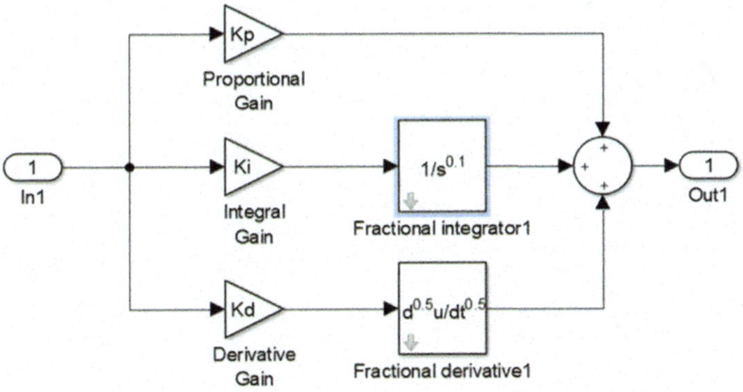

Fig. 3 Fractional order velocity controller

4 Results and Comparison

The simulation results with the nominal parameters of the EV for the two conceived controller is illustrated by Fig. 4.

It's so clear that the FOPI controllers gives a kind of stability to the speed behavior.

Certain circumstances are applied to execute and judge the robustness degree of the FOPI regulator. To begin, temperature change causes the stator resistance to increase by 50% and the inductance to fall by –25% of their nominal values. The friction factor K_t of the wheels thus increases by 50% due to a slick, water-lubricated road. Following that, the EV faces a slope with an angle = 45°. Finally, we attempt to depict the impact of an additive weight by using a weight of 80 kg as an accompanying weight for the conductor.

To signal that, in order to evaluate the speed behavior when the reference takes different situations, the given criteria are carried out for a driving cycle partition. The Fig. 5, presents the results of simulation after application of the test conditions enumerated above.

The simulation results indicate that even at maximum uncertainty range, K_t have no impact on the speed loop performance. Furthermore, the settling time for the FOPI controller is around 1 s. As for the overshoot, 1% is achieved for the FOPI regulator.

The objective of the second test is to assess the controllers' robustness to electrical parameter uncertainties. The electrical parameters are varied according to the following specifications:

- test 1: nominal values
- test 2: $M = 0.8M_e$, $L_d = 0.9L_{dnom}$, $L_q = 0.9L_{qnom}$.
- test 3: $M = 0.8M_e$, $L_d = 0.7L_{dnom}$, $L_q = 0.7L_{qnom}$, $R_s = 1.5R_{snom}$,

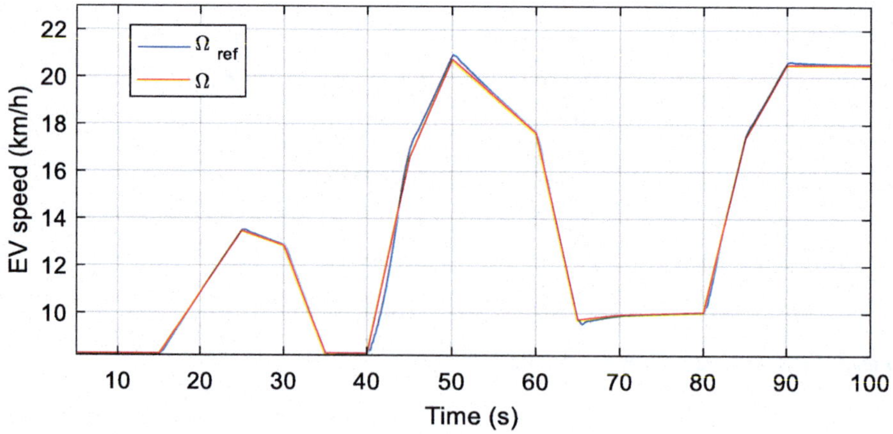

Fig. 4 Performance test under scaled speed sequence

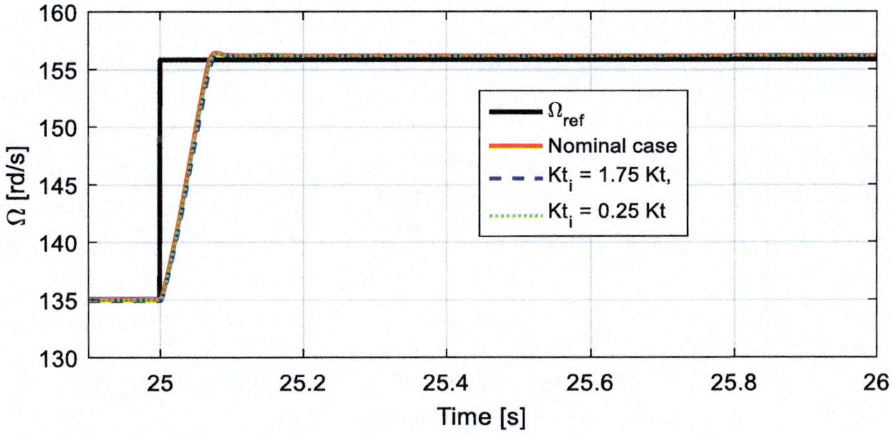

Fig. 5 analysis of the mechanical parameters' robustness

Referring to the Fig. 6, the speed regulated via the FOPI regulator comes to neutralize the slope effect in 25 s, which can preclude the EV skid. While due to the additive mass, the speed moves away from its reference then returns rapidly. So, it's no doubt that the control laws are robust against a supplementary passenger setting in the EV. Finally, when the reference undergoes a sudden variation or the EV encounters a slippery road, the FOPI controller counteracts this action slowly and the speed meets the reference with precision. Consequently, it's to be confirmed that the control laws are robust against different sort of perturbations, insure the steering safety by performing a kind of anti-slip action due to the flexibility of the control components and enhance, inter alia, the EV fault-tolerant. These tangible assets will evidently guaranty the conductor comfort as well as its safety and enhance the EV autonomy.

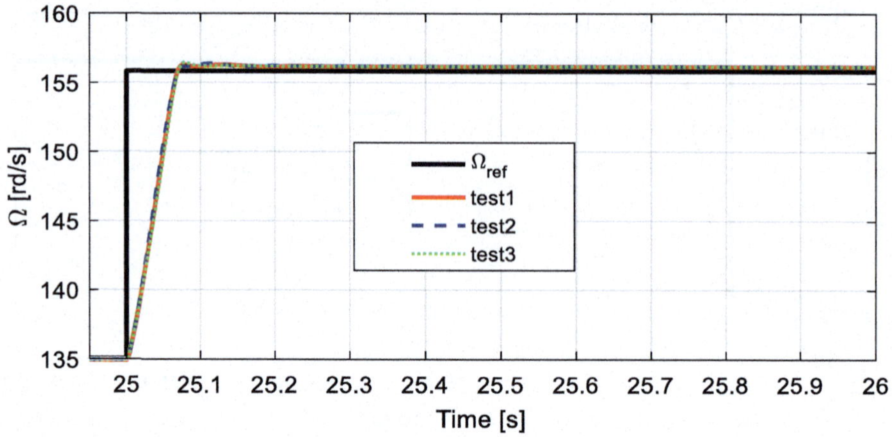

Fig. 6 Robustness tests analysis of the electrical parameters' robustness

5 Conclusion

A FOPI regulator is envisaged in this chapter to govern the dynamic aspect of the studied EV. The obtained results via the conceived control law are persuasive and show good impacts in term of performance and especially in term of robustness. This latter, is sorely demanded and adopted on the traction applications because it can reflect the effectiveness of the control law to preserve the adequacy of the batteries at mid and long-term.

References

1. IEA.: Global EV Outlook 2021. International Energy Agency (2021). https://www.iea.org/reports/global-ev-outlook-2021
2. BloombergNEF.: Electric Vehicle Outlook 2021. BloombergNEF (2021). https://data.bloomberglp.com/professional/sites/24/BNEF-Electric-Vehicle-Outlook-2021.pdf
3. Sperling, D., Gordon, D.: Two Billion Cars: Driving Toward Sustainability. Oxford University Press (2020)
4. Akhade, P., Moghaddami, M., Moghadasi, A., Sarwat, A.: A review on control strategies for integration of electric vehicles with power systems. In: 2018 IEEE/PES Transmission and Distribution Conference and Exposition (T&D), pp. 1–5. Denver, CO, USA (2018)
5. Kong, L., Han, J., Xiong, W., Wang, H., Shen, Y., Li, Y.: A review of control strategy of the large-scale of electric vehicles charging and discharging behavior. In: IOP Conference Series: Materials Science and Engineering, IOP Conference Series: Material Science Engineering, vol. 199, p 012039 (2017)
6. Li, C., et al.: Fuzzy PID control of electric vehicles based on an improved particle swarm optimization algorithm. Energy **225**, 120262 (2021)
7. Zhang, J., et al.: Model predictive control of electric vehicles for energy efficiency improvement. IEEE Trans. Industr. Electron. **68**(11), 10395–10406 (2020)

8. Zhang, Y., et al.: Model predictive control of electric vehicles in an urban environment based on machine learning. Appl. Energy **287**, 116522 (2021)
9. Guo, Q., et al.: Model predictive control of electric vehicles in urban traffic with vehicle-to-infrastructure communication. Appl. Energy **281**, 116067 (2021)
10. Wang, Y., et al.: Adaptive robust control for electric vehicle speed tracking under uncertainties and disturbances. Appl. Energy **279**, 115790 (2020)
11. Li, X., et al.: Sliding mode control of electric vehicles with input constraints based on non-singular terminal sliding mode. Int. J. Control Autom. Syst. **18**(4), 951–962 (2020)
12. Zhou, W., et al.: Robust adaptive control for speed tracking of electric vehicles based on data-driven uncertainties estimation. Appl. Energy **302**, 117670 (2021)
13. Volkswagen ID.3. (2021). Retrieved from https://www.volkswagen-newsroom.com/en/id3-3542
14. Tesla Model 3. (2021). Retrieved from https://www.tesla.com/model3
15. Bilgin, B.: A novel switched reluctance motor based propulsion drive for electric and plug-in hybrid electric vehicles, PHD Thesis Illinois Institute of Technology ProQuest Dissertations Publishing (2011). 3486310
16. Pindoriya, R., Singh, B., Kumar, R., Srivastava, K.: Comparative analysis of permanent magnet motors and switched reluctance motors capabilities for electric and hybrid electric vehicles (2018)

Fractional Order Control of a Grid Connected WindPACT Turbine

Amina Mseddi, Omar Naifar, and Ahmed Abid

Abstract We focused on the grid integration of a wind conversion system WCS based on a Hybrid Excited Synchronous Generator (HESG). We demonstrated an integrated Matlab Simulink platform for modeling the grid, convertors, filters, wind turbine, and generator associated with its power convertors required for command. Following that, we presented the detailed control strategy for both the DC link and the grid side converter. The simulation platform is then used to test the effectiveness of the implemented control strategies in both normal and failure operation modes.

Keywords HESG · WECS · Grid integration · WindPACT turbine

1 Introduction

The integration of renewable energy into the electricity grid causes issues for the various stakeholders involved, including generation, transmission, distribution, and consumers. To effectively integrate energy, a number of issues must be addressed in the following areas:

- Market restructuring issues: bringing all stakeholders together to balance supply and demand.
- Electricity network design and operation: optimization of current infrastructure, extensions and reinforcements...
- Wind energy grid connection: control strategy, energy quality, and wind power plant capacities...

A. Mseddi (✉)
Engineering Preparatory Department, North American Private University, Sfax, Tunisia
e-mail: amina.mseddi.ise@gmail.com

O. Naifar
Control and Energy Management Laboratory, University of Sfax, Sfax, Tunisia

A. Abid
ISE Company, Sfax, Tunisia

A. Ben Makhlouf et al. (eds.), *State Estimation and Stabilization of Nonlinear Systems*, Studies in Systems, Decision and Control 491,
https://doi.org/10.1007/978-3-031-37970-3_22

This work focuses on the creation of the control laws required for the establishment of this link. Indeed, after validating the hybrid generator's applicability in wind conversion systems with an isolated load, [1–3], this paper aims to connect the WindPACT turbine [4, 5] to the electrical grid. Matlab/Simulink is used to implement a medium voltage network (HTA). We begin by modeling the structure of the network. In the Park coordinate system, the connecting device is shown. This rather simplified representation allows for the development of control laws to be facilitated. Then, using "Simpowersystem" tools, a more "realistic" architecture of an HTA network and its connection device will be implemented. The second section of this work is devoted to the development of appropriate regulators to ensure the reliability of this connection. The final section will be devoted to simulation results.

2 Network Structure Modeling

The purpose of this paragraph is to model the electrical assembly ensuring the grid connection of the wind conversion system. The principle assembly is illustrated in Fig. 1 A continuous bus provides the connection between the rectifier and the inverter. The latter ensures the control of the voltage of the DC bus and the exchange of active and reactive power between the turbine and the network. Before being injected into the power grid, the current passes through an *RL* filter. A Phase Locked Loop (PLL) allows instantaneous measurement of voltages and frequencies on the grid side.

Where:

i_{gHESG} The rectified current supplied by the generator

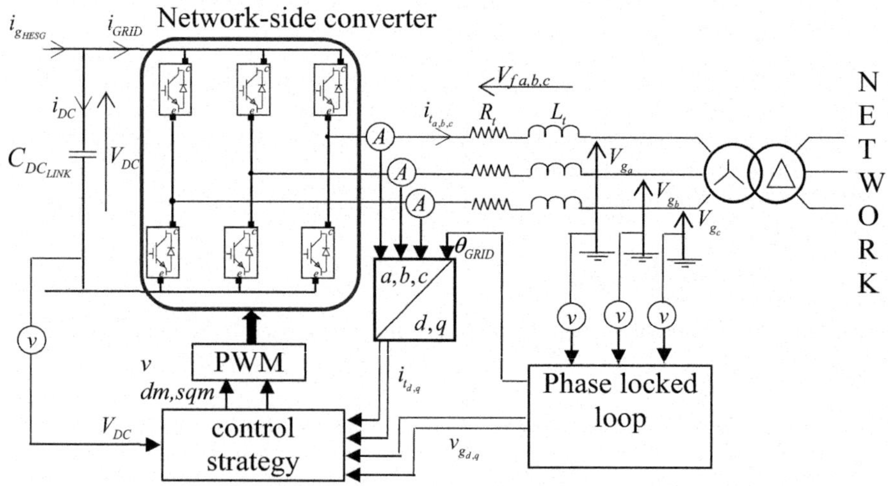

Fig. 1 Electrical diagram of the coupling to the network of the SCE

i_{GRID} Power on the grid side

i_{DC} The current in the continuous bus

$i_{t\,abc}$ The current passing through the filter in the coordinate system (a, b, c)

$i_{t\,dq}$ The current passing through the filter in the coordinate system (d, q)

$In_{g\,abc}$ The voltage on the mains side in the coordinate system (a, b, c)

V_{DC} The voltage at the terminals of the continuous bus

$In_{f\,abc}$ The voltage across the filter in the coordinate system (a, b, c)

$V_{g\,dq}$ The voltage on the mains side in the coordinate system (d, q)

$v_{md,mq}$ The modulating tension

i_{GRID} The instantaneous phase of the network

C_{DCLINK} The capacity of the continuous bus

$\boldsymbol{R_t}$ The resistance of the filter on the network side

$\boldsymbol{L_T}$ The inductance of the filter on the network side.

2.1 *Continuous Bus Modeling*

The rectifier on the HESG side and the inverter on the grid side are connected via a continuous bus (Fig. 1). The latter allows the transfer of power between two sources with different frequencies by decoupling the constant frequency of the network from the variable frequency of the generator.

The mathematical model of the DC bus can be expressed as (1)

$$V_{DC}(t) = \frac{1}{\Delta T}\int_{t_0}^{t_0+\Delta T} \frac{1}{C_{DC_{LINK}}} i_{DC}(t) + v_{DC}(t_0) \tag{1}$$

where is the initial value of the voltage across the capacitor $v_{DC}(t_0)$

The current in the capacitor is equal to the difference of the two modulated currents:

$$i_{DC}(t) = i_{g_{HESG}}(t) - i_{GRID}(t) \tag{2}$$

The DC bus transfer function is given by (3)

$$\frac{V_{DC}(s)}{i_{DC}(s)} = \frac{1}{C_{DC_{LINK}} \cdot s} = \frac{\tau_c}{s} \tag{3}$$

where τ_c is the time constant of the continuous bus which is equal to $1/C_{DCLINK}$.

To be able to send a current into the network, the voltage V_{DC} must satisfy:

$$V_{DC} \geq \sqrt{6} \cdot V_{effa,b,c} \tag{4}$$

where $V_{effa,b,c}$ is the voltage that appears on the filter side. Since the effective voltage at the primary of the connection transformer is 690 V, the nominal voltage of the DC bus is set at 1800 V.

2.2 Power Converter Modeling

The three-phase inverter on the grid side consists of 6 switches formed by IGBTs and antiparallel diodes. To describe its operation, it is assumed that switches can be considered perfect switches.

At the output of the converter, the filter is considered to be a source of current. The latter must never be short-circuited. Therefore, the switching functions on the same inverter arm must be complementary so as to satisfy the relationship (5).

$$f_{1,2,3} + \overline{f_{1,2,3}} = 1 \tag{5}$$

The modulated voltages u $_{12}$ and u_{13} are given by:

$$u_{12} = m_{12} \cdot V_{DC} \qquad u_{13} = m_{13} \cdot V_{DC} \tag{6}$$

where m $_{12}$ and m_{13} are conversion functions defined by (7)

$$m_{12} = f_1 - f_2 \qquad m_{13} = f_1 - f_3 \tag{7}$$

The modulated current injected into the network is a function of the currents of the filter i_{GRID} $i_{t_{a,b,c}}$ such that:

$$i_{GRID} = m_{12} \cdot i_{t_a} + m_{13} \cdot i_{t_b} \tag{8}$$

The continuous equivalent model of the converter facilitates the synthesis of regulators controlling power transit on the grid side. For a balanced three-phase network, one can describe the modulated voltages (at the output of the converter) in the Park reference frame such that:

$$v_{td} = v_{md}\frac{V_{DC}}{2} \qquad v_{tq} = v_{mq}\frac{V_{DC}}{2} \tag{9}$$

where v md and v $_{mq}$ are the modulating voltages and v $_{td}$ and v_{tq} are the direct and quadrature components of the modulated simple voltages defined in (6) [6].

2.3 Filter Modeling

The WCS is connected to the grid via an *RL* filter. Indeed, the current from the inverter is chopped at the modulation frequency. It is therefore necessary to add a filter to attenuate the harmonics of the current. By charging and discharging, during each modulation period, the L_t inductor smoothes the chopped current and transforms it into a sinusoidal current.

In the coordinate system (a, b, c) the filter is modeled by:

$$[V_{t_{a,b,c}}] = R_t \cdot [i_{t_{a,b,c}}] + L_t \frac{d}{dt}[i_{t_{a,b,c}}] + [V_{g_{a,b,c}}] \tag{10}$$

With

$$i_{t_a}(t) = \frac{1}{L_t} \cdot \int v_{l_a}(t) \cdot dt + i_{t_a}(t_0) \quad i_{t_b}(t) = \frac{1}{L_t} \cdot \int v_{l_b}(t) \cdot dt + i_{t_b}(t_0)$$
$$i_{t_c}(t) = -i_{t_a}(t) - i_{t_b}(t) \tag{11}$$

where

$$v_{l_a} = V_{t_a} - V_{g_a} - v_{r_a} \qquad v_{l_b} = V_{t_b} - V_{g_b} - v_{r_b} \tag{12}$$

In Park's (10) becomes:

$$v_{td} = R_t \cdot i_{t_d} + L_t \frac{di_{t_d}}{dt} - L_t \cdot w_s \cdot i_{t_q} + v_{g_d}$$
$$v_{tq} = R_t \cdot i_{t_q} + L_t \frac{di_{t_q}}{dt} + L_t \cdot w_s \cdot i_{t_d} + v_{g_q} \tag{13}$$

where i $_{td}$ and i_{tq} are the direct and quadrature currents passing through the filter and sent to the network. The voltages across the filter are then:

$$v_{fd} = R_t \cdot i_{t_d} + L_t \frac{di_{t_d}}{dt} = v_{td} + e_q - v_{g_d}$$
$$v_{fq} = R_t \cdot i_{t_q} + L_t \frac{di_{t_q}}{dt} = v_{tq} - e_d - v_{g_q} \tag{14}$$

where the coupling voltages e $_d$ and e_q are given by (15)

$$e_q = L_t \cdot w_s \cdot i_{t_q} \quad e_d = L_t \cdot w_s \cdot i_{t_d} \tag{15}$$

Finally, Eq. (14) gives the transfer function of the filter which is written as:

$$\frac{i_{t_q}(s)}{v_{f_q}(s)} = \frac{i_{t_d}(s)}{v_{f_d}(s)} = \frac{1}{L_t \cdot s + R_t} = \frac{1/R_t}{1 + \tau_t \cdot s} \qquad \tau_t = \frac{L_t}{R_t} \tag{16}$$

2.4 Synchronization of the WCS to the Power Grid

Synchronization is one of the most important issues in controlling power electronic equipment connected to the grid. Examples of applications where synchronization is essential include active filters, power factor control, renewable energy production [7].

For normal operation of the electrical network, the frequency must be stable and close to its nominal value. In Europe, the frequency is usually 50 ± 0.1 Hz and it very rarely falls in the range 49–50.3 Hz and in Tunisia it is 50 ± 2.5 Hz. With regard to the voltage limits, each network proposes a voltage tolerance curve in which the voltage to be supplied by the turbine is quantified by its amplitude [8] The voltage and frequency ranges determine the limits within which the turbines should not be disconnected from the grid. It is therefore necessary to detect the parameters of the network voltage during and after the fault, such as the amplitude and phase of the fundamental as well as its positive and negative [9]. Several PLL principles exist in the literature: there are simple principles such as the method of transition to 0 "Zero Crossing Method" or the method of filtering the voltages of the network "Filtering Grid Voltage" but the latter are not reliable especially in case of fault or imbalance in the electrical [10]. The poor performance of conventional methods then leads to the use of other more advanced [9]. The method used in this work is the SRF-PLL "Synchronous Reference Frame PLL" [9]. In order to remove or mitigate the effect of a possible imbalance in the network, the captured voltage passes through a filter thus giving a voltage image of the real voltage. The image voltage is then enslaved. In this way, the controlled oscillator creates a phase synchronized to the network.

2.5 Electrical Network

To study the behavior of our WCS, we propose the architecture of the electrical network illustrated in Fig. 2.

The infinite HTB network is modeled by a three-phase RLE type of short-circuit power $S_{cc\text{-}source}$ 1,300 MVA. The nominal voltage of the $_{HV}$ U source network is equal to 120 kV and the phase shift between the voltage and the current of the source is zero. The rest of the network is composed of:

- A 120 kV/25 kV step-down transformer rated 47 MVA rated "Transformer (2)" in Fig. 2.
- Two three-phase loads, one is 2.6 MW and the other 1.5 MW.
- The WP turbine is connected to the grid by a 690 V/25 kV transformer noted "Transformer (1)" in Fig. 2.

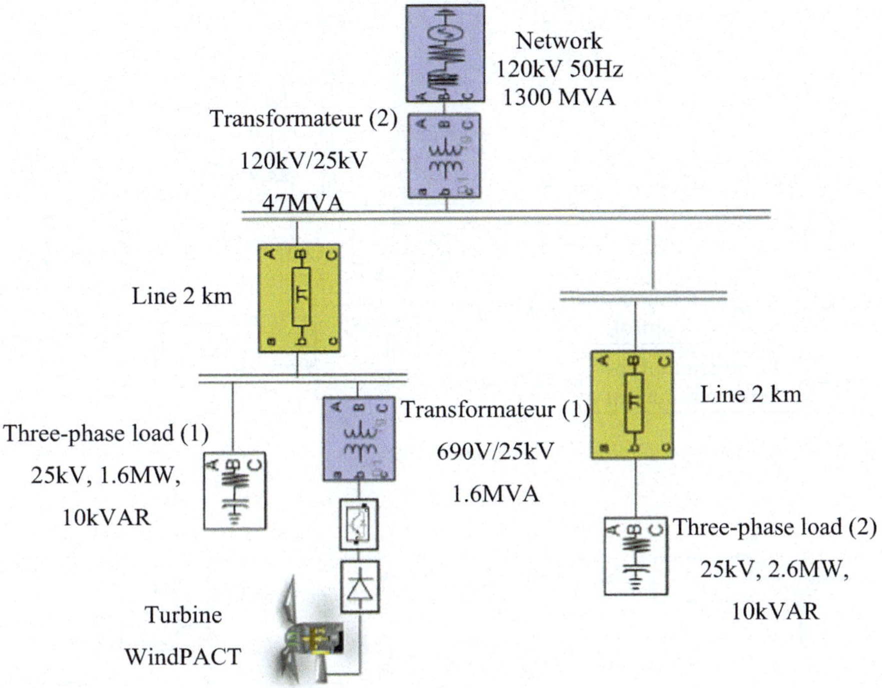

Fig. 2 Network model

All the elements described above are modeled by Matlab's "Simpowersystem" blocks where the block named "WP wind turbine" represents the model of the Wind-PACT 1.5 MW turbine interfaced to the hybrid generator and its power converters (the chopper for the control and the rectifier for the connection to the continuous bus).

3 Network Link Control

Once the system is modeled, we can proceed to the control of the inverter and the continuous bus. The aim is to maintain a constant bus voltage, transfer all the active power produced by the wind turbine to the grid and cancel the reactive power transmitted to have a unit power factor. The control synoptic is shown in Fig. 3.

The control of the network side link is illustrated in Fig. 4 [12, 13].

This control is done through two nested loops. The internal loops are based on the management of filter currents, and the external loops are intended for the management of DC bus voltage and reactive power.

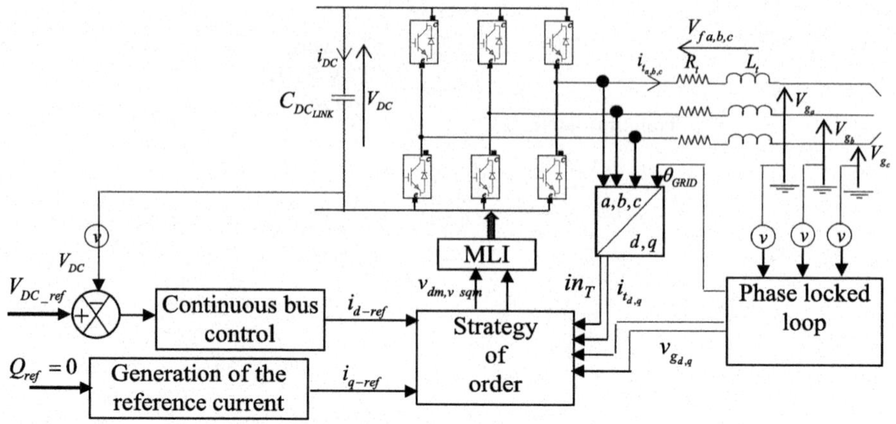

Fig. 3 Block schema of the network link control policy

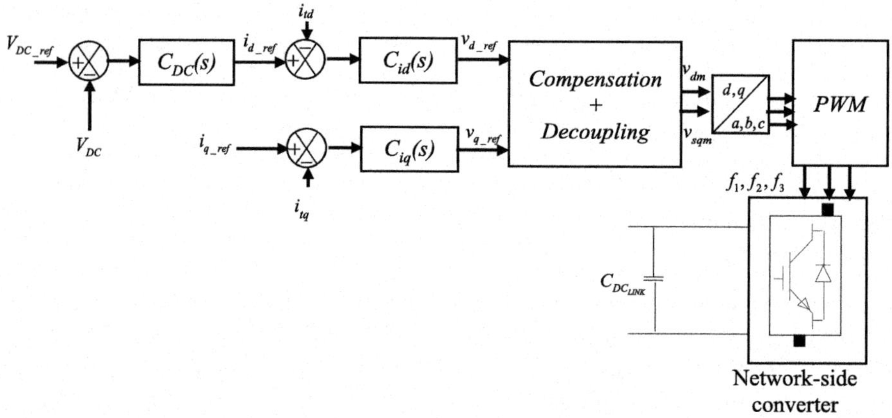

Fig. 4 Network-side converter control loop

3.1 Continuous Bus Control

The adjustment of the continuous bus is carried out by a fractional order PI regulator denoted $C_{DC}(s)$ [12, 14] which generates at output the reference current $i_{d\text{-}ref}$. The response time of the bus's external control loop is usually chosen in the order of a few milliseconds [14]. It must be at least 10 times larger than the inner loop of filter currents. It must also be greater than the calculation time of frequency quantities 50 Hz (10 ms). A response time of 15 ms is adequate for this loop [1]. The time constant of the DC bus corrector is equal to C_{DCLINK} (Fig. 5).

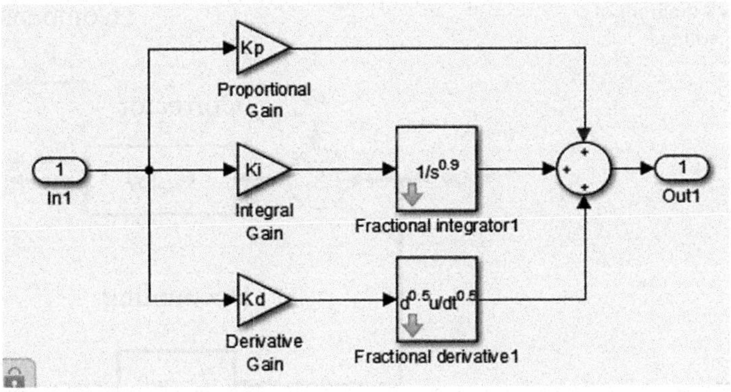

Fig. 5 Fractional order PI controller

The final corrector is given by:

$$C_{DC}(s) = K_{DC} + \left(\frac{1}{T_{DC}s}\right)^{\lambda} \quad with \quad T_{DC} = 0.01\,\mathrm{s} \qquad K_{DC} = 11.4 \quad \lambda = 0.9 \tag{17}$$

3.2 Network Side Converter Control

The grid-side converter is used to control the currents passing through the filter. The command is done in the Park coordinate system which is synchronized with the mains voltages. The modulating voltages v_{dm} and v_{qm} are compared to a triangular carrier of frequency 2,700 Hz to generate the switching frequencies (f_1, f_2, f_3). The carrier frequency is limited to 2.7 kHz to limit the calculation time.

The generation of modulating voltages is shown in Fig. 6. It is done in three actions: correction of the currents of the filter (i_{td} and i_{tq}), compensation of the voltages of the secondary of the transformers (v_{gd} and v_{gq}) and decoupling of the voltages defined in (15) [11].

The quadrature current component is used to regulate the reactive power and the direct component is used for the regulation of the DC bus voltage. The direct current $i_{d\text{-}ref}$ comes from the corrector $C_{DC}(s)$ of the DC bus. On the other hand, the reference current in quadrature is maintained equal to 0. This makes it possible to cancel the reactive power and to have a unit power factor. Indeed, the powers injected into the network can be expressed, in Park's reference frame, by (18) and (19) [13]:

$$P_g = V_{g_d} \cdot i_{t_d} + V_{g_q} \cdot i_{t_q} \tag{18}$$

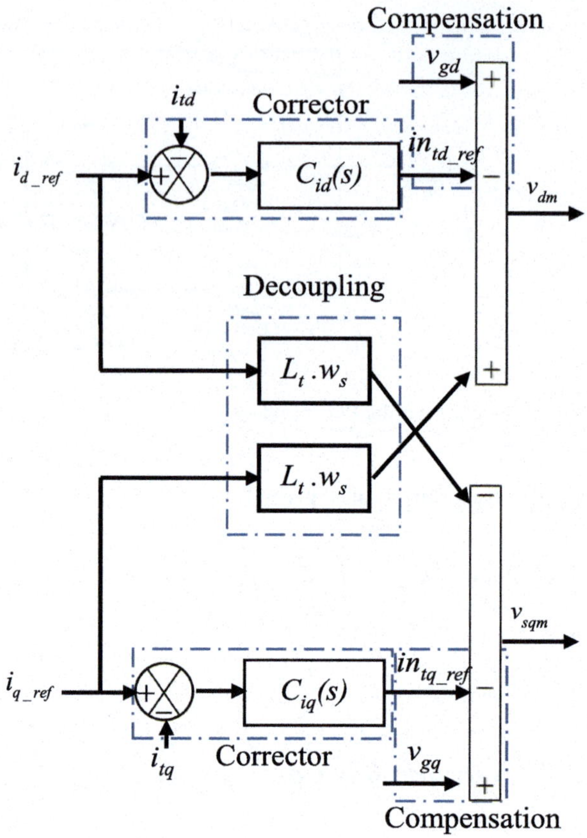

Fig. 6 Block diagram of modulating voltage generation

$$Q_g = -V_{g_d} \cdot i_{t_q} + V_{g_q} \cdot i_{t_d} \tag{19}$$

The Park reference frame being synchronized with the mains voltages, implies that the quadrature component of the voltage of the lattice V_{gq} is zero [14]. We can therefore deduce from equation (19) that:

$$i_{tq-ref} = i_{q-ref} = -\frac{Q_{g-ref}}{V_{g_d}} \tag{20}$$

The control of the currents of the filters is done by the corrector C $_{id}$(s) for the direct current and $C_{iq}(s)$ for the quadrature current. An integral proportional corrector with a response time of 1 ms is usually used for adjusting currents through the filter [1, 14]. The time constant *of* the corrector C $_{id}$(s) (respectively $C_{iq}(s)$) is stopped at the filter time constant and its gain is adjusted to impose the desired response time. The final correctors are then given by (21).

$$C_{id}(s) = C_{iq}(s) = K_{t_{d,q}} + \left(\frac{1}{T_{t_{d,q}} s}\right)^{\lambda} \quad T_{t_{d,q}} = 0.32\,\text{s} \quad K_{t_{d,q}} = 872 \;\; \lambda = 0.1 \qquad (21)$$

4 Simulation Results

In this paragraph, the elaborate control laws are tested. We start first with the validation of the command on the network side alone. Two tests are performed: the first concerns a network in normal operation and the second deals with the case where the network is subject to a fault. Once this command is validated, we move on to a second test where the WindPACT turbine based on a HESG is connected to the grid. The behavior of the network and that of the turbine are observed for normal operation and in case of fault.

4.1 Validation of the Command on the Network Side

4.1.1 Case of a Network in Normal Operation

Figure 7 shows the pattern of the active and reactive power fed into the network. It is zero with small fluctuations of 0.5VAR. The voltage of the DC bus is maintained equal to $V_{DC\text{-}name}$ with a peak of about 50 V when the electrical power is varied in step (Fig. 8). The fluctuations observed in Fig. 8 are due to the calculation step of the Matlab software. The sampling pitch is set to 5 10^{-5} given the complexity of the implemented model. For a stochastic rate of the injected electrical power (Fig. 9). The voltage of the DC bus is stable and equal to its reference value (there is no longer a voltage peak) as shown in Fig. 10.

4.1.2 Case of a Network in Short-Circuit Failure

A constant power equal to 1.5 MW is now injected and the effect of the appearance of certain defects on the behaviour of the continuous bus is being tested. It is noted that for these tests, the sampling pitch is reduced to $^{10-6}$ s, which justifies the smooth

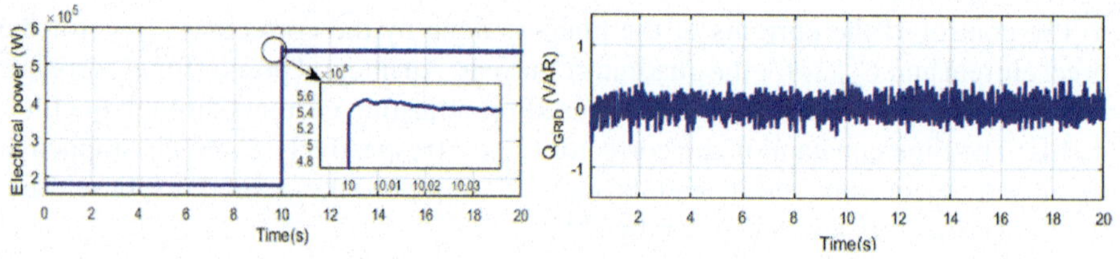

Fig. 7 Profile of the active and reactive power injected into the grid

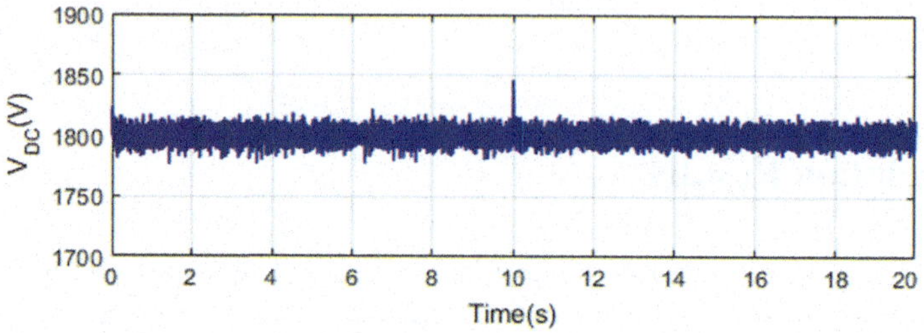

Fig. 8 DC bus voltage

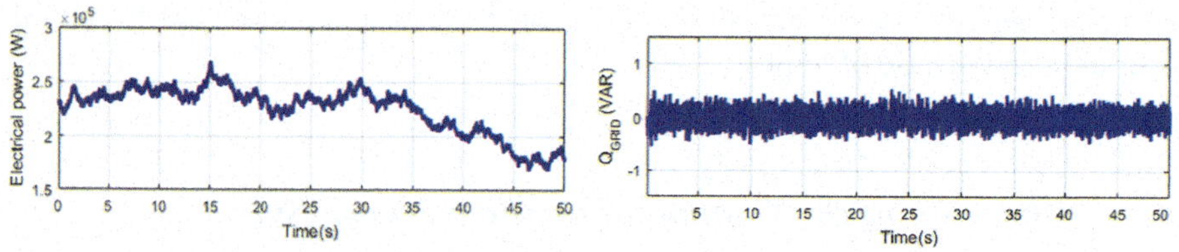

Fig. 9 Profile of the power injected into the grid

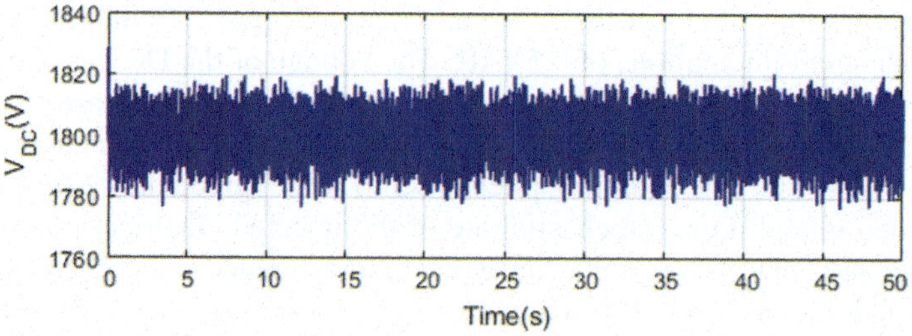

Fig. 10 DC bus voltage

pace of the bus tension. The first fault is a single-phase defect where the phase "a" connected to load (2) of is short-circuited (Fig. 11). Figure 12 shows that the continuous bus is insensitive to this defect: its voltage is maintained equal to $V_{DC\text{-}name}$. A second balanced three-phase fault is applied to the terminals of the load (2). The

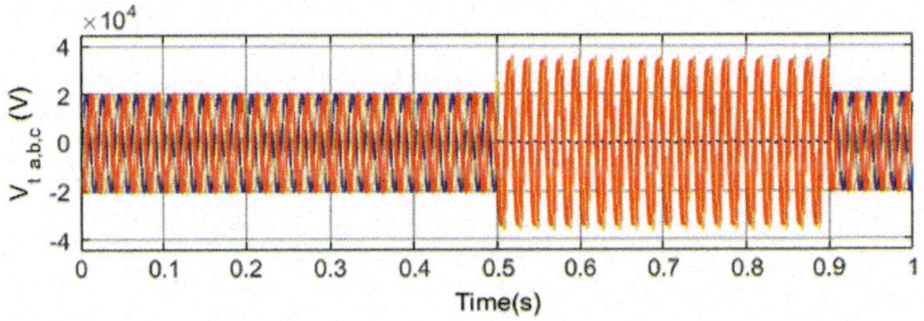

Fig. 11 Rate of voltages at the terminals of the transformer secondary

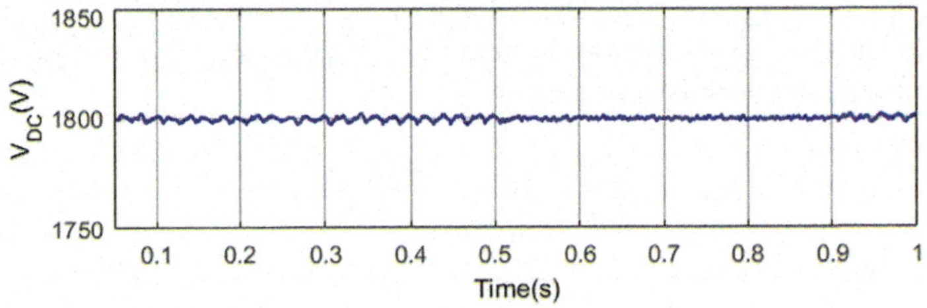

Fig. 12 DC bus voltage

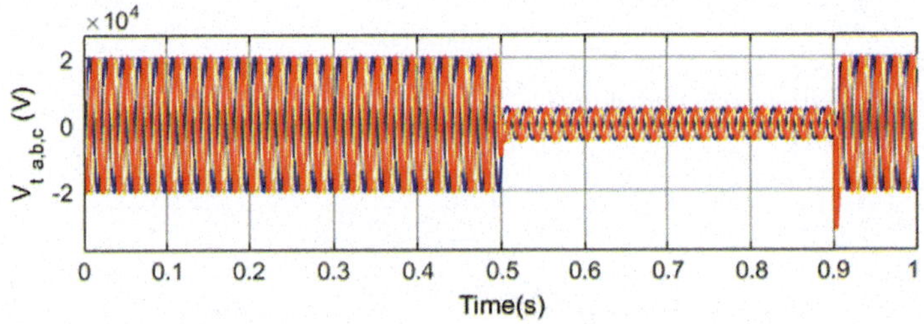

Fig. 13 Rate of voltages at the secondary of the transformer

three phases are grounded for 400 ms (Fig. 13). After this time, a protection is triggered and isolates the load (2) from the rest of the network. This type of defect being more severe, a peak of 50 V is observed at its appearance, as well as at its disappearance (Fig. 14).

4.2 *Validation of the Control Strategy of the Grid-Connected WP Turbine*

The first test concerns normal operation of the electrical network. A stochastic wind of average value 9 m/s is considered (Fig. 15). A good monitoring of the reference

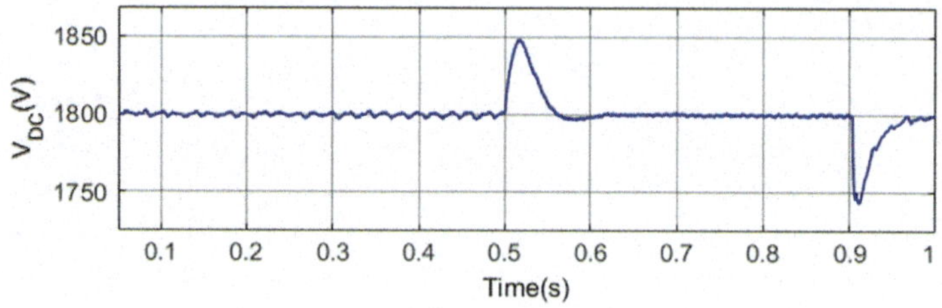

Fig. 14 DC bus voltage

speed is obtained (Figs. 16, 17). The voltage of the DC bus is also constant (Fig. 18) despite the stresses caused by variations in wind speed. Finally, the reactive power injected into the network is zero (Fig. 19).

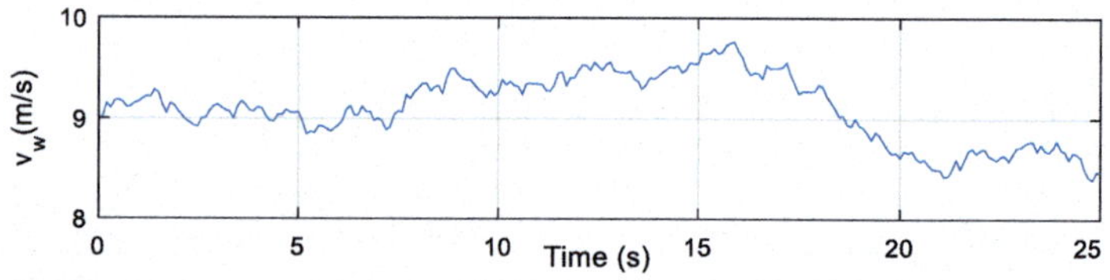

Fig. 15 Wind speed as a function of time

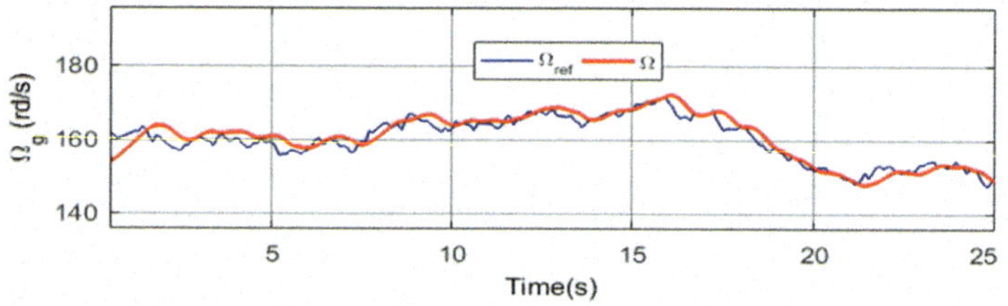

Fig. 16 Rotational speed as a function of time

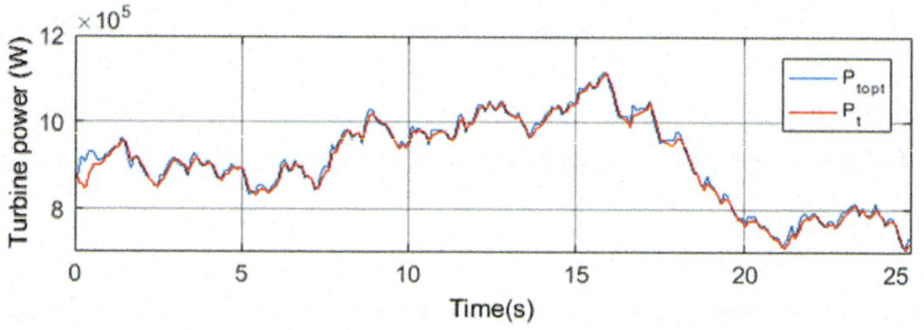

Fig. 17 Turbine power as a function of time

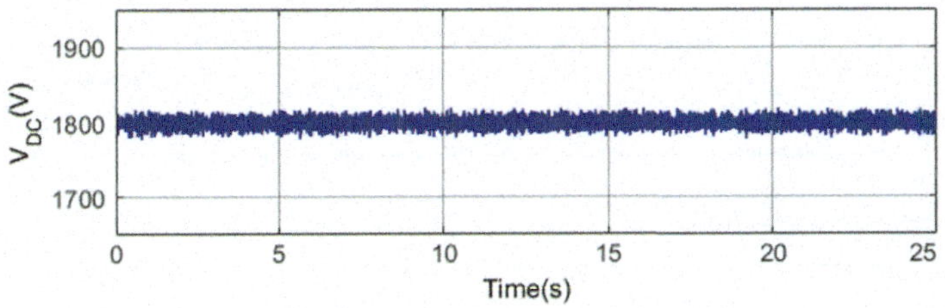

Fig. 18 DC bus voltage

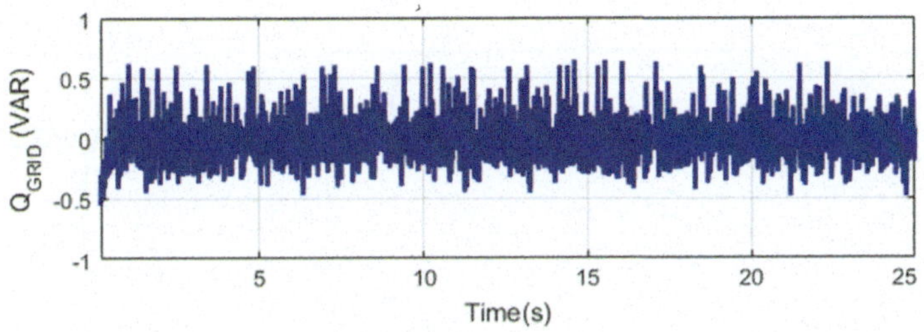

Fig. 19 Reactive power injected into the grid

To test the behavior of the wind system in the face of a three-phase fault balanced on the load (2), a constant wind value of 12 m/s is considered. The defect is applied between 1.5 and 1.9 s. The speed of rotation of the generator is constant (Fig. 20). The behavior of the turbine is not affected, the power extracted from the wind is equal to 1.5 MW (Fig. 21). The voltage of the DC bus is equal to 1800 V with peaks of 10 V when defects appear and disappear (Fig. 22). This shows good resistance during the short circuit. The direct current of the filter (Fig. 23) stabilizes the bus voltage during this fault. This peak is of the order of 0.5% of the value of V_{DCnom}. This value is acceptable by comparing it to that obtained for the same type of defect for a wind turbine with a power of 1.5 MW where the peak obtained is equal to 20% of the nominal value of the voltage V_{Dcnom} [1] which proves the performance of the synthesized regulators.

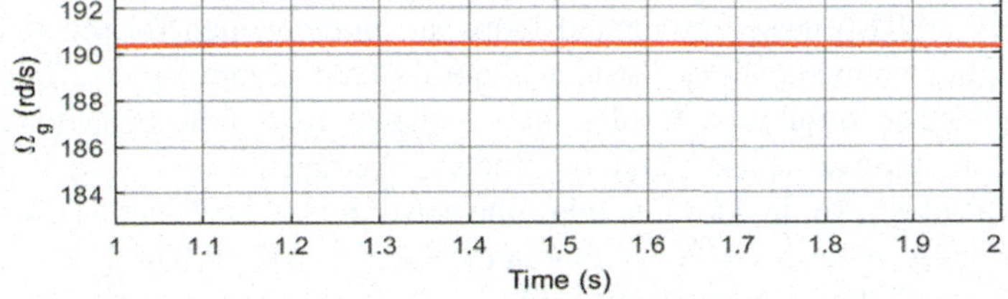

Fig. 20 Rotational speed as a function of time

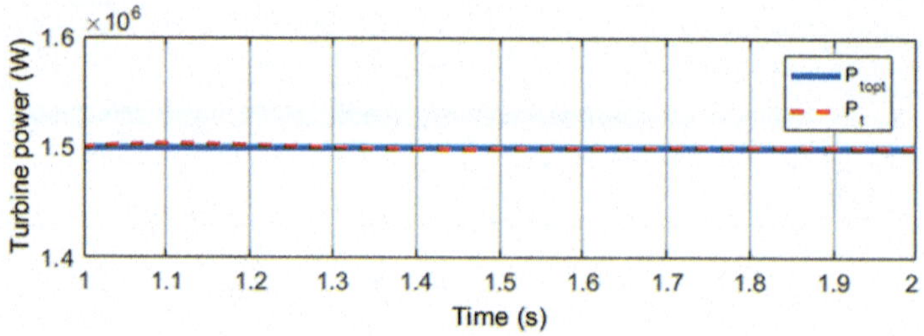

Fig. 21 Turbine power as a function of time

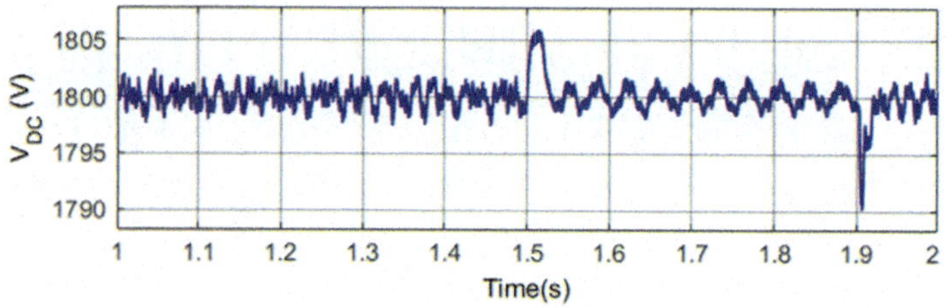

Fig. 22 DC bus voltage as a function of time

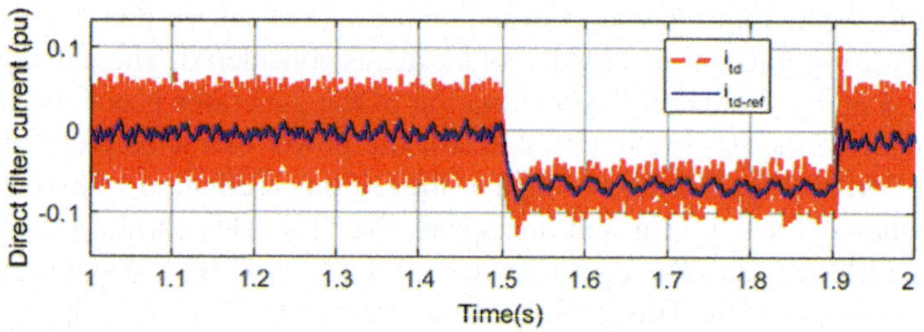

Fig. 23 Direct filter current as a function of time

5 Conclusion

This chapter has focused on the modelling and control of the HESG-based WP turbine grid connection. Simpowersystem's "behavioral" models are used for power grid modeling. Two nested command loops are implemented for the stabilization of the DC bus voltage and for the cancellation of the reactive power injected into the network. The simulation results show satisfactory turbine behavior and good performance of the designed regulators. Indeed, the reactive power injected into the electricity grid is zero. In addition, all of the active power produced is transferred to the grid. Finally, it has been shown that in the event of a disturbance on the grid, the behavior of the WP turbine is unaffected.

References

1. Mseddi, A., Le Ballois, S., Aloui, H., Vido, L.: Robust control of a wind conversion system based on a hybrid excitation synchronous generator: a comparison between H_∞ and CRONE controllers. Math. Comput. Simul. **158**, 453–476 (2019)
2. Mseddi, A., Le Ballois, S., Aloui, H., Vido, L.: Load mitigation and wind power harnessing of a HESG-based wind conversion system. J. Renewable Sustain. Energy **12**(5) (2020)
3. Mseddi, A., Le Ballois, S., Aloui, H., Vido, L.: Robust control of a HESG for a wind energy application. Electric Power Syst. Res. A **168**, 250–260 (2019)
4. Jonkman, J.M., Hayman, G.J., Jonkman, B.J., Damiani, R.R., Murray, R.E.: AeroDyn v15 user's guide and theory
5. Malcolm, D.J., Hansen, A.C.: WindPACT turbine rotor design study, rapport du laboratoire NREL (2006)
6. El Aimani, S.: Modeling of different wind turbine technologies integrated into a medium voltage network, PhD thesis, Ecole centrale de Lille, December 2004. http://l2ep.univ-lille1.fr/fileupload/file/theses/SalmaElAimani.pdf
7. Hsieh, G.C., Hung, J.C.: Phase-locked loop techniques - a survey. IEEE Trans. Indust. Electron. **43**, 609–615 (1996)
8. Zhang, L.D., Bollen, M.H.J.: Characteristic of voltage dips (sags) in power systems. IEEE Trans. Power Deliv. **15**, 827–832 (2000)
9. Nagliero, A., Mastromauro, R.A., Liserre, M., Dell'Aquila, A.: Synchronization techniques for grid connected wind turbines. In: 35th Annual Conference of IEEE Industrial Electronics, Porto, pp. 4606–4613 (2009)
10. Timbus, A., Liserre, M., Teodorescu, R., Blaabjerg, F. : Synchronization methods for three phase distributed power generation systems - an overview and evaluation. In: Power Electronics Specialists Conference, pp. 2474–2481 (2005)
11. Li, S., Haskew, T.A., Williams, K.A., Swatloski, R.P.: Control of DFIG wind turbine with direct-current vector control configuration. IEEE Trans. Sustain. Energy **3**(1), 1–11 (2012)
12. Bingke, Y., Bo, W., Lin, Z., Hesen, L., Yilu, L., Xingpei, J., Dichen, L.: A Novel, stable, and economic power sharing scheme for an autonomous microgrid in the energy internet. Energies **8**, 12741–12764 (2015)
13. Djamel, I., Camara, M.S., Camara, M.B., Brayima, D., Hamid, G.: Permanent magnet synchronous generators for large offshore wind farm connected to grid - comparative study between dc and ac configurations. Int. J. Renewable Energy Res. **4**, 519–527 (2014)
14. Bredard, F.: Modeling and control of a variable speed wind energy conversion chain connected to the grid and using an asynchronous machine, Master's program thesis in engineering, University of Quebec at Rimouski (2014)

Comparative Study Between PI Controller and Fractional Order PI for Speed Control Applied to the Traction System of an Electric Vehicle (EV)

Jalila Kaouthar Kammoun, Amina Mssedi, Moez Ghariani, and Omar Naifar

Abstract This paper proposes a design of a speed control based on a fractional order proportional and integral (PI) applied to the traction system of an electric vehicle (EV). The paper explains also the modeling, control design and simulation, resulting in a numerical simulator composed of detailed models of the major constituent: transmission system, induction motor, control system, and vehicle dynamics. The simulation was refined in MATLAB/Simulink. Simulation results show the efficiency of the designed control.

Keywords Electric vehicle (EV) · Speed control · Fractional order control FO-PI

1 Introduction

Among the most important problems in the world is pollution and overpopulation which has improved the demand for transport [1]. Electric motors consume 50% of the electric power generation, so inmproving their efficiency can have a significant impact on saving electric energy [1]. Digital control techniques that use new power

J. K. Kammoun (✉)
Electronic. Systems and Communication Engineering (Enet'com), Sfax, Tunisia
e-mail: kaouthar.kammoun@enetcom.usf.tn; kaoutharkammoun02@gmail.com

A. Mssedi
Engineering Preparatory Departement, North American Private University Sfax, Sfax, Tunisia

M. Ghariani
Sfax Engineering National School, 3038 Sfax, Tunisia

O. Naifar
Control and Energy Management Laboratory, Sfax Engineering National School, University of Sfax, Sfax, Tunisia
e-mail: omar.naifar@enis.tn

A. Ben Makhlouf et al. (eds.), *State Estimation and Stabilization of Nonlinear Systems*, Studies in Systems, Decision and Control 491,
https://doi.org/10.1007/978-3-031-37970-3_23

electronics devices have evolved to improve the performance of electric motors [2]. Electric machines have made significant progress in their design in recent years.

Three-phase induction motors (IMs) are a popular choice for industrial applications due to their reliability, simplicity, robustness, and efficiency. Frequency converters are commonly employed to control the speed and stop of induction motors with precision. These converters provide a variable frequency power source that allows for the control of the motor's speed. Generally, the use of variable speed drives, such as Variable Frequency Drives (VFDs), has become increasingly popular in recent years due to their ability to control the speed, torque, and flow of motors, which can lead to significant energy savings [3].

Overall, while internal combustion engines (ICE) vehicles are still the dominant technology used in vehicles today, there is a growing interest in alternative technologies like EVs that offer significant advantages over ICE vehicles. ICEs have been the primary technology used in vehicles for over a century, but they have significant drawbacks, including emissions of pollutants and greenhouse gases, dependence on fossil fuels, and limited efficiency. In contrast, electric vehicles (EVs) use electric motors powered by batteries or fuel cells to drive the wheels, which have higher efficiency and lower emissions compared to ICE vehicles [4].

Electric vehicles (EVs) and hybrid electric vehicles (HEVs) are known for their high performance in transforming electrical energy from the battery into mechanical energy for the vehicle to move, with an efficiency of 60–85% [5].

The driving range of EVs is limited by the capacity of the battery, and recharging the battery can take several hours, depending on the charging method used. However, advances in battery technology and the development of fast-charging infrastructure are helping to address these issues and improve the performance of EVs [6].

Mathematical models are an important tool for understanding the energy consumption and improving the performance of electric vehicles [7, 8].

The present work surrounds the process of modeling, control and simulation of an EV. The motor used for the traction system is an induction motor (IM) of 4 kW.

The design and implementation of each component (e.g., transmission system, electric motor, and battery) of an EV is critical to its performance and viability as a clean and effective transportation system. Particular attention was compensated on thedevelopment of IM speed control. This control is done through a proportional and integral (PI) control of fractional order.

Linear proportional–integral (PI) controllers present an excellent tool for controlling the speed of induction machines due to their simplicity and ease of implementation.

Fractional-order PI (FO-PI) controllers are an extension of PI controllers that can provide better performance and stability in controlling the speed of induction machines [2].

The contribution of fractional order algorithms application in cruise control has been manifested recently, with studies demonstrating the effectiveness of these algorithms in controlling the speed of electric vehicles [9–11]. For instance, Hosseinnia et al. presented an improved fractional-order cruise control for a commercial Citroën C3 prototype, which has automatic driving capabilities, at low speeds [9–11].

The content of this paper is organized as follows. First, a brief description of the induction motor model, is presented. Then, the mathematical car model is showed. Besides, the FOPI controller design is presented. Finally, matlab simulation speed control and some conclusions are provided.

2 Three–Phase Induction Motor

In this paper, a three-phase wound rotor induction motor is considered where stationary reference frames are used. Using [12, 13], equations from (1)–(11) are reached. The winding on the *d–q* axis is designated as *ds*, *dr*, *qs*, and *qr*. The direct and quadrature axes of a three-phase induction motor are shown in Fig. 1. Voltage and flux linkage equations can be given as [14]:

$$\mathrm{v}_{\mathrm{s}\alpha} = \mathrm{R}_{\mathrm{s}}\mathrm{i}_{\mathrm{s}\alpha} + \frac{\partial \varnothing_{\mathrm{s}\alpha}}{\partial \mathrm{t}} \tag{1}$$

$$\mathrm{v}_{\mathrm{s}\beta} = \mathrm{R}_{\mathrm{s}}\mathrm{i}_{\mathrm{s}\beta} + \frac{\partial \varnothing_{\mathrm{s}\beta}}{\partial \mathrm{t}} \tag{2}$$

$$0 = \mathrm{R}_{\mathrm{r}}\mathrm{i}_{\mathrm{r}\alpha} + \frac{\partial \varnothing_{\mathrm{r}\alpha}}{\partial \mathrm{t}} - \mathrm{W}_{\mathrm{m}}\varnothing_{\mathrm{r}\beta} \tag{3}$$

$$0 = \mathrm{R}_{\mathrm{r}}\mathrm{i}_{\mathrm{r}\beta} + \frac{\partial \varnothing_{\mathrm{r}\beta}}{\partial \mathrm{t}} - \mathrm{W}_{\mathrm{m}}\varnothing_{\mathrm{r}\alpha} \tag{4}$$

Flux-linkage Equations:

$$\varnothing_{\mathrm{s}\alpha} = \mathrm{L}_{\mathrm{s}}\mathrm{i}_{\mathrm{s}\alpha} + \mathrm{L}_{\mathrm{m}}\mathrm{i}_{\mathrm{r}\alpha} \tag{5}$$

$$\varnothing_{\mathrm{s}\beta} = \mathrm{L}_{\mathrm{s}}\mathrm{i}_{\mathrm{s}\beta} + \mathrm{L}_{\mathrm{m}}\mathrm{i}_{\mathrm{r}\beta} \tag{6}$$

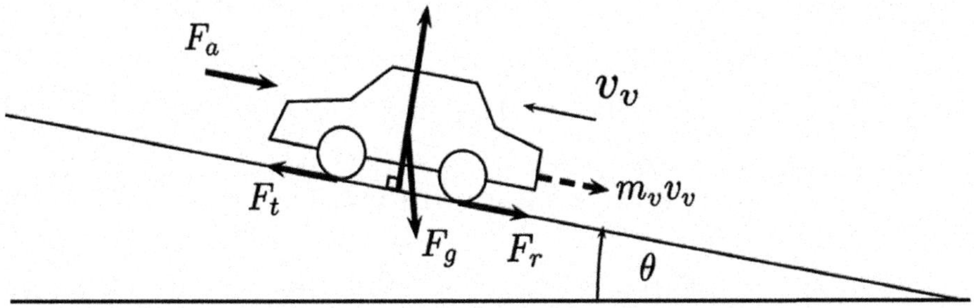

Fig. 1 Forces applied upon vehicle

$$\O_{r\alpha} = L_r i_{r\alpha} + L_m i_{s\alpha} \tag{7}$$

$$\O_{r\beta} = L_r i_{r\alpha} + L_m i_{s\beta} \tag{8}$$

The torque equation is defined as:

$$T_e = P(\O_{s\alpha} i_{s\beta} + \O_{s\beta} i_{s\alpha}) \tag{9}$$

The rotating speed of the stator magnetic field can be given as:

$$N_s = \frac{120 * f}{n} \tag{10}$$

$$\dot{W}_r = \frac{1}{J}[T_e - T_L - f_r w_r] \tag{11}$$

A list of nomenclature used in this work is given in Table 1 at the beginning of the paper and the values of motor parameters are provided in Table 2.

3 Mathematical Car Model

So as to elaborate a mathematical model that describes the dynamics of the vehicle, a balance of forces is generated. The EV is caractirized by diffirent forces the velocity $\upsilon(t)$, m_υ its total mass, $F_t(t)$ the tensile force generated by the contact of the tires with the road and $F_p(t)$ the force of disturbance due to the aerodynamic drag, the friction of bearing and gravity. These forces are shown in Fig. 1 [15].

The differential equation that relates the tensile force $F_t(t)$ with the speed υ_υ (t) of the vehicle are made by applying Newton's second law [15].

$$m_\upsilon\ \upsilon\dot{\upsilon}(t) = F_t(t) - F_P(t) \tag{12}$$

where

$$F_P(t) = F_a(t) + F_r(t) + F_g(t) \tag{13}$$

$$F_a(t) = \frac{1}{2}\rho_a A_f C_{ad}(\upsilon_\upsilon(t) + \upsilon wind(t))^2 \tag{14}$$

$$F_r(t) = sgn(\upsilon_\upsilon(t) + m_v g cos(\theta(t))C_r \tag{15}$$

$$F_g(t) = m_v g\ sin(\theta(t)) \tag{16}$$

Table 1 List of parameters

Symbols	Description
$i_{s\alpha}$	Stator current along the alpha axis
$i_{s\beta}$	Stator current along the beta axis
i_{ds}	Stator current along the direct axis
i_{qs}	Stator current along the quadrature axis
$\emptyset_{s\alpha}$	Stator flux along the alpha axis
$\emptyset_{s\beta}$	Stator flux along the beta axis
$\emptyset_{ds}$	Stator flux along the direct axis
$\emptyset_{qs}$	Stator flux along the quadrature axis
$v_{s\alpha}$	Stator voltages along the alpha axis
$v_{s\beta}$	Stator Voltage along the beta axis
i_{dr}	Rotor current along the direct axis
i_{qr}	Rotor current along the quadrature axis
$\emptyset_{dr}$	Rotor flux along the direct axis
$\emptyset_{qr}$	Rotor flux along the quadrature axis
v_{rd}	Rotor voltages along the direct axis
v_{rq}	Rotor voltages along quadrature axis
σ	Blondel's coefficient
T_e	Electromagnetic torque
T_L	Load torque
W_r	Load torque
n	Number of poles
f	Frequency

Table 2 Motor parameters

Parameter	Symbol	Value	Unit
Resistance of stator	$\mathrm{R_s}$	6.3	Ω
Resistance of rotor	$\mathrm{R_r}$	5.6	Ω
Inductance of stator	$\mathrm{L_s}$	0.6348	H
Inductancc of rotor	$\mathrm{L_r}$	0.6348	H
Inertia coefficient	J	0.1	kg m2
Poles	p	2	_
Power	P	4	kW

Vehicle parameter are presented in Table 3.

The electric motor generates the force $\mathrm{F_t}$ and the torque that's depends on the current flowing through the windings and rotation speed of the motor. Both of the electric motor of the vehicle and the transmission system must be able to defeat the forces due to gravity, wind, friction and inertial effects based on the Eq. 12.

Table 3 Electric vehicle parameters

Symbols	Description
C_d	Drag coefficient
C_r	Coefficient of friction of the bearing
A_f	Frontal area of the vehicle
ρ_a	Density of the air at 20 °C
θ	Angle of the slope
g	Gravity acceleration

4 FOPI Controller Design

Many variations of PID, as well as nonlinear approaches, have been described in the literature. For exemple, PID controllers with integral and derivative of fractional order have been published [16]. Using fractional order controller is due to the enhance on the operating of the controllers outperform the integer order controller as it's confirm at [17].

In effect, fractional-order controller can be used even in the case where the integer order controller is suitable [18, 19]. The fractionalorder proportional integrator (FOPI) controller is a modified version of the existing fraction-order PID and a FOPI used in speed loop of the IM drive is shown in Fig. 2. In these figure ***d*** the load torque disturbance to the speed-loop, ***u*** the actuating signal, and ***n*** the measurement error (in this case it's considred null).

Transfer function described the FOPI controller is presented by Eq. 17.

$$C(s) = K_{p+}K_i \ s^{-\lambda} \tag{17}$$

where $\lambda \in [0,1]$ real integral operator.

K_p, proportional constant.

K_i, integration constant.

The tuning of the FOPI controller involves adjusting the controller parameters (λ, K_p, K_i) to achieve the desired performance specifications, and to overcome the PI.

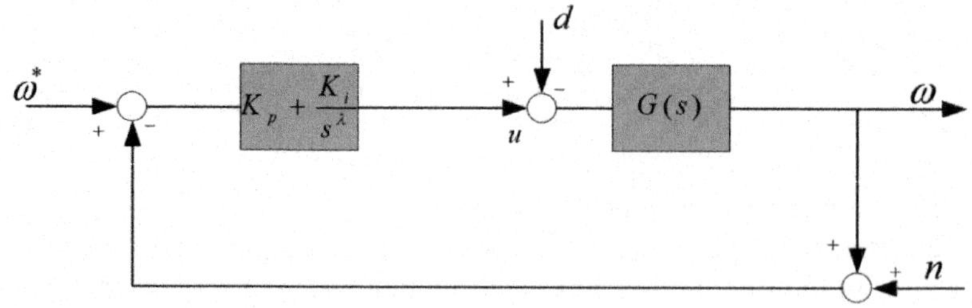

Fig. 2 FOPI controller

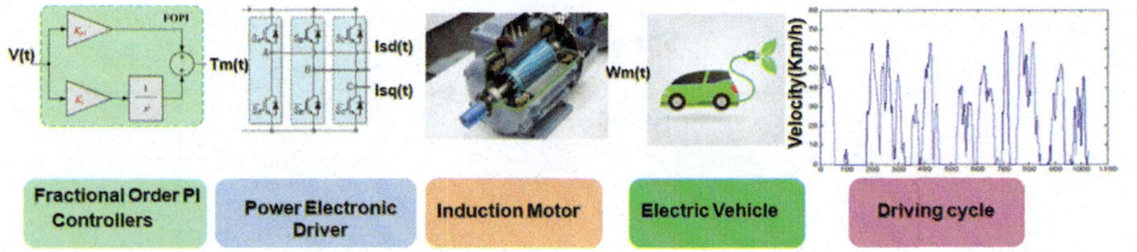

Fig. 3 Block diagram of the control system

Furthermore, there are many ways to regulate the controller basing on the specific application and the desired properties of the system [20, 21].

4.1 *Control System Design*

As the motor torque Tm(t) depends on the currents flowing through the winding, the control loop will not only be closed by the speed measure of the vehicle (or engine), but will also based on the speed of the vehicle's current loop. Figure 3 illustrate the block diagram of the control system.

4.2 *Torque Control Loop*

The electromagnetic torque in 3 phase induction motor ncan be defined as follows:

$$T_{em} = \frac{3}{2}\frac{L_m p}{L_r}\emptyset_{rd} i_{sq} \tag{18}$$

From the expression (18), the couple depends on the component i_{sq}.

According to the flux equation, the quadrature component of the voltage is reformulated as follows:

$$v_{sq} = v_{sqf} + v_{sqc} \tag{19}$$

$$v_{sqf} = R_s i_{sq} + \frac{\sigma L_{sd} \partial i_{sq}}{\partial t} \tag{20}$$

$$v_{sqc} = \sigma L_s w_s i_{sq} + \frac{\sigma L_m \emptyset_{rd}}{L_r} \tag{21}$$

The torque is a function of the current isq which is generated by v_{sqf}, while the v_{sqc} term must be compensated to ensure the decoupling between the two axes. The role

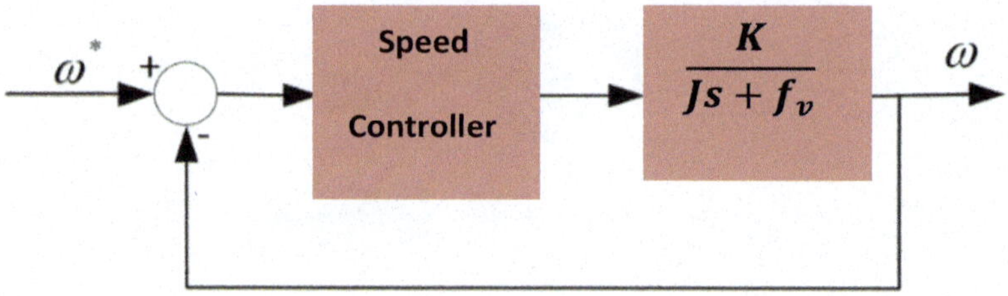

Fig. 4 Simplified speed loop

of the FOPI regulator is to deliver the correct component of v_{sqf} in order to control the torque.

4.3 Speed Control Loop

It is well known that the speed loop for an electrical drive with induction motor with rotational inertia J and viscous coefficient of friction f_v, can be described by the simplified scheme in Fig. 4, where W* is the reference speed, W is the measured speed and K = 1.

The control aim is to regulate the actual quantity with the reference command. Essentially, Particularly a high performance electrical drive must meet needs of speed tracking with minimum exceed as well as good load-torque rejection. To achieve the above requirement the best way is to use a fractional order controller for to its robustness and flexibility [14].

5 Matlab Simulation Speed Control

A SIMULINK block was established to analyze the open loop speed control method using FOPI controller. The SIMULINK block is given below followed by the outcomes (Fig. 5).

The Speed control can be implemented through simulation, including in the system the equations of both the dynamics of the EV and the IM. The electrical vehicle and the induction motor are described using Simulink blocks, which were built specifically for this job. Several profiles were tested like as step and ramp signals as input signals (reference velocities), as seen in Fig. 6.

As was previously mentioned, both controllers were tuned to optimize the Integral of time-weighted absolute error (ITAE) signal of the IM speed. As seen in red cruve the system under a PI controller shows a bigger overshot when a step (0–300 rpm) is applied at 10 s of simulation. However, as is shown in green cruve, the system under an FO- PI controller it has a lower ITAE value.

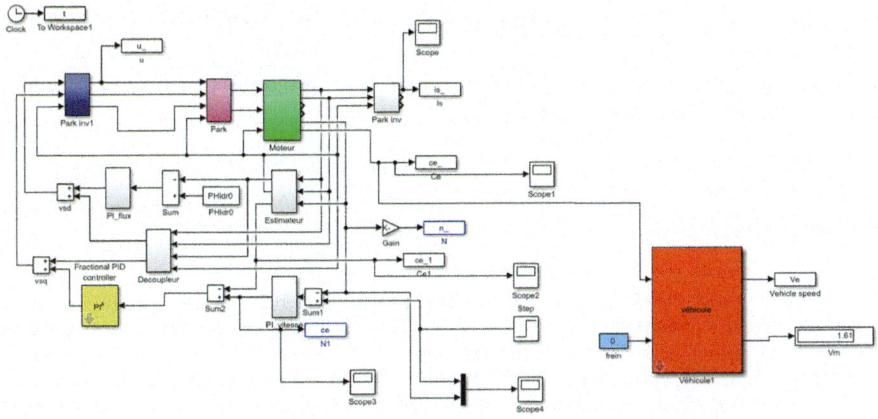

Fig. 5 Simulation of proposed speed control of induction motor

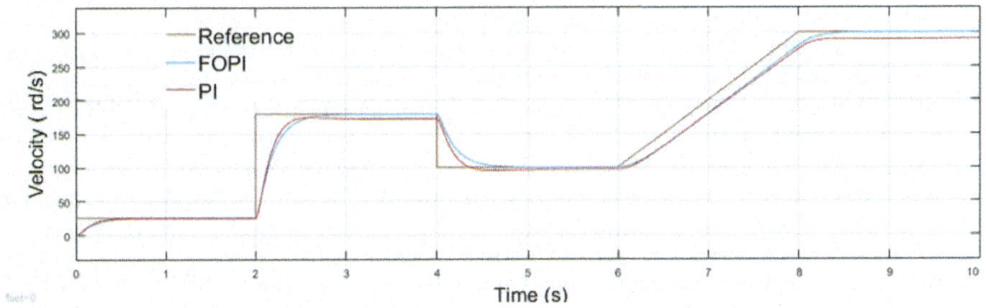

Fig. 6 Step and ramp responses of the system

The fractional order controllers FO-PI is more percise in fact the trending error is nul however in the normal PI regulator it is about 5%. Steped profile is adopted between 2 to 4 s and ramped profile is adopted [6–8 s]. Eventhough the first profile does not represent,a realistic command for electric vehicle but it allows testing the efficiency of the adopted control strategy in a severe environment.

6 Conclusion

This paper presents the application of Fractional Order PI (FOPI) to control the speed of an induction motor in an electric vehicle.

Following presenting a method to design them in the frequency domain, comparisons are carried out with respect to classical PI. Tthe advantages of FOPI as for the quicker dynamical response and decrease of the overshoot in terms of speed tracking and load rejection. Additionally, fractional PI controllers can improve the efficiency of the motor and reduce energy consumption, which is important for extending the driving range of an EV.

Simulation results proves that the elimination of trending error for FO-PI which confirm the goodnes of the method.

References

1. Kosow, I.: Control de Máquinas Eléctricas (in Spanish). Reverte: Cartagena, Colombia (1977)
2. Ponce-Cruz, P., Sampe-Lopez, J.: Máquinas Eléctricas y Técnicas Modernas de Control (in Spanish). AlfaOmega: São Paulo, Brazil (2008)
3. De Almeida, A.T., Ferreira, F.J.T.E., Both, D. : Technical and economical considerations in the application of variable-speed drives with electric motor systems. IEEE Trans. Ind. Appl. **41**, 188–199 (2005). [CrossRef]
4. Emadi, A.: Handbook of Automotive Power and Electronics and Motor Drives. Taylor and Francis, Abingdon, UK (2005)
5. Chan, C.C., Bouscayrol, A., Chen, K.: Electric, Hybrid, and Fuel-Cell Vehicles: Architectures and Modeling. IEEE Trans. Veh. Technol. **59**, 589–598 (2010). [CrossRef]
6. Lukic, S., Emadi, A.: Performance analysis of automotive power systems: effects of power electronic intensive loads and electrically-assisted propulsion systems. In Proceedings of the IEEE Vehicular Techonology Conference (VTC), Vancouver, BC, Canada, 24–28 Sept 2002, pp. 1835–1839 (2002)
7. Mapelli, F.L., Tarsitano, D., Mauri, M.: Plug-In Hybrid Electric Vehicle: Modeling, Prototype Realization, and Inverter Losses Reduction Analysis. IEEE Trans. Ind. Electron. **57**, 598–607 (2010). [CrossRef]
8. Thomas, P., Shanmugam, P.K.: A review on mathematical models of electric vehicle for energy management and grid integration studies. Sci Direct J Energy Storage **55**(Part A 1), 105468 (2022)
9. Hosseinnia, S.H., Tejado, I., Milanés, V., Villagrá, J., Vinagre, B.M.: Experimental application of hybrid fractional-order adaptive cruise control at low speed. IEEE Trans. Control Syst. Technol. **22**(6) (2014)
10. Hosseinnia, H., Tejado, I., Milanés, V., Villagrá, J., Vinagre, B.: Experimental application of hybrid fractional-order adaptive cruise control at low speed HAL Id: hal-01091158. IEEE Trans. Control Syst. Technol. **22**(6) (2014). https://hal.science/hal-01091158
11. HosseinNia, S.H., Tejado, I., Vinagre, B.M., Milanés, V., Villagrá, J.: Low speed control of an autonomous vehicle using a hybrid fractional order controller. In: The 2nd International Conference on Control, Instrumentation and Automation Date Added to IEEE Xplore: 24 Nov 2012
12. Ammar, A., Bourek, A., Benakcha, A.: Sensorless SVM-direct torque control for induction motor drive using sliding mode observers. J. Control. Autom. Electr. Syst. **28**, 189–202 (2017)
13. Chapman, S.J. Electric Machinery Fundamentals. McGraw-Hill, New York, NY, USA (2004)
14. Chen, T.-C., Sheu, T.-T.: Model reference neural network controller for induction motor speed control. IEEE Trans. Energy Convers. **17**, 157–163 (2002)
15. Munoz-Hernandez, G.A., Mino-Aguilar, G., Guerrero-Castellanos, J.F., Peralta-Sanchez, E.: Fractional order PI-based control applied to the traction system of an electric vehicle (EV) mdpi. J. Appl. Sci. (2020)
16. Monje, C.A., Chen, Y., Vinagre, B.M., Xue, D., Feliu, V.: Fractional order systems and controls: fundamentals and applications. Springer, New York, NY, USA (2010)
17. Chen, Y., Petras, I., Xue, D.: Fractional order control-a tutorial. In : American Control Conference, 2009. ACC'09, pp. 1397–1411 (2009)
18. Monje, C.A.: Design methods of fractional order controllers for industrial applications, Ph.D. Thesis, University of Extremadura, Spain, 2006

19. Monje, C.A., Vinagre, B.M., Feliu, V., Chen, Y.: Tuning and autotuning of fractional order controllers for industry applications. Control. Eng. Pract. **16**, 798–812 (2008)
20. Petras, I.: The fractional-order controllers: methods for their synthesis and application (2000). math/0004064
21. Monje, C.A., Chen, Y., Vinagre, B.M., Xue, D., Feliu, V.: Fractionalorder systems and controls: fundamentals and applications. Springer Science & Business Media (2010)

SPRINGER NATURE

GPSR Compliance

The European Union's (EU) General Product Safety Regulation (GPSR) is a set of rules that requires consumer products to be safe and our obligations to ensure this.

If you have any concerns about our products, you can contact us on ProductSafety@springernature.com

In case Publisher is established outside the EU, the EU authorized representative is:

Springer Nature Customer Service Center GmbH
Europaplatz 3
69115 Heidelberg, Germany

The manufacturer's authorised representative in the EU is Springer Nature Customer Service Centre GmbH, Europaplatz 3, 69115 Heidelberg, Germany. If you have any concerns regarding our products, please contact ProductSafety@springernature.com

Printed and bound by CPI Group (UK) Ltd, Croydon, CR0 4YY
01/10/2026
02225902-0016